INTRODUCTION TO ELECTRONICS TECHNOLOGY

Second Edition

JOHN W. McWANE
Hewlett-Packard Company

Breton Publishers
Boston, Massachusetts

PWS PUBLISHERS

Prindle, Weber & Schmidt •♣• Duxbury Press •♠• PWS Engineering •⟁• Breton Publishers •⚙
20 Park Plaza • Boston, Massachusetts 02116

PWS Publishers is a division of Wadsworth, Inc.

Library of Congress Cataloging-in-Publication Data

McWane, John.
   Introduction to electronics technology.

  Prev. ed.: Introduction to electronics and instrumentation. 1981.
  Includes index.
  1. Electronics. I. McWane, John. Introduction to electronics and instrumentation. II. Title.
TK7815.M354 1986   621.381   85-16318
ISBN 0-534-06342-X

Printed in the United States of America
2 3 4 5 6 7 8 9 — 90 89 88 87

ISBN 0-534-06342-X

Sponsoring editor: George J. Horesta
Production supervision: Technical Texts, Inc.
Production editor: Jean T. Peck
Interior and cover design: Sylvia Dovner
Cover photo: Jay Freis/The Image Bank
Composition: A & B Typesetters, Inc.
Cover printing: New England Book Components, Inc.
Text printing and binding: Halliday Lithograph

To Nathaniel H. Frank – teacher, mentor, friend

CONTENTS

PART VI
POWER SUPPLIES   404

**Direct Current Power Supplies in
Electronics   405**

CHAPTER 14
UNREGULATED POWER
SUPPLIES   407

CHAPTER 15
FILTERED POWER
SUPPLIES   447

PREFACE

Introduction to Electronics Technology is the second edition of the text *Introduction to Electronics and Instrumentation*. The new title reflects more a clearer statement of the aims of the book than a change in focus or approach. As stated in the Preface to the first edition, the goal of the text is "an updating of the introductory electronics course to reflect the changes that have taken place in the field and a broadening of the content to make it serve a much wider audience." To that end, *Introduction to Electronics Technology* is designed to prepare the introductory student to an electronics field that utilizes multipurpose, multifunction ICs rather than complex circuit networks filled with discrete components and transistors.

This approach to the introductory course gets the student quickly into the purpose and operation of modern electronic systems with a minimum of the mathematical hurdles found in other introductory texts. Within a few chapters, the student will gain a basic understanding of the dc quantities—voltage, current and resistance—their relation through Ohm's law, and how to analyze simple series and parallel networks. After a brief orientation to electronic signals, the student is ready to tackle operational amplifiers and digital circuits—all before the midpoint of the book. These devices require little mathematical analysis to understand their operation yet produce a variety of electronic signal processing circuits used in today's electronic systems.

As each new concept is introduced, a practical application from an electronic system is given. The operation of analog meters, digital meters, Wheatstone bridges, the 555 timer, constant voltage sources, audio mixers, sine wave generators, and so forth are described as applications of newly developed concepts.

For students who intend to major in electronics, many chapters and sections (starred in the Table of Contents) provide the deeper understanding of the fundamentals and more rigorous development of the concepts that will be needed for subsequent courses. Hence, this text can be used for electronic technology majors and nonmajors alike. For both types of students, the change in content and approach will lead to a more rapid development of the concepts that are relevant to modern electronics technology.

The second edition contains a number of new topics and other additions and revisions. These changes have come from the recommendations of many teachers who have used the text over the past three years. Major topic additions are those in digital electronic circuits. Two chapters have been provided to introduce students to concepts, devices, and applications in this major electronic area.

With this addition, the text now provides comprehensive introductory coverages of all of the major topic areas of electronic technology. Students who complete the material that is not starred in the Table of Contents will have a complete introductory overview of modern electronic technology, while students who complete the starred material are thoroughly prepared, both conceptually and mathematically, for advanced courses in any of the major topic areas of electronics.

The new chapters follow the same developmental approach used in the first edition. Learning objectives are provided at the beginning of each chapter to assist the student in determining the essential concepts and skills to be learned. Many worked-out examples in the text and questions and problems at the end are keyed to the objectives and give students specific examples of the level of performance they should be able to demonstrate.

Throughout the second edition, revisions include many more worked-out examples to give the student a broader perspective of the range of applications of each concept, as well as more questions and problems to give teachers greater latitude in testing student understanding. The mathematical level assumes familiarity with algebra, but algebraic calculations and derivations are kept to a minimum, especially in the introductory-level material (not starred in the Table of Contents). Whenever possible, parallel analyses using graphs are provided to complement the mathematical development.

The text is written in a clear, straightforward style with short simple sentences and paragraphs for ease of student reading and understanding. A parallel, pictorial development of the concepts is also provided. Carefully developed illustrations and captions provide a brief pictorial summary of the key points developed in the text. By looking at the figures and reading the captions, a complete thumbnail summary of the essential concepts and conclusions of the text can be obtained. The second edition also includes major revision and improvement in the quality and clarity of the illustrations throughout.

In summary, *Introduction to Electronics Technology* is a contemporary and comprehensive alternative to the introductory electronics course. Its content has been carefully selected to be of maximum utility in today's electronic environment, and its style and approach have been specifically designed to make the concepts understandable and motivating to a broad range of electronics technology students.

The author would like to acknowledge the contributions of a number of people in the development of this text: Bruce D. Wedlock, for critical guidance on the topical content and analytical techniques appropriate for contemporary electronics programs and for content development of Part II (Basic Circuit Network); Richard J. Duffy, for valuable guidance on topical content and pedagogical approach for the text and for content development of Part III (Time-Varying Signals) and Part VII (Alternating Current, Voltage, and Impedance); Robert F. Coughlin, for content development of Part I (Direct Current, Voltage, and Resistance) and Part VI (Power Supplies); and Frederick F. Driscoll, for content development of Part IV (Operational Amplifiers).

The author wishes to thank George Horesta of Breton Publishers for his interest and support in publishing a book that is a significant departure from conventional practice and for his cooperation in maintaining its features to final publication. Many other individuals have contributed their efforts to this work. The author gratefully acknowledges the help of the staff at Technical Texts, Inc., particularly Jean Peck, who spared no efforts in the production of this revision.

Finally, thanks are due to several firms within the electronics industry for providing information that was helpful in preparing the second edition. The author extends his appreciation to the following firms:

Fairchild Camera and Instrument Company
Hewlett-Packard Company
Littelfuse Company
National Semiconductor Corporation
Raytheon Company
Simpson Electric Company
Wakefield Company

PREFACE
to the First Edition

The 20th Century will surely go down in history as the Age of Electronics. Beginning in the early 1900s, dramatic advances in science and technology have made electronics an increasingly dominant force in modern life. The development of electron tube technology in the 1930s and 1940s created a wave of electronic products that began making inroads into an essentially mechanical society. In the 1950s, the development of semiconductor devices, particularly the transistor, increased electronic system capability and reliability and began a new wave of electronic applications and products that continued to expand into our society.

The 1970s and 1980s are now seeing the latest advances in science and technology—microminiaturization. The result is applications of electronic systems to nearly every aspect of daily life, and an increasing dependence on these sophisticated systems.

Each major technological advance has brought to electronics a new set of knowledge and skills and new requirements in fundamental concepts. The electronics curriculum has responded by adding new topics, deleting outdated ones, and changing its emphasis on basic concepts to adequately prepare students to deal with the new developments.

The introductory electronics course is now in need of another major revision to make it reflect the changes in electronics brought about by microelectronics. A review of today's most widely used introductory electronics texts shows that they were generally written in the 1960s and reflect the state of electronics at that time. Their content and approach prepared students for an electronics field centered largely on transistors and discrete component circuitry. The basic concepts developed included the mathematical analysis and detailed component behavior required for more advanced courses that developed concepts in discrete component circuits.

The "discrete" components of today's electronic system, however, are integrated circuits whose internal behavior is far too complex for even advanced students to comprehend. And the "analysis" skills that must be developed are less concerned with electronic circuits than with electronic systems and subsystems.

The shift in electronics from circuits to systems requires a significant change in the introductory course. For example, greater attention must be paid to input and output elements, including a variety of transducers, control elements, and output display methods, and to the principles and devices for processing electronic information. On the other hand, there is less need for developing mathematical techniques for analyzing complex circuit networks and for delving into the fundamentals of semiconductor behavior.

The expansion of electronics into nearly all fields has also meant an increasing demand for an understanding of electronic fundamentals. Disciplines as diverse as medicine, civil engineering, chemistry and music are therefore adding introductory electronics as a basic requirement in the curriculum. The objectives of these students also require a change in emphasis from that given in most introductory texts. They also have less need for mathematical analysis of complex circuit networks and more need for a practical understanding of the principles of electronic system behavior and the factors that affect their accuracy, reliability, and performance.

The goal of the *Introduction to Electronics and Instrumentation* program is to meet these changing needs — an updating of the introductory electronics course to reflect the changes that have taken place in the field and a broadening of the content to make it serve much wider student population.

This pedagogic approach represents a changing role for the introductory course, making it more motivating and immediately useful to the student, as well as a better preparation for more advanced courses. Most introductory texts, particularly those for students majoring in the field, attempt to develop a deep physical and mathematical foundation of basic concepts and skills with little concern for practical applications of immediate utility.

This course, however, takes the view that a student's first introduction to electronics has a dual objective: the development of a mathematical and theoretical foundation *and* the development of functional literacy. Experience has shown that this is both possible and desirable. While the development of some of the more theoretical topics may have to be deferred to later courses in order to include the more practical ones, the gains in student understanding of the topics that are presented and their motivation for further study far outweigh the seeming decrease in depth of treatment.

In order to serve as broad a student audience as possible, there are no electronic prerequisites, and the concept development assumes no previous experience with electronics. The mathematical level assumes familiarity with algebra, and algebraic calculations and derivations are used when necessary to clarify a concept or describe a system's behavior. However, the algebraic expressions are kept as simple as possible, and rarely does the understanding of a concept demand on following a mathematical argument.

Learning objectives are provided at the beginning of each section to assist the student in determining the essential concepts and skills to be learned. They are written in behavioral terms, indicating what the student should be able to *do* upon completion of that chapter. They are simply stated, however, for maximum student comprehension. Many worked-out Examples in the text and Questions and Problems at the end of each chapter (based on the learning objectives) give students specific examples of the level of performance they are expected to demonstrate.

The text is written in a clear, straightforward reading style with short sentences and paragraphs for ease of student understanding. In addition, in parallel with the text, are figures with captions that provide a brief pictorial summary of the essential points developed in the text. By looking at the figures and reading the captions, a complete thumbnail summary of the essential concepts and conclusions of the text can be obtained. It should be emphasized, however, that the text development is complete and as fully detailed as in a normal text.

This rather revolutionary approach to text development has been carefully designed to aid student study and comprehension in a variety of ways. For the poor reader, or the student who does not learn well from the written word, an alternative form of the text material is available. For the student who simply chooses not to read, or only accesses the text material to solve problems, the figure-caption summary provides both a quick overview of the text content and an efficient guide for locating more detailed information when it is needed.

In summary, this text is a comprehensive and contemporary redesign of the introductory electronics course. Its content has been carefully selected to be of maximum utility in today's electronic environment, and its style and approach have been specifically designed to make the concepts understandable and motivating to a broad range of students.

INTRODUCTION TO ELECTRONICS TECHNOLOGY

Part I

DIRECT CURRENT, VOLTAGE, AND RESISTANCE

DIRECT CURRENT QUANTITIES IN ELECTRONICS

The term **electronics** is derived from **electron,** one of the small, electrically charged particles of matter. The field of electronics is concerned with controlling the flow of electrons in various ways to perform useful functions (Figure 1). In a television set, the flow of electrons reproduces sound and pictures. In calculators, it performs and displays arithmetic calculations. In electronic instruments, it measures and controls a large variety of physical quantities, such as light, temperature, and sound.

To understand electronics in terms of electron flow is quite difficult and, fortunately, is not necessary. The development of modern electronic devices, particularly the integrated circuit, permits us to build and use quite sophisticated instruments without ever referring to or knowing about the electron nature of their operation.

Understanding electronics does require, however, learning a number of concepts that describe the operation of these devices. The most basic concepts are the quantities of electrical current, voltage, and resistance. Electrical **current** is the term that describes the flow of electrons or, more generally, electrical charge in a circuit or device. In order for the charge to flow, an electrical force is required. This force is provided by the **voltage** and is based on the fact that unlike charges attract each other while like charges repel. The ease or difficulty with which the charge can flow in a device is described by its electrical **resistance.** The basic electronic quantities of current, voltage, and resistance can be measured by a versatile device called a multimeter (Figure 2).

Part I concerns only **direct current,** or **dc**—that is, systems in which the current flows only in a single direction. Direct current is contrasted with **alternating current,** or **ac,** in which the current direction changes periodically. The most common example of ac is the electricity suppled by the electric company that appears at wall outlets. Current from ac sources alternates direction 60 times a second.

Most electronic devices—for example, tubes, transistors, and integrated circuits—operate on dc. Thus, TV sets, calculators, radios, computers, and so on, all require dc. Even though they may be plugged into ac outlets, the ac must be converted into dc in order to operate the electronic circuitry. The concepts of voltage, current, and resistance, which you will learn here, all carry over into ac systems, but their description is somewhat more complicated and is discussed later.

DIRECT CURRENT ELECTRONIC SYSTEMS

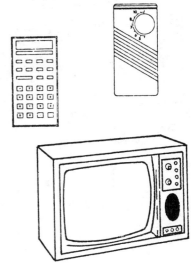

FIGURE 1. Part I explains the basic quantities of direct current (dc) electronics. The devices used in most electronic systems require dc.

MULTIMETER

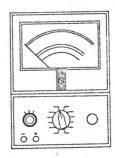

FIGURE 2. The basic electronic quantities are current, voltage, and resistance. These quantities can be measured by a versatile instrument called a multimeter.

REPRESENTING CIRCUIT COMPONENTS

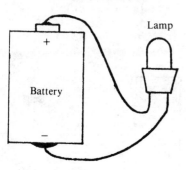

Flashlight Circuit

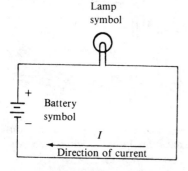

Schematic Diagram

FIGURE 3. Electronic systems are generally illustrated by schematic diagrams in which specific symbols represent the circuit components.

In electronics, a code is used to illustrate the circuits. Instead of drawing pictures of the devices that make up the circuit and the wires that connect them, symbols are used. The symbolic representations of electronic circuits are called **circuit diagrams** or, sometimes, **schematic diagrams** (Figure 3). Each device has its own special symbol, and lines connecting the device to other devices indicate the paths of current flow. The lines may represent actual wires connecting two parts, but in general they only mean that the two parts are electrically connected; for example, they may be touching.

When a new electronic device is introduced in this book, its symbol will be given. The symbols should be learned, as well as the correlation between the symbol markings and the parts of the actual device, because actual working circuits are constructed only from circuit diagrams.

Chapter 1

CURRENT AND VOLTAGE

1.1 OBJECTIVES

Following the completion of Chapter 1, you should be able to:

1. Recognize the units of electrical current, voltage, and power and convert between decimal subunits.
2. Determine the direction of conventional current flow from a circuit diagram, given the voltage source polarity.
3. Identify the schematic symbols and properly correlate symbol markings to components for the following: (a) battery; (b) variable dc voltage source; (c) lamp and alarm; (d) toggle switch (SPST, SPDT, and DPDT); (e) relay; (f) common, chassis, and ground; (g) voltmeter and ammeter; (h) diode; and (i) transistor.
4. Draw a circuit diagram involving a voltage source and series and parallel connections, observing proper polarity.
5. Explain the differences between common, chassis, and ground and observe proper

precautions when dealing with voltage sources.
6. Properly connect an ammeter and a voltmeter in a circuit to measure the current through and the voltage across circuit components.
7. Properly connect an electronic circuit (including a dc power supply) by following a circuit diagram.
8. Accurately read the scale of an analog multimeter for various dc voltage and current settings of the range switch.
9. Use the power relation to calculate the electrical power, current, or voltage of a device, given any two of the quantities.
10. Calculate the time for which a battery can supply various currents, given its ampere-hour rating.

1.2 ELECTRICAL CHARGE

The field of electronics is based on the physical property of **electrical charge.** Electrical charge

5

FORCES ON ELECTRICAL CHARGE

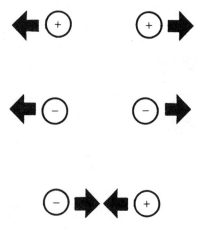

FIGURE 1.1. Electrical charge forms the basis of electronics. Objects with charges of opposite polarity attract each other; objects with the same polarity repel.

OBSERVING FORCES OF ELECTRICAL CHARGE

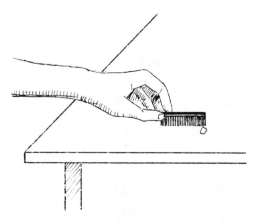

FIGURE 1.2. The forces of electrical charge can be observed in a simple experiment. Rub a plastic comb against cloth and it will pick up a small piece of paper.

is not a quantity we have much "feel" for since it cannot be seen and is not as directly sensed by the human body as are temperature, sound, or light, for example. Though almost everyone has at one time experienced an electrical shock, which is a flow of charge through the body, the feeling is sufficiently unpleasant that most people have never investigated charge through this sort of experience.

Fundamentally, electronics is based on the physics of electrical charge, yet it is not necessary to know much physics to work with electronic devices. For most practical purposes, it is only necessary to know two basic properties. First, charge comes in two distinct forms called positive and negative, or simply + and − . The distinction between the two forms is called **polarity** — positive polarity and negative polarity. Second, if two objects have charges of the *same polarity*, either both positive or both negative, the two objects will *repel each other* (Figure 1.1). The more charge the two objects have, the stronger the force of repulsion. On the other hand, if two objects have charges of *opposite*

polarity, one negative and one positive, the two objects will *attract each other* (Figure 1.1). And again, the more charge the objects carry, the stronger the force of attraction.

While certain effects of charged objects can be observed on a large scale (Figure 1.2), the effects that are important for electronics occur at the microscopic level — that is, between the atomic particles that make up matter. Electrons are the "objects" that carry the negative charge, and protons are the "objects" that carry the positive charge. Though electronics can be described in terms of the behavior of electrons and protons, or systems of electrons and protons, that description is quite complex and unnecessary for most purposes.

However, it is necessary to remember the two properties of electrical charge given earlier because they are the basis of electronics descriptions. For example, two basic quantities of electronics are electrical current and voltage. **Electrical current** is the flow of electrical charge through the devices of electronic systems. It is this flow of electrical charge that causes electronic systems to work. Charge flows because it is attracted to things of opposite polarity. **Electrical voltage** specifies the size of polarity differences and determines the rate at which charge and, therefore, current flow. This chapter describes the two basic electronic quantities of current and voltage in terms of the behavior of electrical charge.

ELECTRICAL CURRENT

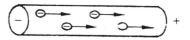

FIGURE 1.3. Electrical current is the flow of electrical charge in materials. The charge is carried by negatively charged electrons that are attracted toward regions of positive polarity.

1.3 ELECTRICAL CURRENT

Electrical current is the flow of electrical charge in matter (Figure 1.3). In electronics, this means a flow through various wires and devices that make up an electronic system. The path (or paths) through which the current flows is called **a circuit.**

Although the flow of charge around the circuit is the essential quantity in electronics, the things that move and thus cause the charge to move are electrons. That is, the electrons contained in some materials—copper, for example—can move around inside the material quite easily. If the material is in the form of a wire and one end of the wire is made positive and the other end negative, the electrons will move toward the positive end. This flow of electrons represents a flow of negative charge through the wire.

Because the electron current flows within the wire, the wire is said to be the **conductor** of the current. Materials in which electrons flow easily are called good electrical conductors, while those in which electrons flow with difficulty or not at all are called poor conductors, or, more specifically, **insulators**. Metals, particularly copper, silver, and gold, are good conductors, while materials such as rubber, ceramic, and mica are insulators.

A third type of material, in which electrons can be made to flow under some conditions but not others, is called a **semiconductor**. Most modern electronic devices, such as transistors,

are based on the properties of semiconductor materials, particularly silicon and germanium.

Units of Electrical Current

Every electron carries exactly the same charge. Therefore, one way to indicate the amount of current flowing in a conductor is to specify the number of electrons that pass a given point in the conductor per second. However, because the *charge* on the electron is its important characteristic, amounts of current are measured by the amount of charge that passes a fixed point per second. Because all electrons have exactly the same charge, the amount of charge flowing is directly related to the number of electrons flowing.

Electrical charge is represented by the letter Q. It is measured in units called **coulombs**, designated C. One coulomb of charge represents a large number of electrons, about 6 billion billion (6,000,000,000,000,000,000). More precisely, in scientific notation, a coulomb is:

$$1 \text{ C} = 6.242 \times 10^{18} \text{ electron charges}$$

Electrical current is defined, therefore, as the rate of flow of coulombs—that is, the number of coulombs that pass a point in the conductor per second. It is symbolized by the letter I and is measured in units of **amperes**. One ampere of current equals a flow of one coulomb of charge past a point during one second. More generally, the current flow can be expressed by

CURRENT FLOW IN AUTOMOBILE TAILLIGHT

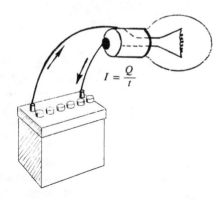

$$I = \frac{Q}{t}$$

FIGURE 1.4. Electrical current is measured in amperes (A). One ampere is a flow rate of 1 coulomb of charge (6×10^{18} electrons) per second. One ampere is about the amount of current flowing in an automobile taillight.

dividing the total charge that passes a point by the time (t) it takes to pass:

$$\text{current (amperes)} = \frac{\text{charge (coulombs)}}{\text{time (seconds)}}$$

or

$$I = \frac{Q}{t}$$

The ampere (often abbreviated *amp*) is symbolized by the letter A. One ampere (1 A) of electrical current is approximately the amount flowing in the taillight of a typical automobile (Figure 1.4). A headlight, on the other hand, conducts about 15 A, while the starter can conduct as much as 180 A when it is turning over the engine. These currents are rather sizable. Most electronic devices conduct much less current. The transistor, which is the principal device used in electronics, requires only about 0.001 A to operate.

Because most electronic circuits carry only fractions of amperes, the smaller units of milliampere (mA) and microampere (μA) are com-

UNITS OF ELECTRICAL CURRENT

$$
\begin{aligned}
1 \text{ mA} &= 0.001 \text{ A } (10^{-3}\text{ A}) \\
1 \text{ }\mu\text{A} &= 0.000001 \text{ A } (10^{-6}\text{ A}) \\
&= 0.001 \text{ mA } (10^{-3}\text{ mA})
\end{aligned}
$$

FIGURE 1.5. Most electronic circuits use small currents: milliamperes (mA) or even microamperes (μA). Conversions between these units have to be done frequently.

GUIDE FOR CONVERTING CURRENT UNITS

To go from:	To:	Multiply by:
mA	A	$\dfrac{1}{1,000}$ or 10^{-3} $\dfrac{A}{mA}$
μA	A	$\dfrac{1}{1,000,000}$ or 10^{-6} $\dfrac{A}{\mu A}$
μA	mA	$\dfrac{1}{1,000}$ or 10^{-3} $\dfrac{mA}{\mu A}$

To go from:	To:	Multiply by:
A	mA	$1,000$ or 10^3 $\dfrac{mA}{A}$
A	μA	$1,000,000$ or 10^6 $\dfrac{\mu A}{A}$
mA	μA	$1,000$ or 10^3 $\dfrac{\mu A}{mA}$

FIGURE 1.6. The table provides a convenient reference for converting between various units of electrical current. Note that either fraction (1/1000) or scientific notation (10^{-3}) can be used.

monly encountered. The **milliampere** is equal to one thousandth of an ampere, or 0.001 A, and the **microampere** is one millionth of an ampere, or 0.000001 A. Figure 1.5 shows the relations between these three units. Conversions between the various units must be done frequently when working with electronic circuits. Figure 1.6 provides a guide to making these calculations. Typical conversions are provided in the following examples.

EXAMPLE 1: A phonograph cartridge conducts 50 μA when the needle is playing on a record. What is the current in A?

Solution: Since amperes are larger units than microamperes, the result must be smaller than 50. From the conversion tables:

$$50 \text{ } \mu A \times \frac{1 \text{ A}}{1,000,000 \text{ } \mu A} = \textbf{0.000050 A}$$

or

$$50 \text{ } \mu A \times \frac{10^{-6} \text{ A}}{\mu A} = \textbf{50} \times \textbf{10}^{-6} \textbf{ A}$$

EXAMPLE 2: A #63 auto-license lamp conducts 0.63 A. What is the equivalent current in mA?

Solution: Since milliamperes are smaller units than amperes, the resulting number should be larger than 0.63. From the conversion tables:

$$0.63 \text{ A} \times \frac{1000 \text{ mA}}{\text{A}} = \textbf{630 mA}$$

or

$$0.63 \text{ A} \times \frac{10^{3} \text{ mA}}{\text{A}} = \textbf{630 mA}$$

Current Direction and Polarity

In circuits that carry direct current, it is important to know in which direction the current is flowing. We might expect that the direction of current flow would be the direction in which the negatively charged electrons flow. Instead, modern electronics uses the conventional current direction.

Conventional current flow is the direction in which an electron would flow if it carried a positive charge. Instead of electrons flowing in the circuit, we speak of the flow of *positive*

DIRECTION OF CONVENTIONAL CURRENT FLOW

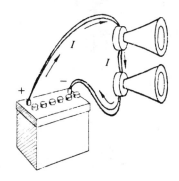

Actual Circuit

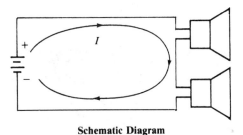

Schematic Diagram

FIGURE 1.7. The direction of conventional current flow is the direction that a positive charge carrier would flow in the circuit – that is, toward points of negative polarity.

charge or, sometimes, *positive charge carriers*, imaginary objects that carry a positive charge. Thus, in conventional current flow, the *arrow points toward the negative end of a conductor* (Figure 1.7).

The reason for the use of the conventional current direction is partly historical and partly mathematical. Before it was known that electrical current was a flow of electrons, it was measured simply as a flow of charge that could be either positive or negative. The results of various experiments could not tell the difference, so the two descriptions were equivalent. In order to standardize discussions, Benjamin Franklin arbitrarily designated current as a flow of posi-

SEMICONDUCTOR DEVICES AND SYMBOLS

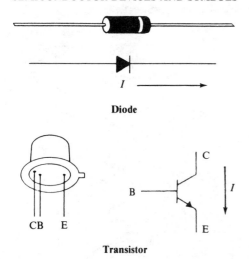

FIGURE 1.8. Arrows on the symbols for semiconductor devices show the direction of flow of positive charge carriers.

tive charge, thus establishing current direction toward negatively charged points.

More recently, as the field of semiconductors developed, the mathematical description of their behavior was sometimes easier if one assumed that positive charge carriers, as well as negatives ones, were flowing. As a result, the electronic symbols for all semiconductor devices, such as diodes and transistors, show the direction of flow of positive charge carriers (Figure 1.8). Because modern electronics is primarily based on semiconductor devices, conventional current direction is used throughout this book.

It should be noted that some textbooks use the **electron current** convention that shows current flow as the direction in which electrons, or negative charge carriers, would flow. The electron current direction is the opposite of the conventional current direction that is used here. In reading other texts, always check to see which current convention it has adopted.

Current Flow in Circuits

In actual circuits, conventional current means that the direction of current flow is the direction in which a positive charge carrier flows through the conductors and devices that make up the circuit. In Figure 1.7, for example, a battery makes one end of a circuit positive and the other end negative. A positive charge carrier flowing around the circuit would flow toward the negative battery terminal. Therefore, the current direction is as shown.

It should be emphasized that current direction, as shown, for example, in Figure 1.7, is determined by the direction in which a positive charge carrier flows *around a complete circuit*, not within the battery. That is, the terminals of the battery or other voltage source establish the polarity difference for the circuit and, thereby, the direction in which positive charge carriers will flow. Thus, the charge carriers are viewed as leaving the positive terminal of the battery and being attracted to the negative terminal via the conductors and any devices that may be in the circuit.

As shown in Figure 1.9, the circuit must be complete, or *closed*, in order for current to flow; there can be no gaps or openings in the path. If the circuit is opened, for example, by a switch, the current immediately stops flowing. Such a circuit is called an **open circuit**, and no dc current will flow. Similarly, any arrangement of devices that allows current to flow in a closed path is called a **closed circuit**.

Devices connected one to the other (in series), as shown in Figure 1.10, make up a **series circuit**. The *same current flows through each device in a series circuit*. This is an important point. The amount of charge flowing around a circuit is everywhere the same. Charge is neither created nor absorbed by any of the devices in a circuit. Therefore, no matter where the current flowing in a series circuit is measured, the value will always be the same.

Two devices connected parallel to each

CURRENT IN OPEN AND CLOSED CIRCUITS

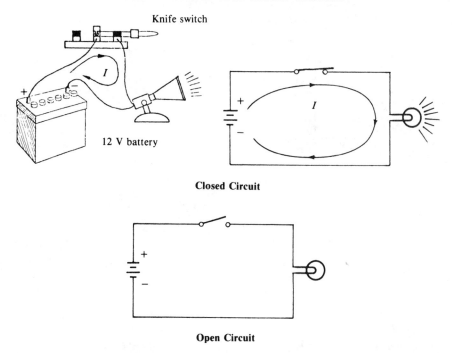

FIGURE 1.9.   Current will only flow in a closed circuit. If the circuit is opened — for example, by a switch — the current immediately stops.

other, as in Figure 1.10, make up what is called a **parallel circuit.** In a parallel circuit, the *current will divide, some current going through each device.* The total current flowing in the circuit remains constant, however, so that in a parallel circuit like Figure 1.10, we always have:

$$I = I_1 + I_2$$

Switches

The opening and closing of an electrical circuit is one of the most fundamental purposes of electronics. In electronic control systems, for example, electronic circuits are used to turn on and off a wide variety of output devices automatically — heaters, lights, motors, pumps,

alarms, and the like. Electronic switches are also the basis of computer operation. Hence, the use of switches to open and close circuits will appear frequently throughout this book.

The simplest type of switch, called a **toggle switch,** is a mechanical one in which moving a lever or pushing a button causes metal contacts to open and close, thus opening or closing an electrical circuit. Figure 1.11 shows the three basic forms of the toggle switch and their circuit symbols.

In the simplest kind of toggle switch, one wire of the circuit is connected to a metal contact, or **pole,** that is part of the movable toggle lever. The other circuit wire is connected to a separate, fixed metal contact, or pole. Moving the toggle lever back and forth causes the

CONNECTION OF CIRCUIT COMPONENTS

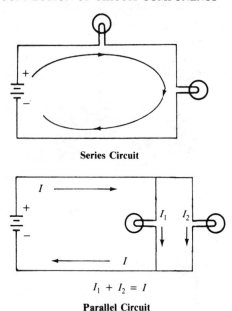

Series Circuit

$$I_1 + I_2 = I$$

Parallel Circuit

FIGURE 1.10. Circuit components can be connected in series or in parallel. In a series circuit, the same current flows through each device. In a parallel circuit, the current divides, but the total current is constant.

TOGGLE SWITCH AND SYMBOLS

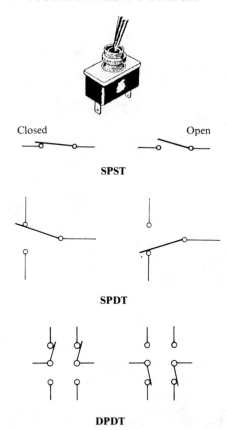

FIGURE 1.11. Opening and closing circuits (switching) is an important function of electronics. The most common switch is a mechanical one in which a lever is used to open and close metal contacts. The common toggle switch has three basic forms: SPST, SPDT, and DPDT.

switch to make or break contact with the fixed pole, thus closing or opening the circuit. Because the toggle of this switch has only one pole and the switch can be thrown to make contact with only one other pole, the switch is called a **single-pole–single-throw** (SPST) switch. Most electrical switches, including those that turn on lights, are SPST switches.

If the toggle has only one pole, but can be thrown in two directions to make contact with either of two other poles, the switch is called **single-pole–double-throw** (SPDT). This type of switch might be used to turn on either a light or an alarm, but not both at the same time.

A third option is the **double-pole–double-throw** (DPDT) switch shown in Figure 1.11. The toggle of this swtich has two poles and can

be thrown in either of two directions to connect each pole to one of two other poles. There are many different ways this switch can be used — for example, to turn on or off two systems simultaneously, to reverse current directions in a circuit, and so on.

There are many varieties of the mechanical switch, including push-button switches, slide switches, and rotary switches. Despite the dif-

REED SWITCH AND SYMBOLS

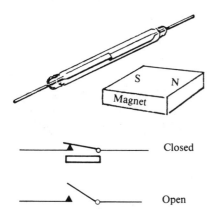

FIGURE 1.12. Another type of mechanical switch is the reed switch. Instead of a lever, a magnet is used to open and close the metal contacts. The reed switch has wide application in burglar alarm systems.

ferent mechanisms for opening and closing the contacts, however, their general classification, basic forms, and schematic symbols are those shown in Figure 1.11.

One interesting type of mechanical switch is the **magnetic reed switch** shown in Figure 1.12. When the magnet is placed next to the reed switch, it attracts one pole of the reed switch to make contact with the other, thus closing the circuit. When the magnet is removed, the reeds separate and the circuit opens. By attaching the magnet to one object — for example, a door — and the switch to another object — for example, the door frame — the switch can be used to indicate whether the two objects are close to each other or apart (door open or closed). This particular switch has widespread application in burglar alarm systems to turn on an alarm when a door or window is opened by an intruder.

While mechanical switches are widely used, the real value of electronics is realized when the electronic circuitry does the switching automatically. The earliest such form of **electronic**

RELAY SWITCH AND SYMBOLS

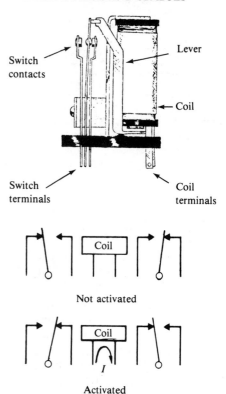

FIGURE 1.13. An early form of an electronic switch is the relay. When a current passes through its coil, the magnetic field produced moves a lever that in turn switches the contacts of a DPDT switch. Thus, a control current in the coil circuit can be used to switch other circuits.

switch is the **relay** shown in Figure 1.13. In this switch, the electrical current from one circuit passes through a special coil of wire in the relay. Current in the coil creates a magnetic field that attracts the lever of the relay switch, causing it to move. The moving lever is the toggle of a DPDT switch, as shown by the relay's schematic diagram. A wide variety of devices can be connected to the relay terminals for turning on or off by an electronic control circuit connected to the coil.

ELECTRICAL VOLTAGE

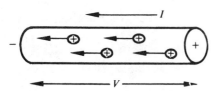

FIGURE 1.14. Voltage specifies the size of a po-larity difference between two points. Its units (volts, V) show that voltage is the energy gained by the charge carriers in moving between the two points.

Because the operation of the relay requires a certain amount of current to activate the lever, typically 10 mA to 100 mA, this type of switch is termed a **current-controlled switch**. It had widespread applications in early electronics, particularly in the early computers. With the development of modern semiconductor devices that require substantially less current and, hence, less electrical power, relay applications have been greatly reduced.

1.4 ELECTRICAL VOLTAGE

For current to flow in a conductor, there must be a difference in polarity between its two ends. The greater the polarity difference, the stronger the force that attracts the charge carriers and the faster they move. The faster the charge carriers move, the greater the flow of charge and the greater the electrical current.

The quantity that measures the size of the polarity difference is called **voltage** and is given the symbol V (Figure 1.14). The greater the voltage across a particular conductor, the greater the current that flows. The unit of voltage is called the **volt** (V) (Figure 1.15).

Voltage is not really a force, however. Rather it is a measure of the amount of *energy transferred to a charge carrier* as it moves

UNITS OF VOLTAGE

$$
\begin{aligned}
1 \text{ mV} &= 0.001 \text{ V } (10^{-3} \text{ V})\\
1 \text{ } \mu\text{V} &= 0.000001 \text{ V } (10^{-6} \text{ V})\\
&= 0.001 \text{ mV } (10^{-3} \text{ mV})
\end{aligned}
$$

FIGURE 1.15. **The unit of voltage is the volt. Most electronic devices operate on about 1 V to 15 V, but incoming signals are often small: millivolts (mV) or microvolts (μV).**

through the conductor. As the charge carriers are attracted from one end of a conductor to another, they pick up energy in the form of motion. This energy is ultimately transferred to the atoms of the conductor and appears as heat.

It should also be noted that two other nearly equivalent terms are sometimes used for the quantity we will call voltage. These terms are potential difference and electromotive force. **Potential difference** expresses the fact that even though a conductor may not be present, a difference in polarity between two points may exist. This polarity *difference* has the *potential* for causing a current to flow, whether a conductor is present or not.

Electromotive force (emf or E) on the other hand, expresses the idea that the polarity difference provides a *force* that will cause *electrons* to *move* (motive). Voltage, potential difference, and electromotive force are terms that are often used interchangeably in books and literature. Throughout this book, however, we will use the term *voltage*.

Units of Voltage

Because voltage is a measure of energy, the volt is defined in terms of energy. The unit of energy that is used in electronics is based on the Standard International (SI) System of units and is called the **joule** (J). A joule of energy is approximately the amount of energy required to lift a 1 kg object (2.2 lb) 10 cm (4 in.) off the ground.

Thus, voltage is defined as the amount of energy (in joules) gained by 1 coulomb of charge carriers in moving along a conductor. That is, 1 volt (V) between the ends of a conductor can cause 1 coulomb (C) of charge carriers to gain 1 joule (J) of energy as they move from one end of the conductor to the other. More generally, the voltage between any two points in a conductor can be calculated by dividing the amount of energy gained by a flow of charge carriers moving between the two points by the amount of charge that moved:

$$\text{voltage (volts)} = \frac{\text{energy (joules)}}{\text{charge (coulombs)}}$$

As with electrical current, a wide range of voltages will be encountered. A typical D cell battery has a voltage of about 1.5 V; an automobile battery has about 12 V, and voltages as high as 20,000 V can be found inside TV sets. Most electronic systems operate in the range of 1 V to 15 V, though many incoming signals, such as the ones from a phonograph pickup, are as low as several millivolts (0.001 V) or even microvolts (0.000001 V).

Because a wide range of voltages is common in electronics, it is necessary to make conversions between these units frequently, particularly between volts, millivolts (mV), and microvolts (μV). The procedure for making the conversions is identical to that presented for current. The conversion guide in Figure 1.6 can be used by simply substituting voltage (V) for current (A). The conversion procedure is illustrated by the following examples.

EXAMPLE 3: A phonograph cartridge has an output of 0.005 V. What is the resulting voltage in millivolts (mV)?

Solution: The larger unit, volts, will equal more ($\times$ 1000) millivolts. Therefore, multiply the number of volts by 1000 to obtain the equivalent voltage in millivolts.

$$0.005 \text{ V} \times \frac{1000 \text{ mV}}{\text{V}} = \textbf{5 mV}$$

EXAMPLE 4: How many volts are in an antenna signal of 50 μV?

Solution: The resulting number of volts must be less than the number of microvolts.

$$50 \text{ μV} \times \frac{10^{-6} \text{ V}}{\text{μV}} = \textbf{50} \times \textbf{10}^{-6} \textbf{ V}$$

EXAMPLE 5: How many millivolts are in 1500 μV?

Solution: The resulting number of millivolts must be less than the number of microvolts.

$$1500 \text{ μV} \times \frac{10^{-3} \text{ mV}}{\text{μV}} = \textbf{1.5 mV}$$

Caution: A voltage difference across an electrical conductor causes a current to flow. The human body is an electrical conductor. If a voltage is placed across two points on the human body, a current will flow, and current flowing in the body can be fatal.

Figure 1.16 shows the results of various amounts of current flowing in the body. Since voltage causes the flow of current, extreme caution should be observed around voltage sources. A voltage as low as 42 V has killed a person. As rule of thumb, *be extremely careful around voltages greater than 30 V.*

EFFECTS OF ELECTRICAL CURRENT

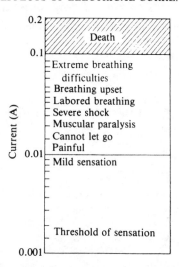

FIGURE 1.16. If a voltage is placed across two points in the human body, a current will flow. Effects of various currents on the human body range from mild sensation to death.

RELATIVE POLARITY AND VOLTAGE

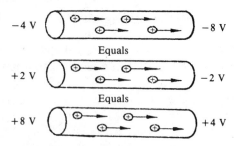

FIGURE 1.17. Polarity and voltage are relative quantities. Positive charge carriers and, therefore, current will flow toward points that are more negative (less positive) than other points.

Common and Ground

We have indicated that electrical current is a flow of positive charge carriers from a region of positive polarity to a region of negative polarity. The greater the polarity difference, the greater the current flow. But polarity, and therefore voltage, is a *relative* quantity. For example, suppose that two ends of a conductor carry a negative voltage, but one is more negative than the other (Figure 1.17). A positive charge carrier would experience a force of attraction from both ends, but the force from the end with the greater negative voltage would be greater. Hence, the positive charge carrier would move toward that end. The end with less negative voltage would therefore appear to repel the positive charge carriers and act as if it had positive polarity. When we use the symbols + and −, therefore, we are referring to the relative polarity of one point to another, rather than to its absolute polarity.

In electronic circuits, there may be many points at which you will want to know the relative voltage, so you will need a reference point. This point is called **common** because it is the common point to which all voltages refer. The voltage at any point in a circuit may be either positive or negative with respect to common, and currents will flow toward points that are more negative. For example, if a point is at $+6$ V with respect to common, the current will flow toward common. If it is at -6 V with respect to common, then it will flow away from common.

When several systems are connected, it is usually necessary to establish a common reference point for all of them. This common point is generally accepted to be the earth or the ground and is called **ground** (GND). It is designated on a circuit diagram by the symbol shown in Figure 1.18. This symbol is often used to indicate common, which, in fact, may or may not be true ground.

Ground means that some point in the system is actually connected to the earth. Generally the connection is done through the water-pipe system of a building because water pipes coming into the building are buried in the ground. In electronic systems that operate from 110 V ac electrical outlets, this connection is made by the plug to the third, or ground, termi-

VOLTAGE REFERENCE SYMBOLS

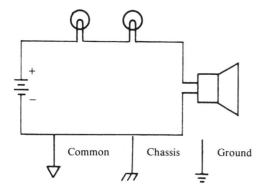

Common   Chassis   Ground

FIGURE 1.18. Because voltage is relative, a point of reference must be selected. This point is called common and is indicated on circuit diagrams by the symbol shown. When a point is connected to the chassis or a water pipe (ground), other symbols are used.

WALL OUTLET SAFETY GROUNDS

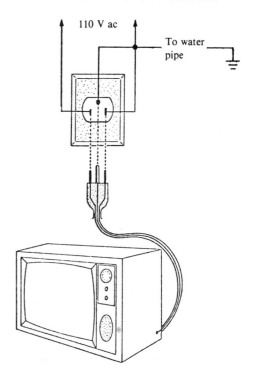

FIGURE 1.19. The third terminal of wall outlets is connected to a water pipe buried in the ground. Any exposed part of an electronic system should be connected to this terminal for safety.

nal of the outlet (Figure 1.19). At some point in the building, this third terminal is connected to a water pipe. One of the other outlet wires is also attached to the ground wire somewhere in the building. Therefore, only one of the three pins is at 110 V ac with respect to the other two.

The use of the earth ground has considerable safety implications. For example, when you stand on the ground, your feet are connected to ground. If you touch something that has a voltage with respect to ground, a current will flow through your body, which could be hazardous. Therefore, any exposed part of an electrical system should be at ground, the same as you. In any approved electrical system, grounding is done by electrically connecting any exposed part of the system to the third terminal of the plug.

Direct Current Voltage Sources

Two types of voltage source are commonly available: one that produces direct current (dc) in a conductor and one that produces alternat-

ing current (ac). In a **dc voltage source**, the relative polarities of the source stay the same. A battery is a typical dc voltage source. One end of the battery is positive and the other end is negative, and they don't change. When a dc voltage source is connected across two ends of a conductor, the current will flow in only one direction through the conductor. Figure 1.20 shows a number of common battery varieties and indicates what voltage they supply and the polarities of each end.

In an **ac voltage source**, however, the polarity changes. The 110 V wall outlet is an example of an ac voltage source. The voltage on the "live" pin alternates from positive to negative

BATTERIES AND SYMBOL

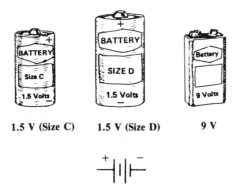

1.5 V (Size C)   1.5 V (Size D)   9 V

FIGURE 1.20.   Batteries are a common dc voltage source. Their symbol shows a fixed polarity.

VARIABLE DC POWER SUPPLY AND SYMBOL

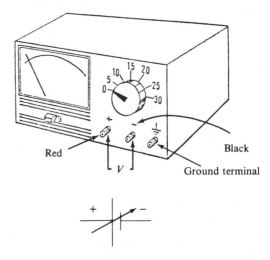

FIGURE 1.21.   Variable dc power supplies are generally used for testing electronic circuits. The voltage appears across the + and − terminals. The ground terminal is used only to establish the relative polarity of the output.

with respect to the other two pins (or ground) at a rate of 60 times per second. When an ac voltage source is connected to a conductor, the current flowing in it will change direction each time the voltage changes.

Almost all electronic systems require dc voltages for their operation. One source of voltage for such systems is, of course, batteries. Because batteries last for a limited time, however, most nonportable electronic systems operate from the electricity supplied by wall outlets. But, the voltage supplied by the wall outlet is ac. Therefore, the first thing that must be done to the ac voltage used to power an electronic system is to convert it to dc. The unit inside the system that does this is called a **dc power supply**.

In studying the behavior of electronic circuits, you will need to supply them with a dc voltage. Because batteries are relatively expensive, you will want to use a dc power supply that operates from a wall outlet. Figure 1.21 shows a typical ac-operated dc power supply used for experimenting with electronic circuits. Just like a battery, it has two terminals labeled + and − . The voltage from the power supply appears across these + and − terminals, which

are connected to the circuit. The current flow is from the + terminal, through the circuit, and then to the − terminal.

Note that there is a third terminal labeled ground. This terminal is connected to the ac plug ground, but the + and − terminals are "floating" with respect to it — that is, neither of them is electrically connected to the ground terminal. This situation lets you use the power supply as either a positive or a negative voltage source with respect to ground.

For example, suppose you wanted to supply a positive voltage to a circuit with respect to ground. You would connect the negative power supply terminal to the ground terminal and the power supply would provide a positive voltage from that terminal with respect to ground.

On the other hand, if you wanted to supply a negative voltage with respect to ground, you would connect the positive terminal to the ground terminal, and the negative terminal would provide a negative voltage with respect to ground.

No matter how the supply is used, the electronic circuit must always be connected across the + and − terminals because that is where the voltage appears. Connecting a circuit across + and ground or across − and ground will produce no current.

For easy, visible reference, a color code is generally used on voltage terminals. The **positive terminal** is always **red** and the **negative terminal** is always **black**. The ground terminal is also generally black, but sometimes a third color, such as green or white, is used to distinguish it from + and −. The ground terminal may also be designated simply by the ground symbol $\perp$.

Most dc power supplies intended for laboratory use are variable, meaning that the voltage appearing at the + and − terminals can be changed by turning a knob. More expensive power supplies also have a meter on them to indicate how much voltage is actually appearing at the + and − terminals. When no meter is present, the terminal voltage has to be measured separately.

Transistor Switch

The relay described earlier as an example of an electronic switch has been largely replaced by semiconductor devices, most notably the transistor. While the transistor has many other applications, one of its earliest and most important is as an electronic switch. The relay was described as a current-controlled switch because it requires a certain current in the coil before the switch will activate. Similarly, the **transistor switch** can be described as a **voltage-controlled switch** because a minimum voltage is required for activation.

A typical transistor and its schematic diagram appear in Figure 1.22. Note that it has three terminals marked B (base), C (collector), and E (emitter). The "switch" is contained in

NPN TRANSISTOR AND SYMBOL

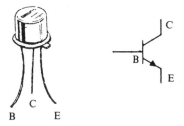

FIGURE 1.22. Electrical voltage can also be used in switching. One common transistor application is as a voltage-controlled switch. A typical transistor has three terminals: collector (C), emitter (E), and base (B).

the junction between the collector and emitter (CE) terminals. Unlike the relay, this switch has no mechanical metal contacts that open and close. Rather, the semiconductor characteristics of the CE junction are such that it will behave either as a good conductor and permit current to pass (switch closed) or as a good insulator and block current (switch open), depending on the voltage at the base terminal.

The base terminal thus acts like the electronic toggle of an SPST switch. If the base voltage V_{BE} is 0 V, the CE junction is open and no current flows; if the base voltage is several volts positive, the CE junction is closed and current flows easily (Figure 1.23). The exact voltage required to open and close the switch depends on the particular transistor, but it is typically several volts.

The physics underlying the properties of the CE junction that permits it to act either as a conductor or an insulator, depending on the base voltage, is beyond the scope of this book. However, the capacity of certain materials, particularly silicon and germanium, to conduct electricity under certain circumstances is the basis for the term *semiconductors*.

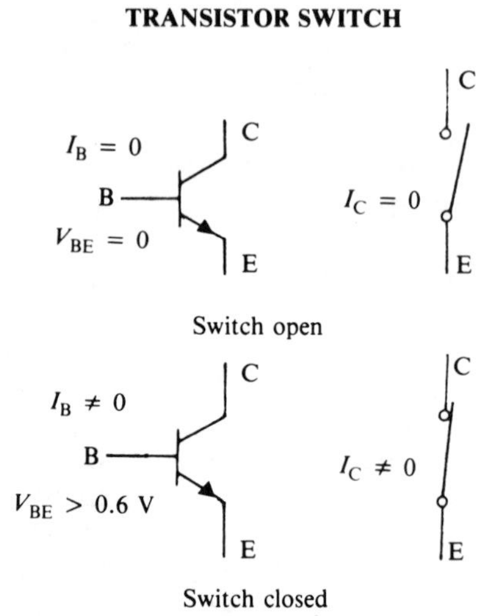

TRANSISTOR SWITCH

$I_B = 0$

B

$V_{BE} = 0$

C

E

$I_C = 0$

C

E

Switch open

$I_B \neq 0$

B

$V_{BE} > 0.6$ V

C

E

$I_C \neq 0$

C

E

Switch closed

FIGURE 1.23.   In a transistor switch, the base terminal acts as the control element that opens and closes a semiconductor junction between the collector and emitter terminals When the base voltage V_{BE} is 0, the CE junction acts like an insulator, and the switch is open. When V_{BE} is several volts positive, the CE junction acts like a conductor, and the switch is closed.

1.5 MEASUREMENT OF CURRENT AND VOLTAGE

Types of Meters

Electrical current and voltage are measured by meters called, respectively, **ammeters** (abbreviation of ampere-meter) and **voltmeters**. The schematic symbols for these meters are shown in Figure 1.24. There are two types of electrical meters in common use today: analog meters and digital meters. Their names reflect the way in which they register the quantity being measured.

On **digital meters,** the voltage (or current) appears as lighted numerals, or *digits*. If the voltage changes, the numerals change to indicate the new value. This type of meter is fairly

easy to read, particularly if the voltage is constant, and the possibility of reading error is small.

On **analog meters,** a needle moves across a marked scale. The measured value is indicated by the position of the needle over the scale. The name *analog* is applied because the needle motion is "analogous," or similar, to the changes in the measured quantity. For constant values of voltage or current, this type of meter is somewhat more difficult to read than the digital type. However, when the quantity is changing, the analog meter will often give a better indication of the value at any time because its scale is fixed. On a digital meter, a changing quantity may cause the numerals to change too rapidly to be read.

In this book, the analog meter is used exclusively in the illustrations for two reasons. First, they are less expensive than digital meters, so you are much more likely to use them. Second, the reading of analog meters requires some explanation and some practice in order to get accurate values.

Proper Meter Connections

An ammeter is essentially a flow meter because it measures the amount of charge flowing through the circuit. As with most flow meters, the *current must flow through the ammeter* to be measured. Therefore, the circuit must be opened where the current is to be measured and the ammeter inserted in the circuit (Figure 1.25).

An ammeter is also sensitive to the direction of the current. It has two terminals marked positive (red) and negative (black), or sometimes *common* instead of negative. Only when the current flows into the positive terminal and out of the negative terminal does the meter read. Therefore, the ammeter indicates both the amount of current and the direction.

A voltmeter, on the other hand, measures *differences* in voltage between various points.

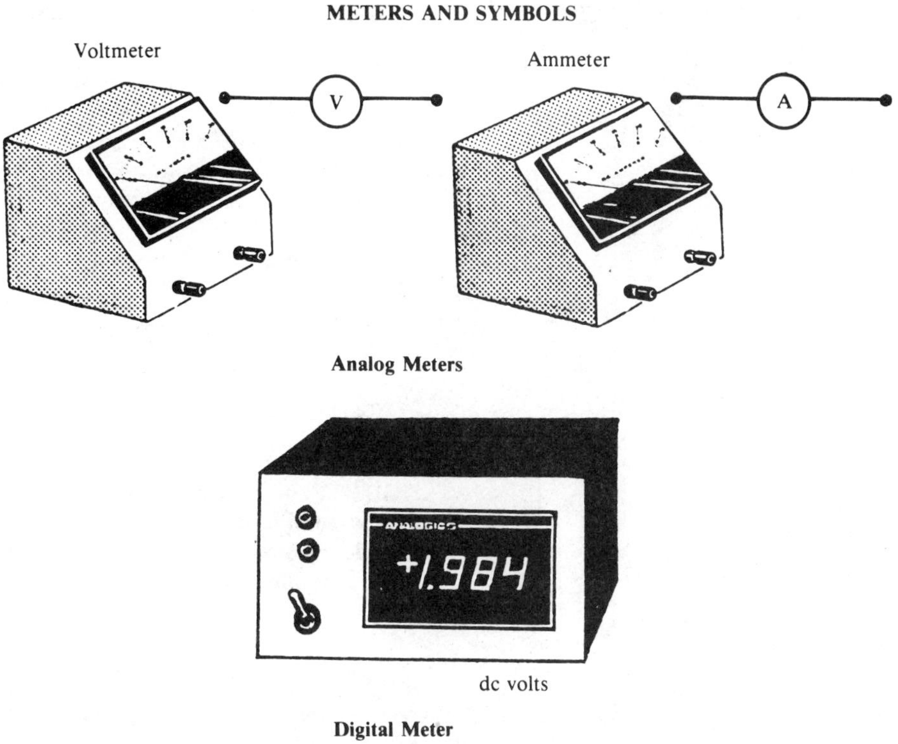

METERS AND SYMBOLS

Voltmeter

Ammeter

Analog Meters

dc volts

Digital Meter

FIGURE 1.24. Current is measured by ammeters and voltage by voltmeters. These meters may be either the digital type that registers the values as lighted numbers or the analog type that uses a needle to indicate values on a marked scale.

Therefore, a *voltmeter is connected across two points in a circuit* where a voltage difference is to be measured. The voltmeter is also sensitive to polarity. It has positive and negative (or common) terminals, and only when the positive terminal is connected to the point of higher voltage does the meter read. Thus, the voltmeter indicates both the amount of voltage and the relative polarity.

The Multimeter

The meter most commonly used in electronics is a multipurpose meter, or **multimeter**. It measures both current and voltage, depending on

how it is connected to the circuit. It can also be used to measure electrical resistance that is described in the next section, as well as ac currents and voltages. The ac quantities are described later, but the measurement techniques are essentially the same as for dc.

Figure 1.26 shows five characteristics shared by all multimeters:

— **Terminals,** to connect the meter to the circuit;
— **Function switch,** to select either ac or dc quantities for measurement;
— **Range switch,** to select the proper quantity and the range of values that can be

AMMETER AND VOLTMETER CONNECTIONS

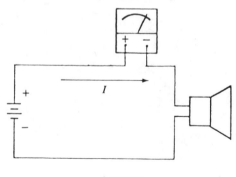

Ammeter

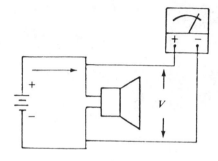

Voltmeter

FIGURE 1.25. An ammeter is a flow meter, so it must be inserted into the circuit. A voltmeter, on the other hand, measures differences in voltage, so it is connected across two circuit points.

measured—that is, at full-scale needle deflection;
- **Needle and scales**, to indicate values;
- **Zero adjustment**, to set the needle to exactly zero.

This type of meter has only one needle to indicate all the quantities it can measure, but numerous scales. Each scale has markings corresponding to the quantity and range being measured—that is, to the different settings of the function and range switches.

Figure 1.27 shows a typical multimeter

MULTIMETER COMPONENTS

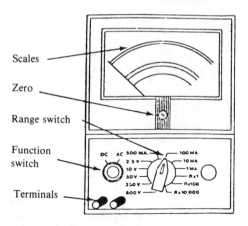

FIGURE 1.26. The multimeter is the basic instrument used in testing electronic circuits. It can measure dc current and voltage, as well as resistance and ac quantities. All multimeters have the five components shown.

MULTIMETER SCALES

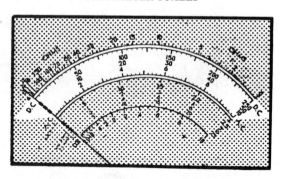

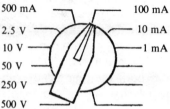

FIGURE 1.27. Because the multimeter has only one needle, many scales are required. The scale shown unshaded is used for reading dc voltage and current. Each set of numbers is also used with more than one range switch position.

EXAMPLE 6

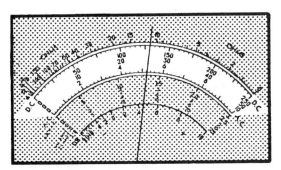

FIGURE 1.28. When the needle is in the position shown, the correct readings for various range switch positions are the ones given in the solution to this example.

EXAMPLE 7

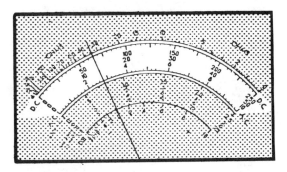

FIGURE 1.29. Determine the correct scale numbers and readings for the different range switch positions given in this example.

scale used to measure dc current and voltage. One set of markings corresponds to three sets of numbers, 0 to 10, 0 to 50, and 0 to 250. Each of these sets of numbers is used with more than one setting of the range switch. In the meter shown in Figure 1.27, they are:

Set of Numbers	Range	Switch Positions
0–10	1 mA	Current
	10 mA	Current
	100 mA	Current
	10 V	Voltage
0–50	500 mA	Current
	50 V	Voltage
	500 V	Voltage
0–250	2.5 V	Voltage
	250 V	Voltage

Because there are so many scales and because a single scale is used for several ranges, a multimeter is somewhat complicated to read. It requires some practice to get an accurate reading, and you must always double-check to be sure that you have read the proper scale correctly.

The following example gives the dc values indicated by the single needle position of Figure

1.28 for different range settings. Example 7 gives you an opportunity to test yourself.

EXAMPLE 6: If a meter needle is positioned as shown in Figure 1.28, what are the correct scale numbers and readings for the different range switch positions given below?

Solution:

Range Switch Position	Correct Scale Numbers	Correct Reading
250 V	dc 0–250	130 V
50 V	dc 0–50	26 V
10 V	dc 0–10	5.2 V
2.5 V	dc 0–250	1.3 V
500 mA	dc 0–50	260 mA
100 mA	dc 0–10	52 mA
1 mA	dc 0–10	0.52 mA

EXAMPLE 7: If a meter needle is positioned as shown in Figure 1.29, fill in the correct scale numbers to be used and the correct scale reading.

Range Switch Position	Correct Scale Numbers	Correct Reading
250 V		
50 V		
10 V		
2.5 V		
500 mA		
100 mA		
1 mA		

Solution: The correct answers are given in the footnote below.*

1.6 ELECTRICAL POWER

So far you have learned about two basic quantities of electronic circuits: current and voltage. These two quantities together determine a third important quantity for electrical systems: **energy**, or more specifically, the rate at which electrical energy is being used.

The **rate** of energy usage means how rapidly the energy is being used—for example, the amount of energy being used each second. Energy rate is called **electrical power** and is designated P.

Units of Electrical Power

If electrical energy is specified in joules, electrical power is specified in units called **watts** (W).

*Solution to Example 7:

Range Switch Position	Correct Scale Numbers	Correct Reading
250 V	dc 0–250	65 V
50 V	dc 0–50	13 V
10 V	dc 0–10	2.8 V
2.5 V	dc 0–250	0.65 V
500 mA	dc 0–50	130 mA
100 mA	dc 0–10	28 mA
1 mA	dc 0–10	0.28 mA

TYPICAL POWER RATINGS

Device	Power Rating (W)
Range	up to 16,000
Clothes dryer	5,600
Air conditioner	up to 2,080
Heater	up to 1,650
Dishwasher	1,400
Toaster	1,200
Iron	1,000
Color TV	420
Clothes washer	400
Refrigerator	320
Hi fi	230
Black and white TV	205
Fan	200
Tape recorder	60
Radio	30
Shaver	11
Clock	2

FIGURE 1.30. Most electrical devices are rated by their power consumption in watts, which is the rate at which they use energy in joules per second.

One watt is a rate of energy usage of one joule per second. Because the rate of energy usage may vary from one second to the next, a more useful quantity is the *average* electrical power—that is, the average rate at which energy is used. The average electrical power is defined as the total energy used divided by the time during which it is used:

$$\text{power (watts)} = \frac{\text{energy (joules)}}{\text{time (seconds)}}$$

Because the total energy used by a device depends on how long it is operated, most devices are given a power rating in watts. This is particularly true of household devices, such as light bulbs, refrigerators, ranges, TVs, and so on (Figure 1.30). Light bulbs use tens to hundreds of watts in producing light, and an elec-

ELECTRIC METER

FIGURE 1.31. Electric companies charge for energy usage. Electric meters add up the product of power consumption in kilowatts and time in hours to give the total energy used in kilowatt hours.

tric range may use several thousand watts in cooking.

Because electric motors are used for various types of mechanical tasks (drilling, pumping, blowing, and so forth), they are sometimes rated by their mechanical power *output* in **horsepower** (hp), rather than by their electrical power *input* in watts. The two are directly related: 1 hp = 746 W. Therefore, a 5 hp electric motor will require an electrical power of:

$$5 \text{ hp} \times 746 \text{ W/hp} = 3730 \text{ W}$$

To determine how much energy a device has used during any period of time, simply multiply its power rating (in watts) by the length of time (in seconds, s). For example, a 100 W bulb (100 J/s) operated for 1 h (3600 s) will use:

$$100 \text{ W} \times 3600 \text{ s} = 360{,}000 \text{ J}$$

Electric companies charge for the *total* energy used (Figure 1.31). Because the joule is a relatively small quantity of energy, they use a larger unit, the **kilowatt hour** (kWh). Electric rates are set at so many cents per kilowatt hour. The kilowatt hour is a unit of energy because it is the power used (in kilowatts, kW) multiplied by the time during which it was used (in hours, h):

$$1 \text{ kW} = 1000 \text{ W}$$
$$1 \text{ h} = 3600 \text{ s}$$

Therefore, 1 kWh is equal to an energy of:

$$1 \text{ kWh} = 1000 \text{ W} \times 3600 \text{ s} = 3{,}600{,}000 \text{ J}$$

A typical home in the United States uses electrical energy at an average rate of about 800 W or 0.8 kW. At $0.09/kWh, this represents an average cost of about $0.07 every hour of the day.

Power Rating of Electronic Devices

The electrical power used by a device is an important consideration not only because electrical energy can be expensive, but also because the power used by a device appears as heat. If this heat is too great, the temperature of the device can get too high and the instrument may be destroyed. As a result, essentially every electronic device has a maximum power rating that should not be exceeded.

Calculating Electrical Power

The maximum power rating of a device may be given in watts, but it is just as likely that it will be given as maximum operating voltage and current. This is so because the power used by an electrical device is determined both by the voltage across it and the current through it. Electrical power is the product of these two quantities and is expressed by the **power relation**: The power *P* used by a device is found by multiplying the voltage *V* across the device by the current *I* flowing through it. In equation form:

power = voltage × current

watts = volts × amps

$$P = V \times I$$

EXAMPLE 8

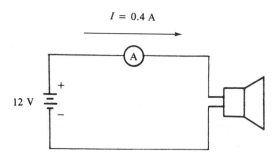

$$P = 12 \text{ V} \times 0.4 \text{ A} = 4.8 \text{ W}$$

FIGURE 1.32. The electrical power used by a device is the voltage across it times the current through it. This is called the power relation: $P = V \times I$.

EXAMPLE 8: An alarm has a measured voltage of 12 V and the current flowing through it is 0.4 A. How much power does the alarm use (Figure 1.32)?

Solution: Using the power relation:

$$P = V \times I = 12 \text{ V} \times 0.4 \text{ A} = \textbf{4.8 W}$$

The electrical energy used by an electronic circuit is provided by the voltage source, either a battery or a dc power supply. Because all the electrical power used by a circuit must come from the voltage source, the power relation can also be used to calculate the power drawn from a source. We simply measure the voltage across the source terminals and the current flowing out of one terminal, and then multiply the two together (Figure 1.33).

EXAMPLE 9: A transistor radio has a 9 V battery as a voltage source. The current flowing from the battery when the radio is

TYPICAL APPLICATION OF POWER RELATION

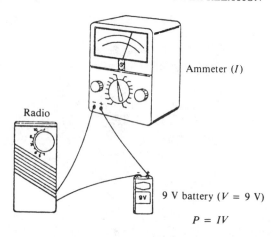

FIGURE 1.33. One application of the power relation is to calculate the power drain from a voltage source.

operating is measured to be 150 mA. What is the power supplied by the battery?

Solution: To use the power relation, the current must be in amperes. Therefore, 150 mA must first be converted to amperes. From Figure 1.6, we get the conversion factor:

$$I = 150 \text{ mA} \times \frac{10^{-3} \text{ A}}{\text{mA}} = 0.15 \text{ A}$$

Using the power relation:

$$P = V \times I = 9 \text{ V} \times 0.15 \text{ A} = \textbf{1.4 W}$$

Because most of the electrical energy used by an electronic device is converted into heat, we must often be careful that a device doesn't get too hot and become damaged. Many electronic devices have a power rating that tells how much power can safely be *dissipated* or used by the device without damage. The power relation

EXAMPLE 10

$P_{max} = 5\ W$

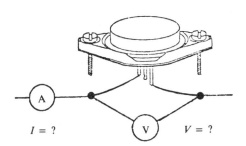

$I = ?$ $V = ?$

FIGURE 1.34. Many devices have a power rating that tells how much power they can dissipate without damage. The power relation can be used to calculate safe operating currents and voltages from this rating – for example, V and I for the transistor shown.

can be used to calculate safe operating voltages and currents for such devices.

EXAMPLE 10: A transistor has a maximum power rating of 5 W when operated at 15 V. How much current can safely flow through the device (Figure 1.34)?

Solution: The power relation states:

$$P = V \times I$$

Since we want to know the current I, we divide both sides of the equation by V:

$$\frac{P}{V} = \frac{VI}{V}$$

or

$$I = \frac{P}{V}$$

Using the values for P and V:

$$I = \frac{5\ W}{15\ V} = 0.33\ A$$

UNDERSTANDING ELECTRICAL POWER

$$I = \frac{Q}{t}$$

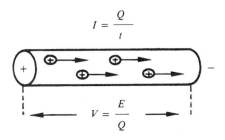

$$V = \frac{E}{Q}$$

FIGURE 1.35. The reason that the product of voltage and current represents power follows from their definitions. Current is the amount of charge flowing per second, and voltage is the energy transferred to each charge.

Why does the product of voltage and current represent the rate of electrical energy usage? This can be understood from the definitions given for voltage and current. Current was defined as the rate at which charge flows through a device, while voltage was defined as the amount of energy transferred to the charge carrier as it moves (Figure 1.35). Therefore, if the amount of charge moving through a device per second (the current) is multiplied by the amount of energy transferred to each charge (the voltage), the result will be the rate at which energy is transferred to the charge carriers per second. This, of course, is power P:

$$I \times V = \frac{\text{charge flowing}}{\text{time (seconds)}} \times \frac{\text{energy transferred}}{\text{charge flowing}}$$

$$= \frac{\text{energy transferred}}{\text{time (seconds)}} = P$$

or

$$IV = \frac{Q}{t} \times \frac{E}{Q} = \frac{E}{t} = P$$

Power Rating of Batteries

The source of power for an electronic system is its voltage source. When the source is a dc power supply operating from a 110 V outlet, the length of time that the supply can provide power is essentially unlimited. When it is a battery, however, the battery life is an important consideration. The **battery life** for a given application is the length of time it can supply a given current at its rated voltage. The term that specifies this is called the battery **capacity**, or sometimes its **current rating**.

The current rating is given in **ampere hours** (Ah), which specify the amount of time the battery can supply a given current at its rated voltage. For example, if a battery has a current rating of 1 Ah, it can supply a current of 1 A for 1 h before its voltage will start to fall below its rated value.

The amount of time a battery can supply other currents can also be determined from this rating. To determine how long a battery can supply a given current, we simply divide the rating by the amount of current supplied:

$$\text{time (hours)} = \frac{\text{current rating (ampere hours)}}{\text{current supplied (amperes)}}$$

EXAMPLE 11: A battery has a current rating of 3 Ah. How long can it supply a current of 200 mA?

Solution: Because the equation requires that the current supplied be specified in amperes, 200 mA must first be converted. From Figure 1.6, we get:

$$I = 200 \text{ mA} \times \frac{10^{-3} \text{ A}}{\text{mA}} = 0.2 \text{ A}$$

$$\text{time (hours)} = \frac{3 \text{ Ah}}{0.2 \text{ A}} = \textbf{15 h}$$

BATTERY CAPACITIES

Battery Size	Voltage (V)	Current Range (mA)	Capacity (Ah)
AAA	1.5	0–20	0.30
AA	1.5	0–25	1.0
C	1.5	0–80	2.7
D	1.5	0–150	5.7
6	1.5	0–1500	47.0
509	6.0	0–250	9.0
216	9.0	0–15	0.4

FIGURE 1.36.   Battery life is specified by its capacity in ampere hours (Ah). The table gives the capacities of a number of common battery sizes.

Figure 1.36 shows the capacities of various types of batteries. The capacity depends to some extent on the amount of current drawn from the battery. When a large current is drawn, the capacity is generally lower than for small currents. Values given are for fresh batteries at 70°F. The operating schedule is 2 h/day. The cutoff voltage is 0.8 V/1.5 V cell for all of the cells.

The ampere-hour rating of a battery is actually a measure of how much energy the battery can supply. Because the current is supplied at the battery voltage, which is essentially constant, multiplying the current rating of a battery by its voltage rating gives the total energy it can supply in watt hours (Wh). For example, a 12 V car battery with a current rating of 40 Ah can supply an energy of:

$$12 \text{ V} \times 40 \text{ Ah} = 480 \text{ Wh (about } 2 \times 10^6 \text{ J)}$$

A 1.5 V D cell flashlight battery with a rating of 4 Ah can supply an energy of only:

$$1.5 \text{ V} \times 4 \text{ Ah} = 6 \text{ Wh (about } 2 \times 10^4 \text{ J)}$$

1.7 QUESTIONS AND PROBLEMS

1. Classify the following as insulators, conductors, or semiconductors: paper, aluminum, air, brass, wood, silicon, glass, steel, vacuum, and germanium.

2. In the circuits diagrammed in Figure 1.37, show the direction of conventional current flow through each device and the polarity of each device connection. If no current flows through a device, write $I = 0$.

3. Use a battery and lamps to draw a closed circuit that contains both series and parallel connections. Indicate the direction of conventional current flow and the polarity of each device.

4. Complete the circuits shown in Figure 1.38, drawing in the proper meters and connections to measure the current and voltage for each device. For each meter, use the proper electrical symbol and show the meter terminal polarity. Use the minimum number of meters to measure the values.

5. Draw wires connecting the devices shown in Figure 1.39 to build the circuit shown in the schematic diagram.

6. Referring to Figure 1.40, fill in the correct scale number to be used and the correct scale reading.

Range Switch Position	Correct Scale Numbers	Correct Reading
500 V		
2.5 V		
10 V		
50 V		
250 mA		
100 mA		
1 mA		

7. Perform the following unit conversions:
   (a) 150 mA = _____ A
   (b) 0.020 A = _____ mA
   (c) 2000 μA = _____ mA
   (d) 2000 mV = _____ V
   (e) 5 mV = _____ μV
   (f) 0.012 A = _____ μA

8. Automobiles use 12 V batteries. If an automobile taillight is required to draw 630 mA, how

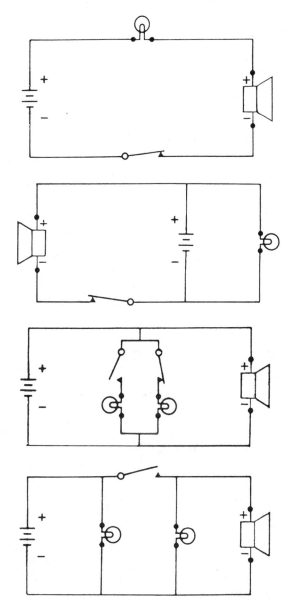

FIGURE 1.37.

much electrical power P does it use? If the total current drawn from the battery is measured as 13.6 A, how much power P is supplied by the battery?

9. The specifications for a transistor indicate that the maximum power it can dissipate at 5 V is

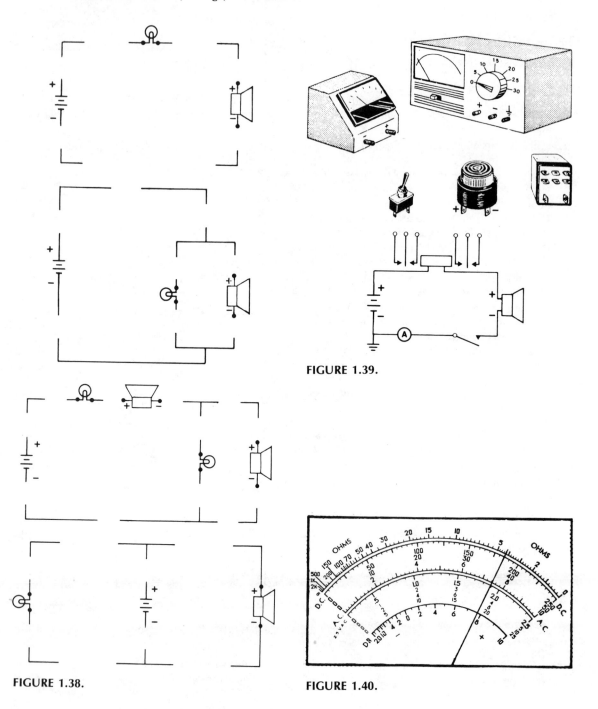

FIGURE 1.39.

FIGURE 1.38.

FIGURE 1.40.

730 mW. What is the maximum current I_{max} that can be drawn by the transistor? Express your answer in milliamperes.

10. A typical electric range has a power consumption of 14,000 W. What current I is flowing through the range if the wall outlet voltage is 115 V? What current I would flow if the specified voltage were 240 V? Which unit would use less energy: a 115 V unit or a 240 V unit?

11. A certain electric motor is designed to operate at 115 V. On the back of the motor, the specifications state that 1 A of current is required by the motor. How much would it cost to operate this motor continuously for 24 h at $0.03/kWh?

Another manufacturer claims that his motor is cheaper to use; it requires 0.3 A at 240 V. Is the claim accurate?

12. A radio requires 20 mA at 1.5 V. Which battery will provide the longest radio playing time: an AAA or an AA cell? What is the approximate playing time for each cell? (Use Figure 1.36.)

13. Calculator manufacturer A makes the model XX machine that requires 5 mW from an AAA battery. Manufacturer B makes the SuperXX that requires 15 mW from an AA battery. If the AA and AAA batteries cost 50 cents each, which calculator costs less to use for 100 h?

Chapter

2

RESISTANCE

2.1 OBJECTIVES

Following the completion of Chapter 2, you should be able to:

1. Recognize the basic units for electrical resistance and convert between power-of-ten equivalents.
2. Use the resistance relation to calculate the resistance of a uniform conductor or insulator, given its resistivity ρ, length l, and cross-sectional area A.
3. Determine the resistance of a length of AWG wire, given the AWG wire resistance table.
4. Identify the resistance, tolerance, and power rating of a resistor by its color code and physical dimensions.
5. Determine the range of possible resistance values of a stock resistor based on its tolerance.
6. Properly use the ohmmeter portion of a multimeter to measure the resistance of a circuit component.

7. Determine the temperature of a thermistor from its resistance and R–T graph.
8. Determine the illumination on a photoconductor from its resistance and R–I graph.

2.2 ELECTRICAL RESISTANCE

Batteries come in a relatively small number of voltage values, most commonly, 1.5 V, 6 V, and 9 V. Yet they are able to supply current to a wide range of electronic systems. These systems draw a wide range of currents, from microamperes or less to tenths of amperes. Because the current may be different, even though the voltage is the same, it is apparent that each device must have some property that determines how much current it will draw.

The property that determines the magnitude of current flow is called **electrical resistance**; it is a measure of how difficult it is for charge carriers to move through the device or

ELECTRICAL RESISTANCE

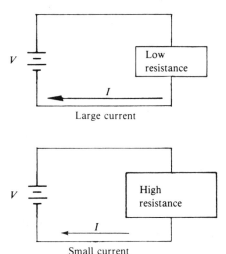

FIGURE 2.1. Electrical resistance determines how much current will flow for a given voltage. With a low resistance, a large current flows. With a high resistance, a small current flows.

UNIT OF ELECTRICAL RESISTANCE

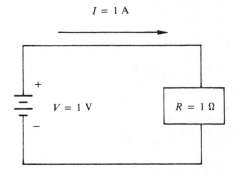

FIGURE 2.2. Electrical resistance is measured in ohms (Ω). If a device has a resistance of 1 Ω, 1 V across it will produce a current of 1 A.

RANGE OF RESISTANCE VALUES

$$
\begin{aligned}
1{,}000\ \Omega &= 1\ \text{k}\Omega \\
1\ \Omega &= 0.001\ \text{k}\Omega\ (10^{-3}\ \text{k}\Omega) \\
1{,}000{,}000\ \Omega &= 1\ \text{M}\Omega \\
1{,}000\ \text{k}\Omega &= 1\ \text{M}\Omega \\
1\ \Omega &= 0.000001\ \text{M}\Omega\ (10^{-6}\ \text{M}\Omega)
\end{aligned}
$$

FIGURE 2.3. Electronic devices have a wide range of resistance values, from ohms to megohms (MΩ).

material. If there is a *low resistance*, the charge carriers can move easily, and a given voltage will produce a *large current*. On the other hand, if there is a *high resistance*, the charge carriers move with difficulty, and a *small current* will flow (Figure 2.1).

Units of Electrical Resistance

Because electrical resistance is a property of the device itself, it is a fixed quantity that can be measured and then stamped on the device for reference. Resistance is indicated by R and is measured in units called **ohms.** The symbol for ohms is the Greek letter omega, which is written Ω. Thus, to indicate that a device has a resistance of 10 ohms, we write $R = 10\ \Omega$.

One ohm is defined as the amount of resistance a device has if 1 V placed across it will cause 1 A of current to flow (Figure 2.2). For larger or smaller resistances, of course, less or

more current would flow. If, for 1 V, 1 Ω produces 1 A, then 10 Ω will produce 0.1 A, 0.1 Ω will produce 10 A, and so on. This concept is discussed in more detail in the next section.

Like the ranges of current and voltage, the range of resistance values encountered in electronic devices is wide. Most often, the resistances will be large: hundreds, thousands, or even millions of ohms. Therefore, the larger resistance units of **kilohm**, kΩ (1000 Ω), and **megohm**, MΩ (1,000,000 Ω), are frequently used. The relationship between these units is shown in Figure 2.3. Sample conversions are given in the following two examples.

EXAMPLE 1:   If a device has a resistance of 2200 Ω, what is its resistance in kilohms?

Solution: Since kilohms is the larger unit, the answer should be smaller than the original number of ohms. Therefore, we multiply the number of ohms by 1 kΩ/1000 Ω:

$$2200 \ \Omega \ \times \ \frac{1 \ \mathrm{k}\Omega}{1000 \ \Omega} = \mathbf{2.2 \ k\Omega}$$

EXAMPLE 2:   How many ohms are there in a resistance of 1.8 MΩ?

Solution: There are 10^6 Ω in a megohm, so multiply by 10^6 Ω/MΩ:

$$1.8 \ \mathrm{M}\Omega \ \times \ \frac{10^6 \ \Omega}{\mathrm{M}\Omega} = \mathbf{1.8 \times 10^6 \ \Omega}$$

Resistivity of Materials

The electrical resistance of a device depends on a number of factors. The most important is the type of material of which the device is made. Materials that give a *low* electrical resistance are called **conductors** and conduct large currents. Most metals, such as gold, silver, copper, and aluminum, have a low electrical resistance and are used in parts of electronic circuits where the current must flow easily. Copper, particularly, is used for wires that carry the current from the voltage source to various circuit devices and in switches that open and close circuits.

Materials that give a relatively *high* electrical resistance and, therefore, conduct small currents are called **insulators.** Rubber, plastic, and porcelain are examples of insulators. These materials are used where a high resistance to current flow is required. For example, wires are insulated with a thin coat of rubber or plastic to prevent accidental contact with other conductors. Also, the metal terminals of voltage sources are recessed inside a plastic case to prevent accidental contact with the hand.

RESISTIVITY ρ

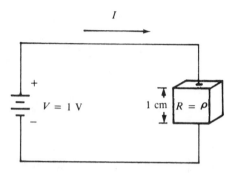

FIGURE 2.4.   The resistance properties of materials are specified by their resistivity ρ, which is the resistance of a cube of the material that measures 1 cm on a side.

The quantity that specifies the relative electrical resistance of a material is called **resistivity.** The symbol for resistivity is the Greek letter rho, ρ, pronounced *rō*. The resistivity of a material is defined as the resistance of a cube of the material 1 cm on a side (Figure 2.4). A 1 cm cube is selected simply as a standard shape for comparing various materials.

Figure 2.5 gives the resistivities of a number of different materials, both conductors and insulators. Also given is the current that would flow if 1 V were placed across the cube. As the list shows, the difference in the electrical properties of conductors and insulators is enormous. One volt across a 1 cm cube of silver results in a current of nearly 1,000,000 A, while a cube of hard rubber gives less than 0.000000000000001 A. This is a difference of nearly 10^{21}, or a billion billion. It explains why a very long wire of copper conducts current easily, while a thin layer of rubber between two wires prevents the wires from shorting.

Resistance of Wires

The resistance of a 1 cm cube of a material may not seem like a very practical number since most devices, wires for example, come in other

RESISTIVITY OF MATERIALS

Conductors	ϱ (Ohm-cm)	I at 1 V
Silver	1.5×10^{-6}	6.7×10^5 A
Copper	1.7×10^{-6}	5.9×10^5 A
Gold	2.4×10^{-6}	4.2×10^5 A
Aluminum	2.8×10^{-6}	3.6×10^5 A
Iron	10.0×10^{-6}	1.0×10^5 A
Lead	22.0×10^{-6}	0.45×10^5 A
Nichrome	100.0×10^{-6}	0.1×10^5 A
Insulators		
Wood	10^{10}–10^{13}	10^{-10}–10^{-13} A
Glass	10^{11}–10^{16}	10^{-11}–10^{-16} A
Porcelain	10^{14}	10^{-14} A
Hard rubber	10^{15}–10^{18}	10^{-15}–10^{-18} A

FIGURE 2.5. The table gives the resistivity of a number of materials used in electronics, both conductors and insulators. Also given is the current that would flow through a 1 cm cube of the material at 1 V.

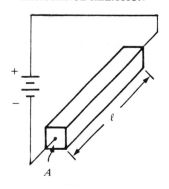

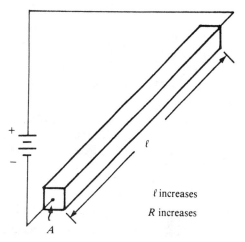

ℓ increases
R increases

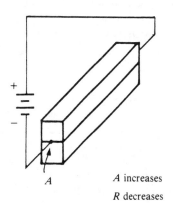

A increases
R decreases

shapes. However, by multiplying this number by a factor that takes into account the shape of an actual device, we can obtain the device resistance.

The *shape factor* takes into account two geometrical aspects of resistance to flow: the length of the flow path l and its cross-sectional area A. These aspects are fairly easy to understand. Resistance to flow is a measure of the difficulty that a charge carrier experiences in moving through a material. Therefore, it is reasonable to expect that the farther the charge carrier has to travel in the material, the greater the resistance it experiences. This situation can be summed up by the following rule:

Resistance increases as the path length l increases.

Resistance also determines the amount of flow. The less the resistance, the greater the flow. If two wires are placed between two points instead of one (Figure 2.6), it is reasonable to expect the amount of flow to increase. Putting

FIGURE 2.6. The resistance of different shapes can be calculated by multiplying the resistivity by a shape factor l/A. This expression is called the resistance relation $R = \rho(l/A)$. The shape factor accounts for the fact that resistance increases with length l, but decreases with cross-sectional area A.

two wires instead of one has essentially increased the cross-sectional area of the conductor path while decreasing the resistance between the two points. Therefore, it can be stated:

Resistance decreases as the cross-sectional area A increases.

The dependence of resistance on the geometrical quantities of cross-sectional area and length can be mathematically expressed by the following **resistance relation:**

$$R = \rho \left(\frac{l}{A}\right)$$

This relation states that the resistance R of any wire is the resistivity ρ of the wire material times a shape factor l/A. For long wires, l is large and R is large. For wires of large cross-sectional area, A is large and R is small.

If the length l is measured in cm and the area A in cm^2, then the resistivity ρ has units of **ohm-cm**. These are the units for resistivity given in Figure 2.5.

The resistance relation can be used to calculate the resistance of any shape of conductor (or insulator). The only restrictions are that the conductor be all of the same material with a uniform resistivity ρ and have the same cross-sectional area A along its entire length l. Two example calculations are given below for the shapes in Figures 2.7 and 2.8.

EXAMPLE 3: What is the resistance of a copper wire 100 cm long with a diameter d of 0.03 cm (Figure 2.7)?

Solution: Using Figure 2.5, we find that the resistivity of copper is:

$$\rho = 1.7 \times 10^{-6} \text{ ohm-cm}$$

The cross-sectional area of a cylindrical wire of diameter d is:

$$A = \frac{\pi}{4}d^2 = \frac{\pi}{4}(0.03 \text{ cm})^2$$
$$= 7.1 \times 10^{-4} \text{ cm}^2$$

Using the resistance relation:

$$R = 1.7 \times 10^{-6} \text{ ohm-cm} \times \frac{10^2 \text{ cm}}{7.1 \times 10^{-4} \text{ cm}^2}$$
$$= 0.24 \ \Omega$$

EXAMPLE 4: What is the resistance of a square sheet of hard rubber 4 cm square on a side and 0.003 cm thick (Figure 2.8)?

Solution: Figure 2.5 gives the resistivity of hard rubber as 10^{15} ohm-cm to 10^{18} ohm-cm, depending on the particular variety.

Assuming the worst case for an insulator, let us take:

$$\rho = 10^{15} \text{ ohm-cm}$$

The cross-sectional area of a square rubber sheet of side length a is:

$$A = a^2 = (4 \text{ cm})^2 = 16 \text{ cm}^2$$

Using the resistance relation:

$$R = 10^{15} \text{ ohm-cm} \times \frac{0.003 \text{ cm}}{16 \text{ cm}^2}$$
$$= 2 \times 10^{11} \ \Omega$$

Standard Wire Sizes

The calculation of resistance using the resistivity relation is rarely required in practical electronics. It was introduced principally to illustrate the way in which the resistance of a device depends on three quantities: type of material, cross-sectional area, and length. This de-

EXAMPLE 3

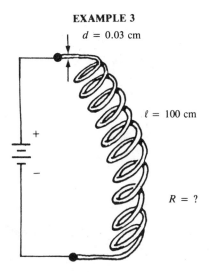

FIGURE 2.7. The resistance of conductor wires can be calculated using the resistance relation. In this example, what is the value of R?

EXAMPLE 4

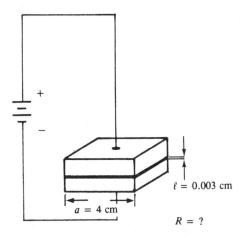

FIGURE 2.8. The resistance of insulating materials can also be calculated using the resistance relation. Here, a thin sheet of hard rubber acts as an insulator between two metal plates. In this example, what is the value of R?

pendence is most directly related to the conductor wires used to connect devices together in electronic circuits.

Conductor wires come in a wide variety of standard sizes, or *gages*. Each gage represents a different diameter and thus has a different resistance per unit length. Figure 2.9 gives the standard American Wire Gage (AWG) size designations for solid round copper wire, the type used in most electronic circuits. The gage wire most commonly used in electronics is about #24. Also given in Figure 2.9 is the diameter of the wire and the resistance in $\Omega/1000$ ft. Note that *as the* AWG # *increases, the diameter of the wire decreases, but the resistance increases*.

The basis of the AWG sizes is that the cross-sectional area is cut in half for every increase in three gage numbers. From the resistance relation, we see that halving the area doubles the resistance. Therefore, the resistance per 1000 ft is approximately doubled for every increase of three gage numbers.

The dependence of resistance on length is illustrated by the wire resistance given in $\Omega/1000$ ft, which indicates that the resistance of a

particular length of wire is directly proportional to its length. To determine its resistance, first determine its diameter or gage "number," then look up its resistance in $\Omega/1000$ ft, and then multiply this number by the number of feet of wire.

EXAMPLE 5: What is the resistance of 8 in. of #32 copper wire?

Solution: First, 8 in. must be converted to feet:

$$8 \text{ in.} \times \frac{1 \text{ ft}}{12 \text{ in.}} = 0.67 \text{ ft}$$

Figure 2.9 gives the resistance of #32 copper wire as 164.1 $\Omega/1000$ ft. Therefore, the resistance is:

$$R = \frac{164.1 \ \Omega}{1000 \text{ ft}} \times 0.67 \text{ ft} = \mathbf{0.11 \ \Omega}$$

AWG WIRE SIZES

AWG#	Diameter (in.)	Ω/1000 ft at 20°C
0000	0.4600	0.0490
000	0.4096	0.0618
00	0.3648	0.0780
0	0.3249	0.0983
1	0.2893	0.1240
2	0.2576	0.1563
3	0.2294	0.1970
4	0.2043	0.2485
5	0.1819	0.3133
6	0.1620	0.3951
7	0.1443	0.4982
8	0.1285	0.6282
9	0.1144	0.7921
10	0.1019	0.9989
11	0.09074	1.260
12	0.08081	1.588
13	0.07196	2.003
14	0.06408	2.525
15	0.05707	3.184
16	0.05082	4.016
17	0.04526	5.064
18	0.04030	6.385
19	0.03589	8.051
20	0.03196	10.15
21	0.02845	12.80
22	0.02535	16.14
23	0.02257	20.36
24	0.02010	25.67
25	0.01790	32.37
26	0.01594	40.81
27	0.01420	51.47
28	0.01264	64.90
29	0.01126	81.83
30	0.01003	103.2
31	0.008928	130.1
32	0.007950	164.1
33	0.007080	206.9
34	0.006305	260.9
35	0.005615	329.0
36	0.005000	414.8
37	0.004453	523.1
38	0.003965	659.6
39	0.003531	831.8
40	0.003145	1049.0

FIGURE 2.9.   Conductor wires come in standard sizes with a specific number of Ω/1000 ft of length. The table gives the diameter and resistance for solid round copper wire. Example 5 shows how to calculate the resistance using the table.

RESISTORS AND SYMBOLS

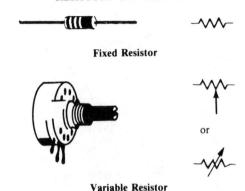

Fixed Resistor

or

Variable Resistor

FIGURE 2.10.   Resistors are used in electronic circuits to control current. Resistors come in two forms: fixed and variable.

2.3 RESISTORS

For a given voltage, it is the resistance that determines how much current will flow in a circuit. Because electronics is largely concerned with controlling current flow, resistors are extremely important circuit devices. They come in a wide variety of values and are inserted in various parts of electronic circuits to establish desired amounts of current.

Resistors come in two forms: fixed and variable. **Fixed resistors** have a specific amount of resistance that cannot be changed. The resistance of **variable resistors**, on the other hand, can be changed, generally by turning a shaft. Typical examples are shown in Figure 2.10, along with their electronic symbols. Note that the resistor symbol illustrates that the current path through it is difficult.

Fixed Resistors

Fixed resistors come in a variety of forms suitable for different applications. They can be classified by three basic characteristics: resistance value, power rating, and type of material and construction.

TYPES OF FIXED RESISTORS

Carbon Composition: $R = 1\ \Omega$ to 22 MΩ
$P_{max} = 2$ W

Carbon Film: $R = 1\ \Omega$ to 300 MΩ
$P_{max} = 2$W

OHMITE

Metal Film: $R = 1\ \Omega$ to 150,000 MΩ
$P_{max} = 1$ W

Wire-Wound: $R = 0.1\ \Omega$ to 100 kΩ
$P_{max} = 1000$ W

FIGURE 2.11. Fixed resistors are classified by type of material and construction, resistance value, and power rating. The four most common types include carbon composition, carbon film, metal film, and wire-wound.

— **Resistance value,** which indicates the resistance in ohms under normal operating conditions;
— **Power rating** in watts, which tells how much power can be dissipated by the resistor without permanent change in the resistance value;
— **Type of material and construction,** which represents different manufacturing techniques.

The four most common types of fixed resistors are shown in Figure 2.11, along with an approximate range of resistance values and maximum power ratings.

CARBON COMPOSITION RESISTORS

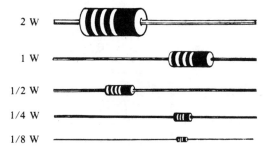

FIGURE 2.12.   The most common resistor in electronics is of carbon composition. Its power rating can be easily identified by its physical size.

Carbon composition resistors are the most widely used type of resistor because of their low cost and availability. They are constructed by molding two lead wires into contact with a carbon composition resistance material inside an insulating shell. The power rating of carbon composition resistors can be easily identified by their physical size, as shown in Figure 2.12.

High-precision resistors are generally of the **metal film** type. These resistors are characterized by extremely accurate resistance values that do not change with changing conditions, such as time or temperature. They are made by depositing, under vacuum, a thin conducting film of metal or metallic oxide on an insulator, such as ceramic or glass. Leads are then brought out from the metal film. **Carbon film resistors** are made in essentially the same way, but have somewhat inferior performance characteristics.

Resistors that can dissipate large amounts of power are generally **wire wound.** Wires of conducting metal are wound on vitreous enamel or ceramic forms that can withstand high temperatures.

Color Code for Identification

As shown in Figure 2.12, carbon resistors have several circular bands around them. The different-colored bands represent the resistance

COLOR CODE FOR CARBON RESISTOR VALUES

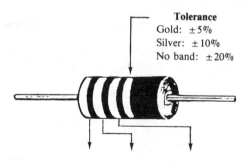

Tolerance
Gold: ±5%
Silver: ±10%
No band: ±20%

Color	1st Digit	2nd Digit	Multiply by:
Black	0	0	1
Brown	1	1	10
Red	2	2	100
Orange	3	3	1,000
Yellow	4	4	10,000
Green	5	5	100,000
Blue	6	6	1,000,000
Violet	7	7	10,000,000
Gray	8	8	100,000,000
White	9	9	1,000,000,000
Gold			0.1
Silver			0.01

FIGURE 2.13. The resistance value and tolerance of a carbon resistor can be determined from the color of the circular bands around it. The table gives the color code for these bands.

value. The translation of the color code is shown in Figure 2.13, where each color represents a particular number.

The first two bands indicate the first two digits of the resistance value. The third band identifies the number of zeros after the second digit, or a multiplier in powers of 10. For example, if the first three bands are green, blue, and orange, the resistance would be:

$$R = 5 \quad 6 \quad \times \quad 10^3 \, \Omega = 56,000 \, \Omega$$
$$\text{green} \quad \text{blue} \quad \text{orange}$$

The fourth band indicates the **tolerance** of the resistor — that is, the amount by which the ac-

tual value may differ from the indicated, or *nominal* value. The actual value of the resistor may be higher or lower than the nominal value by the percent given. For example, if the tolerance digit in the preceding equation were silver, the actual value of the resistor could be ±10% of the nominal value, or:

$$\text{tolerance of } R = \pm 10\% \times R$$
$$= \pm 0.1 \times 56,000 \, \Omega$$
$$= \pm 5,600 \, \Omega$$

Therefore the actual resistance could be anywhere from:

$$56,000 + 5,600 = 61,600 \, \Omega$$

to

$$56,000 - 5,600 = 50,400 \, \Omega$$

If the fourth band were missing, the tolerance would be ±20%.

It is important to learn the color code in order to identify quickly the value of a resistor without having to measure it. A common way of remembering the code is to learn a jingle in which the first letter of each word is the same as the first letter of the color. An example is the following:

"Black	*Bears*	*Raid*	*Our*	*Young*
Black	Brown	Red	Orange	Yellow
0	1	2	3	4
Garden	*But*	*Violet*	*Goats*	*Won't."*
Green	Blue	Violet	Gray	White
5	6	7	8	9

Carbon resistors also come in stock values that are generally quite low in cost, a few cents apiece. When designing a circuit, it is important to select resistance values from stock values in order not to have to pay a premium price for special values. The stock resistor values are given in Figure 2.14. The values are selected so that the tolerance limits of adjacent values overlap. This choice guarantees that from a

STOCK RESISTANCE VALUES

Ohms					Kilohms		Megohms	
0.10	**1.0**	**10**	**100**	**1000**	**10**	**100**	**1.0**	**10.0**
0.11	1.1	11	110	1100	11	110	1.1	11.0
0.12	**1.2**	**12**	**120**	**1200**	**12**	**120**	**1.2**	**12.0**
0.13	1.3	13	130	1300	13	130	1.3	13.0
0.15	**1.5**	**15**	**150**	**1500**	**15**	**150**	**1.5**	**15.0**
0.16	1.6	16	160	1600	16	160	1.6	16.0
0.18	**1.8**	**18**	**180**	**1800**	**18**	**180**	**1.8**	**18.0**
0.20	2.0	20	200	2000	20	200	2.0	20.0
0.22	**2.2**	**22**	**220**	**2200**	**22**	**220**	**2.2**	**22.0**
0.24	2.4	24	240	2400	24	240	2.4	
0.27	**2.7**	**27**	**270**	**2700**	**27**	**270**	**2.7**	
0.30	3.0	30	300	3000	30	300	3.0	
0.33	**3.3**	**33**	**330**	**3300**	**33**	**330**	**3.3**	
0.36	3.6	36	360	3600	36	360	3.6	
0.39	**3.9**	**39**	**390**	**3900**	**39**	**390**	**3.9**	Bold
0.43	4.3	43	430	4300	43	430	4.3	figures
0.47	**4.7**	**47**	**470**	**4700**	**47**	**470**	**4.7**	are
0.51	5.1	51	510	5100	51	510	5.1	10%
0.56	**5.6**	**56**	**560**	**5600**	**56**	**560**	**5.6**	values.
0.62	6.2	62	620	6200	62	620	6.2	
0.68	**6.8**	**68**	**680**	**6800**	**68**	**680**	**6.8**	
0.75	7.5	75	750	7500	75	750	7.5	
0.82	**8.2**	**82**	**820**	**8200**	**82**	**820**	**8.2**	
0.91	9.1	91	910	9100	91	910	9.1	

FIGURE 2.14. Low-cost carbon resistors come in the stock values listed in the table. All values are available in 5% tolerance; bold figures, in 10% tolerance.

stock assortment, you will be able to find a resistor of any desired value.

Variable Resistors

The control of many electronic systems is achieved by controlling the current that flows in some part of the circuit. In turn, controlling the current is achieved by varying the resistance of that part of the circuit. Resistors whose value can be changed are called variable resistors, or, more commonly, **potentiometers** (sometimes shortened to *pot*). A somewhat older name for the variable resistor is **rheostat.** Nearly every control knob on an electronic system is connected to a potentiometer.

The electronic symbol for a potentiometer, shown in Figure 2.15, indicates its general construction. A potentiometer consists of a fixed resistor with connections at each end and a *wiper* that can be moved along the fixed resistor to make contact at any point. Thus, the effective resistance between one end and the wiper contact can be varied from near 0 Ω, when the wiper is at one end, to the full value of the resistor when the wiper is at the other end. (Generally, the zero end will not go exactly to 0 in order to prevent accidental shorting of a circuit when operating the potentiometer at the low end.)

The construction of a typical, general-purpose, low-power potentiometer is shown in Figure 2.16. The fixed resistor is usually made of carbon and is in the form of an arc of about 300°. The wiper is a metal contact attached to a shaft. When the shaft is rotated, the wiper can be moved around the full extent of the arc. Depending on which end of the fixed resistor is

VARIABLE RESISTORS AND SYMBOL

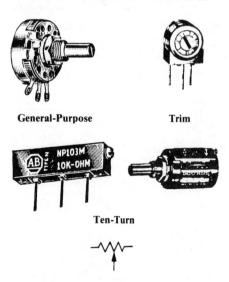

General-Purpose   Trim

Ten-Turn

FIGURE 2.15. Variable resistors, or potentiometers, come in a variety of shapes for different applications. In each case, a wiper moves along a fixed resistor.

GENERAL-PURPOSE POTENTIOMETER

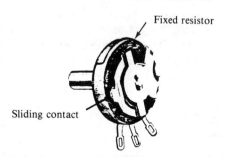

Fixed resistor

Sliding contact

FIGURE 2.16. In a typical general-purpose carbon pot, the shaft rotates a metal contact along a fixed, circular carbon resistor.

connected to the circuit, a full clockwise rotation can represent either 0 Ω or the full value of resistance.

Potentiometers of this kind come in different forms, distinguished by how the resistance changes with shaft rotation. In **linear pots** (Fig-

MEASURING CHANGE IN RESISTANCE

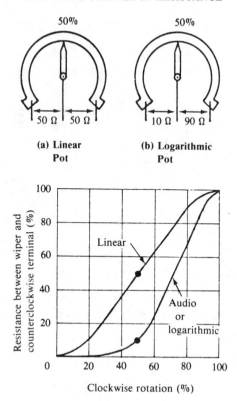

(a) Linear
    Pot

(b) Logarithmic
    Pot

(c) **Resistance versus Rotation for Carbon Element Potentiometers**

FIGURE 2.17. The change in resistance with shaft rotation can be linear or logarithmic. The graph shows the resistance versus rotation for the two cases.

ure 2.17a), the change in resistance is proportional to the shaft rotation. For example, if the potentiometer has a fixed resistance of 100 Ω and the wiper is rotated halfway, the resistance between either end and the wiper is 50 Ω.

More common, however, is the **logarithmic** or **audio pot** (Figure 2.17b). The variation of resistance with shaft rotation for an audio pot is shown in Figure 2.17c. Note that the first 50% of shaft rotation produces only a 10% change in resistance, while the second 50% produces the remaining 90%.

WIRE-WOUND POTENTIOMETERS

FIGURE 2.18. Special wire-wound pots are required for power greater than about 2 W. Two common types are illustrated here.

RESISTANCE BOXES

Resistance Substitution Box

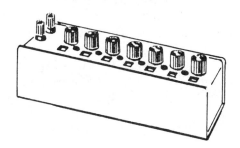

Resistance Decade Box

FIGURE 2.19. Resistance substitution boxes are often used in testing circuit performance. They consist of stock 5% resistors that can be connected to a circuit with a switch. Resistance decade boxes contain precision resistors of integer values. They are used for precision measurements.

Two other forms of the basic potentiometer are shown in Figure 2.15. One is the **trim pot** that is generally used in a circuit where we must accurately set, or trim, a certain resistance to optimize the performance after the circuit has been assembled. The shafts of these pots must be turned with a screwdriver.

Another type has the wiper attached to a sliding, rather than a rotating, shaft. The **sliding pot** is becoming more fashionable in some modern electronic systems.

When very precise resistance variations are required, potentiometers that require ten complete turns of the shaft to cover the full extent of the fixed resistor are used. A typical **ten-turn pot** is also shown in Figure 2.15. Because it takes ten times the number of turns to cover the same resistance range, very precise resistance adjustments can be made with this type of pot.

All these potentiometers come in a wide range of resistance values, from 0 Ω to 50 Ω to 0 Ω to 50 MΩ, and for power ratings of ½ W up to 2 W. (Note: If only a fraction of the total pot resistance is used, then the power rating of the pot must be reduced by that same fraction.)

For power greater than 2 W, special **wire-wound potentiometers** are used. Two common types are shown in Figure 2.18. On one, the wiper is a sliding metal clamp that can be tightened with a nut and bolt to make contact along the resistor wire. On the other, the contact is made with a spring-loaded metal contact that is turned by a shaft.

Substitution and Decade Boxes

In building and studying electronic circuits, you will constantly be changing resistance values to see how the circuit behaves or to optimize its performance. To make this process easier, special **resistor substitution boxes** are available. These boxes consist of a wide range of resistance values that can be connected by a switch, as shown in Figure 2.19. The two terminals of the box are connected across the two circuit points where the resistance is to be varied. As the switch position is changed, a different resistor is substituted between the box terminals.

Generally, the resistors in the substitution box are stock resistors with only 5% or, perhaps, 10% tolerance. When more precise resistance values are required, a **resistance decade box** is used (Figure 2.19). This box is constructed of selected precision resistors with a tolerance of 1% or better. The box contains several switches, each representing a different power of ten, or *decade,* of resistance. Each switch has ten positions for each of the integer values from 0 to 9. Thus, you can dial an exact value of resistance from 0 up to the number of decades provided by the box. If there are five decade switches, the range extends from 0 Ω to 99,999 Ω.

Decade boxes are used primarily for accurate circuit design and special measurement applications. They can be quite expensive, depending on how many decades they provide and the precision of the individual resistors. Substitution boxes, on the other hand, are relatively inexpensive and can even be constructed from a multiple-position switch and stock resistors. They are also quite handy for designing and testing electronic circuits.

2.4 MEASURING RESISTANCE

Ohmmeter

In most situations, resistance is measured with an instrument called an **ohmmeter.** Generally, it is part of a multimeter. Because resistance is a passive quantity, its measurement is somewhat different from that of current and voltage. A voltage must be placed across a resistor, causing a certain current to flow in the circuit. For a fixed voltage, this current is a measure of the resistance.

An ohmmeter contains a battery that is in

RESISTANCE MEASUREMENT

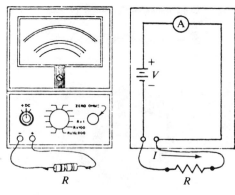

Ohmmeter Circuit   Schematic Diagram

FIGURE 2.20. Resistance is measured with an ohmmeter, which is generally part of a multimeter. The simplified circuit diagram shows that an ohmmeter places a battery in series with its ammeter and the unknown resistor.

series with its own ammeter. The resistance to be measured is connected in series, as shown in Figure 2.20.

Reading an Ohmmeter

Because the current that flows in a device is directly related to its resistance (for a fixed voltage), the meter scale can be directly marked in resistance. Figure 2.21 shows a typical resistance scale from an ohmmeter. Note that the scale has two unusual characteristics:

— *Resistance that increases from right to left*, rather than, as is usual, from left to right;
— *A nonlinear scale*—that is, a scale for which equal divisions do not mean equal changes in resistance. Near the zero end of the scale, the tick marks represent tenths of units, while near the high end, they represent thousands.

OHMMETER RESISTANCE SCALE

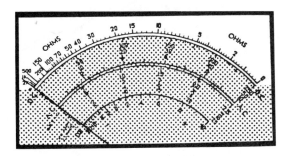

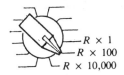

FIGURE 2.21.   The resistance scale of an ohmmeter is difficult to read because it is nonlinear and resistance increases from right to left. The range switch is a multiplier type. The proper switch position puts the needle near mid-scale.

EXAMPLE 6

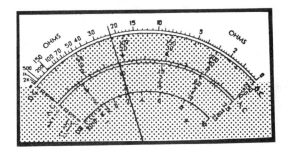

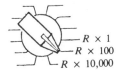

FIGURE 2.22.   When the meter needle and range switch are in the positions shown, the correct resistance reading is that determined in the example.

Due to these two factors, accurate reading of the ohms scale of a multimeter requires some practice. Always be mindful that the scale increases from right to left and be alert to the values of the scale divisions in the region of the scale you are reading. Even experienced people check themselves when reading an ohms scale.

While the scale of an ohmmeter covers the full range of resistance values—0 to ∞—it is difficult to read the scale accurately near the high-resistance end. As on a voltmeter or an ammeter, the most accurate region is near the middle-to-right end of the scale. Therefore, ohmmeters also contain a **range switch** that lets you change the range of the reading to bring the needle near mid-scale.

A typical range switch is shown in Figure 2.21. Its labels are multipliers—that is, the numbers that must be multiplied by the scale reading to give the measured resistance in ohms. The following example illustrates a typical reading.

EXAMPLE 6:   What is the resistance measured by the ohmmeter shown in Figure 2.22?

Solution: The meter needle is one division to the left of 20. Scale divisions in this region represent 2 Ω of resistance because there are five divisions between 20 and 30. Since the scale increases to the left, the scale value is:

$R = 22\,\Omega$

The range switch is on $R \times 100$. Therefore, the scale value must be multiplied by 100 to obtain the measured value:

$$\begin{aligned} \text{measured resistance} &= R \times 100 \\ &= 22\,\Omega \times 100 \\ &= \textbf{2200}\,\Omega \end{aligned}$$

The following example tests your skill in reading the ohms scale of a multimeter.

EXAMPLE 7

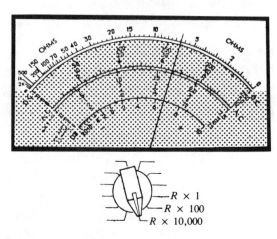

FIGURE 2.23. To test your skill, determine the correct resistance value for the needle and switch positions shown.

EXAMPLE 7: What is the resistance measured by the meter shown in Figure 2.23?

Solution: The correct answer is given in the footnote below.*

When using an ohmmeter to measure resistance, you must also keep in mind the basic circuit diagram (Figure 2.20) because it places some restrictions on how the ohmmeter can be used. For example, if a particular device is wired into a circuit, you cannot simply place the test leads of the ohmmeter across the device and expect to read its resistance. The remainder of the circuit must be taken into account. If there is a voltage source somewhere in the circuit, the current it produces will also flow through the ohmmeter and give an erroneous reading. Even without voltage sources present,

---
*Solution to Example 7: $R = $ **65,000 Ω**

OHMMETER CONNECTION

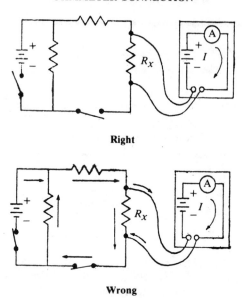

Right

Wrong

FIGURE 2.24. In using an ohmmeter to measure the resistance of a device in a circuit, be certain there are no other voltage sources or resistance paths.

when the ohmmeter test leads are across the device, they are also across the rest of the circuit (Figure 2.24). Therefore, the resistance you read is some unknown combination of the device and the resistance of the rest of the circuit.

In using an ohmmeter to measure resistance, observe the following rule:

Always be sure the device is free of other voltage sources and other possible current paths.

Techniques for measuring the resistance of devices already wired into circuits will be discussed in Chapter 3.

Another point to remember is that an ohmmeter contains a battery that has a limited capacity. If the ohmmeter leads are left across a low resistance or, in the extreme case, touching each other (0 resistance), a large current will be drawn from the battery and it will run down. Therefore, two other rules are:

Never leave an ohmmeter permanently connected to a resistor.

Always switch the multimeter range switch off resistance after using it as an ohmmeter.

Testing Circuit Continuity

While the principal purpose of an ohmmeter is to measure resistance, there is a related application that is quite useful. Because the ohmmeter contains a battery in series with an ammeter, it can be used to test whether a current path is continuous. If you want to determine whether a particular path is open or closed, you simply put the test leads on the ohmmeter across the two ends of the path. If the ohmmeter needle moves, the path is complete, if not, it is an open current.

A continuity tester has many applications. For example, you can identify the terminals of a switch by placing an ohmmeter across successive pairs to see which are connected for each switch position. For multiconductor wire, you can use an ohmmeter to locate the two ends of each wire. You will probably find yourself using an ohmmeter more often as a continuity tester than to measure resistance.

Resistance Bridges

As Figure 2.21 indicates, the accuracy of an analog ohmmeter scale is limited to about two digits. For most work, this accuracy is sufficient, and the multimeter is by far the most common method for measuring resistance in such cases. A digital ohmmeter provides somewhat better accuracy. It has an accuracy of three or more digits, depending on the meter.

When extremely accurate resistance measurement is required, devices called **resistance bridges** are used. These bridges are special circuits composed of very accurately known resis-

tors and very sensitive meters. With a good resistance bridge, you can make extremely precise resistance measurements, to seven or even eight digits.

The design and use of resistance bridge circuits is an important topic and is covered elsewhere in the book.

2.5 RESISTIVE TRANSDUCERS

An important component of any electronic measuring instrument is the device that serves as an interface between the physical system being measured and the electronic circuitry. These devices, called **transducers,** convert, or *transduce,* changes in the physical variable (light, sound, temperature, and so on) into a change in some electrical variable (for example, voltage, current, or resistance). The electronic circuitry can then process the electrical variable in a wide variety of ways.

Probably the widest class of transducers includes those that produce resistive changes, called **resistive transducers.** They are constructed from materials that have been specifically selected because their resistance changes in some predictable way with some change in their environment. By measuring the resistance of a resistive transducer, you can determine the value of the physical quantity.

An understanding of the behavior of transducers is critical to the proper use and construction of electronic measuring instruments. A brief introduction is given here to two specific resistive transducers: the thermistor that measures temperature, and the photoconductor that measures light. These devices have a wide range of practical applications, including electronic thermometers and light meters.

Thermistor

A **thermistor** is a bead of special material whose resistance changes greatly with tempera-

THERMISTORS AND SYMBOL

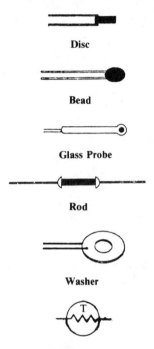

Disc

Bead

Glass Probe

Rod

Washer

FIGURE 2.25. Transducers are key components in electronic instrumentation. They convert physical changes into electrical changes that can be electronically measured. The several types of thermistors shown here are used for measuring temperature.

THERMISTOR R-T CURVE

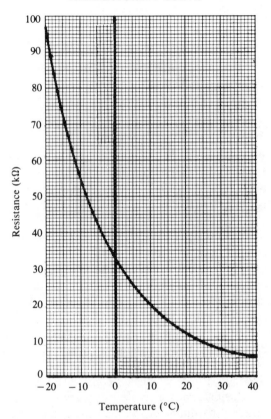

Temperature (°C)

FIGURE 2.26. The resistance of a thermistor depends strongly on its temperature. By measuring the resistance of a thermistor, you can determine its temperature simply by referring to its R-T curve.

ture. Figure 2.25 shows several different types of thermistors, along with the electronic symbol. The T on the symbol indicates that its resistance changes with temperature.

The variation in resistance of a typical thermistor with temperature, its **R-T curve**, is shown in Figure 2.26. Note that as the thermistor *temperature increases,* its *resistance decreases* rapidly. At 0°C, for example, the resistance of this thermistor is 32,650 Ω, while at room temperature (about 25°C), it is only 10,000 Ω, and at 40°C, it is only 5330 Ω: This general behavior—that is, decreasing resistance with increasing temperature—is characteristic

of semiconductor materials, from which thermistors are made.

To measure temperature with a thermistor, simply attach it to an object whose temperature you want to measure. When the temperature of the object changes, the temperature of the thermistor changes and, hence, its resistance changes. These resistance changes can be processed by electronic circuitry.

One common application is in electronic hospital thermometers. A small thermistor attached to a probe is placed in the patient's mouth. The circuitry of the thermometer quickly converts the resistance of the thermistor into a digital reading of the patient's temperature.

Photoconductor

A resistive transducer widely used for measuring light illumination is the **photoconductor.** Figure 2.27 shows typical photoconductors, along with the electronic symbol. The Greek letter lambda, λ, indicates that the resistance changes with light.

The snakelike part of the photoconductor is a special semiconductor material whose *resistance decreases* as the amount of *light increases.* The two leads of the photoconductor are attached to the ends of the snake. A clear plastic window protects the material from dirt and damage.

The calibration graph relating light illumination to resistance for a typical photoconductor is shown in Figure 2.28. Note that the scales of both axes are *logarithmic.* Each major division represents an increase by a factor of 10. A logarithmic scale is needed so that a wide range

PHOTOCONDUCTORS AND SYMBOL

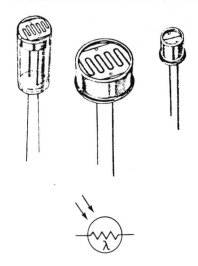

FIGURE 2.27.   The photoconductor is a resistance transducer used for measuring illumination.

of values can be covered on a single graph. Care must be taken in reading this scale, as in reading the resistance scale of an ohmmeter, because equal divisions do not represent equal changes in the quantity.

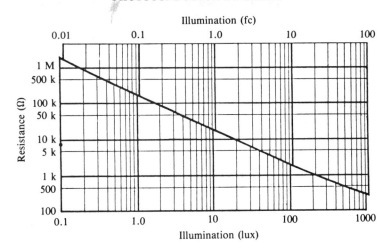

PHOTOCONDUCTOR *R-I* GRAPH

FIGURE 2.28.   The resistance of a photoconductor varies widely with illumination. In order to cover the whole range, the axes of the *R–T* graph are logarithmic. Two illumination scales, the lux (SI unit) and the footcandle (English unit), are shown here.

EXAMPLE 8:   An ohmmeter measures the resistance of a CL5M9M photoconductor (Figure 2.28) to be 8000 Ω. What is the illumination I of the photoconductor in lux and in footcandles (fc)?

Solution: On the resistance scale, 8000 Ω occurs approximately at the point shown on the vertical resistance scale. A vertical line downward from the intersection of 8000 Ω and the graph crosses the illumination scale (in lux) at about:

$I = 22$ lux

A vertical line upward crosses the illumination scale (in footcandles) at about:

$I = 2.1$ fc

APPROXIMATE VALUES OF ILLUMINANCE NATURAL ILLUMINATION

Sky Condition	Lux
Direct sunlight	120,000
Full daylight (not direct sunlight)	15,000
Overcast day	1,000
Twilight	10
Full moon	0.1
Night sky (clear, but moonless)	0.001

FIGURE 2.29.   Units of illumination indicate how much light is falling on a given surface. The table shows typical illuminations in lux for various natural light conditions.

The units of illumination might be unfamiliar to you. Two scales of units are given in Figure 2.28: the **lux**, which is the Standard International (SI) unit, and the **footcandle** (fc), which is commonly used in the United States. These are related by a ratio of about 10 to 1. Specifically:

10.76 lux = 1 fc

These units of illumination are a measure of how much light is falling on, or illuminating, a surface. Figure 2.29 gives the illumination in lux of a number of naturally occurring situations. Note the wide range of illuminations that occur in nature and to which our eyes can adjust. Compare this range with the range of resistance values of the photoconductor (Figure 2.28).

Because photoconductors respond over a broad range of illumination, they have many applications. The light-sensing elements in light meters and cameras, for example, are generally photoconductors. Another application is in switching or counting as an object moving past a photoconductor interrupts a light beam.

2.6 QUESTIONS AND PROBLEMS

1. For each of the carbon resistors shown in Figure 2.30, indicate the colors that would appear on each band.
2. For each of the carbon resistors pictured in Figure 2.31, indicate the value of the resistor and its power rating.
3. What are the maximum and minimum values for each of the resistors in Problem 1?
4. An ohmmeter reading is shown in Figure 2.32. What is the resistance value if the range switch is on:
    (a) $R \times 1$?
    (b) $R \times 100$?
    (c) $R \times 10,000$?
5. The two ends of a reel of AWG #22 wire are brought out as shown in Figure 2.33. How would you use an ohmmeter to:
    (a) Tell if the wire was broken inside the reel?
    (b) Estimate the length of wire on the reel?
6. The resistance-versus-temperature (R–T) curves of five devices (A, B, C, D, and E) are given in

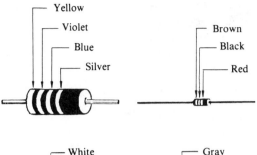

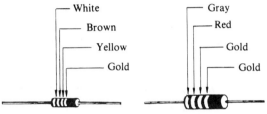

FIGURE 2.31.

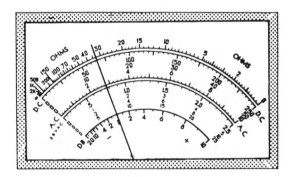

FIGURE 2.32.

FIGURE 2.33.

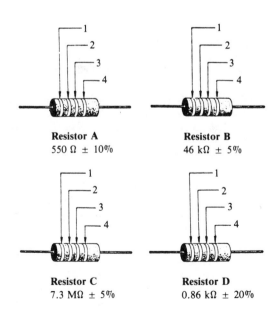

Resistor A
550 Ω ± 10%

Resistor B
46 kΩ ± 5%

Resistor C
7.3 MΩ ± 5%

Resistor D
0.86 kΩ ± 20%

FIGURE 2.30.

Figure 2.34. Which curve or curves represents the R–T characteristics of:

(a) A carbon resistor whose resistance varies very little with temperature?

(b) A device whose resistance increases with an increase in temperature?

(c) A device whose resistance decreases when temperature increases?

(d) A thermistor?

7. Using Figure 2.34 for guidance:

(a) Complete the data shown below by entering in column 3 the resistance that corresponds to the temperature given for each device in column 2.

1 Device	2 Temperature	3 Resistance
A	0°C	
A	50°C	
B	0°C	
B	50°C	
C	0°C	
C	50°C	
D	0°C	
D	50°C	
E	0°C	
E	50°C	

(b) Enter in column 3 the temperature that corresponds to the resistance given in column 2.

1 Device	2 Resistance	3 Temperature
A	3 kΩ	
A	5 kΩ	
B	3 kΩ	
B	5 kΩ	
C	5 kΩ	
C	10 kΩ	
D	22 kΩ	
D	30 kΩ	
E	27 kΩ	
E	30 kΩ	

8. If device C in Figure 2.34 were at body temperature, what resistance would an ohmmeter register?

9. Device B in Figure 2.34 is in contact with your

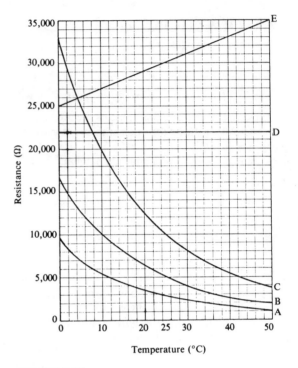

FIGURE 2.34.

body. The thermistor's resistance is measured at about 17 kΩ. Describe the condition of your body.

10. Refer to the resistance-versus-illumination (R–I) characteristic curves in Figure 2.35 to answer the following questions:

(a) As you go from left to right on the horizontal axis, is the illumination getting lighter or darker?

(b) Which characteristic curves are those of a photoconductor?

(c) Which characteristic curves are those of a carbon resistor?

(d) Which characteristic curve has the best dark-to-light ratio?

(e) Locate and draw on the graph a vertical line that corresponds to an illumination of 2 lux.

(f) Locate and draw on the graph a vertical line that corresponds to an illumination of 2 fc.

(g) Would it be easier to hide in an illumination of 1000 lux or 0.1 lux?

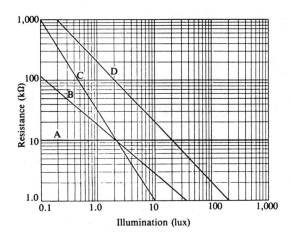

FIGURE 2.35.

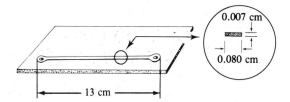

FIGURE 2.36.

11. Perform the following unit conversions:
    (a) 127 MΩ = _____ Ω
    (b) 258 Ω = _____ kΩ
    (c) 4.6 kΩ = _____ MΩ
    (d) 3.78 MΩ = _____ kΩ
12. A printed circuit board has a copper conductor pattern 13 cm long, 0.080 cm wide, and 0.007 cm thick (Figure 2.36). What is its resistance?

What would the resistance be if the conductor were made of aluminum?
13. Which is the better insulator: a square sheet of hard rubber 10 mm square on a side and 0.3 mm thick or a square sheet of porcelain 10 mm square on a side and 1.3 mm thick? (Use Figure 2.5.)
14. What is the length of AWG #40 wire left on a spool if its resistance is 500 Ω?
15. In order to finish the construction of a circuit, you need to build a resistor from wire. What length of AWG #37 wire is required to make a resistance of 29 Ω?
16. You have a spool with approximately 66 ft of wire on it. You measure the resistance of the spooled wire to be 0.044 kΩ. What is the AWG # of the wire?

Chapter

3

OHM'S LAW

3.1 OBJECTIVES

Following the completion of Chapter 3, you should be able to:

1. Use Ohm's law to calculate the voltage across, current through, or resistance of a circuit component, given any two of these quantities.
2. Properly use the consistent units of volts, milliamperes, and kilohms for voltage, current, and resistance, respectively, in calculations using Ohm's law.
3. Record the results of calculations to the proper number of significant digits.
4. Use the power relation to calculate the power dissipated by a circuit component, given its resistance and either the voltage across or the current through it.
5. Properly use the consistent set of units of milliwatts, volts, milliamperes, and kilohms for power, voltage, current, and resistance, respectively, in calculations using the power relation.
6. Recognize the $V-I$ characteristics of a resistor, diode, and LED.

7. Interpret the circuit performance of a diode or LED from its $V-I$ characteristic.

3.2 OHM'S LAW

Chapter 2 stated that for a given voltage, the resistance of a device determines the current. By changing the resistance of a circuit, we can change the current, and this changing current can be utilized in various ways—for example, to activate a switch, which, in turn, can turn on a light, register a count, or activate an alarm; or to measure the resistance of a transducer, which, in turn, can measure temperature or illumination. These examples represent the power of electronics. By selecting the proper electronic components, we can build relatively simple systems to do many useful things.

But how do we determine which components to use? For example, what resistance should the thermistor or photoconductor have? What current should the ammeter read? What should the value of the voltage source be? One

ELECTRONIC THERMOMETER CIRCUIT

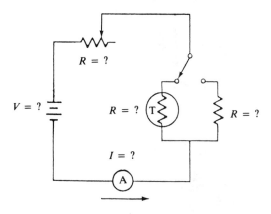

FIGURE 3.1. Essential to designing electronic circuits is the ability to calculate voltage, current, and resistance correctly. In this section, you will learn some of the basic methods.

way of determining these values is by trial and error: Simply try different values until the circuit works. However, it would be much more convenient to calculate the proper values.

Because resistance determines the current for a given voltage, we need to establish a relationship among the three basic electrical quantities: voltage, current, and resistance. This relationship is called **Ohm's law:**

$$V = IR$$

Note that it expresses the proper relationship between resistance and current for a fixed voltage. When R increases, I must decrease in order for V to remain constant.

This simple relation is probably the most basic and often used mathematical expression in electronics. It permits us to calculate the proper values for components in a circuit. For example, when making an electronic thermometer, we can use Ohm's law to calculate the proper thermistor resistance for a given ammeter and voltage source (Figure 3.1). It can also

help to calculate the proper battery voltage for a given relay and so on.

In this chapter, a number of typical applications of Ohm's law are given to suggest the many types of situations in which it can be used to solve circuit problems. In these examples, different forms of the relation are used, depending on the unknown quantity being calculated. These forms are:

To determine voltage:	$V = IR$
To determine current:	$I = \dfrac{V}{R}$
To determine resistance:	$R = \dfrac{V}{I}$

All three forms should be learned because each quantity will be calculated frequently. Remember, however, that it's the *same basic relation*, regardless of the form. Ohm's law is simply written differently depending on the unknown quantity to be calculated (Figure 3.2).

Consistent Units

The use of Ohm's law for calculations requires careful attention to the use of consistent units. The basic set of units is the one previously introduced:

V = volts (V)
I = amperes (A)
R = ohms (Ω)

With Ohm's law, the proper calculations can be made only if all the values of voltage, current, and resistance are first converted to these basic units.

However, there is an equivalent set of consistent units that makes calculations simpler. This set of units takes into account the fact that

OHM'S LAW

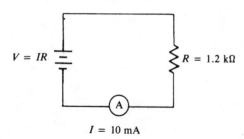

$V = IR$ $R = 1.2\ k\Omega$

$I = 10\ mA$

Unknown Voltage

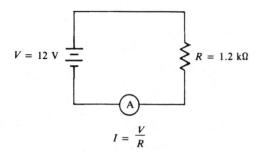

$V = 12\ V$ $R = 1.2\ k\Omega$

$I = \dfrac{V}{R}$

Unknown Current

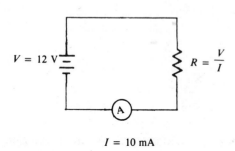

$V = 12\ V$ $R = \dfrac{V}{I}$

$I = 10\ mA$

Unknown Resistance

FIGURE 3.2. The relationship among voltage, current, and resistance for a device is given by Ohm's law: $V = IR$. This basic relationship can be used to calculate any one of the quantities if the other two are known.

UNITS FOR OHM'S LAW CALCULATIONS

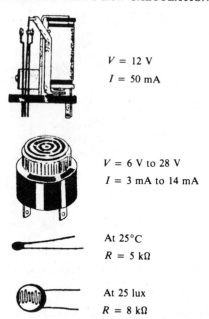

$V = 12\ V$
$I = 50\ mA$

$V = 6\ V$ to $28\ V$
$I = 3\ mA$ to $14\ mA$

At 25°C
$R = 5\ k\Omega$

At 25 lux
$R = 8\ k\Omega$

FIGURE 3.3. The voltage, current, and resistance for most electronic devices occur as volts, milliamps, and kilohms. These units can be used directly in Ohm's law, thus minimizing conversions to other units.

V = volts (V)
I = milliamperes (mA)
R = kilohms (kΩ)

Correct results will also be obtained from Ohm's law if values are expressed in these units.

The reason that these units are equivalent can easily be demonstrated. Because 1 A = 1000 mA while 1 Ω = 0.001 kΩ, the fact that current and resistance occur as a product in Ohm's law means that the factors 1000 and 0.001 nullify each other:

$V = IR$
V = A × Ω = 1000 mA × 0.001 kΩ
  = mA × kΩ

Because most values of current and resistance

in electronics, voltages are generally in the range of 1 V to 10 V, currents are in the range of a few milliamperes, and resistances are in the range of kilohms (Figure 3.3). These values also make up a set of consistent units for Ohm's law:

$V = 15.2$ V

$I = 3$ mA

Calculate R.

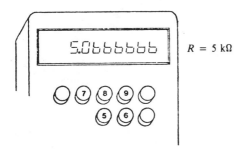

$R = 5$ kΩ

FIGURE 3.4. When using a calculator to calculate circuit values, be sure to round the answer back to the proper number of significant digits.

in electronic circuits are in the mA and kΩ range, using this set of units requires much less conversion to other units. Therefore, this set of consistent units is used throughout this book and is summarized by the following rule:

For calculations with Ohm's law, all voltages must be in V, *all currents in* mA, *and all resistances in* kΩ.

Significant Digits

The following rule applies to significant digits:

The number of significant digits in the calculated unknown quantity cannot exceed the smaller of the two known quantities.

For example, if V is 15.2 V (three significant digits) and I is 3 mA (one significant digit), then the calculated resistance can only be accurate to one significant digit. If a calculator is used in making the calculation, be sure to round off the reading to the proper number of significant digits before recording the answer. (Figure 3.4).

3.3 APPLICATIONS OF OHM'S LAW

Determining Resistance by Measuring Current

As explained in Chapter 2, electrical resistance can be determined by calculation, using the resistance relation $R = \rho(l/A)$, or by measurement, using an ohmmeter. The calculation method is practical only for simple shapes, such as wires. The ohmmeter measurement, on the other hand, can be used for almost any device and is by far the more common method. The ohmmeter places a fixed voltage across the device and then uses the resulting current as a measure of resistance. Ohm's law is the basis from which to determine the relation between current and resistance for this measurement.

EXAMPLE 1: An ohmmeter places a voltage of 1.52 V across an unknown resistor, and its ammeter records a current of 0.27 mA (Figure 3.5). What is the value of the resistance?

Solution: Using Ohm's law, calculate the resistance:

$$R = \frac{V}{I} = \frac{1.52 \text{ V}}{0.27 \text{ mA}} = \textbf{5.6 k}\boldsymbol{\Omega}$$

Note that the units of current used in the equation are mA. Therefore, the calculated resistance is in kΩ.

Note also that the resistance is given only to two significant digits because the current is given to only two significant digits.

EXAMPLE 2: Suppose you want to make a simple ohmmeter from a 1.6 V battery and an ammeter that measures from 0 μA to 100 μA (Figure 3.6.) What is the range of resistance values that can be measured?

EXAMPLE 1

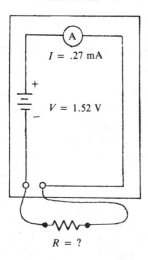

$$I = .27 \text{ mA}$$

$$V = 1.52 \text{ V}$$

$$R = ?$$

FIGURE 3.5. Ohm's law provides the basis for determining the value of an unknown resistor in an ohmmeter circuit. In this example, what is the value of R?

EXAMPLE 2

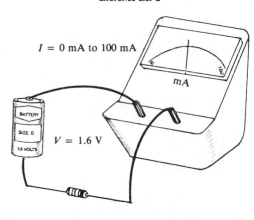

$$I = 0 \text{ mA to } 100 \text{ mA}$$

$$V = 1.6 \text{ V}$$

Range of R = ?

FIGURE 3.6. A simple ohmmeter can be constructed from a battery and an ammeter. The range of resistance values that it can measure is limited, however. In this example, what is the range of R?

Solution: The smallest resistance that can be measured will be determined by the largest current that can be measured, or 100 μA. Similarly, the largest resistance that can be measured will be determined by the smallest measurable current. Assuming the smallest readable current is 1 μA, the measurable current range is:

$$I = 1 - 100 \, \mu A$$

The current range should be converted to mA:

$$I_{min} = 1 \, \mu A \times \frac{1 \text{ mA}}{1000 \, \mu A} = 0.001 \text{ mA}$$

$$I_{max} = 100 \, \mu A \times \frac{1 \text{ mA}}{1000 \, \mu A} = 0.10 \text{ mA}$$

The range of resistance values that will produce these currents, for a fixed voltage of

1.6 V, can be calculated from Ohm's law:

$$R_{max} = \frac{V}{I_{min}} = \frac{1.6 \text{ V}}{0.001 \text{ mA}} = 2000 \text{ k}\Omega$$

$$R_{min} = \frac{V}{I_{max}} = \frac{1.6 \text{ V}}{0.10 \text{ mA}} = 16 \text{ k}\Omega$$

Note that the number of significant digits is larger for the minimum resistance than for the maximum resistance. The reason is that a current of 100 μA can be read to two significant digits, while a current of 1 μA can be read to only one significant digit. Recall the scale of an ohmmeter. The low-resistance (right) end is much more accurate than the high-resistance (left) end (Figure 3.7).

Determining Current by Measuring Voltage

In Chapter 1, we measured the current in a circuit by opening the circuit and inserting an am-

UNDERSTANDING THE OHMMETER SCALE

FIGURE 3.7.   The calculation of Example 2 shows why the scale of an ohmmeter increases from right to left and why it is more accurate at its low (right) end.

meter. This is practical if the circuit can be easily opened. But if the circuit is soldered together, as it is inside a TV, it is not very convenient to have to unsolder the circuit, make the measurement, and then resolder the circuit again. With Ohm's law, we can calculate the current in such situations. First, locate a resistor in the circuit loop where the current is to be measured. From its color code, determine its value. Then, by measuring the voltage across it, calculate the current in that part of the circuit.

EXAMPLE 3

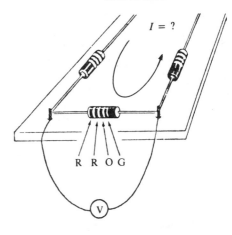

$V = 69.3\ \text{mV}$

FIGURE 3.8.   With Ohm's law, you can determine the current in a circuit without opening it. By measuring the voltage across a known resistor, the current can be calculated. In this example, what is the value of *I*?

EXAMPLE 3:   What is the current in the circuit shown in Figure 3.8 if the voltage across the fixed resistor is 69.3 mV?

Solution: The value of the resistance can be determined from the resistor color code:

$$
\begin{array}{cccc}
\text{red} & \text{red} & \text{orange} & \text{gold} \\
R = \quad 2 & 2 & \times \quad 10^3 & \pm 5\% \\
\end{array}
$$

$$= 22\ \text{k}\Omega \pm 5\%$$

The measured voltage must be converted to proper units:

$$V = 69.3\ \text{mV} \times \frac{1\ \text{V}}{1000\ \text{mV}} = 0.0693\ \text{V}$$

Using Ohm's law, we find the current:

$$I = \frac{V}{R} = \frac{0.0693\ \text{V}}{22\ \text{k}\Omega} = 0.0032\ \text{mA}$$

Note that the number of significant digits in the answer is two. However, the answer is accurate only to the tolerance of the fixed resistor, or $\pm 5\%$. Thus, the current is also accurate to $\pm 5\%$:

$$
\begin{aligned}
0.0032\ \text{mA} \pm 5\% = \\
\mathbf{0.0032\ mA \pm 0.0002\ mA}
\end{aligned}
$$

The accuracy of this method of determining current is generally limited by the uncertainty in the value of the fixed resistor. If several resistors are present in the same loop, choose the most precise one (smallest tolerance). If this is not sufficiently accurate for some reason, then the circuit may have to be opened and the ammeter inserted as before.

3.4 OTHER RELATIONS FOR ELECTRICAL POWER

In Chapter 1, the mathematical relation for calculating the power used by an electrical device was described as the product of the voltage across it and the current through it:

$$P = V \times I$$

In many cases, we know the resistance of a device and the voltage across it, but not the current; or perhaps the current, but not the voltage. With Ohm's law, we can calculate the unknown current or voltage and then use the power relation given above. However, it is also possible to derive two other forms of the power relation so that the intermediate calculation is not necessary.

Known Resistance and Current

Substituting the form of Ohm's law for voltage in terms of resistance into the power relation, we get:

$$P = V \times I$$

and

$$V = I \times R$$

Therefore:

$$P = (I \times R) \times I$$

or, equivalently:

$$P = I^2 R$$

This equation states that for a given voltage across a device, the power used by it increases as the square of the current. If we double the current, the power used by the device will increase by a factor of four (Figure 3.9).

The reason for the double dependence on current can be understood from Ohm's law. According to Ohm's law:

ELECTRICAL POWER: $P = I^2R$

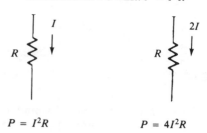

$$P = I^2R \qquad P = 4I^2R$$

FIGURE 3.9. The electrical power used by a device is proportional to the square of the current through it. Double the current and the power increases by a factor of 4.

$$V = IR$$

If the current through a device is increased, there is a proportional increase in voltage. Because power is a product of current and voltage, an increase in current is represented twice in the power expression.

Known Resistance and Voltage

Substituting the form of Ohm's law for current in terms of resistance into the power relation, we get:

$$P = V \times \frac{V}{R}$$

or, equivalently:

$$P = \frac{V^2}{R}$$

This equation indicates that the power also increases as the square of the voltage. The reason for this is the same as that given for the current.

Note also that the power increases *inversely with resistance*. This latter point is important. For a fixed voltage, as the resistance decreases, the power increases. Thus, at a given voltage, a

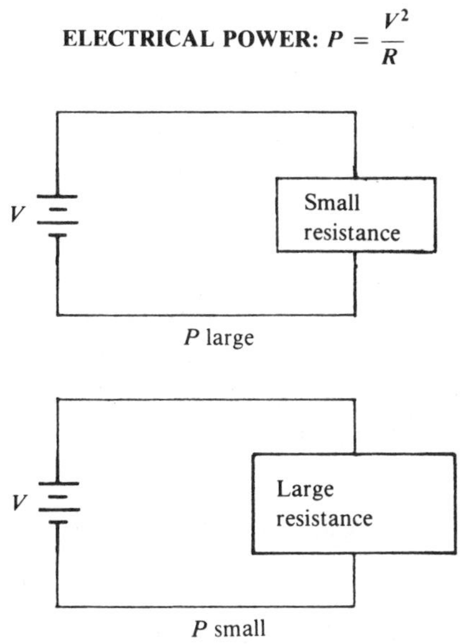

ELECTRICAL POWER: $P = \dfrac{V^2}{R}$

FIGURE 3.10. The electrical power used by a device is proportional to the square of the voltage across it, but it is inversely proportional to its resistance. Thus a low-resistance device uses more power than a high-resistance device for the same voltage.

low-resistance device will use more power than a high-resistance device (Figure 3.10).

Like Ohm's law, the power relation should be learned and understood because calculating power is frequently required. For example, most devices carry a maximum power rating that cannot be exceeded without damaging them. The resistors of Chapter 2 are typical examples. In selecting components for a circuit, always check to be certain that they will be operated at voltage and current levels within their power ratings.

Consistent Units

The basic unit for power is the watt. However, the power used by most electronic devices is generally in the milliwatt range. In fact, low power consumption is very often an important design consideration, particularly when a system is to be battery powered. If current is expressed in mA and resistance in kΩ, the power calculated from the power relation will be in mW. Therefore, the set of consistent units we will use for power is:

P = milliwatts (mW)
V = volts (V)
I = milliamperes (mA)
R = kilohms (kΩ)

As with Ohm's law, be certain that the values substituted into the power relations are in these consistent units and that the answer is expressed to the proper number of significant digits.

3.5 APPLICATIONS OF THE POWER RELATION

Determining Power from Known Resistance

EXAMPLE 4: The minimum current required to activate a relay is 30 mA. The relay resistance is 240 Ω. What is the minimum electrical power required by the relay (Figure 3.11)?

Solution: Because the current and resistance are given, we should use the I–R form of the power relation:

$P = I^2R$

Substituting the current in mA and the resistance in kΩ into the power relation gives:

$P = (30 \text{ mA})^2 \times 0.24 \text{ k}\Omega = \mathbf{220 \text{ mW}}$

EXAMPLE 4

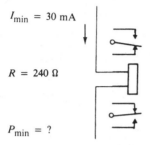

I_{min} = 30 mA

R = 240 Ω

P_{min} = ?

FIGURE 3.11. The I–R form of the power relation can be used to determine the minimum power required to activate the relay. In this example, what is the value of P_{min}?

EXAMPLE 5: A 160 Ω resistor must be placed in series with the relay in order to have the proper voltage across it. What wattage rating should this resistor have if the voltage across it is 8.0 V (Figure 3.12)?

Solution: The electrical power dissipated by the resistor can be calculated from the V–R form of the power relation:

$$P = \frac{V^2}{R}$$

Using the values of V and R,

$V = 8$ V
$R = 0.16$ kΩ

we get:

$$P = \frac{(8 \text{ V})^2}{0.16 \text{ k}\Omega} = \frac{64}{0.16} = \textbf{400 mW}$$

The standard ratings of resistors from Figure 2.12 are $1/8$ W, $1/4$ W, $1/2$ W, 1 W, and 2 W. Expressed in mW, these are 125 mW, 250 mW, 500 mW, 1000 mW, and 2000 mW. In order for the actual power dissipated by the resistor to be less than its rated power, a 500 mW, or $1/2$ W, resistor should be used.

EXAMPLE 5

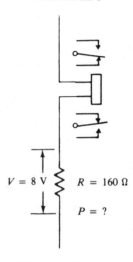

$V = 8$ V R = 160 Ω

P = ?

FIGURE 3.12. The V–R form of the power relation can be used to calculate the proper wattage rating for the series resistor. In this example, what is the value of P?

Determining Maximum Operating Voltage and Current

The calculation of power is most often required for determining maximum allowable operating voltages and currents for devices with a fixed resistance. In these cases, the device has a given maximum power rating, and you must calculate acceptable currents or voltages using the power relation. This calculation requires taking square roots and is most easily done with a calculator. Example calculations are given next.

EXAMPLE 6: A resistor is rated at $1/8$ W maximum power and has a resistance of 1800 Ω. What is the maximum voltage that can be applied to this resistor (Figure 3.13)?

Solution: The solution requires the V–R form of the power relation:

$$P = \frac{V^2}{R}$$

Substitution of the values in consistent units gives:

$$125 \, \text{mW} = \frac{V^2_{\text{max}}}{1.8 \, \text{k}\Omega}$$

or

$$V^2_{\text{max}} = 125 \, \text{mW} \times 1.8 \, \text{k}\Omega = 225 \, \text{V}^2$$

Taking the square root gives:

$$V_{\text{max}} = \sqrt{225 \, \text{V}^2} = \textbf{15 V}$$

EXAMPLE 7:   A circuit contains a $1/4$ W resistor of 2200 Ω. What is the maximum allowable current that can be used in this circuit (Figure 3.14)?

Solution: The solution of this problem requires the I–R form of the power relation:

$$P = I^2 R$$

Substituting the values of P_{max} and R in consistent units gives:

$$250 \, \text{mW} = I^2_{\text{max}} \times 2.2 \, \text{k}\Omega$$

$$I^2_{\text{max}} = \frac{250 \, \text{mW}}{2.2 \, \text{k}\Omega} = 114 \, \text{mA}^2$$

Taking the square root gives:

$$I_{\text{max}} = \sqrt{114 \, \text{mA}^2} = \textbf{11 mA}$$

The maximum current that can be passed through this resistor without exceeding its power rating is only 11 mA.

These examples are only a few of the many calculations that require the different forms of the power relation. The problems at the end of the chapter will give additional experience in using the power relation.

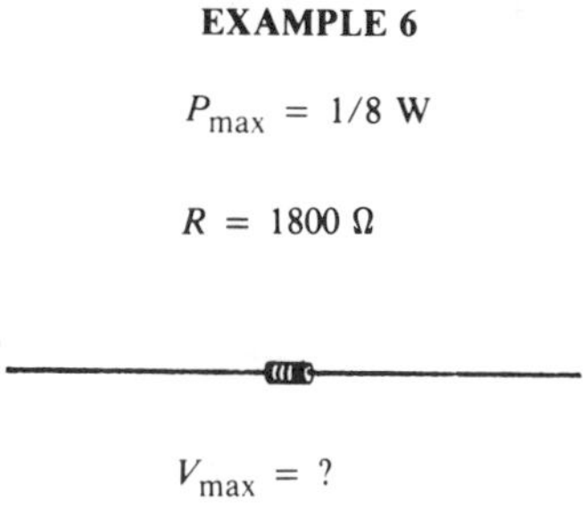

EXAMPLE 6

$$P_{\text{max}} = 1/8 \, \text{W}$$

$$R = 1800 \, \Omega$$

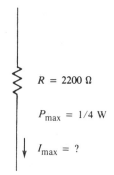

$$V_{\text{max}} = \, ?$$

FIGURE 3.13. The power relation can be used with the power rating of a device to determine maximum operating voltages. In this example, what is the value of V_{max}?

EXAMPLE 7

$$R = 2200 \, \Omega$$

$$P_{\text{max}} = 1/4 \, \text{W}$$

$$I_{\text{max}} = \, ?$$

FIGURE 3.14. The power relation can also be used to determine maximum operating currents. In this example, what is the value of I_{max}?

3.6 V–I CHARACTERISTICS

For any device that obeys Ohm's law, it is a simple matter to find the current or voltage or power for a particular situation. The preceding problems are examples. But how can voltage or current be found for devices that are not resistors?

Basically, a resistor is a device that can be described by Ohm's law — that is, one for which voltage is proportional to current. If the current is doubled, so is the voltage. Many devices, such as diodes and transistors, cannot be described so simply. Changing the current will *not* produce a proportional change in the voltage. In fact, doubling the current may change the voltage by only a small percentage; or, perhaps, a small change in current will produce a large voltage change. In addition, for different devices, current and voltage may vary in completely different ways. There is simply no relation like Ohm's law to describe these devices.

A possible solution is to tabulate current and voltage for each device. Then, given any value for current or voltage, the corresponding value may be found from the tables. A far more convenient scheme is to graph the *V–I* data. With a graph for each device, we can see at a glance exactly how current and voltage are related. These individual graphs are called *V–I* **characteristics**, and they are used for predicting circuit performance in much the same way that Ohm's law is used for resistors.

The *V–I* characteristics of a resistor (or any other purely resistive device) are rarely given as a graph because Ohm's law expresses their behavior so much more simply. However, it is worthwhile to look at the graphical *V–I* characteristic of a resistor as a step toward understanding the more complicated *V–I* characteristics of other devices.

Resistor

The *V–I* characteristic of a resistor is just a graphical presentation of Ohm's law:

$$V = IR$$

According to this expression, the current *I* is directly proportional to the voltage *V*, and the constant of proportionality is the resistance *R*.

As a graphical *V–I* characteristic, Ohm's law is a straight line, as shown in Figure 3.15. Any straight-line relationship is called **linear** and the term *linear behavior* will frequently be used to describe this type of relationship between two quantities.

V-I CHARACTERISTIC OF A 1 kΩ RESISTOR

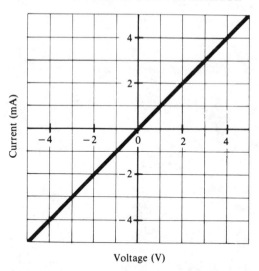

FIGURE 3.15.   An important performance specification of an electrical device is its *V–I* characteristic – that is, the current that will flow in the device for all possible voltage values. The graph shown is the *V–I* characteristic for a 1 kΩ resistance and is just a graphical representation of Ohm's law, *V = IR*.

Two important features of the *V–I* characteristic of a resistor should be noted (Figure 3.15). First, it is a *linear relationship* that passes through 0 to negative values. With 0 V across a device, no current will flow. As a positive voltage is applied, the current increases in direct proportion to the voltage. The negative region expresses the fact that if the polarity of the voltage is reversed (*V* negative), then the current direction also reverses. For a resistor, the linear relationship between voltage and current in the negative direction is identical to that in the forward direction. Therefore, a resistor can be connected either way in a circuit and it will behave in the same way. This is not the case with all devices.

Second, observe the *slope of the line*. The slope is determined by the value of the resistance. *The greater the resistance, the flatter the*

V-I CHARACTERISTICS OF THREE RESISTANCES

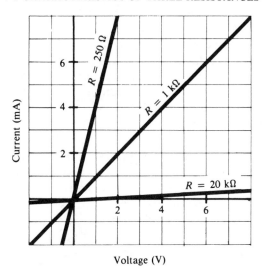

FIGURE 3.16. The **V–I** characteristics for different resistors show two important features: They are straight lines (linear), and the slopes of the lines decrease with increasing values of resistance.

slope. This situation is illustrated in Figure 3.16. Note that for a very large resistance, the line is quite flat, which agrees with the observation that the larger the resistance, the smaller the current for a given voltage. A perfectly horizontal line would indicate an infinite resistance—for example, an open circuit. No matter how much the voltage is increased across an infinite resistance, no current will flow.

Similarly, *the smaller the resistance, the steeper the slope*. This situation graphically expresses the observation that large currents will flow through devices that have a very small resistance. A vertical line along the *y*-axis would mean a short circuit, or 0 resistance.

The features of a resistor's *V–I* characteristic and their meaning are important because they help to explain the circuit behavior of other devices. In summary, these features are:

— *Straight line*, which means that current increases directly (linearly) with voltage;

DIODES AND SYMBOL

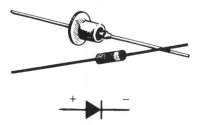

FIGURE 3.17. Modern electronics is based on the **V–I** characteristics of semiconductor devices, especially the diode. Shown are two typical diodes and the symbol for a diode. The symbol arrow points in the direction of current flow when the diode is forward biased – that is, with the polarity shown.

— *Flat slope*, which indicates small changes in current for large changes in voltage *or* large changes in voltage for small changes in current;
— *Steep slope*, which indicates large changes in current for small changes in voltage *or* small changes in voltage for large changes in current.

Diode

Most of modern electronics has been developed around the *V–I* characteristics of **semiconductors.** These are a special class of materials—particularly silicon and germanium—that conducts current easily under some circumstances but not under others. By careful and special construction using these materials, a whole variety of modern devices has been developed that will perform the many functions that comprise modern electronics.

The basic semiconductor device is the **diode.** The typical diode looks very much like a resistor (Figure 3.17), but its *V–I* characteristic is quite different. Figure 3.18 shows the general *V–I* characteristic of a diode. It consists of two basic regions: a flat one that extends to negative voltages and a very steep one at positive voltages.

V-I CHARACTERISTIC OF A SILICON DIODE

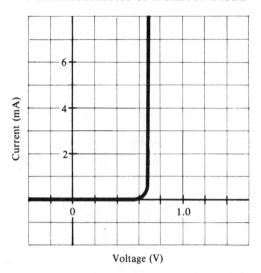

FIGURE 3.18.   The *V–I* characteristic of a diode shows that it operates much like a one-way electronic valve. Below 0.6 V, the diode acts like an open circuit, and no current will flow. Above 0.6 V, the diode acts like a closed circuit, and current flows easily. The transition voltage of 0.6 V is characteristic of a silicon diode.

The transition occurs at a small positive voltage that is typically 0.2 V to 2 V depending on the diode material. For a silicon diode, the transition occurs at about 0.6 V; for a germanium diode, it is about 0.2 V. The material of which a diode is made can generally be determined simply by measuring its transition voltage.

The behavior of a diode in a circuit can be understood from its *V–I* characteristic. With 0 V across the diode, the *V–I* curve is at 0, indicating that no current will flow. The slope of the line is very flat up to the transition voltage of about 0.6 V, indicating that even if the voltage were increased, very little current would flow. At 0.6 V, the *V–I* curve changes abruptly to a steep slope, indicating that the device will suddenly conduct large currents — that is, very

small changes in voltage will produce very large changes in the current, above 0.6 V.

For a negative voltage, the curve is very flat and near 0 current. Even if the negative voltage across the diode were made large, essentially no current would flow. The small current that does flow at negative voltage is termed the *leakage current* and is generally on the order of microamperes or less.

Another way of describing a diode's behavior is to say that below the transition voltage, a diode acts as an open circuit, and above the transition voltage, it acts as a closed, or short, circuit. Consequently, the diode is often referred to as a one-way electronic valve that lets current pass in one direction but not the other. This special *V–I* characteristic has many applications.

The symbol for the diode (Figure 3.17) indicates in which direction the diode will pass current. The arrow points in the direction of current flow when the terminal voltages are as shown. Most diodes also have a white line around one end corresponding to the bar on the symbol. When this end is negative, the diode will conduct current above the transition voltage. In this connection, the diode is said to be **forward biased.** When this end is positive, the diode is said to be **back biased**, or **reverse biased**, and only the small leakage current will flow.

The one-way nature of a diode is easy to observe if the diode is connected to an ohmmeter. Because the ohmmeter contains a battery, one of its terminals is at a higher potential than the other. Therefore, if the ohmmeter causes the diode to be reverse biased, no current will flow. Reversing the leads causes the ohmmeter needle to deflect. For an unmarked diode, this simple ohmmeter test is a convenient way to determine the polarity.

Does a reading on the ohmmeter mean that a diode has resistance? The answer is no. Resistance has meaning only in terms of Ohm's law. It is the proportionality constant between voltage and current. For a device for which cur-

LIGHT-EMITTING DIODE AND SYMBOL

FIGURE 3.19. A common diode that most people have seen is the light-emitting diode (LED) that emits light when it conducts. Shown are a typical LED and its electronic symbol.

rent and voltage are not proportional, resistance has no meaning. The number indicated on the ohmmeter is simply the ratio of the voltage divided by the current, or V/I. As can be seen in Figure 3.18, a slight change in voltage may produce a very large change in the current, and this ratio will change. Thus, the ohmmeter reading depends on the exact point on the V–I characteristic at which the measurement is made.

Light-Emitting Diode

A special type of diode that most people have seen is the **light-emitting diode**, or LED. This diode has the additional property that when it conducts, it emits light. A typical LED and its symbol are shown in Figure 3.19.

To date, LEDs can be constructed to emit only certain colors—for example, red, green, and yellow. Red LEDs are the most widely used because they give the most light for a given power consumption. The red LED is therefore a common element used for the individual segments of the seven-segment numerals of electronic digital displays.

Zener Diode

Another type of diode that has several important applications is the **zener diode**. A typical

ZENER DIODE AND SYMBOL

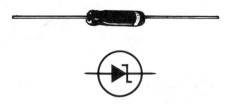

FIGURE 3.20. Another diode with important electronic applications is the zener diode. Shown is a typical case style and its electronic symbol.

V-I CHARACTERISTIC OF A ZENER DIODE

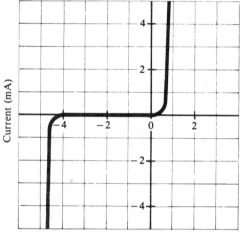

FIGURE 3.21. The V–I characteristic of a zener diode shows a reverse breakdown voltage of about –4.6 V. This so-called zener voltage is extremely constant and is commonly used as a reference voltage in electronic circuits—for example, regulated power supplies. Zener diodes have a wide range of zener voltages—from a few volts to several hundred volts.

zener diode and its symbol are shown in Figure 3.20.

The V–I characteristics of the zener diode are shown in Figure 3.21. Of primary interest is the very steep portion of the curve at a specific negative voltage. This behavior is called **reverse**

breakdown. With a negative (or reverse) voltage on the diode, it acts like an open circuit (essentially 0 reverse current) up to some value of voltage and then "breaks down," becoming, effectively, a short circuit. This breakdown voltage is called the **zener voltage**, and it can vary from a few volts to several hundred volts, depending on the device.

Once a reverse breakdown begins, the re-

verse current can change over a wide range, while the reverse voltage remains essentially constant. This property makes the zener diode very useful as a source of constant voltage — that is, the zener diode can provide a wide range of currents, all at the same voltage. Almost all regulated voltage sources employ a zener diode to maintain constant voltage.

3.7 QUESTIONS AND PROBLEMS

1. Suppose you have three 120 V heaters with internal resistances R of 10 Ω, 12.5 Ω, and 20.5 Ω. Which heater will dissipate the most electrical power P_{max}?

2. The voltage across a resistive heater producing 100 W is increased by a factor of three. How much power P will the heater now dissipate?

3. A resistor is connected to a constant-current power supply. If you wanted to increase the power produced, would you increase or decrease the value of the resistor?

4. Suppose you have a diode and cannot tell in which direction to connect it in a circuit in order to have it forward biased. Describe or illustrate a simple method to determine the proper circuit connection.

5. On the graph shown in Figure 3.22, draw the V–I characteristic of a 0.4 kΩ resistor, a germanium diode, and a light-emitting diode. Label the curves.

6. Suppose a device has a V–I characteristic like that shown in Figure 3.23. In what region does it act like a resistor?

In the following problems, be sure to use consistent units (V, mA, kΩ) and to express the answer to the proper number of significant digits.

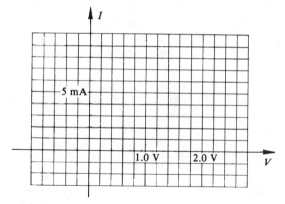

FIGURE 3.22.

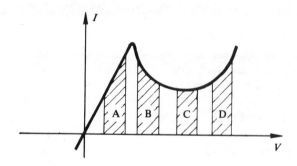

FIGURE 3.23.

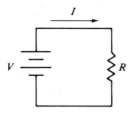

V	I	R
6.5 V		820 Ω
15 V	10.6 mA	
	0.06 A	39 kΩ
1.2 V		5600 Ω
	150 mA	0.39 MΩ
0.8 V	1.053 mA	

FIGURE 3.24.

7. If the voltage, current, and resistance values for the circuit shown in Figure 3.24 are those given in the table, find the unknown quantity.

8. If the voltage drop across a resistor is measured to be 64 mV and the current through it is 43 μA, what is the value of the resistor R?

9. If an ohmmeter is constructed from a 9 V battery and a 0 μA to 50 μA ammeter, what is the range of resistance R it can measure? Assume the ammeter can be read to an accuracy of 0.5 μA and has no internal resistance.

10. A thermometer is made by placing a 1.52 V battery in series with a UUA41J1 thermistor (Chapter 2, Figure 2.26) and a 0 μA to 50 μA ammeter with an internal resistance $R_m = 1000$ Ω. If the ammeter records a current of 27 μA, what is the temperature of the thermistor?

11. Suppose you want to know the current through a transistor in a portable radio. A resistor in series with the transistor has the color code blue, gray, orange, and silver. The voltage across it is measured to be 23 mV. What is the range of possible current values I through the transistor? If the last color were gold, what would the range of possible current values be?

12. If a thermistor draws 237 μA at 9.0 V, what is its resistance R? How much electrical power P is it using? If the thermistor is the UUA41J1, what is its temperature (Chapter 2, Figure 2.26)?

13. If a thermistor has a dissipation constant of 1 mW/°C and a minimum resistance of 5.3 kΩ, what is the maximum voltage V_{max} that can be supplied to it to be sure that its self-heating does not exceed 1°C?

14. A $1/2$ W resistor has the color code green, blue, yellow, and gold. What is the maximum current I_{max} it can carry?

15. If the power output of a 12 Ω dc heater is 500 W, what is the voltage V across the heater?

16. A power supply can deliver 100 mA at 15 V. What is its power rating P_{max}?

17. A circuit contains a $1/2$ W resistor of 0.38 MΩ. What is the maximum allowable current I_{max} that the resistor can carry?

18. A circuit requires exactly 5 V for proper operation. Unfortunately, the power supply must be located 500 ft away from the circuit. The only wire available is AWG #18. What voltage V must the power supply provide to have exactly 5 V at 200 mA? At 20 μA? Find the AWG # of the wire that must be used if the only power supply available is 5.1 V and a current of 200 mA is required at exactly 5 V.

19. Suppose you have a 15 V power supply available and want to power a device rated at 6.0 V with an internal resistance of 156 Ω. Draw the circuit that you would use, showing component values. If you use a resistor, be sure to specify a stock value (Chapter 2, Figure 2.14) and its power rating.

Part

II

BASIC CIRCUIT NETWORKS

NETWORK THEOREMS

In all technologies, there are physical laws that govern the behavior of the parameters, or quantities, that describe a device under study. For example, in the study of the flight of an aircraft, equations can be written to express the lift and drag forces on the plane as a function of its speed. In optics, the magnification of a lens system can be expressed in terms of the spacing and focal lengths of the individual lenses.

In electronics, the quantities that describe a system's behavior are the voltages and currents that exist throughout the system. In Part II, some of the basic laws that govern the relationship of voltage and current in electronic circuits are described. These circuits are often referred to as **networks** because their schematic diagram resembles an intricate "net" of electronic components woven together by connecting wires.

Although these laws are expressed as mathematical equations, the emphasis is on their use as "tools" to understanding how electronic networks and devices work. Thus, the network laws should be viewed as the "wrenches" and "screwdrivers" that allow us to "disassemble" a complex network into simpler components in order to understand how the circuit accomplishes its assigned task.

One of these tools has already been introduced: Ohm's law. Ohm's law states that for a resistor, the voltage at the resistor terminals is proportional to the current that the resistor carries. Expressed as an algebraic equation, Ohm's law is:

$$V = IR$$

For the circuit shown in Figure 1, $V = 9$ V and $R = 1$ kΩ. Therefore, we can find the current I by rewriting Ohm's law as:

$$I = \frac{V}{R}$$

and substituting the known values of V and R. Thus:

$$I = \frac{9 \text{ V}}{1 \text{ k}\Omega} = 9 \text{ mA}$$

Note that when the numerical values are substituted, voltages are in units of *volts*, resistance is in units of *kilohms*, and the resulting current is given in *milliamperes*. As explained in Part I, these units make up a consistent set of units, and using them makes it unnecessary to convert to the basic units of volts,

AN ELECTRICAL CIRCUIT

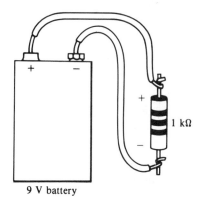

9 V battery

Pictorial Diagram

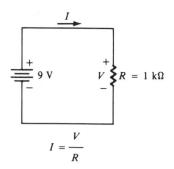

Schematic Diagram

FIGURE 1. The behavior of electronic systems is described by the voltages and currents that occur in the circuit. This part describes some of the basic network laws that govern these two quantities. The simplest of these is Ohm's law, $V = IR$.

ohms, and amperes before carrying out the calculations. The use of this consistent set is continued in the following chapters.

In Part II, several types of basic electronic networks are explored that occur repeatedly in electronic systems. It is important to memorize some of the properties of these networks so that they will be readily available when analyzing a circuit. Several network theorems are also examined. These theorems will allow us to analyze more complex circuit connections. Keep in mind that these network theorems are not abstract mathematical concepts, but are the basic tools for understanding the operation of electronic circuits.

Chapter 4

SERIES AND PARALLEL NETWORKS

4.1 OBJECTIVES

Following the completion of Chapter 4, you should be able to:

1. State Kirchhoff's current law (KCL) and write an equation that describes currents entering and leaving a node.
2. Identify parallel circuit connections in a circuit diagram and calculate their equivalent resistance, given circuit resistance values.
3. Calculate the current through each branch of a current divider, given the applied voltage and the resistance of each branch of the divider.
4. Identify the basic components of a D'Arsonval meter movement and explain how current produces needle deflections.
5. Explain the basic D'Arsonval meter characteristics of accuracy, sensitivity, and resistance and give typical values.
6. Design an ammeter circuit for a specific current range using a D'Arsonval meter with a given current sensitivity and resistance.

7. State Kirchhoff's voltage law (KVL) and write an equation that describes the voltage drops around a series loop.
8. Identify series circuit connections in a circuit diagram and calculate their equivalent resistance, given circuit resistance values.
9. Calculate the voltage across each component of a voltage divider, given the applied voltage and the resistance of each segment of the divider.
10. Design a voltmeter circuit for a specific voltage range using a D'Arsonval meter with a given current sensitivity and resistance.
11. Explain the ohms-per-volt rating of a D'Arsonval voltmeter and from it calculate the maximum circuit resistance across which the voltmeter can be connected, maintaining a specific accuracy.
12. Explain the basic digital voltmeter characteristics of sensitivity, accuracy, and resistance and give typical values.
13. Design a voltmeter circuit with a specific voltage range using a digital voltmeter with

FIGURE 4.1. Electrical current is the flow of charge through an electrical conductor. It is the product of the number of charged particles that pass a given point per second times the charge carried by each particle.

FLOW OF CHARGE THROUGH AN ELECTRICAL CONDUCTOR

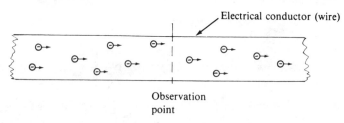

Observation point

FIGURE 4.2. The law of conservation of charge states that the amount of charge in any region remains constant. Thus, the charge passing point B must be the same as that passing point A.

LAW OF CONSERVATION OF CHARGE

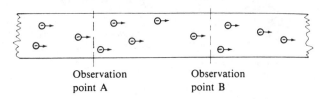

Observation point A   Observation point B

a given voltage sensitivity, accuracy, and meter resistance.

14. Design an ammeter circuit with a specific current range using a digital voltmeter with a given voltage sensitivity, accuracy, and meter resistance.

15. Design an ohmmeter circuit with a specific resistance range using a digital voltmeter with a given voltage sensitivity, accuracy, and meter resistance.

4.2 KIRCHHOFF'S CURRENT LAW

Electrical current in a conductor is produced by the flow of charged particles, usually electrons, through the conductor. One way of picturing current is shown in Figure 4.1. As the electrons in the wire move to the right, they cross a plane called the observation point (shown as a dashed line). The electrical current produced by this flow is equal to the number of electrons that cross the observation point in 1 s multiplied by

the charge carried by each electron. Thus, the magnitude of an electrical current is directly proportional to the number of electrons that cross the observation point per second.

Conservation of Charge

An important property of electrons moving in a conductor is that they obey the **law of conservation of charge**, which means that there can be no change in the amount of electrical charge within the electrical conductor. At any point, the net charge can neither increase nor decrease. Thus, if we had two observation points along the conductor, the number of electrons crossing observation point A in 1 s would be exactly the same as the number crossing observation point B (Figure 4.2).

The KCL Equation

Let's now examine what happens if we have a junction of wires called a **node.** The region of the node is defined by three observation points: A, B, and C. Electrons enter the node at point

CONSERVATION OF CHARGE AT A NODE

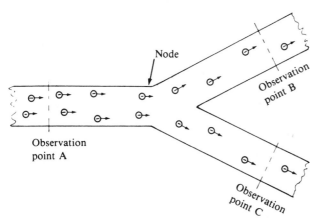

FIGURE 4.3. Where there is a node, the law of conservation of charge requires that the total charge entering the node equals the total charge leaving the node. Thus, the sum of the charge passing points B and C must equal the charge passing point A.

A and leave at points B and C. According to the law of conservation of charge, there can be no buildup of charge in the node. Hence, the number of electrons entering at point A must exactly equal the number leaving at B and C (Figure 4.3)

Since we know that the magnitude of the current is proportional to the number of electrons, we can use the law of conservation of charge to write an equation for the behavior of electrical currents at a node. This equation is called **Kirchhoff's current law** (KCL). Kirchhoff's current law states that *the algebraic sum of the currents entering a node is zero*. Because arrows indicating the direction of the currents will not all point toward the node, we establish the following rule:

Currents that point toward a node have a positive sign, and currents that point away from a node have a negative sign.

Thus, for Figure 4.4, KCL requires:

$$+I_1 - I_2 - I_3 = 0$$

KCL may be expressed equivalently as:

$$I_1 = I_2 - I_3$$

To see how useful KCL is in analyzing the

KIRCHHOFF'S CURRENT LAW

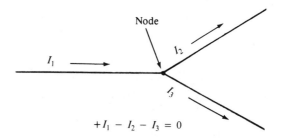

$$+I_1 - I_2 - I_3 = 0$$

FIGURE 4.4. The behavior of current at a node is described by Kirchhoff's current law (KCL): The algebraic sum of currents entering a node equals zero, where currents entering a node are positive and currents leaving are negative.

behavior of electronic networks, we will examine some circuits that require KCL for their solution.

4.3 PARALLEL CIRCUITS

Let us examine the behavior of the connection of the battery and two resistors shown in Figure 4.5. This particular arrangement is called a **parallel circuit** because the two resistors in the schematic diagram form parallel lines connected at the ends. Note that the node in the schematic diagram can be large because the resistors and their connections are spaced far enough apart to provide room for the component values.

EXAMPLE 1: What is the total current drawn from the battery of Figure 4.5?

Solution: Because the 9.0 V battery is connected directly across both resistors, we can find the currents I_1 and I_2 directly from Ohm's law:

$$I_1 = \frac{9.0\ \text{V}}{12\ \text{k}\Omega} = 0.75\ \text{mA}$$

$$I_2 = \frac{9.0\ \text{V}}{4.0\ \text{k}\Omega} = 2.25\ \text{mA}$$

Note that in substituting numerical values into the equations, the consistent set of units volts and kilohms was used. Thus, the results are in milliamperes.

Although we know the current in each resistor, we do not know the current being drawn from the battery I. To find this current, we use KCL. Writing KCL at node 1 in accordance with the rule of negative sign for currents pointing away from the node, we have:

$$+I - I_1 - I_2 = 0$$

We can rewrite this equation to solve for I:

$$I = I_1 + I_2$$

and substitute for I_1 and I_2 the values previously calculated. Thus:

$$I = 0.75 + 2.25 = \textbf{3.0 mA}$$

Equivalent Resistance

Now that we know the current I, we are in a position to consider a new question: How do the

EXAMPLE 1

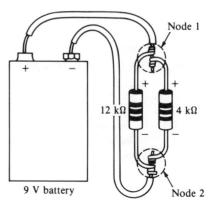

Pictorial Diagram

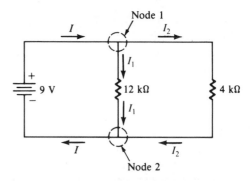

Schematic Diagram

FIGURE 4.5. When two resistors are connected at both ends, they form a parallel circuit. The total current drawn by the parallel combination can be determined by KCL.

individual values of two resistors in parallel combine to form an equivalent resistance? Consider the arrangement shown in Figure 4.6. At the top of the figure is the original circuit, with the resistors enclosed by a dashed box. We know that the battery voltage is 9 V and the circuit current is 3 mA. At the bottom of the figure is a second circuit with the same battery voltage and current, but with an unknown resistance R_{eq}.

EXAMPLES 2 AND 3

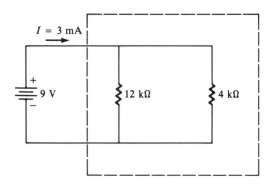

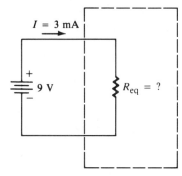

FIGURE 4.6. The current produced by two resistors in parallel can be produced also by a single equivalent resistor. The equivalent series resistor of this circuit is calculated in Examples 2 and 3.

EXAMPLE 2: What is the value of R_{eq} that will produce the same current as the parallel combinations of 4 kΩ and 12 kΩ?

Solution: Using Ohm's law, we can find R_{eq}:

$$R_{eq} = \frac{V}{I} = \frac{9 \text{ V}}{3 \text{ mA}} = 3 \text{ k}\Omega$$

Thus, a 3 kΩ resistor will produce the same current as the parallel combination of a 4 kΩ and a 12 kΩ resistor. We can say, therefore, that the *equivalent resistance* R_{eq} of the parallel combination is 3 kΩ.

CALCULATING EQUIVALENT RESISTANCE OF TWO PARALLEL RESISTORS

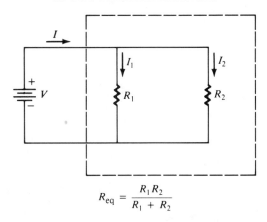

$$R_{eq} = \frac{R_1 R_2}{R_1 + R_2}$$

FIGURE 4.7. Finding the equivalent resistance of two parallel resistors is often required in circuit analysis. It can be calculated from Equation 6.

Let us now consider a more general case and develop an equation for the equivalent resistance of a parallel connection (Figure 4.7). For this circuit, the voltage across each resistor is V. Thus, we can find I_1 and I_2 from Ohm's law:

$$I_1 = \frac{V}{R_1} \quad \text{and} \quad I_2 = \frac{V}{R_2} \tag{1}$$

Using KCL at the node, we get:

$$I - I_1 - I_2 = 0 \tag{2}$$

Solving for I:

$$I = I_1 + I_2 \tag{3}$$

If we substitute Equations 1 and 3, we get:

$$I = V\left(\frac{1}{R_1} + \frac{1}{R_2}\right) = V\left(\frac{R_1 + R_2}{R_1 R_2}\right) \tag{4}$$

From Ohm's law, we know the equivalent resistance will be given by:

$$R_{eq} = \frac{V}{I} \tag{5}$$

Therefore, solving Equation 4 for V/I, we obtain the *equivalent resistance of two resistors connected in parallel*:

$$R_{eq} = \frac{R_1 R_2}{R_1 + R_2} \qquad (6)$$

Parallel combinations of resistors are very common in electronic circuits, so this is an important and useful relationship and should be memorized.

Note also that Equation 6 derives directly from Equation 4, in which it is seen that resistors in parallel add by their inverses:

$$\frac{1}{R_{eq}} = \frac{1}{R_1} + \frac{1}{R_2} + \cdots \qquad (7)$$

This relation applies to any number of resistors in parallel.

> **EXAMPLE 3:** What is the equivalent resistance of the parallel connection of 12 kΩ and 4 kΩ shown in Figure 4.6?
>
> **Solution:** Using Equation 6, we write:
>
> $$R_{eq} = \frac{12 \text{ k}\Omega \times 4 \text{ k}\Omega}{12 \text{ k}\Omega + 4 \text{ k}\Omega} = \frac{48}{16} = 3 \text{ k}\Omega$$
>
> This answer checks with the result in Example 2.

An important property of the equivalent resistance of a parallel connection is that the equivalent resistant is always *less* than either of the two resistors that form the parallel combination. Since parallel resistor combinations are so common in circuit analysis, the algebraic result of Equation 6 is often denoted:

$$\frac{R_1 R_2}{R_1 + R_2} = R_1 \parallel R_2 \qquad (8)$$

This form simplifies writing the expression in complex calculations.

Current Divider Property

The parallel connection has wide practical application as a **current divider** — that is, current coming into a node divides, with some current passing through each leg of the parallel connection. To see how a current divider works, we first use KCL to relate the current I in the parallel network to the current in each resistor:

$$I = I_1 + I_2 \qquad (9)$$

Because the voltage across each resistor is the same, we also have:

$$I_1 R_1 = I_2 R_2$$

or

$$I_2 = I_1 \left(\frac{R_1}{R_2} \right) \qquad (10)$$

Substituting Equation 10 into Equation 9 to eliminate I_2, we find:

$$I = I_1 \left(1 + \frac{R_1}{R_2} \right) = I_1 \left(\frac{R_1 + R_2}{R_2} \right) \qquad (11)$$

Thus I_1 as a *fraction* of the input current I is:

$$I_1 = \frac{R_2}{R_1 + R_2} I \qquad (12)$$

If we had eliminated I_1 between Equations 9 and 10, we would have found:

$$I_2 = \frac{R_1}{R_1 + R_2} I \qquad (13)$$

CURRENT DIVIDER

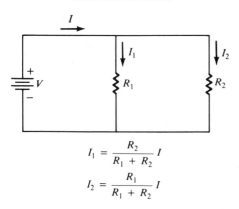

$$I_1 = \frac{R_2}{R_1 + R_2} I$$

$$I_2 = \frac{R_1}{R_1 + R_2} I$$

FIGURE 4.8. An important application of a parallel connection is the current divider. The fraction of the current that flows through each resistor is equal to the opposite resistor divided by their sum (Equations 12 and 13).

These two equations express the current divider property of the parallel resistors: *The fraction of the current in each resistor is equal to the value of the opposite resistor divided by their sum* (Figure 4.8). Note that each of the divided currents I_1 and I_2 must be *less* than the input current I. Since we will use the current divider in several measurement applications, these equations should be memorized.

4.4 D'ARSONVAL AMMETER

Although KCL provides a mathematical means of calculating the current in any circuit element, it is often more practical to measure directly an unknown current in the actual circuit. To accomplish this, an ammeter is required. (See Chapter 1, Section 1.5.) There are several types of ammeters in general use, depending on the specific application. In this section, attention is focused on the **D'Arsonval meter** that is the most common type of moving-needle meter used to measure direct current (Figure 4.9).

D'ARSONVAL AMMETER

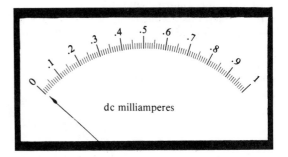

dc milliamperes

FIGURE 4.9. The ammeter is used for measuring current. The most common type of ammeter is based on the D'Arsonval meter movement.

Force on Current-Carrying Wire in a Magnetic Field

The basis of operation of the D'Arsonval meter is the force produced between a wire carrying a dc current and the magnetic field produced by a permanent magnet. The existence of this force can be demonstrated by means of the experimental arrangement shown in Figure 4.10.

To illustrate the force on the permanent magnet produced by a current in the wire, a magnetic compass is placed on the cardboard next to the wire. The needle of the compass is simply a permanent magnet placed on a bearing so it can rotate easily. With the switch open (no current in the wire), the compass is placed so that the needle points at the wire. With the switch closed, the needle will rotate to a position at right angles to its original position. If the switch is reopened, the needle will return to point at the wire again. Since the presence of the current in the wire causes the compass needle to rotate, there must be a force between the wire and the needle that depends upon the current.

Actually, the force results from the interaction of two magnetic fields. One field is produced by the permanent magnet of the compass

FIGURE 4.10. The basis of the D'Arsonval meter movement is the force that exists between a current-carrying wire and a magnetic field. This force can be demonstrated with a magnetic compass and the arrangement shown here.

PRINCIPLE OF D'ARSONVAL METER MOVEMENT

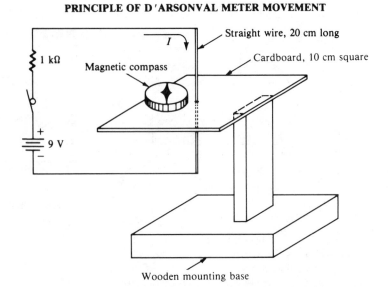

Straight wire, 20 cm long

Cardboard, 10 cm square

Magnetic compass

1 kΩ

I

9 V

Wooden mounting base

needle; the other, by the current in the wire. These fields interact, with the result that the compass needle, which is free to rotate, aligns the direction of its magnetic field with the direction of that produced by the wire. By closing the switch and moving the compass around the wire, we can determine the direction of the wire's magnetic field.

The force of interaction between the two magnetic fields is a very important phenomenon. Not only can it be used to measure current, as will be explained next, but it is the basis of operation of all rotating electric motors and generators.

Meter Operation

The D'Arsonval meter has three basic parts controlling its operation:

— A **permanent magnet** that provides a uniform magnetic field;
— A **coil** that carries the current to be measured;

— A **spring** that measures the force produced by the measured current.

A sketch of the mechanical arrangement is shown in Figure 4.11.

During operation, the current to be measured passes through the meter coil, producing a magnetic field that interacts with the magnetic field produced by the permanent magnet. The interaction forces the coil to rotate and tighten the spring fastened to the bottom of the coil. The coil will stop at the position at which the force on the coil produced by the current is just balanced by the force produced by tightening the spring.

Thus, the measurement of the electrical current is achieved by comparing the electrical force caused by the current to the mechanical force of the compressed spring. A pointer fastened to the coil and a scale indicate the coil's rest position. By passing known values of current through the coil, we can calibrate the scale to read the current's magnitude directly.

D'ARSONVAL METER MOVEMENT

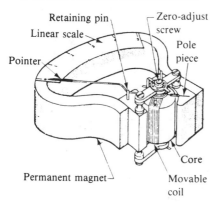

Typical Movement

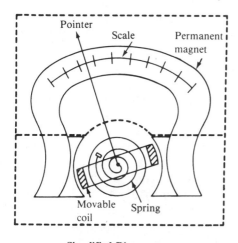

Simplified Diagram

FIGURE 4.11. The D'Arsonval movement has a coil that is free to rotate in the field of a permanent magnet. When the coil carries current, the force between it and the field causes the coil to turn, tightening a spring. The equilibrium rest position of the needle over a calibrated scale indicates the value of the current in the coil.

Meter Accuracy

Because the D'Arsonval meter depends on a mechanical spring to measure the magnitude of the current, its accuracy is affected by the quality of the spring and the friction forces in the

METER ACCURACY

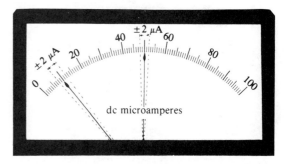

FIGURE 4.12. The accuracy of a meter is specified as a certain percent of the full-scale deflection. A typical value is ±2%. Thus, measurements at the low end can be inaccurate by as much as 20%.

bearings. Typical accuracy is about ±2% of the *full-scale reading* (Figure 4.12). However, when built with extreme care, D'Arsonval meters can achieve full-scale (fs) accuracy of about ±0.1%.

It is important to note that the accuracy of a D'Arsonval meter is always expressed in terms of the full-scale reading, not the reading the pointer indicates. If a meter has an accuracy of ±2% fs and the full-scale reading is 100 μA, then the accuracy of the meter is:

$$\text{accuracy} = \pm 0.02 \times 100 = \pm 2 \ \mu A$$

which means that the reading is in doubt by ±2 μA no matter where the needle is pointing. If, for example, the meter indicates 10 μA, the reading is in doubt by ±20% since 2 μA is 20% of 10 μA.

Meter Sensivity

D'Arsonval meters can be built with a wide range of sensitivities. At the lower end, meters with a full-scale current sensitivity of 50 μA are common, though meters with a sensitivity as low as 5 μA are available. Because of the small

MEASURING CURRENT WITH AN AMMETER

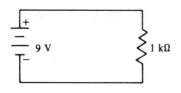

(a) Original Circuit

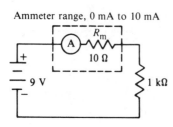

(b) Circuit with Ammeter

FIGURE 4.13. In using an ammeter to measure current, we must be certain that the meter resistance is small compared to that of the circuit. If it is not, the value of the current being measured may be changed.

force that such small currents produce, these meters, having small coils and bearings, are rather delicate.

At the other extreme are meters with full-scale current sensitivities of 5 A. These meters are quite rugged. It is possible to make meters of sensitivity greater than 5 A full scale, but they are rare. Rather, the range is extended by use of shunts, as explained later.

Meter Resistance

A factor that must be considered when using a D'Arsonval ammeter is the meter's internal resistance R_m. When an ammeter is used to measure the current in a circuit, it must be inserted in series with the circuit element whose current it is to measure. This situation is illustrated in Figure 4.13. The original circuit is shown in Figure 4.13a. If we want to measure the current in the 1 kΩ resistor, we must insert an ammeter in series with the 1 kΩ resistor, as shown in Figure 4.13b. According to KCL, all current flowing in the 1 kΩ resistor must also flow in the ammeter.

However, there is another problem that can affect the measurement. Will the ammeter's presence cause the current in the 1 kΩ resistor to change? Clearly, if the resistance of the meter were zero, the currents and voltages in Figure 4.13a and b would be identical. This is the ideal situation. In actuality, the ammeter's resistance is not zero, although it is made as small as possible. Recall that in the D'Arsonval meter the current flows through a coil of wire. This coil has some resistance that must be added to the circuit in Figure 4.13b when the meter is present. If the ammeter resistance is small compared to the rest of the circuit, however, the effect of the meter resistance on the original circuit will also be small and can be neglected.

Generally speaking, the more sensitive the D'Arsonval meter, the higher the meter resistance. Typical values for a 50 μA meter are 2 kΩ, while a 5 A meter has a resistance of about 0.02 Ω (Figure 4.14). For the 0 mA to 10 mA ammeter shown in Figure 4.13b, its resistance is about 10 Ω, which is small compared to the circuit resistance of more than 1 kΩ. Therefore, we can neglect the meter resistance in making the measurement.

Multirange Ammeters

Suppose we want to measure a current that we expect to be in the 1 mA to 10 mA range, but we have only a 100 μA full-scale D'Arsonval meter on hand. At first glance, it might seem that a less sensitive meter must be found. However, there is a way to make the 100 μA meter less sensitive in order to make the measurement. This is accomplished by the use of an external resistor in parallel with the meter. This parallel

TYPICAL D'ARSONVAL AMMETER RANGES

dc Microammeters		dc Milliammeters		dc Ammeters	
Range	Approx. Ω	Range	Approx. Ω	Range	Approx. Ω
0–50	1800	0–1	43.0	0–1	0.050
0–100	1800	0–3	2.0	0–1.5	0.033
0–200	1100	0–5	2.0	0–2	0.025
0–500	90	0–10	10.0	0–3	0.0166
25–0–25	1800	0–15	6.6	0–5	0.010
50–0–50	1800	0–20	5.0	0–10	0.005
100–0–100	1100	0–25	4.0	0–15	0.0033
500–0–500	43	0–50	2.0	0–25	0.0020
		0–100	1.0	0–30	0.0017
		0–150	0.66	0–50	0.001
		0–200	0.50	0–100	10.0
		0–250	0.40	0–150	10.0
		0–300	0.33	0–200	10.0
		0–500	0.20	0–300	10.0
		0–750	0.13	0–500	10.0
		0–1000	0.05	15–0–15	0.0033
				30–0–30	0.0017

FIGURE 4.14. D'Arsonval ammeters are available in a wide range of sensitivities. In general, the more sensitive the meter, the greater its resistance.

resistor is called a **shunt resistor**, or simply a **shunt.**

The arrangement is shown in Figure 4.15. Again notice that the combination of the meter, its internal resistance R_m, and the shunt resistance R_s form a current divider such that a fraction of the current I at the meter terminals will pass through the actual meter as I_1. The remainder will flow through the shunt R_s as I_2 such that $I_1 + I_2$ remains equal to I. The smaller the value of R_s, the larger the fraction of I that will flow in R_s and the smaller the fraction that will flow in the meter.

In the preceding example, it is expected that the current to be measured will be as large as 10 mA. The problem is to determine what value of R_s will result in a full-scale meter reading when $I = 10\,\text{mA}$.

EXAMPLE 4

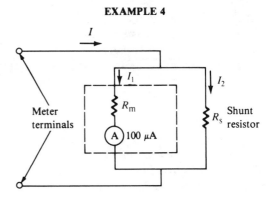

FIGURE 4.15. An ammeter's range can be extended to higher currents by placing a shunt resistor in parallel with the meter. The resulting current divider passes only a certain fraction of the current through the meter coil. Calculation of R_s is given in Example 4.

EXAMPLE 4:   What value shunt resistor is required to make a 100 μA meter movement read 10 mA full scale (Figure 4.15)?

Solution: Since the full-scale reading is 100 μA, or 0.1 mA, we have:

$$I_1 = 0.1 \text{ mA}$$

when $I = 10$ mA. Using the current divider equation, Equation 12, we find:

$$\frac{R_s}{R_s + R_m} = \frac{I_1}{I} = \frac{0.1 \text{ mA}}{10 \text{ mA}} = 0.01$$

Solving for R_s, we find:

$$R_s = 0.01R_s + 0.01R_m$$

or

$$100R_s = R_s + R_m$$

or

$$R_s = \frac{R_m}{99} \qquad (14)$$

Thus, if we placed a resistor whose value is 1/99 of the internal meter resistance R_m in parallel with the terminals of the 100 μA D'Arsonval meter, the result would be a meter that read 10 mA full scale.

Because it is much easier to change the full-scale sensitivity of a meter by adding a shunt rather than by changing the meter coil or the spring, meters are available in only a few basic sensitivities, and special ranges are obtained by the use of shunts. Note that while a shunt can reduce a meter's full-scale sensitivity, *it cannot increase it*. Shunts are widely used in the multimeter, in which several shunts and a switch are employed to obtain several ranges from a single meter movement (Figure 4.16).

CURRENT RANGE SWITCH ON A MULTIMETER

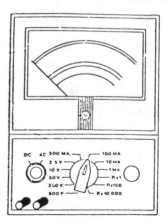

FIGURE 4.16.   The switch that selects the dc current range on a multimeter places various shunt resistors in parallel with the meter. The scale is calibrated to read the different terminal currents, but the coil current remains 0 μA to 100 μA.

KIRCHHOFF'S VOLTAGE LAW

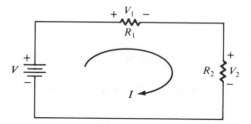

FIGURE 4.17.   The voltages in a circuit are described by Kirchhoff's voltage law (KVL): The sum of the voltages around any closed circuit is zero. KVL is based on the law of conservation of energy.

4.5 KIRCHHOFF'S VOLTAGE LAW

The voltage that exists between two points in an electrical circuit is a measure of the amount of energy a charged particle will gain or lose as it moves between the two points. For example, in Figure 4.17, as a charge carrier moves through the circuit, it will gain or lose energy depending on whether it is attracted to, or repelled by, the potential difference between the terminals of each circuit element. If we assume that the current I represents the flow of positive charge carriers, then these charge carriers will gain energy as they pass through the battery V. As the charge carriers continue through the circuit consisting of R_1 and R_2, they will give up their energy in the form of heat.

Conservation of Energy

If we consider the total energy transferred by a charge carrier as it makes a complete tour through the circuit, the *law of conservation of energy* requires that the charge carrier's energy be the same after the round trip as it was before. In other words, the energy gains and losses during the journey must add up to zero.

The mathematical statement of the law of conservation of energy in an electrical circuit is known as **Kirchhoff's voltage law** (KVL). This law states that *the sum of the voltages around any closed loop in a circuit must be zero.* As with KCL, we must pay strict attention to the polarity assigned to each voltage as we progress around the loop. The following convention will be adopted:

> *The sign of the voltage in* KVL *is the sign of the polarity encountered first while tracing the loop.*

The KVL Equation

Let us examine the expression of KVL for the circuit diagrammed in Figure 4.18. We shall start in the upper-left corner and travel around the circuit in a clockwise direction. As we move clockwise from the starting point, we first meet voltage V_1. Thocpolarity sign for V_1 is $+$. Therefore, we enter $+V_1$ in KVL. We meet and describe $+V_2$ similarly. Finally, as we return to the starting point, we meet V. We have passed

KVL EQUATION FOR A CURRENT LOOP

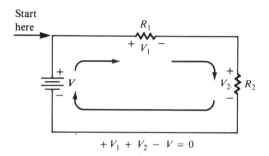

$$+V_1 + V_2 - V = 0$$

FIGURE 4.18. To write the KVL equation for a current loop, you can start at any point, but the sign of each voltage must be the sign of the polarity encountered first when going around the loop.

the $-$ sign, however, so we enter $-V$ in KVL. The complete KVL equation is:

$$+V_1 + V_2 - V = 0$$

It makes no difference where we start in writing KVL as long as we make a complete loop. If we had started at the lower-left corner and traveled counterclockwise, we would have obtained for KVL:

$$-V_2 - V_1 + V = 0$$

Notice that the two equations are identical since one can be obtained from the other by simply multiplying each side by -1 and rearranging the terms.

EXAMPLE 5: What is KVL for the circuit shown in Figure 4.19?

Solution: There are three possible loops in this network. One contains V, R_1, and R_2; one contains R_2, R_3, and R_4; and one contains V, R_1, R_3, and R_4. Let us denote them as loops 1, 2, and 3, respectively.

To write KVL, let us start at the points indicated in the figure and move clockwise:

EXAMPLE 5

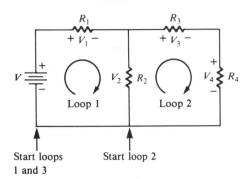

FIGURE 4.19.   For circuits with more than one loop, several KVL equations can be written. Generally, not all of them will be required to completely specify the circuit.

loop 1: $-V + V_1 + V_2 = 0$ **(15)**

loop 2: $-V_2 + V_3 + V_4 = 0$ **(16)**

loop 3: $-V + V_1 + V_3 + V_4 = 0$ **(17)**

The astute mathematician will notice that only two of these three equations are independent. For example, Equation 17 can be obtained by eliminating V_2 between Equations 15 and 16. In general, if KVL is written for all possible loops in a circuit, some redundant equations will result. With practice, you will learn which loops to use and which are necessary.

4.6 SERIES CIRCUITS

Let us now examine in detail the connection of the battery and two resistors shown in Figure 4.20. Note that there is a one-to-one correspondence between the junctions of components and the nodes in the schematic design.

The arrangement of circuit elements in Figure 4.20, in which each element is connected to the next to form a single loop, is called a **series circuit**. Note that in a series circuit, KCL requires that the current be the same in each circuit element. To solve for the circuit's performance, we must find the current I and the unknown voltages V_1 and V_2.

EXAMPLE 6:   What is the total current drawn from the battery of Figure 4.20, and what are the voltages across the two resistors?

Solution: We start by writing KVL. Beginning at the upper-left corner of the circuit:

$$V_1 + V_2 - 9.0 = 0 \qquad \textbf{(18)}$$

Although we do not yet know the value of I, we can express V_1 and V_2 in terms of I by means of Ohm's law:

$$V = RI$$
$$V_1 = 2.7I$$

and

$$V_2 = 1.8I \qquad \textbf{(19)}$$

Notice that consistent units of volts, kilohms, and milliamperes are used in this calculation. Substitution of Equation 19 into Equation 18 yields:

$$2.7I + 1.8I - 9.0 = 0 \qquad \textbf{(20)}$$

Collecting terms and transposing the –9 to the right-hand side produces:

$$(2.7 + 1.8)I = 9.0$$

or

$$I = \frac{9.0 \text{ V}}{4.5 \text{ k}\Omega} = \textbf{2 mA} \qquad \textbf{(21)}$$

Now that the current I is known, we can use Equation 19 to determine the voltages:

$$V_1 = 2.7 \times 2 = \textbf{5.4 V}$$
$$V_2 = 1.8 \times 2 = \textbf{3.6 V} \qquad \textbf{(22)}$$

EXAMPLE 6

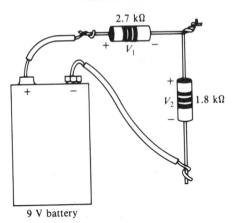

9 V battery

Pictorial Diagram

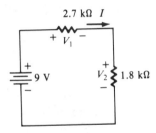

Schematic Diagram

FIGURE 4.20. When two resistors are connected end to end, they form a series circuit and the same current flows through each. Circuit current and voltage can be calculated using KVL and Ohm's law.

EXAMPLE 7

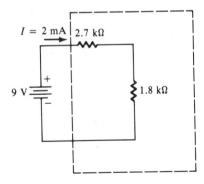

(a) Original Circuit

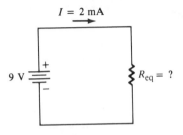

(b) Equivalent Circuit

FIGURE 4.21. The current produced by two resistors in series can be produced also by a single equivalent resistor. R_{eq} for the circuit shown here is calculated in Example 7.

To check our work, we can substitute V_1 and V_2 back into KVL, Equation 18, to see that indeed KVL is satisfied:

$$5.4 \text{ V} + 3.6 \text{ V} - 9.0 = 0$$

Because this equation is correct, we have some confidence that our calculations were performed correctly.

Equivalent Resistance

Since we know the current I, we can find the equivalent resistance of the two resistors in series. Consider the circuit shown in Figure 4.21. The circuit in Figure 4.21a is the original series circuit with the two resistors enclosed by a dashed box. We know the battery voltage is 9 V and the current is 2 mA. Figure 4.21b shows a second circuit in which the battery voltage and current are the same, but the resistor R_{eq} is unknown.

EXAMPLE 7:   What value of R_{eq} will produce the same current as the series combination of 2.7 kΩ and 1.8 kΩ?

Solution: Using Ohm's law, we find for R_{eq}:

$$R_{eq} = \frac{V}{I} = \frac{9.0 \text{ V}}{2 \text{ mA}} = 4.5 \text{ k}\Omega \qquad (23)$$

Thus, a 4.5 kΩ resistor will produce the same battery current as the series combination of a 2.7 kΩ and a 1.8 kΩ resistor. We say, therefore, that the equivalent resistance of the series combination is 4.5 kΩ.

CALCULATING EQUIVALENT RESISTANCE OF
TWO SERIES RESISTORS

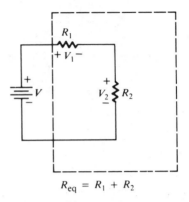

$$R_{eq} = R_1 + R_2$$

FIGURE 4.22.   The equivalent resistance of two resistors in series is simply their sum.

We now consider a more general case and develop the equation for the equivalent resistance of the series combination shown in Figure 4.22. In this circuit, the current in the battery and each resistor is the same and is denoted by I. Using Ohm's law, we find the voltages to be:

$$V_1 = R_1 I$$
$$V_2 = R_2 I \qquad (24)$$

Using KVL, we also find:

$$V_1 + V_2 - V = 0 \qquad (25)$$

If we substitute the equations of Equation 24 into Equation 25, we have:

$$V = (R_1 + R_2)I \qquad (26)$$

From Ohm's law, we know that the equivalent resistance is given by:

$$R_{eq} = \frac{V}{I} \qquad (27)$$

Therefore, we can solve Equation 26 for R_{eq} of a series connection of resistors:

$$\frac{V}{I} = R_1 + R_2$$

The important and useful result states that the *equivalent resistance of two resistors in series is their sum:*

$$R_{eq} = R_1 + R_2 \qquad (28)$$

To check its use, we shall find the equivalent resistance of the series connection of 2.7 kΩ and 1.8 kΩ:

$$R_{eq} = 2.7 + 1.8 = 4.5 \text{ k}\Omega$$

which agrees with the result in Equation 23.

An important property of the equivalent resistance of a series connection is that the equivalent resistance is always *greater* than the resistance of either of the two resistors that form the series connection. Contrast this result with the result for parallel resistance, in which the equivalent resistance is always *less* than the resistance of the two resistors.

EXAMPLE 8: What is the equivalent resistance of the series connection shown in Figure 4.23?

Solution: To find the equivalent resistance, we simply add the values of the two resistors:

$$R_{eq} = 4.7 + 5.6 = \textbf{10.3 k}\Omega$$

EXAMPLE 9: What is the equivalent resistance of the three-resistor circuit shown in Figure 4.24?

Solution: In this case, we have three resistors, so it might appear that Equation 28 does not apply. However, we can attack the problem in two steps. First, let us find the equivalent resistance of the 2.7 kΩ and 3.3 kΩ series combination, which is simply:

$$2.7 \text{ k}\Omega + 3.3 \text{ k}\Omega = 6.0 \text{ k}\Omega$$

We now simplify the circuit, as shown in Figure 4.24. The right-hand circuit now has only two resistors, so we can find R_{eq}:

$$R_{eq} = 1.0 + 6.0 = \textbf{7.0 k}\Omega$$

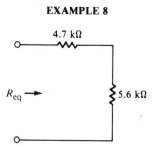

EXAMPLE 8

FIGURE 4.23. Equation 28 can be used to calculate the equivalent resistance of two resistors in series.

Voltage Divider Property

A very useful and widespread application of the series connection is as a **voltage divider** — that is, a given supply voltage is divided across two resistors in series, one fraction appearing across one of the resistors and the remaining fraction appearing across the other. To see how a voltage divider works, we return to the generalized circuit (Figure 4.25). The voltage V_1 can be written in terms of V, and we get the resistors by eliminating I between Equations 24 and 26. Ex-

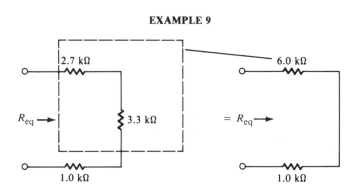

EXAMPLE 9

FIGURE 4.24. Equation 28 can be used to calculate the equivalent resistance of any number of series resistors by successive addition of resistor pairs.

VOLTAGE DIVIDER

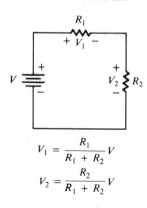

$$V_1 = \frac{R_1}{R_1 + R_2} V$$

$$V_2 = \frac{R_2}{R_1 + R_2} V$$

FIGURE 4.25.   An important application of a series connection is as a voltage divider. The fraction of the voltage that appears across each resistor is equal to its value divided by their sum.

pressed as a fraction of the input voltage V, V_1 is:

$$V_1 = \frac{R_1}{R_1 + R_2} V \qquad (29)$$

Similarly, for V_2 we find:

$$V_2 = \frac{R_2}{R_1 + R_2} V \qquad (30)$$

These important equations express the voltages V_1 and V_2 as fractions of the input voltage V. *The fraction of the voltage across each resistor is equal to its resistance divided by their sum.* Note that this fraction is always *less than one.* These equations are used frequently and should be memorized.

EXAMPLE 10

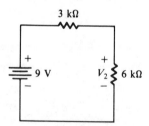

FIGURE 4.26.   Equations 29 and 30 can be used to calculate the voltage across each resistor.

EXAMPLE 10:   What is the voltage V_2 in the circuit shown in Figure 4.26?

Solution: We can find V_2 by using the voltage divider property of the series network. Since V_2 is developed across the 6 kΩ resistor, V_2 is given by:

$$V_2 = \frac{6}{6 + 3} \times 9 = \textbf{6 V}$$

$\qquad\quad$ divider $\qquad$ input
$\qquad\quad$ fraction $\qquad$ voltage

4.7 D'ARSONVAL VOLTMETER

The previous discussion of the D'Arsonval meter described how the meter's needle moves in relation to the current in the meter's coil. Thus, the D'Arsonval meter is fundamentally a current meter. In many instances, however, it is necessary to measure a voltage. The question is, can a D'Arsonval meter be made to indicate voltage? The answer is yes, and this problem is explored next.

EXAMPLE 11

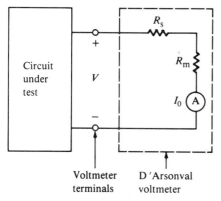

Voltmeter   D'Arsonval
terminals   voltmeter

FIGURE 4.27. A D'Arsonval ammeter can be used to measure voltage by adding a series resistor. The value of R_s for a given ammeter and desired voltmeter range can be calculated from Equation 32.

Voltmeter Design

The problem is to measure voltage when the meter responds only to current. By finding a circuit that produces a current proportional to the voltage to be measured, we can measure the current with a D'Arsonval meter and recalibrate the meter scale to read volts rather than amperes.

We have already encountered the required circuit. It is simply the proportional relationship between voltage and current expressed by Ohm's law. Ohm's law states that:

$$I = \frac{1}{R} V$$

This equation shows that the current in a resistive circuit is proportional to the voltage where the constant of proportionality is simply the circuit's resistance R.

The circuit shown in Figure 4.27 illustrates

how Ohm's law and a D'Arsonval meter can be used to construct a voltmeter. The D'Arsonval voltmeter consists of a meter movement that has an internal resistance R_m and produces a full-scale deflection when the meter current is I_0. In series with the meter movmement is a resistance R_s that is used to set the relationship between the meter needle deflection and the terminal voltage to be measured V. Using Equation 28 for the value of two resistors in series and using Ohm's law to relate V and I, we have:

$$I = \frac{V}{R_m + R_s}$$

The value of R_m is determined by the resistance of the coil of the meter, but the value of R_s can be chosen as required to establish the range of voltmeter operation. The following section shows how this choice is made.

Setting the Voltmeter Range

Suppose we desire a voltmeter whose full-scale voltage is V_0. In Figure 4.27, V_0 must correspond to a current I_0. Substitution into Equation 30 yields:

$$I_0 = \frac{V_0}{R_m + R_s} \tag{31}$$

The only unknown in Equation 31 is R_s, for which we can solve:

$$R_s = \frac{V_0}{I_0} - R_m \tag{32}$$

Thus, given a D'Arsonval meter whose full-scale deflection current is I_0 and internal resistance is R_m and given the need for a voltmeter with a full-scale reading of V_0, we can choose a proper scale resistor R_s.

EXAMPLE 11: What value of series resistor is needed to construct a voltmeter that reads 100 V full scale using a 1.0 mA D'Arsonval meter movement that has a meter resistance of 910 Ω?

Solution: The required scale resistor can be calculated directly using Equation 32. From the problem statement, we know:

$$V_0 = 100 \text{ V}$$
$$I_0 = 1.0 \text{ mA}$$
$$R_m = 0.91 \text{ k}\Omega$$

Note that to use a consistent set of units, we had to convert the value of R_m to kilohms. From Equation 32, the value of R_s is:

$$R_s = \frac{100}{1.0} - 0.91 = \textbf{99 k}\Omega$$

Multirange Voltmeters

In a multirange voltmeter, such as a multimeter, the D'Arsonval movement is combined with several resistors that can be selected by a switch to provide a variety of voltage measurement ranges. This arrangement is shown in Figure 4.28. The values of R_{s1}, R_{s2}, and R_{s3} are chosen to provide a full-scale current of I_0 when the voltmeter terminal voltage is 1 V, 10 V, or 100 V, respectively, according to Equation 32. Thus, by simply changing a switch position, we can change the voltmeter range.

EXAMPLE 12: What values of R_{s1}, R_{s2}, and R_{s3} would be required if a 100 μA meter movement with a meter resistance of 2 kΩ were used in Figure 4.28?

EXAMPLE 12

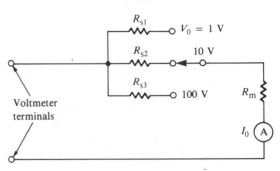

FIGURE 4.28. A multirange voltmeter can be constructed by using a switch to select a different series resistor for each range. Calculations of resistance values are given in Example 12.

Solution: The values of the series resistors are found using Equation 32. For R_{s1}, we have $V_0 = 1$ V; $I_0 = 0.1$ mA; and $R_m = 2$ kΩ. Substituting into Equation 32 yields:

$$R_{s1} = \frac{1.0}{0.10} - 2 = \textbf{8.0 k}\Omega$$

For R_{s2} and R_{s3}, only V_0 changes. Thus:

$$R_{s2} = \frac{10}{0.10} - 2 = \textbf{98 k}\Omega$$

and

$$R_{s3} = \frac{100}{0.10} - 2 = \textbf{998 k}\Omega$$

Maximum Voltmeter Sensitivity

In the foregoing discussion, it was assumed that a series resistor could be found to meet the measurement requirements. However, examination of Equation 32 shows that if V_0 is too small or if I_0 is too large, the value of R_s could be negative! Should this happen, it would be impossible to design a D'Arsonval voltmeter to meet the specifications. This possibility raises the question of how sensitive a voltmeter we can construct using a D'Arsonval movement.

A given meter movement is characterized by the current for full-scale deflection I_0 and the coil resistance R_m. Equation 32 shows that in the most sensitive voltmeter, R_s would just equal zero. Thus, the minimum full-scale voltage (maximum sensitivity) is:

$$V_{0(min)} = I_0 R_m \qquad (33)$$

Earlier we saw that in general, as the current sensitivity I_0 goes down (becomes more sensitive), the meter resistance goes up. A typical value for the resistance of a sensitive 50 μA meter movement is about 2 kΩ (see Figure 4.14). Thus, a typical maximum voltage sensitivity is:

$$\begin{aligned} V_0 &= I_0 \times R_m = 0.05 \text{ mA} \times 2 \text{ k}\Omega \\ &= 0.1 \text{ V} \end{aligned}$$

Thus, practical considerations of meter construction, ruggedness, friction, and so on, limit $V_{0(min)}$ to about 0.1 V. It is unusual to find a multimeter with a full-scale voltage range of less than 0.1 V. Should a more sensitive meter be required, an amplifier must be introduced into the meter to increase the measured voltage. Figure 4.29 shows the most common ranges of voltmeter sensitivity that are available commercially. Note that their internal meter resistance is expressed in *ohms per volt* (Ω/V). This important voltmeter specification is described next.

Measurement Errors – Ohms-per-Volt Rating

An ideal voltmeter will not change the values of voltage and current in the circuit under test when it is connected to the circuit. Stated another way, the ideal voltmeter will draw no current from the circuit being tested. A voltmeter

TYPICAL D'ARSONVAL VOLTMETER RANGES

dc Voltmeters	
Range	Approx. Ω/V
0–1.5	
0–3	
0–5	
0–8	
0–10	
0–15	
0–25	1000
0–30	
0–50	
0–100	
0–150	
0–200	
0–250	
0–300	
0–500	
0–750	
0–1000	
0–1500	2000
0–2000	
0–2500	
0–3000	
0–4000	
0–5000	

FIGURE 4.29. D'Arsonval voltmeters are available in a wide range of sensitivities. Measurements below a few tenths of a volt generally require a voltmeter with an amplifier.

that draws too much current is said to *load* the circuit under test and is considered nonideal. (Figure 4.30).

From the operation of the D'Arsonval voltmeter illustrated in Figure 4.28, it is clear that it is not ideal because the current that operates the meter movement is taken from the circuit under test. Since a D'Arsonval meter movement requires a current in order to operate, the question arises, How much current can the meter draw from the circuit before it loads the circuit and the measurement becomes inaccurate? To answer this question, we must determine how much loading of the circuit is acceptable and

IDEAL AND NONIDEAL VOLTMETERS

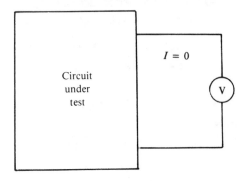

Ideal Voltmeter

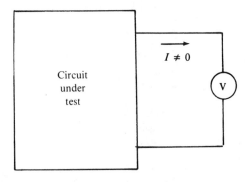

Nonideal Voltmeter

FIGURE 4.30. An ideal voltmeter draws no current from the circuit. Because a D'Arsonval voltmeter requires current for its operation, its accuracy depends on how much current it draws.

then find a method of specifying the meter in terms of the values in the test circuit.

Most D'Arsonval meters are not accurate to more than about ±1% full scale. Thus, a reasonable standard is that the meter should not cause the circuit values to change by more than 1%.

When a D'Arsonval voltmeter is connected to a circuit, an additional resistance path of value $R_s + R_m$ is added in parallel with the circuit terminals across which the voltmeter is to

be measured (Figure 4.31). The addition of the voltmeter will change the circuit in that the original resistor R, shown in the original circuit of Figure 4.31, will appear to have a smaller value, given by:

$$R' = R \parallel (R_s + R_m) \qquad (34)$$

If the value of R' is within 1% of the original value of R, the original currents and voltages will not change by more than 1% and an acceptable measurement can be made. In order for R' to be within 1% of R, the value of $R_s + R_m$ must be greater than $100R$. Thus, for an acceptable measurement, the value of the resistance in the voltmeter must be greater than 100 times the value of the resistor across which the voltage is measured:

$$R_s + R_m \geq 100R \qquad (35)$$

From the previous discussion of the multirange voltmeter, we know that the value of $R_s + R_m$ changes as the voltage range changes. Thus, the voltmeter resistance is not a constant, but varies with the scale switch position. What is constant, however, is the full-scale meter current I_0. From Equation 31, we see that the ratio of $R_s + R_m$ to V_0 is just I_0, which doesn't change as the scale switch is moved:

$$\frac{R_s + R_m}{V_0} = \frac{1}{I_0} \qquad (36)$$

This ratio, which is simply the reciprocal of the full-scale meter current, is known as the **ohms-per-volt (Ω/V) rating** of the voltmeter. Because it is a measure of how much current the voltmeter will draw from the test circuit, it is also termed the **voltmeter sensitivity**. It can be used to determine the value of $R_s + R_m$ as follows. From Equation 36, we see that:

D'ARSONVAL VOLTMETER ACCURACY

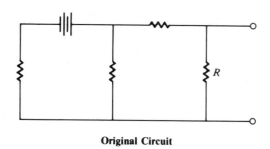

Original Circuit

FIGURE 4.31. The current that is drawn by a voltmeter is determined by how large its resistance is compared to the circuit under test. For 1% accuracy, $R_s + R_m$ should be 100R or more.

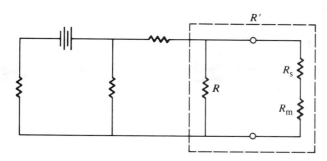

Circuit with D'Arsonval Voltmeter Connected

$$R_s + R_m = \frac{1}{I_0} \times V_0 \quad (37)$$

meter resistance	Ω/V rating	full-scale voltage

Thus, the meter resistance can be found by multiplying the Ω/V rating by the full-scale voltage for which the meter is set.

Note that the rating is ohms per volt (Ω/V) and not kilohms per volt (kΩ/V). Therefore, the consistent unit for current must be amperes, so the 50 μA was converted to 5×10^{-5} A for use in this equation.

EXAMPLE 13: What is the Ω/V rating of a D'Arsonval voltmeter that has a 50 μA full-scale movement (Figure 4.32)?

Solution: The Ω/V rating is the reciprocal of the full-scale current sensitivity. Thus:

$$\text{ohms-per-volt rating} = \frac{1}{5 \times 10^{-5} \text{ A}}$$
$$= 20,000 \ \Omega/\text{V}$$

The importance of the Ω/V rating is that it lets us determine whether or not a voltmeter will draw too much circuit current to give an accurate voltage measurement. As seen by Equation 35, the answer depends on the value of circuit resistance across which the voltmeter will be connected. The following example shows how the maximum acceptable value of circuit resistance can be determined for a meter with a given Ω/V rating.

FIGURE 4.32. The ohms-per-volt (Ω/V) rating is a measure of the loading effect of a voltmeter in measurement. It is used to calculate the maximum circuit resistance across which the meter can be connected.

EXAMPLES 13 AND 14

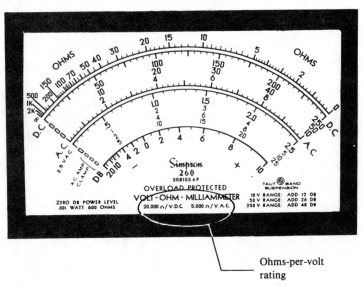

Ohms-per-volt rating

EXAMPLE 14:   What is the largest circuit resistance across which the voltmeter in Example 13 can be connected to measure voltage on the 10 V scale (Figure 4.32)?

Solution: The Ω/V rating for the meter is 20,000 Ω/V. Using Equation 36, we can calculate the meter resistance:

$$R_s + R_m = 20{,}000 \ \Omega/V \times 10 \ V$$
$$= 200{,}000 \ \Omega = \mathbf{200 \ k\Omega}$$

According to Equation 35, a meter with a resistance of 200 kΩ can be used across a circuit resistance of only 2 kΩ or less without causing more than a 1% change in the voltages and currents in the circuit under test. A circuit resistance of 2 kΩ is not a large value! When using simple D'Arsonval voltmeters to measure voltages in electronic circuits, be extremely conscious of the Ω/V rating of the meter. Most voltmeters, such as a multimeter, display their

Ω/V ratings clearly on the scale so that the user can determine whether the meter can be used for the measurement under consideration.

Generally, the greater the Ω/V rating, the more expensive the meter. The value calculated in Example 13 − 20,000 − is typical for most inexpensive D'Arsonval voltmeters. For values greater than about 20,000, we must generally resort to a buffer amplification stage between the circuit under test and the meter. This stage essentially increases the effective meter resistance, thus increasing its Ω/V rating (Equation 35).

4.8 ELECTRONIC MULTIMETERS

The discussion of the D'Arsonval voltmeter stated that this voltmeter has a maximum sensitivity of about 0.1 V full scale and that its low resistance can seriously affect the accuracy of a measurement by drawing current from the cir-

BLOCK DIAGRAM OF ELECTRONIC VOLTMETER

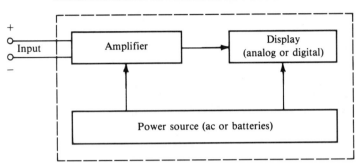

FIGURE 4.33. The limiting characteristics of a D'Arsonval voltmeter (low sensitivity and low resistance) can be improved with an electronic amplifier and external power source. Digital electronic voltmeters using integrated circuits and digital displays have a voltage sensitivity of up to 1 mV full scale and a resistance of 10 MΩ.

cuit under test. Both of these limitations relate to the fact that a D'Arsonval meter is a passive device; it contains no internal source of energy, depending instead on energy (current) from the test circuit. More sensitive meters with higher meter resistance are quite delicate and not suitable for day-to-day use in a laboratory.

In an effort to minimize the limitations of passive-type voltmeters, **electronic voltmeters** have been developed. The block diagram in Figure 4.33 shows the main components of an electronic voltmeter. It uses an internal battery or power supply for its operation, thereby greatly reducing the amount of current that it must draw from the circuit. In addition, an electronic voltmeter may employ amplifier circuits to increase the voltage sensitivity to as much as 1 mV full scale and its resistance to 10 MΩ or more.

Recent advances in integrated circuits and digital displays have made digital electronic voltmeters and multimeters competitive with units employing the D'Arsonval movement. As a result, digital multimeters capable of measuring voltage, current, and resistance are now commonly available. This section focuses attention on the digital voltmeter as the basic meter of an electronic digital multimeter. The design of the basic instrument and the circuit techniques used in making measurements are explored here.

Digital Voltmeter

The fundamental operation of the D'Arsonval meter depends on balancing the force produced by a current-carrying wire in a magnetic field against that of a spring. The measurement is achieved by comparing the electrical force to its mechanical counterpart. The nonlinearity of the spring of a D'Arsonval meter, friction, and other factors make it difficult to obtain an accuracy better than about 1%. This accuracy limitation exists whether or not an electronic amplifier is used to increase the sensitivity and input resistance.

The **digital electronic voltmeter** (DVM), on the other hand, operates by comparing the voltage to be measured with a reference voltage that is much more accurate than the motion of a mechanical spring. Thus, digital voltmeters can achieve accuracies of 0.05% or better with relative ease. The comparison of the unknown voltage to the reference voltage is accomplished by a circuit called an **analog-to-digital (A/D) converter**. The details of the operation of this circuit are beyond the scope of this text. For our purposes, it is necessary only to understand that the measured voltage is compared to a reference voltage and that the result is converted to a digital signal that drives the digital display.

The reference voltage is normally in the range of 6 V to 8 V and generally derives from a

ELECTRONIC DIGITAL VOLTMETER AND BLOCK DIAGRAM

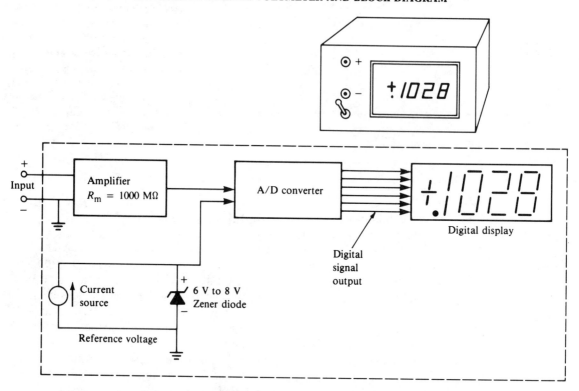

FIGURE 4.34. The electronic digital voltmeter is the basic meter in digital multimeters. Its operation is based on comparing the input voltage to a stable reference voltage derived from a zener diode. An A/D converter transforms the measured voltage to a form suitable for a digital display. The typical DVM has an input resistance of 1000 MΩ and a sensitivity of 0.1 V full scale that is accurate to ±0.05% (±0.1 mV).

zener diode. A zener diode produces a nearly constant voltage when carrying a current in the reverse direction. The accuracy of the measurement can be no better than the accuracy of the reference voltage, so the reference diodes are carefully selected for minimum voltage variations with current, temperature, and life. Most DVMs contain amplifiers to increase the sensitivity of the measurement and to increase the meter resistance to as much as 1000 MΩ. When multirange capability is required, however, the

input resistance is usually on the order of 10 MΩ, as explained later.

A block diagram of the basic parts of a DVM is shown in Figure 4.34. For clarity, the internal power supplies are not shown. The basic DVM can be purchased in a complete unit as the foundation of a digital multimeter. Typical parameters for the basic DVM are 0.1 V full-scale sensitivity with 0.1 mV as the least significant digit, 0.05% accuracy, and an input resistance R_m greater than 1000 mΩ. Let us see

EXAMPLE 15

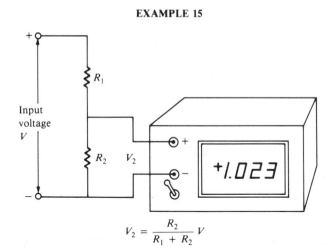

$$V_2 = \frac{R_2}{R_1 + R_2} V$$

FIGURE 4.35. The range of a basic DVM can be extended to higher voltages with a voltage divider. The divider reduces the input resistance from 1000 MΩ to the sum of $R_1 + R_2$. This sum is generally designed to be 10 MΩ.

how to use the basic DVM to construct a multirange digital voltmeter, ammeter, and ohmmeter.

Multirange Digital Voltmeter

Because the meter resistance of the basic DVM is so large, a simple voltage divider network will extend its range, as shown in Figure 4.35. The relationship between the input voltage to be measured V and the voltage at the DVM input terminals V_2 is given by the voltage divider relationship:

$$V_2 = \frac{R_2}{R_1 + R_2} V$$

In writing this equation, it is assumed that R_2 is much less than the 1000 MΩ of the DVM input and, therefore, the loading by the DVM on R_2 can be neglected.

Because the voltage divider ratio $R_2/(R_1 + R_2)$ is less than 1, we can measure voltages V that are greater than the maximum voltage sensitivity V_2 of the DVM−0.1 V, in this case. For example, if the values of R_1 and R_2 are chosen such that $R_2/(R_1 + R_2) = 0.1$, then the input to the basic DVM will be $V_2 = 0.1$ V when $V = 1$

V. Thus, we have extended the range of the meter by a factor of ten.

The penalty paid to extend the meter range is a reduction in the meter resistance. In Figure 4.35, the meter resistance at the V terminals is:

$$R_m = R_1 + R_2$$

While this value can be made very large, practical considerations usually require that R_m be on the order of 10 MΩ in multirange instruments.

EXAMPLE 15: What values of R_1 and R_2 will produce a full-scale reading of 10 V and a meter resistance of 10 MΩ in the circuit of Figure 4.35?

Solution: The basic DVM has a full-scale voltage sensitivity of $V_2 = 0.1$ V that must correspond to $V = 10$ V at the input. Thus, we have for the voltage divider:

$$\frac{R_2}{R_1 + R_2} = \frac{V_2}{V} = \frac{0.1}{10} = 0.01$$

The required meter resistance is given as 10 MΩ. Thus:

$$R_1 + R_2 = 10 \text{ M}\Omega$$

Combining these equations, we find:

$$R_2 = 0.01(R_1 + R_2) = 0.01(10 \text{ M}\Omega)$$
$$= \textbf{0.1 M}\boldsymbol{\Omega}$$

and

$$R_1 = 10 \text{ M}\Omega - R_2 = 10 \text{ M}\Omega - 0.1 \text{ M}\Omega$$
$$= \textbf{9.9 M}\boldsymbol{\Omega}$$

The accuracy of a digital voltmeter with shunt resistors depends on both the accuracy of the DVM itself ($V_2 = \pm0.05\%$) and the accuracy of the shunt resistor values. In order to maintain a measurement accuracy of $V_1 = \pm0.05\%$, the values of the shunt resistors also must be accurate to $\pm0.05\%$:

$$R_1 = 9.9 \text{ M}\Omega \pm 0.05\%$$
$$= (9.900 \pm 0.005) \text{ M}\Omega$$

$$R_2 = 0.1 \text{ M}\Omega \pm 0.05\%$$
$$= (0.10000 \pm 0.00005) \text{ M}\Omega$$

Resistors with specific values and an accuracy of $\pm0.05\%$ are relatively expensive, and they therefore tend to increase the cost of precision DVMs.

To construct a **multirange DVM**, simply extend the voltage divider concept to more resistors, as shown in Figure 4.36. The series resistor string consisting of the 9 MΩ, 900 kΩ, 90 kΩ, and 10 kΩ resistors provides various voltage dividers on the input voltage. As in Example 15, each of these resistors must be accurate to $\pm0.05\%$ to maintain a measurement accuracy of $\pm0.05\%$.

The voltage divider output is selected by the range switch such that voltages larger than 0.1 V are appropriately scaled down before being applied to the basic DVM input. For example, when the range switch is in the 10 V position, the voltage divider consists of:

$$R_1 = 9 \text{ M}\Omega + 900 \text{ k}\Omega = 9.9 \text{ M}\Omega$$
$$R_2 = 90 \text{ k}\Omega + 10 \text{ k}\Omega = 100 \text{ k}\Omega$$
$$= 0.1 \text{ M}\Omega$$

Hence, the input voltage is reduced by a factor of:

$$\frac{V_2}{V} = \frac{R_2}{R_1 + R_2} = \frac{0.1}{9.9 + 0.1} = 0.01$$

Thus, when 10 V are applied to the input, 0.1 V will appear at the basic DVM terminals. If the decimal point in the digital display is moved with a separate switch, the display will appear to read 10.0 V, even though the basic DVM input is only 0.1 V.

Because the basic DVM has such a large resistance, there is no loading on the voltage divider string as the range switch is changed. Thus, the resistance of the multirange DVM is always that of the series string—10 MΩ in the case of the instrument shown in Figure 4.36. Contrast this resistance with the meter resistance of the multirange D'Arsonval voltmeter that changes as the range switch is changed.

Multirange Digital Ammeter

To construct a digital ammeter from a basic DVM, we need only a shunt resistor R_s to convert the current to be measured into a voltage through Ohm's law. This arrangement is shown in Figure 4.37. In this circuit, the DVM input voltage is given by:

$$V = R_s I$$

Since the meter resistance R_m is so large, it is safe to assume that all of the input current I passes through the shunt. For the case in which the full-scale value of V is 0.1 V, then the value of R_s in ohms will be 0.1 divided by the value of the desired full-scale current in amperes (or 100 divided by the desired full-scale current in milliamperes).

MULTIRANGE DVM

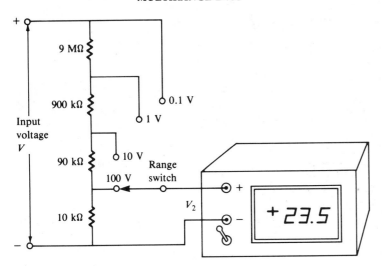

FIGURE 4.36. A basic multi-range DVM can be made by using a switch to tap different points of a voltage divider string. Note that the input resistance remains 10 MΩ for all switch positions. In commercial models, the switch also moves the decimal point in the display to the proper position for each range.

EXAMPLE 16

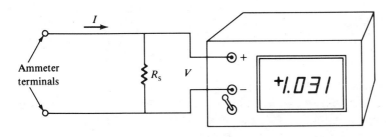

FIGURE 4.37. A basic DVM can be turned into a digital ammeter by adding a shunt resistor. Calculations of shunt resistance values for different current ranges are given in Example 16.

EXAMPLE 16: What value of shunt resistor R_s is required in Figure 4.37 to construct a 10 A full-scale digital ammeter? A 100 mA full-scale digital ammeter?

Solution: The basic DVM has a full-scale voltage sensitivity of 0.1 V. Since we want the corresponding full-scale current to be 10 A, R_s is given by:

$$R_s = \frac{V}{I} = \frac{0.1 \text{ V}}{10 \text{ A}} = \textbf{0.01 Ω}$$

For a 100 mA full-scale meter, we have:

$$R_s = \frac{0.1 \text{ V}}{100 \text{ mA}} = 0.001 \text{ kΩ} = \textbf{1 Ω}$$

A list of values of R_s for various ammeter ranges is given in Figure 4.38. These values have been chosen so that the full-scale voltage measured by the DVM is 0.1 V, and the decimal point has been positioned so that the digital reading will be numerically equivalent to the current being measured. Note that the resist-

DIGITAL AMMETER SHUNT RESISTANCES

Full-Scale Current (I)	Shunt Resistor (R_s)	Full-Scale DVM Display
10 A	0.01000 Ω	10.00 A
1 A	0.1000 Ω	1.000 A
100 mA	1.000 Ω	100.0 mA
10 mA	10.00 Ω	10.00 mA
1 mA	100.0 Ω	1.000 mA
100 μA	1000 Ω	100.0 μA

FIGURE 4.38. The table gives the shunt resistance values necessary to achieve various current ranges with a 0.1 V full-scale DVM. The resistance values must be accurate to ±0.05% to retain the 0.05% accuracy of the meter. Also given are proper decimal point locations for each range.

ance of the digital ammeter for various circuit ranges is comparable to those of D'Arsonval ammeters shown in Figure 4.14.

Again, in order to retain the high accuracy of the digital voltmeter when it is used as a digital ammeter, these shunt resistors must have an accuracy that is at least as good as the DVM (0.05% or better, in most cases). The accuracy of a resistor is also dependent on its temperature, meaning that the I^2R heat generated in the shunt resistor can raise its temperature and

cause its resistance to change. At 10 A, for example, the 0.01 Ω shunt resistor will generate 1 W ($P = I^2R = 10^2 \times 0.01 = 1$ W) that in turn will cause a substantial temperature rise unless adequate steps are taken to remove the heat generated.

To construct a multirange digital ammeter, we need only to provide a set of shunt resistors and a selector switch, as shown in Figure 4.39.

Multirange Digital Ohmmeter

The construction of a digital ohmmeter from the basic DVM is much the same as the construction of a digital ammeter. The digital ammeter operation is based on Ohm's law:

$$V = RI$$

The unknown current is passed through a known resistor R to produce a voltage that is measured, in turn, by the basic DVM. Resistor values are chosen so that the digital display has the same numerical value as the current being measured.

The digital ohmmeter is constructed by interchanging the roles of current and voltage. Using an accurate constant-current source to produce the current I makes the voltage proportional to the unknown value of resistance being

MULTIRANGE DIGITAL AMMETER

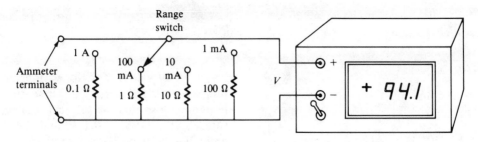

FIGURE 4.39. A multirange digital ammeter can be made by using a switch to select different shunt resistances.

EXAMPLE 17

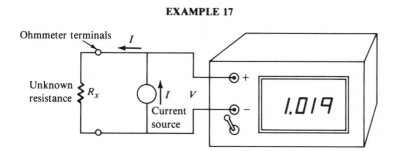

FIGURE 4.40. A DVM can be used to make a digital ohmmeter by adding a constant-current source. Calculations of current source values for various resistance ranges are given in Example 17.

DIGITAL OHMMETER CURRENT SOURCE VALUES

Full-Scale Resistance	Current Source	Full-Scale DVM Display
10 Ω	10.00 mA	10.00 Ω
100 Ω	1.000 mA	100.0 Ω
1 kΩ	100.0 μA	1.000 kΩ
10 kΩ	10.00 μA	10.00 kΩ
100 kΩ	1.000 μA	100.0 kΩ

FIGURE 4.41. The table gives the current source requirements for various ohmmeter ranges with a 0.1 V full-scale DVM. For 0.05% measurement accuracy, the current source must be accurate and stable to ±0.05%.

measured. This arrangement is shown in Figure 4.40.

Because the meter resistance R_m is so large, it is assumed that all the current I flows through the unknown resistance R_x, and none flows into the basic DVM. Clearly, there is a limit to the size of R_x beyond which this assumption fails. If 0.1% error or less is desired, then R_x should be less than 0.1% of $R_m - 1$ MΩ in the case shown in Figure 4.40. Also, the accuracy of the resistance measurement depends directly on the accuracy of the current source used to produce the voltage across R_x.

EXAMPLE 17: What value of current I is required in Figure 4.40 to construct a 10 kΩ full-scale digital ohmmeter?

Solution: The basic DVM has a full-scale voltage of $V = 0.1$ V. Since we want the corresponding full-scale resistance to be $R =$

10 kΩ, the required current is given by:

$$I = \frac{V}{R} = \frac{0.1}{10} = 0.01 \text{ mA} = \mathbf{10 \ \mu A}$$

To produce a multirange ohmmeter, we need only make the current source adjustable by increasing or decreasing the current according to the full-scale resistance range desired. As with the digital ammeter, select the current values so that the full-scale voltage measured by the DVM is 0.1 V and position the decimal point to give the same numerical value as the resistance being measured.

The set of current values for various full-scale resistance ranges is shown in Figure 4.41. Like the current values for the voltmeter and ammeter, these current values must have the same accuracy as the basic voltmeter in order to maintain the measurement accuracy over the whole set of ranges.

4.9 QUESTIONS AND PROBLEMS

1. On the circuits shown in Figure 4.42, label each node and write the KCL equation that describes all currents entering and leaving the node.
2. On the circuits shown in Figure 4.43, label each series connection and calculate its equivalent resistance.
3. For the circuit in Figure 4.44, calculate the values of I_1, I_2 and I for the following values of R_1 and R_2:

   (a) $R_1 = 2.2 \text{ k}\Omega$ and $R_2 = 4.7 \text{ k}\Omega$

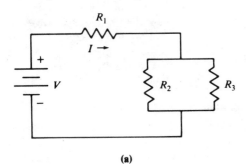

(a)

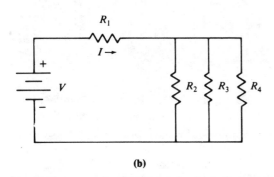

(b)

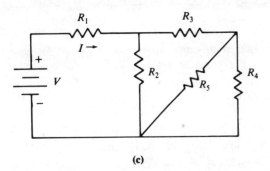

(c)

FIGURE 4.42.

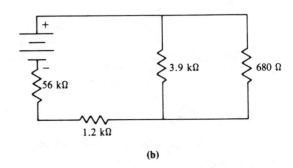

(a)

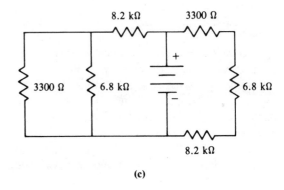

(b)

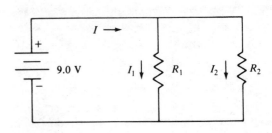

(c)

FIGURE 4.43.

FIGURE 4.44.

(b) $R_1 = 68$ kΩ and $R_2 = 0.33$ MΩ

(c) $R_1 = 100$ kΩ and $R_2 = 1$ kΩ

4. What is the net current and the current through each resistor in the circuit shown in Figure 4.45?

5. If a D'Arsonval meter has a full-scale sensitivity of ±2% fs, what is the accuracy (in current and percent) of the following readings for the full-scale sensitivities given?

Full-Scale Sensitivity	Reading	Accuracy	Accuracy (%)
100 μA	90 μA		
100 μA	20 μA		
100 μA	1.5 μA		
30 mA	22.6 mA		
30 mA	6.1 mA		
1.5 A	1.32 A		
1.5 A	0.22 A		

6. Using the 0 μA to 100 μA meter movement given in Figure 4.14, design a multirange ammeter that has full-scale current sensitivities of 0.1 mA, 0.3 mA, 1.0 mA, and 3.0 mA.

7. On the circuits shown in Figure 4.46, label all series voltage loops and write the KVL for each loop.

8. On the circuit shown in Figure 4.43, label each parallel connection and calculate its equivalent resistance.

9. For the circuit shown in Figure 4.47, calculate the circuit current and the voltage across each resistor using the following values of V, R_1, and R_2:

(a) $V = 1.50$ V; $R_1 = 2.2$ kΩ; and $R_2 = 4.7$ kΩ

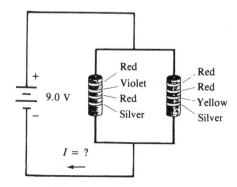

FIGURE 4.45.

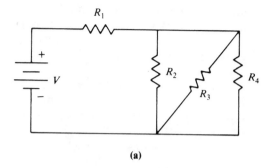

(a)

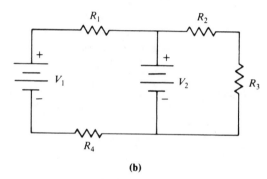

(b)

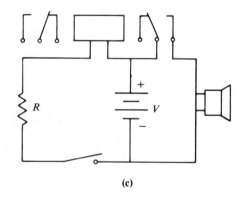

(c)

FIGURE 4.46.

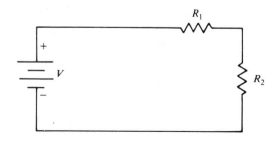

FIGURE 4.47.

(b) $V = 9$ V; $R_1 = 0.680$ MΩ; and $R_2 = 91$ kΩ

(c) $V = 0.010$ V; $R_1 = 22$ kΩ; and $R_2 = 220$ Ω

10. For the circuit in Figure 4.48, what is the total series resistance, the current, and the voltage drop across each resistor?

11. A relay is rated at 12 V and has a resistance of 240 Ω. If only a 20 V power supply is available, design a circuit that will activate the relay, but stay within its rated voltage.

12. What resistance must be placed in series with the UUA41J1 thermistor diagrammed in Figure 4.49 in order to limit the current to 75 μA for a temperature of 30°C? Refer to Chapter 2, Figure 2.26, for the thermistor characteristics.

13. Design a multirange voltmeter using the 0 μA to 50 μA meter movement that has full-scale voltage sensitivities of 0.10 V, 0.30 V, and 1.0 V (see Figure 4.14).

14. If the meter designed in Problem 13 has an accuracy of ±1% full scale, what is the minimum detectable voltage of the voltmeter?

15. What is the Ω/V rating of D'Arsonval voltmeters that have meter sensitivities of 100 μA, 1.0 mA, and 10 mA?

16. What are the largest values of a circuit resistance across which the voltmeters in Question 15 can be connected in order to maintain a measurement accuracy of ±2%?

17. Design a DVM with a full-scale sensitivity of 20 V ± 0.05% and an input resistance of 10 MΩ using a DVM with a voltage sensitivity of 0.1999 V, an input resistance of 1000 MΩ, and an accuracy of ±0.05%.

18. Design a digital ammeter with a full-scale sensitivity of 2 mA ± 0.05% using a DVM with a voltage sensitivity of 0.1999 V, an input resistance of 1000 MΩ, and an accuracy of ±0.05%. What is the ammeter's resistance?

19. Design a digital ohmmeter with a range of 0 MΩ to 1 MΩ ± 0.05%, using a digital voltmeter with a voltage sensitivity of 0.1999 V, an input resistance of 1000 MΩ, and an accuracy of ±0.05%.

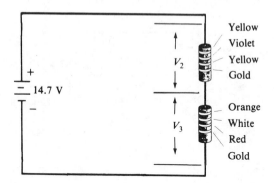

FIGURE 4.48.

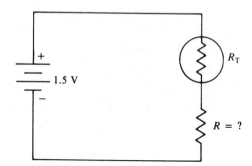

FIGURE 4.49.

Chapter 5

COMPLEX CIRCUIT ANALYSIS

5.1 OBJECTIVES

Following the completion of Chapter 5, you should be able to:

1. Calculate the equivalent resistance R_{eq} of resistor networks consisting of series and parallel combinations.
2. Calculate the net current of a resistor network, given the voltage across the network and its equivalent resistance R_{eq}.
3. Calculate the voltage across and current through any resistor of a resistor network, given the voltage across the network and the net circuit current.
4. Use KVL and KCL to check circuit calculations.
5. Use the method of dimensional analysis to check the dimensional validity of algebraic expressions.
6. Explain the principles of operation of the Wheatstone bridge circuit in resistance measurement.
7. Calculate the value of an unknown resist-

ance in a balanced Wheatstone bridge, given the values of the known resistors.

5.2 OVERVIEW

Chapter 4 introduced the series and parallel connections of resistors, including numerical methods for calculating the equivalent resistance of the two combinations. These basic circuit connections have fundamental electronic applications in dividing circuit currents and voltages to meet the $V–I$ requirements of other circuit components. Also discussed was the use of current and voltage dividers for making multirange voltmeters, ohmmeters, and ammeters from a single D'Arsonval ammeter or digital voltmeter.

The mathematical technique for arriving at these series-parallel results involves two basic electronic laws: Kirchhoff's current law (KCL) and Kirchhoff's voltage law (KVL). In this chapter, these laws are applied to more complex networks that combine the series and parallel connections.

5.3 GENERAL STRATEGY

Before analyzing specific circuits, we must clearly understand what quantities we want to know and then establish a general strategy for determining them. When a circuit network is connected to a voltage source, there are several quantities that might be of interest. One quantity is the total current drawn from the voltage source. If the voltage source is a battery, for example, the current drawn will determine how long the battery will last. Other examples are the voltage across or current through some particular component in the network. It may be desirable to verify that the voltage or current of a device is adequate for its proper operation or, conversely, to make certain that voltage or current ratings are not exceeded.

In other situations, we may want to know the value of resistance necessary in a network to produce a specific voltage or current. This information is particularly useful for circuit design. We may know that we need a voltage or current divider in order to meet the voltage or current requirements of a device, but how do we determine what circuit values to use?

Rather than attacking each possibility separately, Chapter 5 outlines a general strategy for analyzing complex networks that will yield all of the circuit quantities. If only one specific quantity is desired, of course, the whole strategy need not be implemented, only the portion that produces the desired result. Needless to say, there are many shortcuts for obtaining specific quantities, but these will be given after the more general approach.

The circuits we will analyze are shown as a combination of resistors and a voltage source V (Figure 5.1). The circuit components are shown

TYPICAL COMPLEX CIRCUIT NETWORK

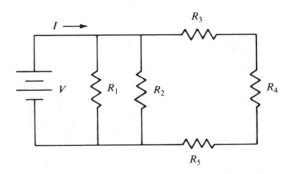

FIGURE 5.1. Chapter 5 explains how to analyze more complex circuits. These circuits are represented by a network of resistors connected to a voltage source V, where the resistors can represent any resistive device, such as a thermistor or relay.

as resistors, but they can be any resistive device, such as a thermistor, photocell, relay, meter movement, or loudspeaker.

To analyze a circuit like the one shown in Figure 5.1, the procedure is as follows:

1. *Determine the equivalent resistance of the resistor network* (Figure 5.2). The collection of resistors in the network draws the same current from the voltage source as a single "equivalent" resistor. The value of this equivalent resistance can be calculated by successively combining series and parallel combinations using the methods learned in Chapter 4.

2. *Calculate the current drawn from the voltage source by the equivalent resistance.* This step, of course, is simply the application of Ohm's law to Figure 5.2.

3. *Use the circuit current (or voltage) to calculate the voltage across (or current through) a specific resistor in the original network.* This procedure may have to be done in several steps by reversing the resistor-combining process of step 1.

**COMPLEX RESISTOR NETWORK REDUCED
TO ITS EQUIVALENT RESISTANCE**

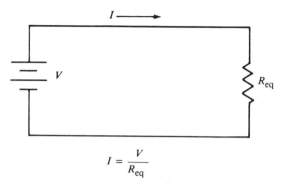

$$I = \frac{V}{R_{eq}}$$

FIGURE 5.2. The general strategy for analyzing a circuit like the one shown in Figure 5.1 is to determine the equivalent resistance of the resistor network and use it to calculate the net circuit current *I* using Ohm's law. In turn, *V* and *I* can be used to determine the voltage across or current through any device in the original resistive network.

5.4 THREE-RESISTOR COMBINATIONS

Let us begin by considering combinations of three resistors. There are four possible ways to connect three resistors, as shown in Figure 5.3. Each of these combinations involves a series, parallel, or combined series-parallel connection that will divide an applied voltage or current in a certain way. The calculation of the voltage across or current through a specific resistor uses essentially the same methods presented in Chapter 4, except that the algebra is considerably more messy. The first step in each case is to determine an equivalent resistance R_{eq} for the combination in order to calculate the circuit current *I* for any applied voltage *V*.

Simple Series Connection

The simplest case is Figure 5.3a, in which three resistors are connected in series. This is an ex-

ample of a three-component voltage divider. We know that two resistors in series, R_1 and R_2, have an equivalent resistance R_{eq}, which is just the sum of the two resistors, as given by Equation 27 in Chapter 4:

$$R_{eq} = R_1 + R_2$$

If we add a third resistor in series, its value will simply add to the value of the sum of the first two. Thus, for the connection in Figure 5.3a, the equivalent resistance is:

$$R_{eq} = R_1 + R_2 + R_3 \tag{1}$$

See Figure 5.4.

This result, of course, can be extended to any number of resistors connected in series — that is, *for any number of resistors connected in series n, the equivalent resistance is simply their sum*:

$$R_{eq} = R_1 + R_2 + R_3 + \cdots + R_n \tag{2}$$

The circuit current *I* for an applied voltage *V* is determined by Ohm's law:

$$I = \frac{V}{R_{eq}} \tag{3}$$

Because the resistors are connected in series, this current flows through each of them.

Knowing the circuit current permits calculations of the voltage across each resistor, again using Ohm's law:

$$V_1 = IR_1$$
$$V_2 = IR_2$$
$$V_3 = IR_3 \tag{4}$$

FIGURE 5.3. The methods for determining the equivalent resistance of simple series and parallel connections can be expanded to more complex networks. Shown are four possible ways of connecting three resistors. The first step in calculating the voltage or current for a specific resistor is to determine the equivalent resistance of the combination.

FOUR WAYS TO CONNECT THREE RESISTORS

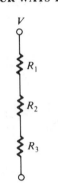

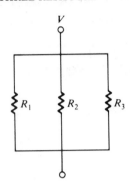

(a) Series Connection

(b) Parallel Connection

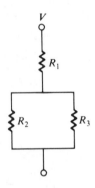

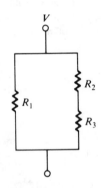

(c) Series-Parallel Connection

(d) Parallel-Series Connection

EXAMPLE 1: If the applied voltage in Figure 5.4 is $V = 15$ V, and $R_1 = 10$ kΩ, $R_2 = 47$ kΩ, and $R_3 = 22$ kΩ, what is the voltage across each resistor?

Solution: Using Equation 1, we find the equivalent resistance of the circuit to be:

$$R_{eq} = R_1 + R_2 + R_3$$
$$= 10 \text{ k}\Omega + 47 \text{ k}\Omega + 22 \text{ k}\Omega = 79 \text{ k}\Omega$$

Using Equation 3, the circuit current is therefore:

$$I = \frac{V}{R_{eq}} = \frac{5 \text{ V}}{79 \text{ k}\Omega} = 0.19 \text{ mA}$$

The voltage across each resistor is:

$$V_1 = IR_1 = 0.19 \text{ mA} \times 10 \text{ k}\Omega = \mathbf{1.9 \text{ V}}$$
$$V_2 = IR_2 = 0.19 \text{ mA} \times 47 \text{ k}\Omega = \mathbf{8.9 \text{ V}}$$
$$V_3 = IR_3 = 0.19 \text{ mA} \times 22 \text{ k}\Omega = \mathbf{4.2 \text{ V}}$$

According to the KVL, the sum of these voltages must equal the applied voltage V:

EXAMPLE 1

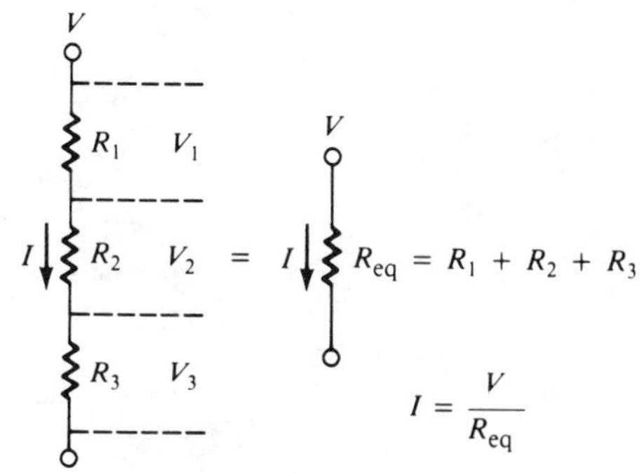

FIGURE 5.4. For three (or more) resistors in series, the equivalent resistance is simply the sum of the three resistance values. R_{eq} can then be used to calculate the circuit current I that in turn, can be used to calculate the voltage across each resistor.

$$V = V_1 + V_2 + V_3$$
$$= 1.9\ V + 8.9\ V + 4.2\ V = 15\ V$$

Applying KVL serves as a good check of this calculation. In this case, since the sum of the voltages equals the applied voltage, we have some confidence that the algebraic calculations were performed correctly.

Simple Parallel Connection

The next most complicated circuit is three resistors in parallel, as shown in Figure 5.3b. This is an example of a three-component current divider. From Equation 6 in Chapter 4, we know that the equivalent resistance of two resistors in parallel is given by:

$$R_{eq} = \frac{R_1 R_2}{R_1 + R_2}$$

or, in shorthand notation:

$$R_{eq} = R_1 \parallel R_2$$

To find the equivalent resistance of three parallel resistors, we simply break the circuit into two smaller parts, as shown in Figure 5.5, and then apply Equation 6 in Chapter 4 twice. Thus, the equivalent resistance of R_1 in parallel with R_2 is:

$$R_{12} = R_1 \parallel R_2 = \frac{R_1 R_2}{R_1 + R_2}$$

and the equivalent resistance of R_{12} in parallel with R_3 is:

EXAMPLE 2

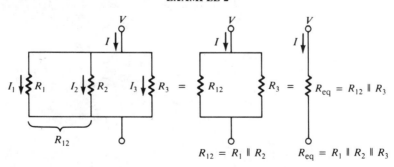

FIGURE 5.5. Three resistors in parallel form a three-component current divider. The equivalent resistance can be found by successive application of the parallel resistor equation. The algebra becomes quite messy, however. Typical circuit calculations are given in Example 2.

$$R_{eq} = R_{12} \parallel R_3 = \frac{R_{12}R_3}{R_{12} + R_3}$$

$$= \frac{R_1R_2R_3/(R_1 + R_2)}{R_1R_2/(R_1 + R_2) + R_3}$$

The final result is:

$$R_{eq} = \frac{R_1R_2R_3}{R_1R_2 + R_1R_3 + R_2R_3} \qquad (5)$$

It is clear that the algebra is becoming a little messy. The use of the shorthand symbol $\parallel$ gives us a considerably more compact form:

$$R_{eq} = R_1 \parallel R_2 \parallel R_3 \qquad (6)$$

As in the previous case, the net circuit current drawn from the voltage source can be calculated using Ohm's law:

$$I = \frac{V}{R_{eq}} \qquad (7)$$

The voltage across each section is, of course, the applied voltage, and the current through each resistor is the applied voltage divided by that resistance value:

$$I_1 = \frac{V}{R_1}$$

$$I_2 = \frac{V}{R_2}$$

$$I_3 = \frac{V}{R_3} \qquad (8)$$

As in the case of the series connection, a check of the calculation can be made by applying the KCL requirement that:

$$I = I_1 + I_2 + I_3 \qquad (9)$$

EXAMPLE 2:   In Figure 5.5, if $V = 15$ V, $R_1 = 10$ kΩ, $R_2 = 47$ kΩ, and $R_3 = 22$ kΩ, what are the net circuit current and the current flowing through each resistor?

Solution: Equation 5 gives the equivalent resistance of the parallel combination:

$$\begin{aligned} R_{eq} &= \frac{R_1R_2R_3}{R_1R_2 + R_1R_3 + R_2R_3} \\ &= \frac{10 \times 47 \times 22}{10 \times 47 + 10 \times 22 + 47 \times 22} \text{ k}\Omega \\ &= \frac{10{,}340}{470 + 220 + 1034} \text{ k}\Omega \\ &= \frac{10{,}340}{1224} \text{ k}\Omega = 6.0 \text{ k}\Omega \end{aligned}$$

Note that as for any parallel combination of resistors, the net equivalent resistance is smaller than the smallest resistor value.

The net circuit current can now be calculated from Ohm's law:

$$I = \frac{V}{R_{eq}} = \frac{15 \text{ V}}{6.0 \text{ k}\Omega} = 2.5 \text{ mA}$$

The current through each resistor is, of course, simply the applied voltage V divided by that resistor:

$$I_1 = \frac{V}{R_1} = \frac{15 \text{ V}}{10 \text{ k}\Omega} = 1.5 \text{ mA}$$

$$I_2 = \frac{V}{R_2} = \frac{15 \text{ V}}{47 \text{ k}\Omega} = 0.32 \text{ mA}$$

$$I_3 = \frac{V}{R_3} = \frac{15 \text{ V}}{22 \text{ k}\Omega} = 0.68 \text{ mA}$$

The accuracy of our calculations can be checked with KCL:

$$\begin{aligned} I &= I_1 + I_2 + I_3 \\ &= 1.5 \text{ mA} + 0.32 \text{ mA} + 0.68 \text{ mA} \\ &= 2.5 \text{ mA} \end{aligned}$$

Again the sum of the currents equals the net circuit current, so we can believe that our calculations are correct.

Series-Parallel Connection

Although Examples 1 and 2 contained three resistors, they were completely series or completely parallel connections. Figure 5.3c represents the series connection of R_1 with the parallel connection of R_2 and R_3. To find the equivalent resistance, we must first look at each of the series parts, as shown in Figure 5.6. The equivalent resistance is the sum of the two series components. In shorthand notation, this can be written:

$$
\begin{aligned}
R_{eq} &= R_1 + (R_2 \parallel R_3) \\
&= R_1 + \frac{R_2 R_3}{R_2 + R_3}
\end{aligned}
\tag{10}
$$

As before, R_{eq} can be used to determine the net circuit current I due to an applied voltage V, and this value, in turn, can be used to determine the current and voltage for each circuit resistor.

EXAMPLE 3: In Figure 5.6, if $V = 15$ V, $R_1 = 10$ kΩ, $R_2 = 47$ kΩ, and $R_3 = 22$ kΩ, what are the voltage across and the current through each resistor?

Solution: Equation 10 gives the equivalent resistance:

$$
\begin{aligned}
R_{eq} &= R_1 + \frac{R_2 R_3}{R_2 + R_3} \\
&= 10 \text{ kΩ} + \frac{47 \text{ kΩ} \times 22 \text{ kΩ}}{47 \text{ kΩ} + 22 \text{ kΩ}} \\
&= 10 \text{ kΩ} + 15 \text{ kΩ} = 25 \text{ kΩ}
\end{aligned}
$$

The net circuit current is therefore:

$$
I = \frac{V}{R_{eq}} = \frac{15 \text{ V}}{25 \text{ kΩ}} = \mathbf{0.60 \text{ mA}}
$$

This answer is the current through resistor R_1. The voltage across R_1 is:

$$
V_1 = IR_1 = 0.60 \text{ mA} \times 10 \text{ kΩ} = \mathbf{6.0 \text{ V}}
$$

EXAMPLE 3

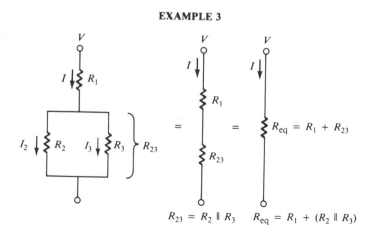

$$R_{23} = R_2 \parallel R_3 \qquad R_{eq} = R_1 + (R_2 \parallel R_3)$$

FIGURE 5.6. The series-parallel connections can be reduced to a single equivalent resistance by first reducing the parallel connection to an equivalent single resistance and then adding the resulting two series resistors. Typical calculations of the voltage and current are given in Example 3.

By KVL, the voltage across the parallel combination $R_2 \parallel R_3$ must be:

$$V = V_1 + V_{23}$$
$$V_{23} = 15\text{ V} - 6.0\text{ V} = \mathbf{9.0\text{ V}}$$

and the currents through resistors R_2 and R_3 are:

$$I_2 = \frac{V_{23}}{R_2} = \frac{9.0\text{ V}}{47\text{ k}\Omega} = \mathbf{0.19\text{ mA}}$$

$$I_3 = \frac{V_{23}}{R_3} = \frac{9.0\text{ V}}{22\text{ k}\Omega} = \mathbf{0.41\text{ mA}}$$

By KCL, the sum of I_2 and I_3 must equal the net circuit current I:

$$I = I_2 + I_3 = 0.19\text{ mA} + 0.41\text{ mA}$$
$$= 0.60\text{ mA}$$

This agreement suggests that our calculations were performed correctly.

Parallel-Series Connection

The final example, Figure 5.3d, is the combination of R_1 in parallel with the series connection of R_2 and R_3. In shorthand notation, this is written:

$$\begin{aligned} R_{eq} &= R_1 \parallel (R_2 + R_3) \\ &= \frac{R_1(R_2 + R_3)}{R_1 + R_2 + R_3} \end{aligned} \qquad \textbf{(11)}$$

See Figure 5.7

Note that in this calculation, the combination of R_2 and R_3 must be made before calculating the parallel combination with R_1. In this regard, the $\parallel$ symbol may be viewed as having the same algebraic behavior as the multiplication sign $\times$.

EXAMPLE 4:   In Figure 5.7, if $V = 15$ V, $R_1 = 10$ kΩ, $R_2 = 47$ kΩ, and $R_3 = 22$ kΩ, what are the voltage across and current through each resistor?

Solution: Equation 11 gives the equivalent resistance:

$$\begin{aligned} R_{eq} &= \frac{R_1(R_2 + R_3)}{R_1 + R_2 + R_3} \\ &= \frac{10\text{ k}\Omega(47\text{ k}\Omega + 22\text{ k}\Omega)}{10\text{ k}\Omega + 47\text{ k}\Omega + 22\text{ k}\Omega} \\ &= \frac{690}{79}\text{ k}\Omega = 8.7\text{ k}\Omega \end{aligned}$$

The net circuit current is:

$$I = \frac{V}{R_{eq}} = \frac{15\text{ V}}{8.7\text{ k}\Omega} = 1.7\text{ mA}$$

The current through R_1 is given by:

$$I_1 = \frac{V}{R_1} = \frac{15\text{ V}}{10\text{ k}\Omega} = \mathbf{1.5\text{ mA}}$$

By KCL, the current through the series connection of R_2 and R_3 must be:

$$I = I_1 + I_{23}$$
$$I_{23} = 1.7\text{ mA} - 1.5\text{ mA} = 0.2\text{ mA}$$

And the voltages across R_2 and R_3 are:

$$V_2 = I_{23}R_2 = 0.2\text{ mA} \times 47\text{ k}\Omega = \mathbf{9.4\text{ V}}$$
$$V_3 = I_{23}R_3 = 0.2\text{ mA} \times 22\text{ k}\Omega = \mathbf{4.4\text{ V}}$$

By KVL, the sum of these voltages should equal the applied voltage V:

$$\begin{aligned} V &= V_2 + V_3 = 9.4\text{ V} + 4.4\text{ V} \\ &= 13.8\text{ V} \neq 15\text{ V} \end{aligned}$$

Note that this result is less than V! It suggests that we have an error in our calculations. However, a check of the calculations

EXAMPLE 4

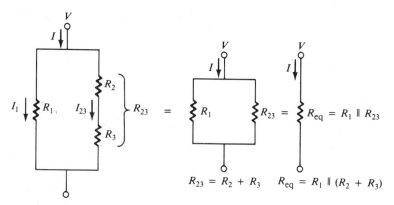

$$R_{23} = R_2 + R_3 \qquad R_{eq} = R_1 \parallel (R_2 + R_3)$$

FIGURE 5.7. The parallel-series resistance is reduced by adding the two series resistors and then calculating the equivalent resistance of the resulting parallel connection. Typical calculations of the voltage and current for each resistor are given in Example 4.

will show that no error has been made. What, then, is the problem?

The problem arises from the use of KCL to get the current through the series resistor. This calculation required a subtraction that led to a reduction in significant figures from two to one. The current through $R_2 + R_3$ was found to be 0.2 mA—only one significant figure.

The accuracy of our calculation can be maintained by finding I_{23} in a different way to get the result in two significant digits. This method requires calculating I_{23} directly:

$$I_{23} = \frac{V}{R_1 + R_2} = \frac{15 \text{ V}}{47 \text{ k}\Omega + 22 \text{ k}\Omega}$$
$$= 0.22 \text{ mA}$$

This gives V_2 and V_3 as:

$$V_2 = I_{23} R_2 = 0.22 \text{ mA} \times 47 \text{ k}\Omega$$
$$= 10 \text{ V}$$

$$V_3 = I_{23} R_3 = 0.22 \text{ mA} \times 22 \text{ k}\Omega$$
$$= 4.8 \text{ V}$$

And checking by KCL gives:

$$V = V_2 + V_3 = 10 \text{ V} + 4.8 \text{ V} = 15 \text{ V}$$

Thus the answer now agrees to two significant digits. This example illustrates the importance of maintaining the proper number of significant digits in calculations and the value of using KVL and KCL to check for calculation errors.

Dimensional Analysis and Error Checking

Sometimes it is necessary to construct the algebraic expression for the equivalent resistance of a complex network. Because the calculations can involve considerable algebraic manipulations, it is desirable to have some way to check the expression for errors that can easily occur.

One technique is to pick specific resistor values that will yield a simple numerical result and then compare the algebraic result with the correct answer. For example, let us check the result of Equation 11. If we pick $R_2 = R_3 = 1$ kΩ and $R_1 = 2$ kΩ for the circuit shown in Figure 5.7 from which we derived Equation 11, then we have $R_2 + R_3 = 2$ kΩ, which is in parallel with the 2 kΩ of R_1. Two 2 kΩ resistors in parallel have an equivalent resistance of 1 kΩ.

Substitution of these values into Equation 11 yields:

$$R_{eq} = \frac{R_1(R_2 + R_3)}{R_1 + R_2 + R_3}$$

$$= \frac{2(1 + 1)}{2 + 1 + 1} \text{ k}\Omega = 1 \text{ k}\Omega$$

This answer checks.

Another error-checking technique that can be applied to algebraic expressions is **dimensional analysis**. In any mathematical expression relating physical quantities, the dimensions of the equation and its subparts must agree with the quantity represented. For example, the dimension of a resistor is its unit, the ohm. Thus, when we express the equivalent resistance of the three resistors in series, as described by Equation 1, all terms in the sum must have the dimension of ohms, and they do:

$$R_{eq} (\Omega) = R_1 (\Omega) + R_2 (\Omega) + R_3 (\Omega)$$

On the other hand, the expression:

$$R_{eq} = R_1 + R_2R_3$$

cannot be correct since the terms R_{eq} and R_1 have the dimension ohms, but R_2R_3 has the dimension ohms × ohms, or ohms². Adding ohms to ohms² is like adding apples and oranges — they don't match.

In Equation 11, each term in the numerator has the dimension ohms², while each term in the denominator has the dimension ohms:

$$R_{eq} = \frac{R_1R_2 + R_1R_3}{R_1 + R_2 + R_3}$$

Thus, each sum appears correct. In addition, the dimension of the numerator divided by the dimension of the denominator must yield the dimension of the expected result, ohms. Thus:

$$R_{eq} = \frac{\text{ohms}^2}{\text{ohms}} = \text{ohms}$$

which is the correct dimension for a resistor.

Dimensional analysis is not a foolproof error check, but it does aid in checking complex algebraic expressions. If there is an error in the dimensional analysis, there is surely an error in the algebra.

EXAMPLE 5: Check the following algebraic expressions for correctness:

(a) $R_{eq} = R_1 + \dfrac{R_2R_3}{R_2 + R_3}$

(b) $R_{eq} = \dfrac{R_1R_2 + R_3(R_2 + R_4)}{R_1R_2 + R_3R_4}$

(c) $R_{eq} = \dfrac{R_1 + R_2R_3}{R_1 + R_2} + R_4$

Solution: Substituting ohms (Ω) for each resistor gives:

(a) $R_{eq} = \Omega + \dfrac{\Omega \times \Omega}{\Omega + \Omega} = \Omega + \dfrac{\Omega^2}{\Omega} = \Omega$

Because R_{eq} has the units of ohms (Ω), this equation is dimensionally correct.

(b) $R_{eq} = \dfrac{\Omega \times \Omega + \Omega(\Omega + \Omega)}{\Omega \times \Omega + \Omega \times \Omega}$

$= \dfrac{\Omega^2 + \Omega^2}{\Omega^2 + \Omega^2} = \dfrac{\Omega^2}{\Omega^2} = 1$

This equation is dimensionally incorrect because the result has no units at all. Hence, it must also be mathematically incorrect.

(c) $R_{eq} = \dfrac{\Omega + \Omega \times \Omega}{\Omega + \Omega} + \Omega$

$= \dfrac{\Omega + \Omega^2}{\Omega} + \Omega = 1 + \Omega + \Omega$

Because the first term is unitless, this expression is dimensionally incorrect and, hence, mathematically incorrect.

EQUIVALENT RESISTANCE OF FOUR-RESISTOR CONNECTIONS

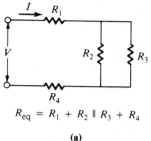

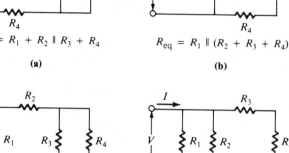

$$R_{eq} = R_1 + R_2 \parallel R_3 + R_4$$

(a)

$$R_{eq} = R_1 \parallel (R_2 + R_3 + R_4)$$

(b)

$$R_{eq} = R_1 \parallel (R_2 + R_3 \parallel R_4)$$

(c)

$$R_{eq} = R_1 \parallel R_2 \parallel (R_3 + R_4)$$

(d)

FIGURE 5.8. The same methods used to reduce three-resistor connections can be used for any number of resistors. For the four-resistor networks shown, the mathematical relations indicate the order of reducing the network to a single equivalent resistance. Knowing R_{eq} allows calculation of the net circuit current I. Knowing V and I allows calculations of the voltage across or the current through any specific resistor.

5.5 FOUR-RESISTOR COMBINATIONS

With three resistors, the algebraic expressions for equivalent resistance became complex. The use of the $\parallel$ symbol saved considerable writing, however. With more than three resistors, this savings is even more necessary. Figure 5.8 shows various combinations of four resistors. The algebraic expressions of some of them are quite unwieldy.

To calculate R_{eq} in each case, follow the same procedure as for three resistors, successively reducing the circuit to simpler circuits by adding series resistors together and calculating the equivalent resistance of parallel combinations. The order in which this reduction takes place is represented mathematically in the R_{eq} expressions shown in Figure 5.8 by the parentheses ( ) and parallel bars $\parallel$. Parenthetical expressions represent series connections whose resistances must be added together before a $\parallel$ combination with it can be made.

Once R_{eq} is determined for the resistor network, the net circuit current can be determined. Knowing the net circuit voltage and current permits calculation of the voltage across and current through each individual resistor. The calculations for the circuits in Figure 5.8 are left as an exercise for the end of the chapter.

Note that these methods can be applied to networks of any number of resistors. The only difficulty is that the more resistors in the network, the more complex and messy the algebra required to analyze it. Fortunately, resistor networks of more than three resistors are rarely used, so the equations in Section 5.4 will solve most practical problems.

5.6 WHEATSTONE BRIDGE

One of the most common and important four-resistor networks is the **Wheatstone bridge**, shown in Figure 5.9. This circuit is extremely useful for measurement of an unknown resistor

WHEATSTONE BRIDGE CIRCUIT

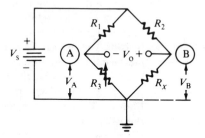

FIGURE 5.9. **An important example of a four-resistor network is the Wheatstone bridge that is used for accurate measurement of unknown resistances. This circuit is essentially two voltage dividers connected in parallel. When $V_o = 0$ V, the bridge is said to be balanced and the unknown resistance R_x is simply related to the known resistances R_1, R_2, and R_3.**

R_x. The importance of the Wheatstone bridge for measuring resistance is that properly designed, it can measure resistance to extremely high accuracy: seven or eight significant digits. When R_x is a resistive transducer, this accuracy translates into a correspondingly high measurement accuracy of the physical quantity. For example, if R_x is a thermistor, temperature changes of microdegrees (10^{-6}°C) are readily measurable, given the accuracy achieved by the Wheatstone bridge.

In the Wheatstone bridge circuit, the values of R_1, R_2, and R_3 are known, and R_x is the unknown resistance to be measured. The operation of the bridge circuit depends on adjusting the values of one (or more) of the known resistances (R_3, in this case) until the bridge output voltage V_o becomes zero. In that condition, the bridge is said to be "in balance," and the unknown resistance can be determined from the values of the known resistances.

Principles of Operation

One simple way to understand the operation of the bridge circuit is to view it as a combination of two voltage divider networks. One voltage divider is formed by R_1 and R_3, and the other is formed by R_2 and R_x. Thus, the voltage at node A, V_A, is given by:

$$V_A = \frac{R_3}{R_1 + R_3} V_s \tag{12}$$

and the voltage at node B, V_B, is given by:

$$V_B = \frac{R_x}{R_2 + R_x} V_s \tag{13}$$

Applying KVL around the loop of V_A, V_o, and V_B, we get:

$$-V_A - V_o + V_B = 0$$

or

$$V_o = V_B - V_A \tag{14}$$

Substituting the previous expressions for V_A and V_B, we obtain:

$$V_o = \left(\frac{R_x}{R_2 + R_x} - \frac{R_3}{R_1 + R_3} \right) V_s \tag{15}$$

The condition for bridge balance is $V_o = 0$. This condition will occur only when the term in the brackets is set equal to zero. Thus:

$$\frac{R_x}{R_2 + R_x} - \frac{R_3}{R_1 + R_3} = 0 \tag{16}$$

or

$$\frac{R_x}{R_2 + R_x} = \frac{R_3}{R_1 + R_3}$$
$$R_x(R_1 + R_3) = R_3(R_2 + R_x)$$
$$R_1 R_x + R_3 R_x = R_2 R_3 + R_3 R_x$$

which reduces to:

$$R_1 R_x = R_2 R_3 \tag{17}$$

or, equivalently, the expression for the unknown resistance R_x:

$$R_x = \frac{R_2 R_3}{R_1} \qquad (18)$$

We see, therefore, that the value of the unknown resistor R_x is given in terms of the three known resistors R_1, R_2, and R_3 when the bridge circuit is balanced. Notice that the actual value of the input voltage source V_s does not appear in the equation for R_x. Hence, *the precise value of V_s does not affect the accuracy of the measurement.* Rather, it merely serves to excite the circuit so that the balance condition can be detected.

Measurement of Unknown Resistance

We have seen how the value of an unknown resistor can be determined from the other three known resistors in a balanced Wheatstone bridge. Let us now look at a practical use of this capability to determine the value of an unknown resistor.

The basic bridge circuit for measuring an unknown resistor is shown in Figure 5.10. The voltage source V_s is a dc power supply. The null detector is a digital voltmeter with a very high input resistance. The digital voltmeter is set on its most sensitive dc scale. The fixed resistors R_1 and R_2 are usually chosen to be of equal value so that in the equation for R_x, Equation 18, they cancel out, giving:

$$R_x = R_3 \text{ for } R_1 = R_2 \qquad (19)$$

WHEATSTONE BRIDGE FOR MEASURING RESISTANCE

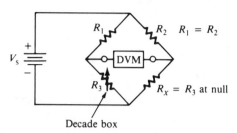

FIGURE 5.10. In a practical Wheatstone bridge, R_1 is made equal to R_2, R_3 is a decade resistance box, and the bridge voltage V_o is measured with a sensitive voltmeter. R_3 is carefully adjusted until the voltmeter reads 0 V, then $R_x = R_3$. The accuracy of the measurement depends on the accuracy of the known resistors R_1, R_2, and R_3 and on the voltage sensitivity of the DVM.

The values of R_1 and R_2 are also chosen to be near the expected value of R_x (within a factor of two or three). It will be shown later that such a choice improves the sensitivity of the null detection.

With V_s applied to the circuit, the measurement procedure is simply to adjust the decade box R_3 until a null ($V_o = 0$ V) is indicated on the DVM. Then, according to Equation 19, the value of R_x is equal to the value of R_3 indicated on the decade box.

5.7 QUESTIONS AND PROBLEMS

1. For the circuits diagrammed in Figure 5.11, calculate R_{eq}, the net circuit current, and the voltage across and current through each resistor for the following circuit values. Check each calculation using KVL and KCL.
   (a) $V = 15$ V, $R_1 = 82$ kΩ, $R_2 = 68$ kΩ, and $R_3 = 0.18$ kΩ
   (b) $V = 1.24$ V, $R_1 = 2.2$ kΩ, $R_2 = 8.2$ kΩ, and $R_3 = 470$ kΩ
   (c) $V = 0.063$V, $R_1 = 12$ kΩ, $R_2 = 4.7$ kΩ, and $R_3 = 910$ kΩ

2. Repeat Problem 1 for the circuit shown in Figure 5.12

3. Repeat Problem 1 for the circuit shown in Figure 5.13.

4. Repeat Problem 1 for the circuit shown in Figure 5.14.

5. Figure 5.15 shows a voltage divider string used to provide decimal intervals of voltage. What value resistors should be used in order to tap voltages of $V_1 = 0.010$ V; $V_1 + V_2 = 0.10$ V; $V_1 + V_2 + V_3 = 1.0$ V; and $V_1 + V_2 + V_3 + V_4$

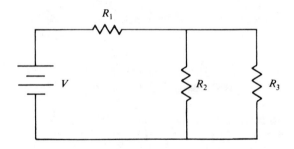

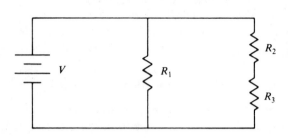

FIGURE 5.14.

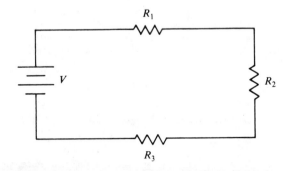

FIGURE 5.11.

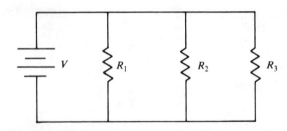

FIGURE 5.12.

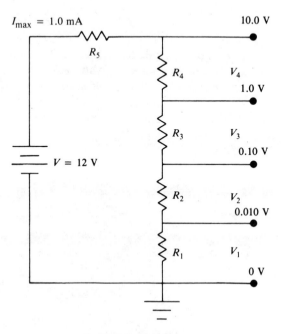

Decimal Voltage Source

FIGURE 5.15.

= 10.0 V. The maximum current drawn from the voltage source should not exceed 1.0 mA.

6. A 12 V automobile battery is used to power a latching alarm circuit consisting of a relay, an alarm, and a solenoid lock, as shown in Figure 5.16. If switch S is closed and then reopened, draw the resulting circuit diagram and the equivalent resistance network. Also, calculate the net current drawn from the voltage source and the current drawn by each device.

7. The operating voltage of two relays is specified as 6.0 V, and they each have a resistance of 560 Ω. Unfortunately, only a 12 V power supply is available. Using the 12 V supply, design a circuit that will power the two relays.

8. Design a circuit that will supply 5 mA to each of three LEDs, using a voltage source rated at 14 V. Assume the voltage across a lighted LED to be 0.6 V.

9. Two relays are available. One is rated at 12 V with a coil resistance of 560 Ω, and the other is rated at 6.0 V with a coil resistance of 250 Ω. Using a 12 V power supply, design a circuit that will power both relays.

10. Check the following expressions for dimensional correctness:

   (a) $R_{eq} = \dfrac{R_1^2(R_2 + R_3)}{R_1 R_2 + R_2 R_3 + R_3 R_1}$

   (b) $R_{eq} = \dfrac{(R_1 + R_2)(R_3 + R_4)}{(R_1 + R_2) + (R_3 R_4)}$

   (c) $R_{eq} = \dfrac{R_1(R_2 + R_3)R_4}{(R_1 + R_2) + (R_3 + R_4)}$

   (d) $V = \dfrac{R_1 R_2}{R_1 + R_2} I$

   (e) $V_1 = \dfrac{R_1}{R_1 + R_2} V_2$

   (f) $I = V\left(\dfrac{R_1}{R_1 + R_2}\right)$

   (g) $V_1 = \dfrac{R_1 R_2}{R_1 + R_2} V_2$

11. Derive expressions for R_{eq} for the circuits shown in Figure 5.8.

12. Calculate R_{eq} and the net circuit current of each of the circuits in Figure 5.8, given the following circuit values: $V = 12.0$ V, $R_1 = 82$ kΩ, $R_2 = 22$ kΩ, $R_3 = 0.18$ MΩ, and $R_4 = 47$ kΩ.

13. Calculate the voltage across and current through each of the resistors in Figure 5.8 using the results of Problem 12.

14. Calculate the value of the unknown resistance in the balanced Wheatstone bridge circuit shown in Figure 5.17, given the following values of source voltage and known resistors. Be sure to express your answer to the proper number of significant digits.

   (a) $V_s = 15$ V, $R_1 = 1.00$ kΩ, and $R_3 = 1.00$ kΩ and $R_3 = 2.234$ kΩ

   (b) $V_s = 6.0$ V, $R_1 = 5003$ Ω, $R_2 = 5007$ Ω, and $R_3 = 12,038$ Ω

   (c) $V_s = 1.5$ V, $R_1 = 105$ Ω, $R_2 = 107$ Ω, and $R_3 = 27.63$ Ω

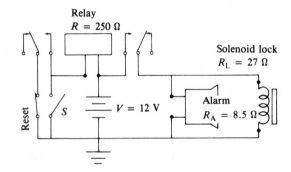

Latching Alarm–Lock System

FIGURE 5.16.

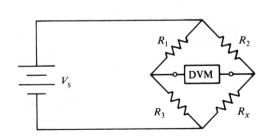

FIGURE 5.17.

Chapter

6

THEVENIN'S THEOREM AND SUPERPOSITION

6.1 OBJECTIVES

Following the completion of Chapter 6, you should be able to:

1. State Thevenin's theorem and draw the Thevenin equivalent network for any pair of terminals in a given resistive network.
2. Describe how to determine the open circuit voltage and Thevenin resistance of a resistive network experimentally.
3. State the procedure for calculating the open circuit voltage and Thevenin resistance of a resistive network.
4. Determine the voltage across any element of a complex resistive network using the principles of Thevenin's theorem.
5. State the principle of superposition and describe how to use it to determine the voltage across a given pair of terminals in a multisource network.
6. Determine the voltage across any element of a complex, multisource resistive network using the principle of superposition.

7. State the basis of the binary number system and convert numbers between it and the decimal number system.
8. Explain how data are expressed electronically in digital form and the basic requirements of digital-to-analog conversion.
9. Draw an R–$2R$ ladder network for up to four bits of digital data, explain how it acts as a digital-to-analog converter, and express the analog output voltage in terms of the digital inputs.
10. State Norton's theorem and draw the Norton equivalent network for any pair of terminals in a given resistive network.
11. Describe how to determine the short-circuit current experimentally.
12. State the procedure for calculating the short-circuit current of a resistive network.
13. Determine the voltage across any element of a complex resistive network using the principle of Norton's theorem.

6.2 OVERVIEW

The previous chapter discussed methods for determining the voltage across any resistor in a complex network of series and parallel combinations. The general tactic is to reduce the resistor network to a single equivalent resistor, use that to find the net circuit current, and then reconstruct the network, successively finding the voltage and current for each resistor. As the networks get more and more complex, however, the algebraic manipulations get messier and messier, keeping track of the calculations gets tougher, and the probability of making an error increases. Add to this the possibility of having more than one voltage source, or even current source, and the problem grows even more difficult.

In the analysis of most circuit networks, however, we are generally concerned only with a single pair of terminals. Although the overall network may be quite complex, the problem's solution only requires knowing the behavior at one terminal pair, often referred to as the output terminals. If we want only the behavior at the output terminals, it would seem a nuisance to have to calculate the voltages and currents for the entire circuit.

Fortunately, there is a method that greatly simplifies the calculations. This method is based on a mathematical tool called **Thevenin's theorem.** For situations involving more than one source of voltage or current, there is a second mathematical tool that also simplifies the calculations: **superposition.** These two techniques are examined next.

6.3 THEVENIN'S THEOREM

Thevenin's theorem can be used to determine the voltage and current at any pair of terminals of a complex resistive network, as shown in Figure 6.1. The network, consisting of resistors,

TYPICAL COMPLEX CIRCUIT NETWORK

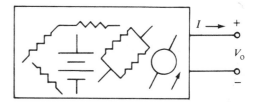

FIGURE 6.1. For a complex network such as this one, we generally want to find only the behavior at the output terminals. Thevenin's theorem simplifies the calculations. If there is more than one voltage or current source, the principle of superposition further simplifies the problem.

voltage sources, and current sources, can be any connection of any size to which a single pair of terminals is connected.

Thevenin Equivalent Network

According to Thevenin's theorem, *the behavior of a complex resistive network at a given terminal pair is identical to that of a single voltage source in series with a single resistor* (Figure 6.2). This Thevenin equivalent network consists of a voltage source V_{oc} in series with a **Thevenin equivalent resistance** R_{Th} and the output terminal pair. The Thevenin equivalent resistance is the effective resistance of the circuit as seen at the output terminals.

The voltage V_{oc} is called the **open circuit voltage** and is the *voltage that would be measured at the output terminals if they were an open circuit*—that is, without a load resistor. If the output terminals were open circuited, no output current would flow and there would be no voltage drop across the Thevenin equivalent resistance R_{Th}. Hence, the voltage V_{oc} would also appear at the output terminals. In an actual circuit, the value of V_{oc} is the voltage that would be measured by a voltmeter with very

**COMPLEX NETWORK REDUCED TO ITS
THEVENIN EQUIVALENT**

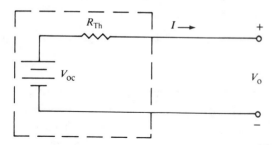

FIGURE 6.2. By Thevenin's theorem, a complex network is equivalent to a single voltage source and a series resistor. The voltage source, called the open circuit voltage, is the voltage at the output terminals when they are open circuited. The resistance is called the Thevenin equivalent resistance.

high input resistance, such as a DVM, that would draw negligible current from the circuit.

It should be understood, however, that the equivalency of the Thevenin network holds only at a given pair of output terminals. In a really complex network, many voltages and currents in its elements change as the terminal variables are changed, but the Thevenin equivalent network tells us nothing about this behavior. It governs only the terminal behavior. Behavior at a different pair of terminals can be described by a different Thevenin equivalent network with different values of V_{oc} and R_{Th}.

Open Circuit Voltage and Thevenin Resistance

The question now becomes: How do we determine the values of V_{oc} and R_{Th} so that we can construct the Thevenin equivalent network of a complex circuit? There are several ways, and we shall examine some of them next.

The most straightforward method is to perform a series of measurements to determine the values for V_{oc} and R_{Th}. It has been pointed out already that V_{oc} is the open circuit voltage mea-

sured at the terminal pair. This value can be found in the laboratory by using a DVM or, if time-varying sources are present, an oscilloscope. Note that the Thevenin equivalent works for time-varying sources as well as dc sources.

To measure R_{Th}, we can apply a test voltage V_T to the output terminals and measure the resulting current, as Figure 6.3 illustrates. Using KVL, we have:

$$-V_{oc} - IR_{Th} + V_T = 0$$

or

$$R_{Th} = \frac{V_T - V_{oc}}{I} \tag{1}$$

Since we know the value of V_T and can measure the values of V_{oc} and I, we can use Equation 1 to calculate R_{Th}. Determination of V_{oc} and R_{Th} by actual measurement is particularly useful when the circuit for which we desire the Thevenin equivalent contains nonlinear devices, such as transistors and op amps.

In simple circuits that contain only resistors and independent voltage and current sources, we can also calculate V_{oc} and R_{Th} using the network analysis tools developed earlier, such as series and parallel connections and voltage and current dividers. Because V_{oc} is the voltage at the output terminals when they are open circuited, we simply open the output terminals by removing any load connection and calculate the voltage that appears there following normal circuit analysis techniques. Thus, the plan for calculating V_{oc} is:

1. Open-circuit the output terminals by removing any load connection.
2. Calculate the open circuit voltage V_{oc} at the output terminals.

With the output terminals open, the calculation of V_{oc} is generally fairly simple.

Calculation of R_{Th} is aided by some network simplification. If we examine Equation 1, we see that if V_{oc} were zero, then the terminal

PROCEDURE FOR MEASURING THEVENIN EQUIVALENT VALUES

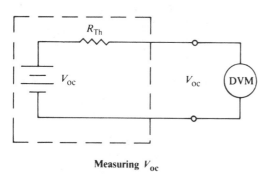

Measuring V_{oc}

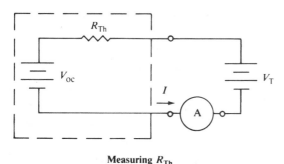

Measuring R_{Th}

FIGURE 6.3. The open circuit voltage can be measured by a high input resistance voltmeter, such as a DVM. The Thevenin resistance can be measured by (1) applying a test voltage to the output terminals, (2) measuring the resulting current, and (3) calculating the resistance using KVL.

behavior of the network would be that of a single resistor R_{Th}. Equation 1 then would become a simple Ohm's law expression with one unknown quantity, R_{Th}.

But, V_{oc} is not zero because of the voltage and current sources inside the network. If all of the voltage and current sources were made equal to zero, however, then V_{oc} would also be zero! A voltage source that is zero is a short circuit. Similarly, a current source that is zero is an open circuit. Thus, the plan for calculating R_{Th} becomes:

1. Set all independent voltage sources equal to zero by replacing them with short circuits.
2. Set all independent current sources equal to zero by replacing them with open circuits.
3. Calculate the resistance R_{Th} at the terminal pair of the modified network.

Knowing V_{oc} and R_{Th}, we can calculate the voltage across any real load resistor simply from the Thevenin equivalent network. The following examples illustrate how the above-mentioned procedure is used in typical circuits.

EXAMPLE 1: What is the voltage across the load resistor in the circuit diagrammed in Figure 6.4a?

Solution: The first step in finding the Thevenin equivalent network is to find V_{oc}. To do this, we remove the load resistor R_L to open-circuit the output terminals, as shown in Figure 6.4b. When the output terminals are open circuited, $I = 0$, and there is no voltage across R_3. From Figure 6.4b, we see that V_{oc} is related to the supply voltage V_s through the voltage divider relation. Thus:

$$V_{oc} = \frac{R_2}{R_1 + R_2} V_s \qquad (2)$$

Next, we must find the value of R_{Th}. Since there is only one voltage source V_s, the circuit that yields R_{Th} is found by replacing V_s with a short circuit, as shown in Figure 6.4c. The Thevenin resistance R_{Th} is then found by calculating the equivalent resistance at the terminal pair in Figure 6.4c. This is a series-parallel combination that results in:

$$R_{Th} = R_3 + R_1 \parallel R_2$$
$$= R_3 + \frac{R_1 R_2}{R_1 + R_2} \qquad (3)$$

EXAMPLE 1

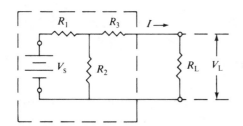

(a) Original Circuit: $V_L = ?$

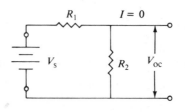

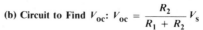

(b) Circuit to Find V_{oc}: $V_{oc} = \dfrac{R_2}{R_1 + R_2} V_s$

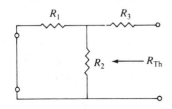

(c) Circuit to Find R_{Th}: $R_{Th} = R_3 + \dfrac{R_1 R_2}{R_1 + R_2}$

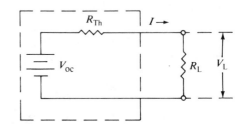

(d) Thevenin Equivalent Network: $V_L = \dfrac{R_L}{R_{Th} + R_L} V_{oc}$

FIGURE 6.4. To calculate the open circuit voltage, open the output terminals, as in (b), and calculate the resulting voltage. To calculate the Thevenin resistance, short-circuit the voltage source, as in (c), and calculate the equivalent resistance of the remaining network.

The resulting Thevenin equivalent network is shown in Figure 6.4d with the load resistor reconnected. At the terminal pair, the circuit characteristics of Figure 6.4d will be identical to the real circuit of Figure 6.4a. The voltage V_L across the load resistor R_L is simply found from the voltage divider equation:

$$V_L = \frac{R_L}{R_{Th} + R_L} V_{oc} \qquad (4)$$

To be more specific about the use of Thevenin equivalent networks, let us insert the following numerical values for the elements in the circuits shown in Figure 6.4a:

$V_s = 10 \text{ V}$
$R_1 = 1 \text{ k}\Omega$
$R_2 = 10 \text{ k}\Omega$
$R_3 = 2.7 \text{ k}\Omega$
$R_L = 5.6 \text{ k}\Omega$

Using these values in Equations 2 and 3 gives:

$$V_{oc} = \frac{10 \text{ k}\Omega}{10 \text{ k}\Omega + 1 \text{ k}\Omega} 10 \text{ V} = 9.1 \text{ V}$$

$$R_{Th} = 2.7 \text{ k}\Omega + \frac{1 \text{ k}\Omega \times 10 \text{ k}\Omega}{1 \text{ k}\Omega + 10 \text{ k}\Omega}$$

$$= 3.6 \text{ k}\Omega$$

EXAMPLE 2

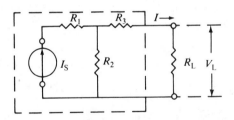

(a) Original Circuit: V_L = ?

(b) Circuit to Find V_{oc}: $V_{oc} = I_s R_2$

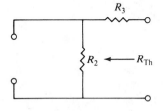

(c) Circuit to Find R_{Th}: $R_{Th} = R_2 + R_3$

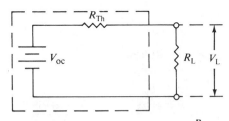

(d) Thevenin Equivalent Network: $V_L = \dfrac{R_L}{R_{Th} + R_L} V_{oc}$

FIGURE 6.5. Thevenin's theorem can also be used for circuits containing current sources. To determine the Thevenin resistance, however, a current source must be open circuited, as in (c).

Equation 4 gives the voltage across the load resistor:

$$V_L = \frac{5.6 \text{ k}\Omega}{3.6 \text{ k}\Omega + 5.6 \text{ k}\Omega} \, 9.1 \text{ V} = \textbf{5.5 V}$$

EXAMPLE 2: What is the voltage across the load resistor in the circuit shown in Figure 6.5a?

Solution: To find the value of V_{oc}, we remove the load resistor to open-circuit the output terminals, as shown in Figure 6.5b.

When the terminals are open circuited, there is no current through resistor R_3, and the voltage across the output terminals V_{oc} is the voltage across resistor R_2 due to the constant current I_S:

$$V_{oc} = I_S R_2 \tag{5}$$

To find the value of R_{Th}, we must open-circuit the constant-current source, as shown in Figure 6.5c. In this case, the equivalent resistance at the output terminals is given by:

$$R_{Th} = R_2 + R_3 \tag{6}$$

The resulting Thevenin equivalent network is

shown in Figure 6.4d. The load voltage is the same as in Example 1, Equation 4:

$$V_L = \frac{R_L}{R_{Th} + R_L} V_{oc}$$

Using the same circuit values as in Example 1 and a constant-current source of:

$$I_S = 1.2 \text{ mA}$$

and substituting into Equations 5 and 6 gives values for the Thevenin equivalent network:

$$V_{oc} = 1.2 \text{ mA} \times 10 \text{ k}\Omega = 12 \text{ V}$$
$$R_{Th} = 10 \text{ k}\Omega + 2.7 \text{ k}\Omega = 12.7 \text{ k}\Omega$$

The load voltage can now be found from Equation 4:

$$V_L = \frac{5.6 \text{ k}\Omega}{12.7 \text{ k}\Omega + 5.6 \text{ k}\Omega} 12 \text{ V} = \textbf{3.7 V}$$

EXAMPLE 3:   What is the voltage across the load resistor in the circuit shown in Figure 6.6a?

Solution: To find the value of V_{oc}, we remove the load resistor, as shown in Figure 6.6b. The open circuit voltage V_{oc} is then the voltage across resistor R_4 due to the source voltage V_s. This value is given by the following voltage divider expression:

$$V_{oc} = \frac{R_4}{R_3 + R_4} V_s \tag{7}$$

EXAMPLE 3

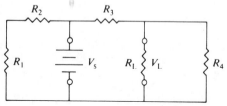

(a) Original Circuit: $V_L = ?$

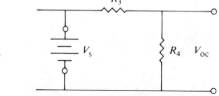

(b) Circuit to Find V_{oc}: $V_{oc} = \dfrac{R_4}{R_3 + R_4} V_s$

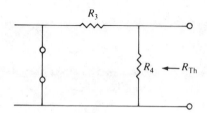

(c) Circuit to Find R_{Th}: $R_{Th} = \dfrac{R_3 R_4}{R_3 + R_4}$

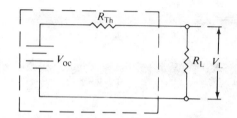

(d) Thevenin Equivalent Network: $V_L = \dfrac{R_L}{R_{Th} + R_L} V_{oc}$

FIGURE 6.6.   Another typical circuit whose analysis Thevenin's theorem simplifies.

To find R_{Th}, the voltage source V_s is short circuited, as shown in Figure 6.6c. This procedure shorts out both resistors R_1 and R_2, which can be ignored in the calculation. Thus, the equivalent resistance as seen at the output terminals is simply R_3 in parallel with R_4:

$$R_{\text{Th}} = \frac{R_3 R_4}{R_3 + R_4} \tag{8}$$

The resulting Thevenin equivalent network is shown in Figure 6.6d. The load voltage is given by Equation 4:

$$V_L = \frac{R_L}{R_{\text{Th}} + R_L} V_{\text{oc}}$$

Circuit values may be substituted to obtain specific results.

6.4 APPLICATIONS

In order to better appreciate how Thevenin's theorem simplifies network analysis, we will work two examples involving two common types of networks: the ladder and the unbalanced Wheatstone bridge. Both of these networks require a moderate amount of tedious and error-prone algebra when attacked by the standard node or loop analysis techniques. The network analysis tools we have developed, however, provide a simple and direct approach to the solutions.

Ladder Network

The ladder network shown in Figure 6.7 is a typical circuit configuration. The name *ladder* comes from its geometrical shape. Actually, the correct term is **unbalanced ladder** because one side of the ladder is simply a conductor without any circuit elements between the "steps." The **balanced ladder** is rare, so *ladder* is usually as-

APPLYING THEVENIN'S THEOREM TO THE LADDER NETWORK

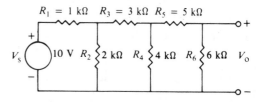

FIGURE 6.7. A typical application of Thevenin's theorem is the analysis of the ladder network shown. The process can be applied to ladders of any length.

sumed to mean an unbalanced ladder, like the one shown in Figure 6.7. A ladder network can be short—three or four circuit elements—or long—eight, ten, or more elements. Of course, the longer the ladder, the more involved the calculations.

To find the output voltage V_o in Figure 6.7, a technique will be developed that can be used for ladder networks of any size. It is important to understand the general approach so that it can be applied to any ladder network.

The first step in the analysis is to notice that the voltage source and the 1 kΩ, 2 kΩ, and 3 kΩ resistors form a network identical to the one in Figure 6.4 that we used in Example 1 to illustrate reduction to a Thevenin equivalent. Therefore, we can use the result of Example 1 to replace the original voltage source and the 1 kΩ, 2 kΩ, and 3 kΩ resistors by their Thevenin equivalent, as shown in Figure 6.8a, thereby reducing the size of the network. The solutions for V_{oc} and R_{Th} are found from Equations 2 and 3 as follows:

$$V_{\text{oc}} = \frac{R_2}{R_1 + R_2} V_s = \frac{2}{1+2} \times 10 \text{ V}$$
$$= 6.67 \text{ V} \tag{9}$$

$$R_{\text{Th}} = R_3 + \frac{R_1 R_2}{R_1 + R_2} = 3 + \frac{1 \times 2}{1+2}$$
$$= 3.67 \text{ k}\Omega \tag{10}$$

THEVENIN ANALYSIS OF LADDER NETWORK

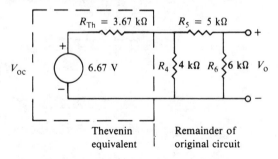

(a) First Reduction of Ladder Network

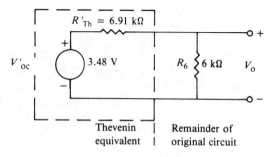

(b) Second Reduction of Ladder Network

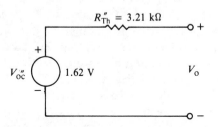

(c) Thevenin Equivalent of Ladder Network

FIGURE 6.8. Applying Thevenin's theorem to the ladder network involves successive reduction of a network of three resistors connected to a voltage source, as shown in (a) and (b). The final Thevenin equivalent is shown in (c).

The next step involves repeating the process by finding the Thevenin equivalent of Figure 6.8a for the 6.67 V voltage source and the 3.67 kΩ, 4 kΩ, and 5 kΩ resistors. Again, the net-

work is identical to that in Figure 6.4, so we simply repeat the process:

$$V'_{oc} = \frac{R_4}{R_{Th} + R_4} V_{oc} = \frac{4}{3.67 + 4} 6.67 \text{ V}$$
$$= 3.48 \text{ V} \tag{11}$$

$$R'_{Th} = R_5 + \frac{R_{Th} R_4}{R_{Th} + R_4} = 5 + \frac{3.67 \times 4}{3.67 + 4}$$
$$= 6.91 \text{ k}\Omega \tag{12}$$

The second reduction of the ladder, using the results of Equations 11 and 12, is shown in Figure 6.8b. At this point, if V_o were the only desired result, we could use the voltage divider equation to find V_o. However, we can also repeat the iteration once more, thereby finding both V_o and the Thevenin resistance at the output terminals of the original ladder network. In this case, however, the value of R_3 in Figure 6.4 is equal to zero. The calculations are:

$$V''_{oc} = \frac{R_6}{R'_{Th} + R_6} V'_{oc} = \frac{6}{6.91 + 6} 3.48 \text{ V}$$
$$= 1.62 \text{ V} \tag{13}$$

$$R''_{Th} = 0 + \frac{R'_{Th} R_6}{R'_{Th} + R_6} = \frac{6.91 \times 6}{6.91 + 6}$$
$$= 3.21 \text{ k}\Omega \tag{14}$$

The final network is shown in Figure 6.8c.

The open circuit output voltage V_o is equal to the value of V''_{oc} in Equation 13, and the output resistance is given by R''_{Th} in Equation 14. The voltage across a load resistance R_L connected across the output terminals is given by the voltage divider expression, Equation 4:

$$V_L = \frac{R_L}{R''_{Th} + R_L} V''_{oc}$$

If $R_L = 1.0 \text{ k}\Omega$, the load voltage is:

$$V_L = \frac{1.0}{3.2 + 1.0} 1.6 \text{ V} = 0.38 \text{ V}$$

Unbalanced Wheatstone Bridge

We saw in Chapter 5 how the Wheatstone bridge can be used to measure the value of an unknown resistance, which is only one of this circuit's applications. When one of the bridge resistors is a transducer that changes resistance in response to a physical quantity—light, temperature, pressure, position, strain, and so on— the bridge circuit forms the basis of many measurement and control systems. This chapter examines the bridge in more detail to see how it performs when unbalanced, which is its condition in most applications. In addition, the Thevenin equivalent network is used to simplify the analysis.

The Wheatstone bridge circuit is shown in Figure 6.9a. The resistor R_t denotes the resistance of a transducer that will change in proportion to the physical property the transducer detects. Thus, the bridge will not be balanced in most cases. Rather, as the physical quantity varies, R_t will vary, thus varying V_o. The value of V_o provides a signal voltage that can be related to the physical quantity measured by the transducer. The relationship between R_t and the physical quantity for a transducer will generally be given and our problem will be to find the relationship between V_o and R_t for the unbalanced bridge.

In order to relate V_o to R_t, we first find the Thevenin equivalent of the circuit in Figure 6.9a. The equation for V_{oc} is the same as Equation 15 in Chapter 5, except that R_t replaces R_x and V_{oc} replaces V_o. Thus:

$$V_{oc} = \left(\frac{R_t}{R_2 + R_t} - \frac{R_3}{R_1 + R_3} \right) V_s \quad (15)$$

By combining the two terms in parentheses, we get:

$$V_{oc} = \frac{R_1 R_t + R_3 R_t - R_2 R_3 - R_3 R_t}{(R_2 + R_t)(R_1 + R_3)} V_s$$

THEVENIN ANALYSIS OF UNBALANCED WHEATSTONE BRIDGE

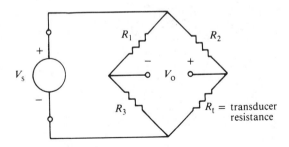

(a) Wheatstone Bridge Circuit

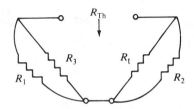

(b) Equivalent Circuit with V_s Short Circuited

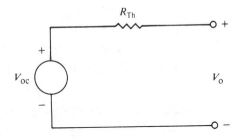

(c) Thevenin Equivalent of Unbalanced Bridge:

$$V_{oc} = \frac{R_1 R_t - R_2 R_3}{(R_2 + R_t)(R_1 + R_3)} V_s$$

$$R_{Th} = R_1 \parallel R_3 + R_2 \parallel R_t$$

FIGURE 6.9. Thevenin's theorem also simplifies analysis of the unbalanced Wheatstone bridge. This circuit is particularly valuable when the unknown resistance R_t is a transducer. Then, the bridge output voltage V_o provides a direct measure of the physical variable that the transducer measures. When a high input resistance DVM is used to measure V_o, $V_o = V_{oc}$.

which results in:

$$V_{oc} = \frac{R_1 R_t - R_2 R_3}{(R_2 + R_t)(R_1 + R_3)} V_s \qquad (16)$$

As a check, we see that $V_{oc} = 0$ (bridge balance) when:

$$R_1 R_t = R_2 R_3$$

or

$$R_t = \frac{R_2 R_3}{R_t} \qquad (17)$$

which agrees with the result derived for the balanced bridge equation, Equation 18 in the previous chapter.

To find R_{Th}, we set $V_s = 0$. Replacing V_s by a short circuit, we get the circuit shown in Figure 6.9b. In this circuit, we see that R_{Th} is the series connection of two pairs of parallel-connected resistors:

$$\begin{aligned} R_{Th} &= R_1 \parallel R_3 + R_2 \parallel R_t \\ &= \frac{R_1 R_3}{R_1 + R_3} + \frac{R_2 R_t}{R_2 + R_t} \end{aligned} \qquad (18)$$

The resulting Thevenin equivalent of the Wheatstone bridge is shown in Figure 6.9c. Two special situations here deserve mention. First, the Thevenin equivalent network has a common terminal between the negative terminal of V_{oc} and the negative output terminal. However, there was no common terminal between V_s and V_o in the original bridge circuit. Its presence in Figure 6.9c is due to the mathematical model we have used to represent the original circuit. The resulting analysis will be correct, but, of course, there is no common ground between the real source and the output terminals.

Second, both V_{oc} and R_{Th} contain the variable R_t. As R_t changes, both V_{oc} *and* R_{Th} will change. Thus, for every different value of R_t,

new values of V_{oc} and R_{Th} must be calculated in order to use the Thevenin equivalent circuit. In the following discussion, some assumptions will be made in order to simplify these calculations.

In any specific application, we must determine values for R_1, R_2, and R_3, for use in the bridge. Usually, we choose them to equal the value of R_t at some midpoint in its expected range of values. For example, if a transducer for the measurement of temperature over the range of 0°C to 100°C produces a resistance variation of 500 Ω to 1500 Ω, we would usually choose $R_1 = R_2 = R_3 = 1$ kΩ. This decision is not as arbitrary as it might seem. When all resistances are the same, the bridge output is the most sensitive to changes in R_t. Thus, the choice of equal resistors is actually the optimum design choice as well as a means of simplifying the algebra.

If we denote the midrange value of R_t as R_0, then we can substitute R_0 for R_1, R_2, and R_3 in Equations 17 and 18 and simplify the expressions. The results are:

$$V_{oc} = \frac{R_t - R_0}{2(R_0 + R_t)} V_s \qquad (19)$$

$$R_{Th} = R_0 \left(\frac{1}{2} + \frac{R_t}{R_0 + R_t} \right) \qquad (20)$$

Notice that when $R_t = R_0$ — the midpoint of the range of R_t — the bridge is balanced ($V_{oc} = 0$) and $R_{Th} = R_0$. Recall, too, that the open circuit voltage V_{oc} is also the output voltage of the bridge when it is measured by a high input resistance voltmeter, such as a DVM.

6.5 PRINCIPLE OF SUPERPOSITION

Thevenin's theorem can be effectively used with a variety of complex resistive networks, including those with more than one current and/or voltage source. When there is more than one source, however, another mathematical tool helps greatly to simplify circuit calculations: the **principle of superposition** (Figure 6.10). It is stated as follows: For a network composed solely of resistive elements and independent voltage and current sources, *the voltage across any pair of terminals is equal to the algebraic sum of the contributions of each of the sources.* Stated mathematically:

$$V_L = A_1 V_1 + A_2 V_2 + \cdots \\ + B_1 I_1 + B_2 I_2 + \cdots \quad (21)$$

where V_L is the voltage across a load resistor R_L connected to the output terminals. The Vs and Is are, respectively, the independent voltage and current sources of the network. The As and Bs are constants that express the fractional contribution to V_L due to each source and are determined by the particular network construction.

According to the principle of superposition, we can set all of the sources equal to zero (voltage sources short circuited and current sources open circuited) except one—for example, V_1 in Equation 21—and calculate the contribution to V_L due to V_1—call it V_{L_1}:

$$V_{L_1} = A_1 V_1 \quad (22)$$

Similarly, we can set all the sources but V_2 equal to 0 and calculate the contribution to V_L due to V_2:

$$V_{L_2} = A_2 V_2 \quad (23)$$

and so on, for each of the sources:

$$V_{L_3} = B_1 I_1 \\ V_{L_4} = B_2 I_2 \quad (24)$$

Methods for calculating the contributions V_{L_1}, V_{L_2}, V_{L_3}, V_{L_4}, . . . are the ones we used in Chapter 5 or Thevenin's theorem, whichever

PRINCIPLE OF SUPERPOSITION

Multisource network

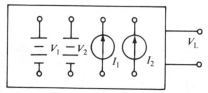

$$V_L = A_1 V_1 + A_2 V_2 + B_1 I_1 + B_2 I_2$$

FIGURE 6.10. For circuits with more than one voltage (or current) source, the principle of superposition further simplifies calculations. By superposition, the voltage across any pair of terminals is equal to the algebraic sum of the contributions of each source.

seems easier for a particular case. The net value of the load voltage, then, will be the sum of the separate voltage contributions:

$$V_L = V_{L_1} + V_{L_2} + V_{L_3} + V_{L_4} + \cdots \quad (25)$$

which is identical to Equation 21 with the expressions of Equations 22 to 24 substituted in. Note that when adding separate voltage contributions, *the proper polarity of each voltage contribution must be carefully observed* in order to obtain the net load voltage.

The principle of superposition generally simplifies analyzing circuits containing more than one source, as the following examples illustrate.

EXAMPLE 4: What is the voltage across the load resistor in the circuit of Figure 6.11a?

Solution: We begin by setting voltage source $V_2 = 0$ (a short circuit) and determining the voltage V_{L_1} across R_L due to voltage source V_1, as shown in Figure 6.11b. V_{L_1} is the voltage across the parallel combination R_L and R_2 ($R_L \parallel R_2$) that is determined by the voltage divider of $(R_L \parallel R_2)$ in series with R_1. This is expressed by:

$$V_{L_1} = \frac{R_L \parallel R_2}{R_1 + (R_L \parallel R_2)} V_1$$

or

$$V_{L_1} = \frac{R_L R_2/(R_L + R_2)}{R_1 + [R_L R_2/(R_L + R_2)]} V_1 \quad (26)$$

Similarly, the contribution to V_L of V_2 with V_1 set to zero can be found as shown in Figure 6.11c. It is seen to be:

$$V_{L_2} = \frac{R_L \parallel R_1}{R_2 + (R_L \parallel R_1)} (-V_2)$$

or

$$V_{L_2} = -\frac{R_L R_1/(R_L + R_1)}{R_2 + [R_L R_1/(R_L + R_1)]} V_2$$

$$(27)$$

Note that the polarity of the contribution of V_2 is opposite that of V_1; hence, the negative sign. The total voltage is given by Equation 25:

$$V_L = V_{L_1} + V_{L_2}$$

If we assume the following values for the circuit parameters, what is V_L?

$V_1 = 9.0$ V
$V_2 = 12$ V
$R_1 = 2.7$ kΩ
$R_2 = 10$ kΩ
$R_L = 5.6$ kΩ

Using Equations 26 and 27:

$$V_{L_1} = \frac{(5.6 \times 10)/(5.6 + 10)}{2.7 + [(5.6 \times 10)/(5.6 + 10)]} 9.0 \text{ V}$$

$$= \frac{3.6 \text{ k}\Omega}{2.7 \text{ k}\Omega + 3.6 \text{ k}\Omega} 9.0 \text{ V}$$

$$= 0.57 \times 9.0 \text{ V} = 5.1 \text{ V}$$

EXAMPLE 4

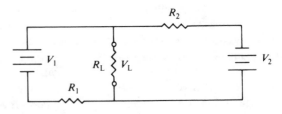

(a) Original Circuit

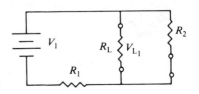

(b) Setting $V_2 = 0$: $V_{L_1} = \dfrac{R_L \parallel R_2}{R_1 + (R_L \parallel R_2)} V_1$

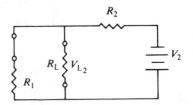

(c) Setting $V_1 = 0$: $V_{L_2} = \dfrac{R_L \parallel R_1}{R_2 + (R_L \parallel R_1)} (-V_2)$

FIGURE 6.11. When using superposition, we set all sources but one equal zero (b) and calculate the terminal voltage due to the remaining source. This is done for each source separately (c). The net terminal voltage is the algebraic sum of the contributions (Equation 25).

$$V_{L_2} = -\frac{(5.6 \times 2.7)/(5.6 + 2.7)}{10 + [(5.6 \times 2.7)/(5.6 + 2.7)]} 12 \text{ V}$$

$$= \frac{1.8 \text{ k}\Omega}{10 \text{ k}\Omega + 1.8 \text{ k}\Omega} 12 \text{ V}$$

$$= -0.15 \times 12 \text{ V} = -1.8 \text{ V}$$

Using Equation 25 for V_L gives:

$$V_L = 5.1 \text{ V} - 1.8 \text{ V} = \mathbf{3.3 \text{ V}}$$

EXAMPLE 5

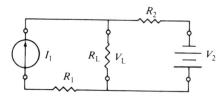

(a) Original Circuit

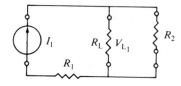

(b) Setting $V_2 = 0$: $V_{L_1} = \dfrac{R_2 R_L}{R_2 + R_L} I$

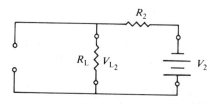

(c) Setting $I_1 = 0$: $V_{L_2} = \dfrac{R_L}{R_L + R_2}(-V_2)$

FIGURE 6.12.   When current sources are present, we set them equal to zero by open-circuiting them.

$$V_{L_1} = I_{L_1} R_L = \frac{R_2}{R_2 + R_L} I R_L$$

$$= \frac{R_2 R_L}{R_2 + R_L} I \qquad (28)$$

Next we set the current source equal to zero (open circuit) and determine the contribution to V_{L_2} due to V_2 (Figure 6.12c). This is simply a voltage divider, but again attention must be paid to polarity:

$$V_{L_2} = \frac{R_L}{R_L + R_2}(-V_2) = -\frac{R_L}{R_L + R_2} V_2$$

The total voltage V_L is simply the sum of V_{L_1} and V_{L_2} as given by Equation 25:

$$V_L = V_{L_1} + V_{L_2}$$

6.6 APPLICATION: DIGITAL-TO-ANALOG CONVERTER

We now consider a more complex example of the use of superposition: the R–$2R$ ladder network, a circuit that is widely used as the basis of **digital-to-analog (D/A) converters.** The purpose of a D/A converter is to change a variable expressed in digital form—that is, a binary number—into a corresponding analog signal. For example, the electronic processing of data by most computers and microprocessors is done in digital (binary) form, but the output display of that data may be desired in an analog form, such as a meter reading. The conversion of the data expressed as a series of digital voltages to a corresponding analog voltage can be accomplished by a resistive network called an R–$2R$ ladder.

EXAMPLE 5:   What is the voltage across the load resistor in the circuit shown in Figure 6.12a?

Solution: As before, we begin by setting the voltage source equal to zero (short circuit) and determining the voltage across R_L due to the current source (Figure 6.12b). The voltage across R_L is determined by the current through R_L as a result of the current divider of R_L and R_2. It is given by:

Binary Number System

Before analyzing the *R-2R* ladder, a brief summary of the binary number system and the way it is used to express information in digital form is given here. The binary number system is similar to the decimal system, except that each digit has only one of two values—0 or 1—rather than one of the ten values of the decimal system—0 to 9. The result is that a binary number is a series of 1s and 0s rather than a series of numbers ranging from 0 to 9.

In the decimal number 783, for example, the first digit, 3, tells the number of units; the second digit, 8, tells the number of tens; and the third digit, 7, tells the number of hundreds. Hence, the number 783 represents seven hundreds, eight tens, and three units. The total number is found by adding these together:

$$
\begin{aligned}
7 \times 100 &= 700 \\
8 \times 10 &= 80 \\
3 \times 1 &= \underline{3} \\
&\ 783
\end{aligned}
$$

Of course, this operation is unnecessary because we immediately recognize the meaning of 783.

In the binary number system, each digit is called a **bit**, and the sequence of bits stands for the number of 1s, 2s, 4s, 8s, 16s, 32s, and so on, rather than the 1s, 10s, 100s, 1000s, and so forth, in the decimal system. Thus, the sequence 110 represents (reading from right to left) zero 1s, one 2, and one 4. To convert the binary number 110 to its decimal equivalent, we must perform this addition:

$$
\begin{aligned}
0 \times 1 &= 0 \\
1 \times 2 &= 2 \\
1 \times 4 &= \underline{4} \\
&\ 6
\end{aligned}
$$

Hence, 110 in binary stands for the decimal number 6. Figure 6.13 gives all of the eight possible combinations of a three bit binary num-

BINARY NUMBER SYSTEM

Binary Number	Decimal Equivalent
001	1
010	2
011	3
100	4
101	5
110	6
111	7

FIGURE 6.13. Digital electronic systems are based on the binary number system. The binary digits stand for the number of 1s, 2s, 4s, 8s, 16s, and so on, and each digit (or bit) has only two possible values: 0 or 1. The table gives the decimal equivalents of all three bit binary numbers.

ber. Take a moment to see how each was determined.

Note that with three bits of binary data, we can express only eight decimal numbers (0 to 7). Contrast this with the decimal number system in which three digits can express any of 1000 numbers (0 to 999). In binary, the expression of 999, for example, requires ten bits:

Binary 1111100111	Decimal 999
$1 \times 512 =$	512
$1 \times 256 =$	256
$1 \times 128 =$	128
$1 \times 64 =$	64
$1 \times 32 =$	32
$0 \times 16 =$	0
$0 \times 8 =$	0
$1 \times 4 =$	4
$1 \times 2 =$	2
$1 \times 1 =$	1
	999

Thus, the binary system is much less efficient for expressing numbers. On the other hand, it requires only two digits, 0 and 1. Electronically, this system is very efficient because 0 and 1 can correspond to two voltage levels: low and high. In digital electronic systems, *low* is generally defined as 0 V and *high* as +5 V.

The more bits, of course, the larger the decimal number that can be represented. Most microprocessor systems have eight bits and can be used to represent decimal numbers 0 to 256. Thus, a given analog signal can be expressed to an accuracy of one part in 256, or ±¼%, which is sufficiently accurate for most signal-processing applications. More bits obviously permit greater accuracies.

Digital-to-Analog Conversion

Digital electronic systems, such as microprocessors, express and process data in binary form. To transmit the data to and from various parts of the system, each bit uses a separate line. Thus, an eight bit system requires eight transmission lines and eight sets of terminals. The voltage at each terminal is either high (+5 V) or low (0 V), depending on whether that bit is 1 or 0.

The problem of digital-to-analog conversion, then, is to convert the digital signal coming in on eight terminals into a two-terminal analog output voltage whose amplitude is proportional to the binary number expressed by the highs and lows on the eight incoming terminals.

Figure 6.14 shows this process for three input digital terminals. Also shown is a typical sequence of input digital signals and the corresponding output analog voltage. For signal *a*, the input digital signal is the binary number 010 with terminals S_1 low (0 V), S_2 high (5 V), and S_3 (0 V). The D/A converter converts these three input voltages into an output voltage of +2 V that represents the decimal equivalent 2. Similarly, for signal *c*, the input digital signal is the binary number 110, and the required output is +6 V, corresponding to 6. Take a moment to study how the D/A converter produces the correct analog output for each of the corresponding input digital signals.

D/A CONVERTER

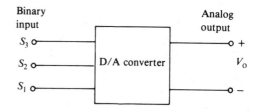

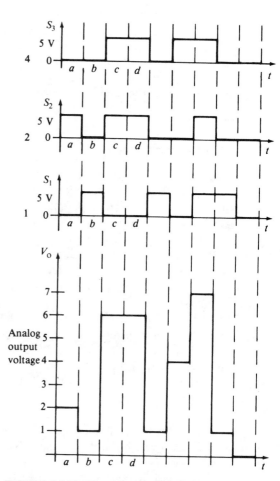

FIGURE 6.14. To transmit digital data electronically, a separate line is used for each bit, and 0 typically represents 0 V and 1 represents +5 V. Digital data can be converted to their analog equivalents using a D/A converter. The figure shows three bit digital data coming into a D/A converter and the resulting analog, decimal output.

R-2R LADDER NETWORK

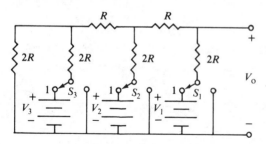

FIGURE 6.15. The *R-2R* ladder network shown can be used as a D/A converter. Three bit input digital data are represented by voltage sources V_1, V_2, and V_3, each of which can be made 0 or 1 by switches S_1, S_2 and S_3. The resulting analog output appears as V_o.

R–2R Ladder Network

The conversion of a digital signal expressed on several terminals to a corresponding analog signal expressed on two output terminals can be accomplished by a relatively simple resistor network called an *R–2R* **ladder.** Figure 6.15 shows the *R–2R* ladder for a three bit input signal, such as the one in Figure 6.14. The ladder network can be extended to accommodate as many bits as required.

The name *R–2R ladder* comes from the ladderlike shape of the network and the fact that the vertical resistors of the "rungs" are two times the values of the horizontal resistors. The actual values of the resistors are not important so long as the 1 to 2, or *R–2R*, ratio is maintained.

In Figure 6.15, the output analog voltage is shown as V_o at the right-hand terminals. The digital input signals are shown as switch positions to a series of voltage sources. When the switch is connected to the 0 terminal (ground), the binary value is zero; when it is connected to the 1 terminal ($+V_1$), the binary value is one. Thus, a digital input to the circuit involves setting the positions of the switches S_1, S_2, and S_3. In practice, the switches represented as mechan-

ical contacts in Figure 6.15 would be replaced by electronic switches made from bipolar or field-effect transistors. For simplicity, however, the circuit of Figure 6.15 is analyzed here with the input signal determined by the switch positions.

Note also that three separate batteries are shown to provide the $+V$ voltage for each of the 1 switch positions. Of course, a single battery or dc supply would be sufficient to operate the circuit with all of the $+V$ terminals connected together.

Analysis of Figure 6.15 requires determining the output voltage V_o for each of the eight combinations of 0 to 1 switch positions. The analysis is done for the combination 111, but the results can be generalized to all eight. The mathematical tool used is the principle of superposition because it is the best suited for multisource circuits. Also, Thevenin's theorem is used. These tools make the analysis of Figure 6.15 relatively straightforward and illustrate how these tools are applied to modern circuits.

The problem, then, is to determine the analog output for the binary number 111, as shown by the switch positions in Figure 6.15. We must determine the output voltage V_o due to the three voltage sources V_1, V_2, and V_3. According to Figure 6.15, this value should correspond to an analog output of seven. Using the principle of superposition, we first set $V_2 = V_3 = 0$ and determine the contribution to V_o due to V_1. Then we set $V_1 = V_3 = 0$ and determine the contribution to V_o due to V_2.

Figure 6.16a shows the circuit redrawn for the first step, $V_2 = V_3 = 0$ (short circuited). V_o will be given by the voltage source V_1 and the voltage divider of $2R$ with the equivalent resistance of the resistor network to the left of the dashed line. To find this equivalent resistance, we start at the left-hand side. The two $2R$ resistors in parallel have an equivalent resistance of:

$$R_{eq} = \frac{2R \times 2R}{2R + 2R} = \frac{4R^2}{4R} = R \qquad (29)$$

ANALYSIS OF *R-2R* LADDER USING SUPERPOSITION

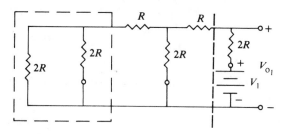

(a) Circuit to Find V_{o_1}

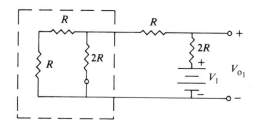

(b) Reduction of (a)

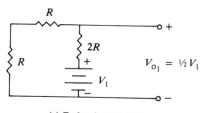

(c) Reduction of (b)

FIGURE 6.16. Analysis of the *R–2R* ladder can be simplified with the principle of superposition. In (a) voltage sources V_2 and V_3 are set to zero (short circuit) and the contribution to V_o due to V_1 calculated. (b) and (c) show the successive combining of the *R–2R* resistive network to determine V_{o_1}.

This is in series with a resistor R to form a net resistance of $2R$, again in parallel with a resistance $2R$ (Figure 6.16b). Again, this parallel combination of two $2R$ resistors produces a net resistance of R that is in series with a resistance R to produce a net equivalent resistance to the left of the dashed line of:

$$R_{eq} = 2R \qquad (30)$$

The resulting circuit is shown in Figure 6.16c. The value of V_{o_1} is the voltage across the $2R$ equivalent resistance, which is given by the voltage divider:

$$V_{o_1} = \frac{2R}{2R + 2R}V_1 = \frac{1}{2}V_1 \qquad (31)$$

Notice that in the final result, the actual value of the resistance R is not involved, only the value of V_1. Note also that the R–$2R$ ratio of resistance values made calculation of the equivalent resistance of the network extremely simple.

By a similar method, we can find the contribution to V_o due to source V_2 with $V_1 = V_3 = 0$ (Figure 6.17a). The plan for solving this circuit is:

1. Find the Thevenin equivalent network of the portion in the dashed box.
2. Combine the Thevenin equivalent with the remaining network.
3. Apply the voltage divider relation to determine V_{o_2}.

The Thevenin equivalent network for the part of Figure 6.17a in the dashed box can be found with the aid of Figures 6.16b and 6.16c. The comparison shows that the value of open circuit voltage at the terminals AB will be:

$$V_{oc} = \frac{1}{2}V_2$$

The Thevenin equivalent resistance is found by short-circuiting V_2 and determining the equivalent resistance of two $2R$ resistors in parallel, which we know to be equal to R by Equation 29. Thus, the Thevenin equivalent network is as shown in Figure 6.17b. Using the voltage divider relation to get the voltage V_{o_2} across the $2R$ resistor, we find:

$$V_{o_2} = \frac{2R}{2R + R + R}\frac{1}{2}V_2 = \frac{1}{4}V_2 \qquad (32)$$

ANALYSIS OF R-$2R$ LADDER USING THEVENIN'S THEOREM

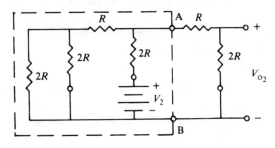

(a) Circuit to Find V_{O_2}

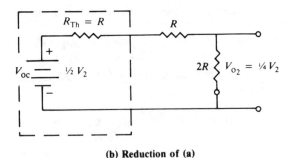

(b) Reduction of (a)

FIGURE 6.17. When $V_1 = V_3 = 0$ (a), Thevenin's theorem can be used to determine V_{O_2}. The resulting Thevenin equivalent network is shown in (b). The final expression for V_0 is given by Equation 34.

The procedure for finding V_{O_3}, the output resulting from S_3 in the 1 position, is similar to that used to find V_{O_1} and V_{O_2}. The result is:

$$V_{O_3} = \tfrac{1}{8} V_3 \tag{33}$$

To find the value of V_0 due to all three sources, we use the principle of superposition:

$$\begin{aligned} V_0 &= V_{O_1} + V_{O_2} + V_{O_3} \\ &= \tfrac{1}{2} V_1 + \tfrac{1}{4} V_2 + \tfrac{1}{8} V_3 \end{aligned} \tag{34}$$

In the preceding calculations, the values found corresponded to the case in which the switches are in the 1 position. If a switch were in the 0 position, the corresponding contribution to the output would be zero. The two pos-

sible switch positions (0 and 1) can be included in Equation 34 by adding a prefix S to each voltage to denote whether that switch input is 0 or 1:

$$V_0 = \tfrac{1}{2} S_1 V_1 + \tfrac{1}{4} S_2 V_2 + \tfrac{1}{8} S_3 V_3 \tag{35}$$

Let us also set each of the voltages to the same value V. Then:

$$V_1 = V_2 = V_3 = V \tag{36}$$

and

$$V_0 = (\tfrac{1}{2} S_1 + \tfrac{1}{4} S_2 + \tfrac{1}{8} S_3) V$$

or

$$V_0 = (S_1 + \tfrac{1}{2} S_2 + \tfrac{1}{4} S_3) \tfrac{1}{2} V \tag{37}$$

The output voltage V_0 of the circuit in Figure 6.15 for any of the eight combinations of three bit binary input signals can be determined from Equation 37 by simply inserting the appropriate values of S_1, S_2, and S_3. For example, suppose we want to find the analog output for the binary number:

$$S_1 S_2 S_3 = 001$$

Substitution gives:

$$V_0 = (0 + 0 + \tfrac{1}{4}) \tfrac{1}{2} V = \tfrac{1}{8} V$$

If we let $V = 8$ V, the output will be:

$$V_0 = 1 \text{ V}$$

which is the decimal equivalent of the binary number 001. Then, for $V = 8$ V, the general equation describing Figure 6.15 becomes:

$$V_0 = (S_1 + \tfrac{1}{2} S_2 + \tfrac{1}{4} S_3) \times 4 \tag{38}$$

For the binary number:

$$S_1 S_2 S_3 = 101$$

Equation 38 yields:

$$\begin{aligned} V_0 &= (1 + 0 + \tfrac{1}{4}) \times 4 \\ &= \tfrac{5}{4} \times 4 = 5 \text{ V} \end{aligned}$$

**COMPLEX NETWORK REDUCED TO ITS
NORTON EQUIVALENT**

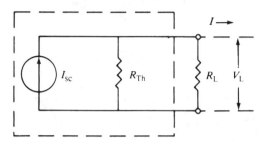

FIGURE 6.18. For some circuits, Norton's theorem will simplify analysis. By Norton's theorem, a complex network is equivalent to a single current source and a parallel resistor. The current source, called the short-circuit current, is the current at the output terminals when they are short circuited. The resistance is the Thevenin equivalent resistance.

which is the decimal equivalent of the binary number 101. A substitution of all the possible combinations of a three bit number would show that Equation 38 will yield the decimal equivalent for each. As indicated earlier, this network can be extended to accommodate any number of binary inputs to yield a D/A conversion.

6.7 NORTON'S THEOREM

We have seen that a linear network of some complexity can be simplified considerably by the use of Thevenin's theorem. The result is the Thevenin equivalent network that consists of a single voltage source V_{oc} and a single series resistor R_{Th}.

Norton Equivalent Network

Another theorem that is similar to Thevenin's, but is simpler to use for certain types of circuits, is **Norton's theorem** that states that *the behavior of a complex resistive network at a given terminal pair will be identical to that of a*

**PROCEDURE FOR MEASURING NORTON
EQUIVALENT VALUES**

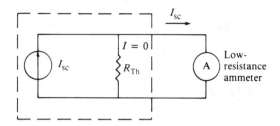

FIGURE 6.19. The short-circuit current can be measured by a low-resistance ammeter connected to the output terminals. The Thevenin resistance is measured the same way as in Figure 6.3.

single current source in parallel with a single resistor. The resulting network is called the Norton equivalent network, shown in Figure 6.18. The parallel resistance R_{Th} is the Thevenin resistance used in the Thevenin equivalent network — that is, the effective resistance of the circuit measured at the output terminals. Hence, the value of R_{Th} is found by the same methods described for the Thevenin equivalent network.

Short-Circuit Current

The current source I_{sc} is called the **short-circuit current** and is the *current that would pass through the output terminals if they were short circuited* — that is, if any load resistor were replaced with a short circuit. When the output terminals are short circuited, no current flows through the Thevenin resistance, but instead passes through the output short circuit.

In an actual circuit, the value of I_{sc} is the current that would be measured by a very low resistance ammeter connected to the output terminals (Figure 6.19). A note of caution, however: If I_{sc} is to be measured by applying a short circuit to the output terminals, we must be certain that none of the components in the circuit will be damaged by the high currents (and, consequently, large power dissipation) that may occur.

The value of I_{sc} can be calculated by the following procedure:

1. Short-circuit the output terminals.
2. Calculate the current I_{sc} that flows through the short circuit.

This method, of course, is comparable to finding the value of V_{oc} using Thevenin's theorem. When we know I_{sc}, we can find the current through a load resistor R_L by the current divider relation:

$$I_L = \frac{R_{Th}}{R_{Th} + R_L} I_{sc} \qquad (39)$$

The load voltage is:

$$V_L = I_L R_L = \frac{R_{Th} R_L}{R_{Th} + R_L} I_{sc} \qquad (40)$$

According to Norton's theorem, both of the circuits shown in Figure 6.20 will produce identical behavior at the output terminals. If the two circuits are equivalent, then we should be able to relate the open circuit voltage V_{oc} of the Thevenin equivalent network to the short-circuit current I_{sc} of the Norton equivalent network. We can see this by writing the circuit equations for each network. For Figure 6.20a, KVL yields:

$$V_{oc} - IR_{Th} - V_o = 0$$

or

$$V_o = V_{oc} - IR_{Th} \qquad (41)$$

For Figure 6.20b, KCL yields:

$$I_{sc} = I + \frac{V_o}{R_{Th}}$$

EQUATING THEVENIN AND NORTON NETWORKS

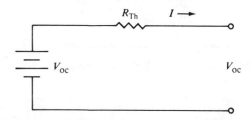

(a) Thevenin Equivalent Network: $V_0 = V_{oc} - IR_{Th}$

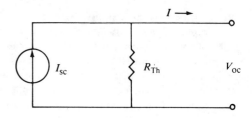

(b) Norton Equivalent Network: $V_0 = I_{sc}R_{Th} - IR_{Th}$

FIGURE 6.20.   Because the Thevenin and Norton equivalent networks both produce the same terminal behavior, they can be related. Equation 43 relates the Thevenin open circuit voltage to the Norton short-circuit current.

or

$$V_o = I_{sc}R_{Th} - IR_{Th} \qquad (42)$$

Equating Equations 41 and 42 gives:

$$V_{oc} = I_{sc}R_{Th} \qquad (43)$$

Thus, if we know the values of V_{oc} and R_{Th} for the Thevenin equivalent network, we can quickly obtain the Norton equivalent network by solving for I_{sc} in Equation 43.

In summary, we can represent the output terminal behavior of a resistive circuit with either the Thevenin or the Norton equivalent network. The choice of which network to use in a particular application depends on which will most simplify the calculations. In general, if the remaining network is a series connection, the calculations will be simpler with the Thevenin equivalent. Conversely, if the remaining network is a parallel combination, then the Norton equivalent will usually make the algebra simpler. Note, however, that either equivalent will yield the correct solution, provided that Equation 43 is satisfied.

The discussion of Norton's theorem concludes with two example calculations. The first will be the circuit of Figure 6.4, for which we earlier found the Thevenin equivalent network (see Example 1). Comparison of the two results provides a test of Equation 43.

EXAMPLE 6: What is the voltage across the load resistor in the circuit shown in Figure 6.21a?

Solution: The calculation of R_{Th} is the same for both equivalent networks, so we can use the results found in Example 1. Short-circuiting the voltage source (Figure 6.21b) and solving for the equivalent resistance gives Equation 3:

$$R_{Th} = R_3 + \frac{R_1 R_2}{R_1 + R_2}$$

Combining terms:

$$R_{Th} = \frac{R_1 R_2 + R_1 R_3 + R_2 R_3}{R_1 + R_2} \tag{44}$$

To find I_{sc}, we must short-circuit the output terminals, as shown in Figure 6.21c. The current I_{sc} through resistor R_3 is determined

EXAMPLE 6

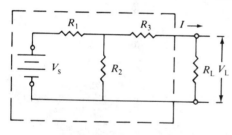

(a) Original Circuit: V_L = ?

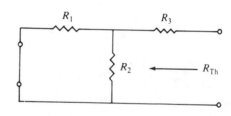

(b) Circuit to Find R_{Th}: $R_{Th} = R_3 + \dfrac{R_1 R_2}{R_1 + R_2}$

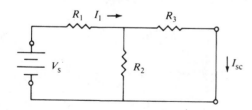

(c) Circuit to Find I_{sc}: $I_{sc} = \dfrac{V_s R_2}{R_1 R_2 + R_1 R_3 + R_2 R_3}$

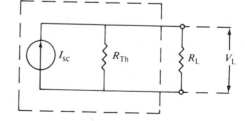

(d) Norton Equivalent Network: $V_L = \dfrac{R_{Th} R_L}{R_{Th} + R_L} I_{sc}$

FIGURE 6.21. To calculate the short-circuit current, we short the output terminals as in (c) and calculate the resulting terminal current. The Thevenin resistance is calculated in the same way as the Thevenin equivalent network.

by the net circuit current I_1 and the current divider of R_2 and R_3. Thus, we must first determine the value of I_1, which is the source voltage V_s divided by the series combination of R_1 and $R_2 \parallel R_3$:

$$I_1 = \frac{V_s}{R_1 + (R_2 \parallel R_3)}$$

or

$$I_1 = \frac{V_s}{R_1 + [R_2 R_3/(R_2 + R_3)]} \qquad (45)$$

The desired current I_{sc} is related to I_1 through the current divider found by R_2 and R_3:

$$I_{sc} = \frac{R_2}{R_2 + R_3} I_1 \qquad (46)$$

Substituting Equation 46 into Equation 45 and simplifying terms gives:

$$I_{sc} = \frac{V_s R_2}{R_1 R_2 + R_1 R_3 + R_2 R_3} \qquad (47)$$

The resulting Norton equivalent network is shown in Figure 6.21d.

According to Equation 43, the product of R_{Th} and I_{sc} should give the open circuit voltage V_{oc} of the Thevenin equivalent network:

$$V_{oc} = I_{sc} R_{Th} = \frac{V_s R_2}{R_1 R_2 + R_1 R_3 + R_2 R_3}$$
$$\times \frac{R_1 R_2 + R_1 R_3 + R_2 R_3}{R_1 + R_2}$$

$$= \frac{R_2}{R_1 + R_2} V_s \qquad (48)$$

Comparison with the results of Example 1, Equation 2, shows that they agree.

A further check is to substitute the circuit values used in Example 1:

$$V_s = 10 \text{ V}$$

$$R_1 = 1 \text{ k}\Omega$$
$$R_2 = 10 \text{ k}\Omega$$
$$R_3 = 2.7 \text{ k}\Omega$$
$$R_L = 5.6 \text{ k}\Omega$$

Substitution into Equation 47 gives the short-circuit current:

$$I_{sc} = \frac{10 \text{ V} \times 10 \text{ k}\Omega}{(1 \times 10) + (1 \times 2.7) + (10 \times 2.7)}$$
$$= \frac{100}{39.7} = 2.5 \text{ mA}$$

In Example 1, substitution of these values gave the Thevenin resistance:

$$R_{Th} = 3.6 \text{ k}\Omega$$

The load voltage can now be found from Equation 40:

$$V_L = \frac{3.6 \text{ k}\Omega \times 5.6 \text{ k}\Omega}{3.6 \text{ k}\Omega + 5.6 \text{ k}\Omega} 2.5 \text{ mA}$$
$$= 2.2 \text{ k}\Omega \times 2.5 \text{ mA} = \textbf{5.5 V}$$

which is the same as in Example 1.

EXAMPLE 7: What is the load voltage for the circuit shown in Figure 6.22a?

Solution: This is the circuit from Example 3 that gave for the Thevenin resistance (Equation 8):

$$R_{Th} = \frac{R_3 R_4}{R_3 + R_4}$$

The short-circuit current is found by replacing the load resistor R_L with a short circuit. This shorts out resistor R_4, which can then be ignored (Figure 6.22c). The short-circuit current, then, is the current through resistor R_3, which is simply the source voltage divided by R_3:

$$I_{sc} = \frac{V_s}{R_3} \quad \text{(49)}$$

The resulting Norton equivalent network is shown in Figure 6.22d.

These results can be compared to the Thevenin equivalent network by using equation 43:

$$V_{oc} = I_{sc}R_{Th} = \frac{V_s}{R_3} \times \frac{R_3 R_4}{R_3 + R_4}$$

$$= \frac{R_4}{R_3 + R_4} V_s$$

Note that this result agrees with Equation 7.

The load voltage can now be found from Equation 40:

$$V_L = \frac{R_{Th} R_L}{R_{Th} + R_L} I_{sc}$$

Substitution of circuit values will give the desired result.

In conclusion, it should be noted that a knowledge of Thevenin's theorem and the principle of superposition is sufficient to analyze most circuit problems. Norton's theorem and several other theorems not discussed here can be helpful or provide shortcuts for certain prob-

EXAMPLE 7

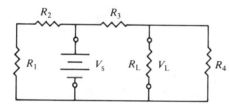

(a) Original Circuit: $V_L = ?$

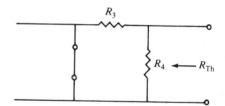

(b) Circuit to Find R_{Th}: $R_{Th} = \dfrac{R_3 R_4}{R_3 + R_4}$

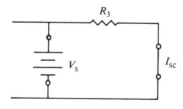

(c) Circuit to Find I_{sc}: $I_{sc} = \dfrac{V_s}{R_3}$

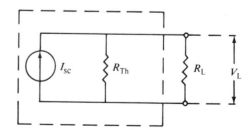

(d) Norton Equivalent Network: $V_L = \dfrac{R_{Th} R_L}{R_{Th} + R_L} I_{sc}$

FIGURE 6.22. Like Thevenin's theorem, Norton's theorem can be used with a variety of networks.

lems, but they are not a necessary tool to have at your fingertips.

As you have also seen, it is essential to have a thorough knowledge of voltage and current dividers (presented in Chapter 4) and basic methods of combining series and parallel connections (presented in Chapter 5) in order to implement Thevenin's theorem and the principle of superposition. In fact, these simple techniques will solve the majority of practical circuit problems.

6.8 QUESTIONS AND PROBLEMS

1. Draw the Thevenin equivalent network for terminals AB of each of the circuits shown in Figure 6.23.
2. Determine expressions for the open circuit voltage V_{oc} and Thevenin resistance R_{Th} of each of the Thevenin equivalent networks in Problem 1.
3. Use Thevenin's theorem to calculate the voltage across the load resistance R_L of each of the circuits shown in Figure 6.23, given the following circuit values: $V = 9.0$ V, $I = 6.0$ mA, $R_1 = 10$ kΩ, $R_2 = 8.2$ kΩ, $R_3 = 4.7$ kΩ, $R_4 = 2.2$ kΩ, and $R_L = 10$ kΩ.
4. Determine expressions for the voltage across the load resistance R_L of each of the circuits shown in Figure 6.24.
5. Calculate the voltage across the load resistance R_L of each of the circuits shown in Figure 6.24, given the following circuit values: $V_1 = 9.0$ V, $V_2 = 6.0$ V, $I = 6.0$ mA, $R_1 = 10$ kΩ, $R_2 = 8.2$ kΩ, $R_3 = 4.7$ kΩ, $R_4 = 2.2$ kΩ, and $R_L = 10$ kΩ.
6. Express the decimal equivalent of each of the following binary numbers: (a) 10011101; (b) 00111101; (c) 11111111; (d) 11011011; (e) 10101010; (f) 01010101.
7. Express the binary equivalent of each of the following decimal numbers: (a) 225; (b) 172; (c) 46; (d) 85; (e) 235; (f) 197.
8. Draw an R–$2R$ ladder network that will convert four bits of digital data into an analog output voltage and derive an expression for this output voltage in terms of the "highs" and "lows" of the digital inputs. (Note: The results derived in the text for a three bit ladder network can be assumed and used as starting point.)
9. Draw the Norton equivalent network for terminals AB of each of the circuits shown in Figure 6.23.
10. Determine expressions for the short-circuit current I_{sc} and Thevenin resistance R_{Th} of each of the Norton equivalent networks in Problem 9. Compare your results with the results of Problem 2.
11. Use Norton's Theorem to calculate the voltage across the load resistor R_L of each of the circuits shown in Figure 6.23 for the following circuit values and compare your results with the results of Problem 3: $V = 9.0$ V, $I = 6.0$ mA, $R_1 = 10$ kΩ, $R_2 = 8.2$ kΩ, $R_3 = 4.7$ kΩ, $R_4 = 2.2$ kΩ, and $R_L = 10$ kΩ.

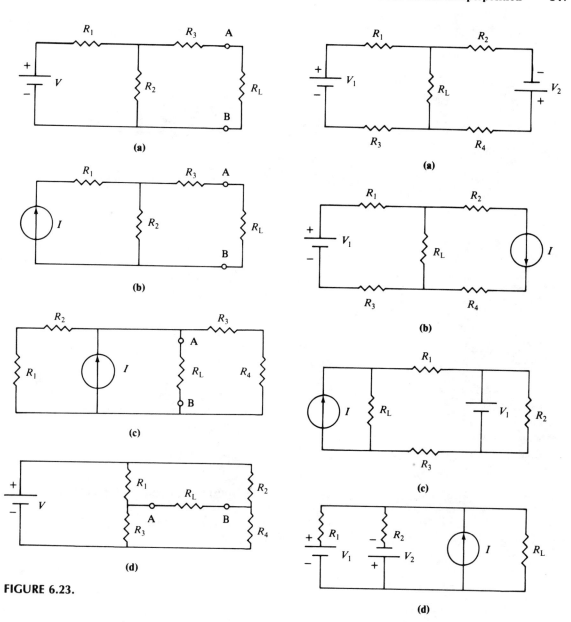

FIGURE 6.23.

FIGURE 6.24.

Part III

TIME-VARYING SIGNALS

TIME-VARYING QUANTITIES IN MEASUREMENT

The sound of a siren, vibrations of an earthquake, velocity of the wind, a flash of lightning, pressure in a compressor, and the temperature outside your house are all quantities that change with time. The ability to measure such quantities as they change is an important part of understanding and controlling our environment.

In nearly every case, these measurements are made electronically. A transducer converts the change in physical quantity — for example, sound, force, pressure, light, or temperature — into a change in an electrical quantity — for example, voltage, current, or resistance — that an electronic system can process to perform some useful task.

In the simplest case, the electronic system simply gives a readout of the measured quantity. In more sophisticated systems, however, the changing physical data have many other uses. Computers use changing weather information as input data to calculate weather patterns; automobiles use engine and road information to control fuel consumption; and there are many other examples.

Because every physical quantity changes with time in one way or another, the measurement of time-varying signals is an important part of electronics and instrumentation (Figure 1). As a result, time-varying electrical signals are introduced quite early in this book so that you can use them in measurement applications.

Part III concentrates primarily on the electronic aspects of time-varying signals rather than on changes of physical quantities or on the transducers that convert them into electrical signals. The latter are extremely important concepts, however, in designing a particular electronic system, so they are discussed too.

FIGURE 1. Nearly every physical quantity changes with time. These changes can be electronically measured and recorded, as in the figure example. The measurement of time-varying electrical signals is a major part of electronics and is described in Chapter 7.

EXAMPLE OF QUANTITIES THAT CHANGE WITH TIME

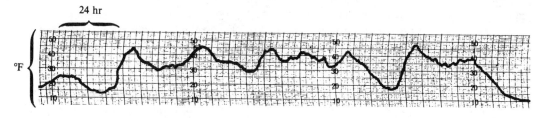

Temperature for a Week in January at Lincoln, Nebraska

Chapter

7

DESCRIBING AND MEASURING TIME-VARYING SIGNALS

7.1 OBJECTIVES

Following the completion of Chapter 7, you should be able to:

1. Identify time-varying signals by their waveforms, including step signals, pulses, pulse trains, and periodic signals (square wave, sine wave, triangle wave, and sawtooth wave).
2. Explain and give mathematical expressions for the terms used to describe time-varying signals, including peak amplitude, peak-to-peak amplitude, pulse height, rise time, fall time, pulse width, period, frequency, duty cycle, average voltage, rms voltage, and slew rate.
3. Determine the amplitude and rise (or fall) time of a step signal from its waveform.
4. Determine the pulse height, pulse width, rise time, and fall time of a pulse signal from its waveform.
5. Determine the period, frequency, duty cycle, and average voltage of a pulse train from its waveforms.

6. Determine the peak amplitude, peak-to-peak amplitude, period, frequency, average voltage, and slew rate of a periodic signal from its waveform.
7. Determine the average voltage, rms voltage, and electrical power of a sine wave voltage, given the voltage waveform (or current waveform) and the circuit resistance.
8. Identify each of the controls of a function generator and explain its purpose in determining the characteristics of the output waveform.
9. Determine the voltage across any load resistance connected to the output of a function generator, given its control settings and output resistance.
10. Determine the maximum signal slew rate that can be measured by an instrument, given its response time (or frequency response) and full-scale voltage sensitivity.
11. Select the proper instrument, according to type, range, and response time, for measuring a given time-varying signal.

12. Explain the operation of a cathode-ray oscilloscope in displaying a periodic time-varying signal, including the operation of the cathode-ray tube, vertical amplifier, and horizontal sweep circuit.
13. Identify each of the controls of a dual-trace oscilloscope and explain its use in determining the display of an input waveform.
14. Determine the amplitude and time characteristics of a waveform displayed on an oscilloscope, given the front panel control settings for a given type of connector probe.

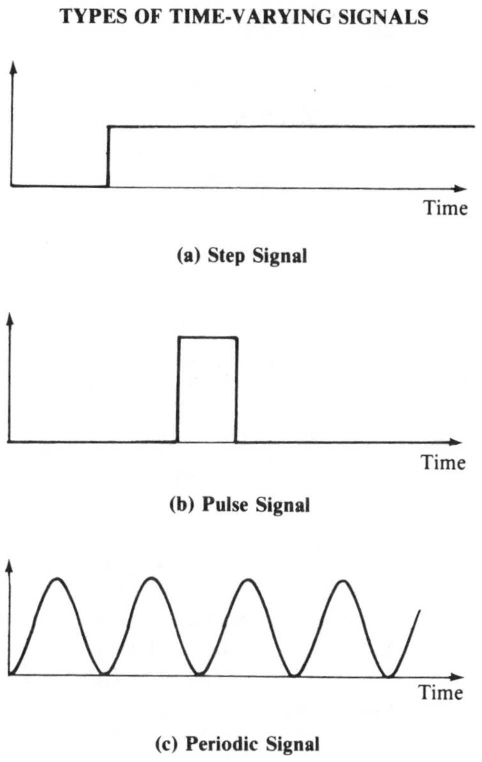

7.2 TYPES OF TIME-VARYING SIGNALS

The study of time-varying signals begins by categorizing them into certain general types. Because the range of possible ways in which signals can change is so broad—some slow, some fast, some small, some large, some simple, and some complicated—there will be many signals that do not exactly fit these categoric descriptions. However, the categories will serve as a basis for describing the signals and their characteristics whether or not they are perfectly accurate.

The basic types of time-varying signals are called *step signals*, *pulse signals*, and *periodic signals* (Figure 7.1). For each one, there is a perfect, or idealized, form. But, in general, most real signals only approximate the idealized form. Hence, a variety of terms must be used to describe the idealized and the approximated forms of each signal type.

After the types and descriptions of time-varying signals have been introduced, the remainder of this section describes the basic instruments used to measure time-varying signals—meters, chart recorders, and oscilloscopes—and the function generator that electronically generates nearly ideal time-

TYPES OF TIME-VARYING SIGNALS

(a) Step Signal

(b) Pulse Signal

(c) Periodic Signal

FIGURE 7.1. Time-varying signals can be approximately divided into three categories: step, pulse, and periodic. The examples shown are idealized; most real signals are not nearly so exact. Also, many complex signals do not fit any of the three categories.

varying signals. The function generator–produced signals are used to test the ability of an electronic circuit to respond to different types of time-varying signals.

Step Signals

The simplest example of a time-varying signal is the step signal shown in Figure 7.2. A **step signal** is produced whenever a quantity makes a single change in value. The electrical signal then goes from one steady value to another.

Ideally, a step signal rises—or falls—

STEP SIGNALS

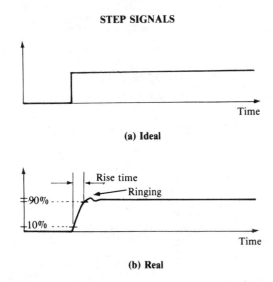

(a) Ideal

(b) Real

FIGURE 7.2.   The simplest time-varying signal is a step signal. An ideal step signal changes instantaneously from one constant value to another. A real signal requires a finite amount of time. The rise time measures the time needed to go from 10% to 90% of the new value.

instantaneously, as shown in Figure 7.2a. Any real signal, of course, always requires some time to change its value. The time needed is called the **rise time** — or **fall time** — of the signal, depending on the direction of the step change.

The rise time is defined as the time it takes to go from 10% to 90% of its final value. The fall time is similarly defined, except that the direction of the step change is negative.

An ideal signal also goes exactly to the new value and stays there. A real signal, however, may overshoot the mark and take a while to settle down to the new value. This type of behavior is called **overshoot** or **ringing** when it takes more than one oscillation to come to the new value.

Figure 7.3 shows two typical signals that can be classified as step signals. In Figure 7.3a, the temperature of a water bath increases after a temperature control system has been set to a

higher value. The time scale may be minutes or even hours depending on the size of the bath.

In Figure 7.3b, the signal measured is the speed of an automobile during a collision. The time scale for the speed to fall to zero is on the order of seconds or less. Obviously, the shorter the time, the more serious the accident. Note also that the signal overshoots to a negative velocity, which indicates that the automobile bounces backward after the collision.

Pulse Signals

Pulse signals are produced whenever a system goes temporarily to a new value but then returns to its original state. An ideal pulse signal is a perfect rectangle (Figure 7.4a). The quantity goes instantaneously to a specific new value, remains there for a specific length of time, and then returns instantaneously to its original value. The difference between the original value and the new value is called the **pulse height**, and the time that it stays at the new value is called the **pulse width.**

A real pulse, of course, takes a certain amount of time to reach the new value and to return to the original value. Also, the new value may or may not be constant (Figure 7.4b). In this case, the pulse height is the maximum of the new value, and the pulse width is measured between the 50% rise and fall points.

Figure 7.5 shows some typical pulse signals. The specific shapes are different for each signal, but they each have characteristics that are common to pulse signals. They all begin at a certain level, undergo a level change, and then return approximately to the level where they began. In the first example, (Figure 7.5a), the signal is the barometric pressure during a severe storm. Figure 7.5b is an electrical signal produced by a human heart beat called an electrocardiogram (ECG). Figure 7.5c is a signal produced by an electronic device called a unijunction transistor. The time scales for the three signals are vastly different, as are their shapes.

EXAMPLES OF STEP SIGNALS

FIGURE 7.3. The two examples of time-varying signals shown can be classed as step signals even though they are not ideal.

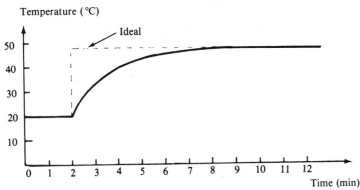

(a) Temperature of a Water Bath

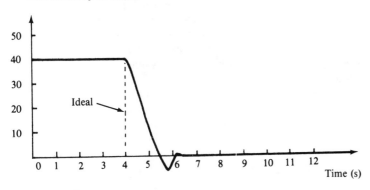

(b) Speed of an Automobile during a Collision

For each, however, we can define an approximate pulse height and an approximate pulse width.

Pulse signals are extremely important in electronics, particularly in digital electronics, which depends totally on high- and low-voltage pulses. Pulse signals are also important in measurement. In the simplest case, the mere existence of a pulse signal, regardless of its exact height or width, can be used in a practical way. In Figure 7.6, for example, a light beam is directed across a conveyor belt of the kind used on a production line. As objects travel along the conveyor belt, they block and then unblock the light beam. The photodetector output produces a corresponding pulse that can cause a counter to advance one unit.

The pulse height is determined by the brightness of the light. The pulse width is determined both by the length of the object on the belt and by the conveyor's speed. The smaller the object and the faster the speed, the shorter the width of the pulse. But, if the circuit is designed only to count, the pulse height and width are relatively unimportant.

In a camera, however, the pulse height and

PULSE SIGNALS

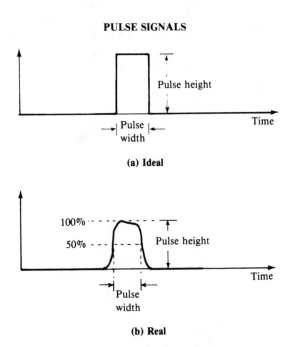

(a) Ideal

(b) Real

FIGURE 7.4. Ideal and typical real pulse signals are shown here. For the ideal pulse, the height and width are exact values, but for a real pulse, they must be approximated.

TYPICAL PULSE SIGNAL MEASUREMENT ON A PRODUCTION LINE

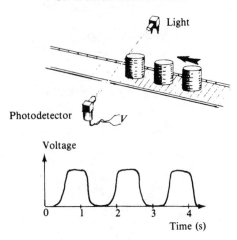

FIGURE 7.6. Pulse signals have widespread application in electronic measurement. In the example shown, the pulse registers the passage of an object on an assembly line. In this case, the pulse height and width are relatively unimportant.

EXAMPLES OF PULSE SIGNALS

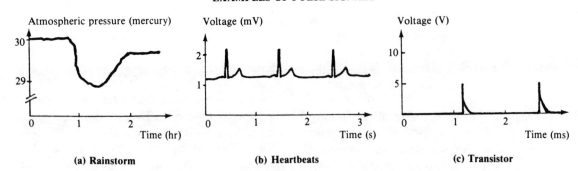

(a) Rainstorm   **(b) Heartbeats**   **(c) Transistor**

FIGURE 7.5. The examples of pulse signals shown here have different shapes, but have characteristics common to all pulse signals. In each case, the signal starts at one value, goes to a new value, and then returns approximately to the original value.

**TYPICAL PULSE SIGNAL MEASUREMENT
IN A CAMERA**

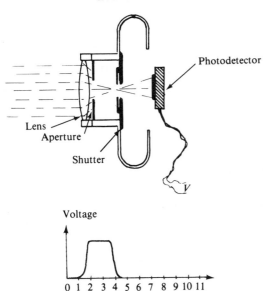

FIGURE 7.7. In a camera, the pulse width measures the time the camera shutter is open (the shutter speed), and the pulse height represents the intensity of the light striking the film. The light intensity is determined both by the source of light and by the camera setting.

width are extremely important. As an example, consider the camera arrangement shown in Figure 7.7. The camera lens is pointed toward an illuminated source, and a photodetection circuit is placed in the plane of the film to measure the light incident on the film.

Whenever the shutter is opened and closed, a pulse is produced by the detector. The pulse width is determined by the time the shutter is open, and the pulse height is fixed by the amount of light striking the detector. For setting shutter speeds, a measurement of the pulse width is important. But, to adjust the lens aperature, only the pulse height is important.

A series of similar pulses that occur in succession, as shown in Figures 7.5b and 7.5c, is called a **pulse train**. The pulse train can be irregular, as shown in Figure 7.8a, where the pulse width, height, and spacing differ from one pulse to the next. Or, the pulses can be identical and evenly spaced, as shown in Figure 7.8b.

Both kinds of pulses might be produced by objects on the conveyor belt in Figure 7.6. Objects of different sizes (variable pulse height) and uneven spacing on the belt (variable pulse spacing) would produce the irregular pulse train of Figure 7.8a, while identical objects evenly spaced would produce the periodic pulse train of Figure 7.8b.

Periodic Signals

When all of the pulses in a pulse train are identical, as well as evenly spaced, the signal is called periodic. A **periodic signal** is one in which every detail of the signal repeats itself after a specific time interval called the **period** T. The period need not be timed from any specific point on the signal. The only requirement is that for any given point, the amplitude will be exactly the same one period later.

The pulse train shown in Figure 7.8b is periodic, but so is the signal shown in Figure 7.9. Unlike a series of pulses that can be thought of as identical signals equally spaced, the signal of Figure 7.9 is continuous. There is no obvious way to break it into a series of individual pulses. However, the signal is periodic: Each point of the signal repeats itself after a time T.

The period T is the time it takes for a periodic signal to repeat. An equivalent way to express the same information is to say how frequently the signal is repeated. This is called the **frequency** f, and it tells how many times the signal repeats in one unit of time, generally each second.

FIGURE 7.8. A series of pulses occurring one after the other is called a pulse train. A pulse train can be irregular, in which the pulse height, width, and spacing vary, or it can be periodic, in which the pulse height, width, and spacing are identical.

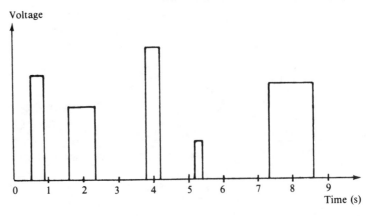

PULSE TRAINS

(a) Irregular

(b) Periodic

PERIODIC SIGNAL

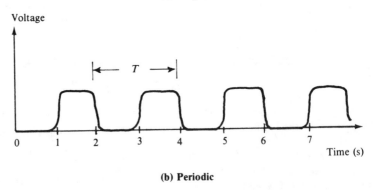

FIGURE 7.9. In general, a periodic signal is one that repeats itself exactly after a time called the period T. For any given point on the signal, the amplitude will be exactly the same one period later.

The frequency f is simply the reciprocal of the period T:

$$f = \frac{1}{T} \tag{1}$$

If the period T is measured in seconds, then the frequency is measured in cycles per second, or **hertz** (Hz).

Any periodic physical event can produce a periodic electrical signal as long as the appropriate transducer is available. These signals can take any shape and have any period or amplitude, but as long as every feature is repeated

SQUARE WAVE

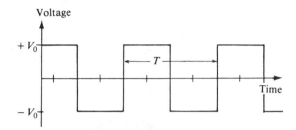

FIGURE 7.10.   The simplest periodic signal is a square wave that is a pulse train whose pulse width equals the time between pulses. The signal changes abruptly from a high voltage $+V_0$ to a low voltage $-V_0$ and back again at equal time intervals.

SINE WAVE

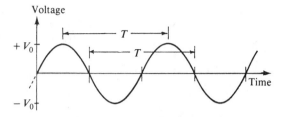

FIGURE 7.11.   The most common periodic wave is the sine wave. It varies continuously from $+V_0$ to $-V_0$ and back to $+V_0$ during each period.

TRIANGLE WAVE

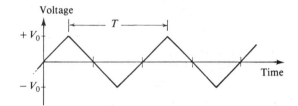

FIGURE 7.12.   Another common periodic wave is called a triangle wave. It varies linearly between $+V_0$ and $-V_0$ and back again.

$$V(t) = V_0 \sin\left[2\pi\left(\frac{t}{T}\right)\right] \qquad (2)$$

after each period, the signals are periodic. There is no point in trying to list all the possibilities—they are endless. On the other hand, several types of periodic signals occur so frequently in electronics that you should be familiar with them.

Square Wave: The simplest of the periodic signals is called a **square wave** (Figure 7.10). This signal is essentially a train of pulses whose pulse width is identical to the time between pulses. Thus, the signal is at a high voltage level for the same length of time that it is at a low voltage level. Frequently, one of these levels will be 0 V, but this is not a necessary condition for a square wave.

Sine Wave: The most common and important of all the periodic signals is the **sine wave.** Whenever anything vibrates, be it the earth's surface during an earthquake, the plucked string of a musical instrument, or the water surface of a wave, the motion is approximately sinusoidal. Other examples are the motion of a spring, the swing of a pendulum, and sound. When you hear a pure note, you are experiencing air pressure that is changing sinusoidally.

A typical sinusoidal signal is shown in Figure 7.11. The voltage at any time $V(t)$ can be written mathematically as:

where the quantity V_0 is the maximum amplitude.

Note whenever the time t increases—or decreases—by an amount T, the quantity in parentheses changes by 2π and the sine function takes on the same value. Thus, the sine function is clearly periodic with a period T. This relationship will be described in more detail later.

Triangle Wave: The signal drawn in Figure 7.12 is also periodic. Its shape, or waveform, is different from that of a sine function, but as shown in the figure, every point on the wave is repeated after one period. Because of its

SAWTOOTH WAVE

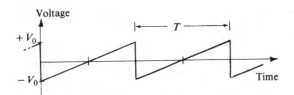

FIGURE 7.13. A periodic signal commonly used in television and oscilloscope circuits is the sawtooth wave. It varies linearly from $-V_0$ to $+V_0$ and then drops abruptly back to $-V_0$.

trianglelike shape, this signal is called a **triangle wave.**

Sawtooth Wave: Most examples of triangle waves are symmetric—the time needed for the signal to rise is the same as the time needed for it to fall. The asymmetric version, shown in Figure 7.13, has a voltage that increases at a steady rate and then returns quickly to its original value. Because its shape resembles the teeth of a saw, this waveform is called a **sawtooth wave.** Sawtooth signals are of particular importance for the operation of an oscilloscope, a topic we will discuss later.

Modulated Signals

A periodic signal is regular. There are no variations in either amplitude or frequency. Every detail of the signal is produced over and over again. However, real signals may not be like that. They may vary in amplitude, or frequency, or both.

Signals that have a constant frequency but vary in amplitude are called **amplitude modulated** (AM). All real oscillating systems are amplitude modulated because eventually their amplitude decreases with time. The tremor of an earthquake, the swing of a pendulum, and the ring of a bell all repeat many times before they stop. The graph in Figure 7.14a shows the

AMPLITUDE-MODULATED WAVES

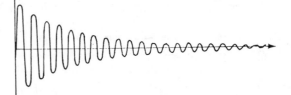

(a) Sine Wave Decaying in Amplitude

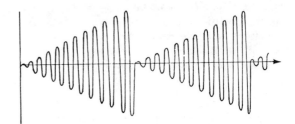

(b) Sine Wave Modulated by a Sawtooth Wave

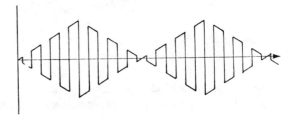

(c) Square Wave Modulated by a Triangle Wave

FIGURE 7.14. Waves, such as the ones shown here, whose frequency is constant but whose amplitude changes are called amplitude modulated (AM).

typical decay in amplitude of a real oscillating system.

Other examples of amplitude modulation are shown in Figures 7.14b and 7.14c. In Figure 7.14b, a sine wave is modulated by a sawtooth wave, and in Figure 7.14c, a square wave is modulated by a triangular wave. Note that the frequencies of both the primary wave and the modulator wave do not change.

FREQUENCY-MODULATED WAVE

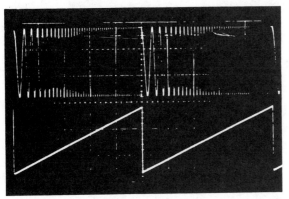

Frequency-
modulated
sine wave

Sawtooth
signal
voltage

FIGURE 7.15. Waves whose amplitude is constant but whose frequency varies are called frequency modulated (FM). A common example of FM is in radio and TV transmission. Here, a sine wave carrier frequency is modulated over a range of frequencies by a sawtooth signal voltage.

Signals that have a constant amplitude but a frequency that changes are said to be **frequency modulated** (FM) (Figure 7.15). The most common example of FM signals is in communication—radio and TV. In this case, a high-frequency carrier signal (approximately 100 MHz) is intentionally modulated with the program signal (approximately 100 Hz to 10 kHz).

The resulting FM signal is broadcast by a transmitter and received by an FM tuner. The FM tuner separates the program signal from the carrier signal and sends it on to the signal amplifier. The reason for using frequency modulation is that the high-frequency carrier wave can be more accurately transmitted than the lower-frequency program signals. In Figure 7.15, a sine wave carrier signal is modulated over a range of frequencies by a signal voltage that is a sawtooth wave. In communications, of course, the signal voltage is much more complicated.

In the most general case, both the amplitude and frequency can vary. Descriptions of these signals get quite complicated, however, and are reserved for advanced texts. Modulated signals, both AM and FM, are also used primarily in the communications field, so discussion of these signals here will be rather limited.

Of more importance to us are the general terms that are used to describe the characteristics of step signals, pulses, and periodic signals.

7.3 DESCRIBING TIME-VARYING SIGNALS

In the overview of the different types of time-varying signals, many of the technical terms used to describe them were ignored. In this section, these signals are treated in more detail, along with the terms that describe them. We will assume that the signal is a voltage because that is what most electronic circuits depend on. However, the descriptions apply equally well to electrical current or any other physical quantity—sound, pressure, temperature, and the like.

Step Signals

Generalized step signals for a step increase and a step decrease are shown in Figure 7.16. Note that they start at one voltage V_1 and end at a different voltage V_2, neither of which needs to be zero.

STEP SIGNALS

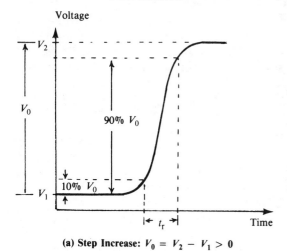

(a) Step Increase: $V_0 = V_2 - V_1 > 0$

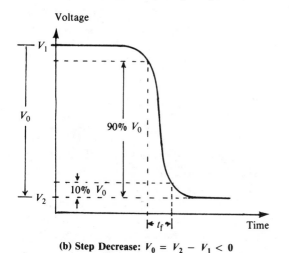

(b) Step Decrease: $V_0 = V_2 - V_1 < 0$

FIGURE 7.16. Two quantities are used to describe a step signal: the amplitude of the change from one value V_1 to a new value V_2 and the time required to make the change, either a rise time or a fall time.

Amplitude: The first quantity used to describe a step signal is the change in voltage. This quantity is simply the difference between V_1 and V_2 and is called the **amplitude** of the step signal V_0. In equation form:

$$V_0 = V_2 - V_1 \qquad (3)$$

If V_2 is greater than V_1, as in Figure 7.16a for a step increase, then V_0 is positive. If V_2 is less than V_1, as in Figure 7.16b for a step decrease, then V_0 is negative. If the value of V_1 or V_2 varies somewhat or has an ac signal superimposed on it, then the value must be approximated. Generally, the estimated average value is sufficient for most purposes.

Rise and Fall Times: The second important quantity describing a step signal is the time it takes to change values. For a step increase, this quantity is called the *rise time* (Figure 7.16a), and for a step decrease, it is called the *fall time* (Figure 7.16b). Because the exact way in which a signal rises or falls can vary widely, an exact quantity is difficult to determine. In Figure 7.16, for example, it is difficult to give an exact time when the signal begins or completes the step. As a result, an approximate rule of thumb is used.

The **rise time** is the time required to go from 10% to 90% of the final value. Similarly, the **fall time** is the time required to go from 90% to 10% of the initial value. These times are illustrated in Figure 7.16. The following example shows how they can be determined.

EXAMPLE 1: What is the amplitude and rise time of the step signal shown in Figure 7.17?

Solution: The initial and final voltages are approximated by the dashed line in the figure. They are:

$V_1 \approx 1.4$ V
$V_2 \approx 6.9$ V

The step amplitude V_0 can be calculated

EXAMPLE 1

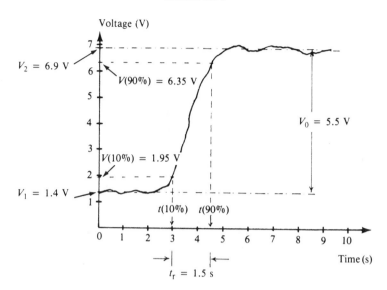

FIGURE 7.17. The rise time is the time taken to go from 10% to 90% of the new amplitude. While this value can be calculated, an estimate made by looking at the graph is generally sufficient.

from Equation 3:

$$V_0 = V_2 - V_1 = 6.9 \text{ V} - 1.4 \text{ V}$$
$$= \textbf{5.5 V}$$

The rise time is the time required to go from 10% to 90% of V_0. Thus, we must first calculate $0.10V_0$ and $0.90V_0$:

$$0.10V_0 = 0.10(5.5 \text{ V}) = 0.55 \text{ V}$$
$$0.90V_0 = 0.90(5.5 \text{ V}) = 4.95 \text{ V}$$

To determine the value on the voltage axis that corresponds to these points, we add the value of V_1:

$$V(10\%) = V_1 + 0.10V_0$$
$$= 1.4 \text{ V} + 0.55 \text{ V} = 1.95 \text{ V}$$
$$V(90\%) = V_1 + 0.90V_0$$
$$= 1.4 \text{ V} + 4.95 \text{ V} = 6.35 \text{ V}$$

These points are shown in Figure 7.17. The

corresponding times are found on the time axis to be:

$$t(10\%) = 3.0 \text{ s}$$
$$t(90\%) = 4.5 \text{ s}$$

Thus, the rise time:

$$t_r = t(90\%) - t(10\%)$$
$$= 4.5 \text{ s} - 3.0 \text{ s} = \textbf{1.5 s}$$

A similar process is used to find the fall time of a decreasing step function.

This procedure for finding an exact value for the rise or fall time is generally not warranted in most applications because normally only an approximate, or order-of-magnitude, value is required. Hence, a rough estimate made by looking at the graph and sketching in some lines is sufficient for most purposes.

PULSES

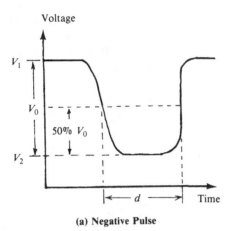

(a) Negative Pulse

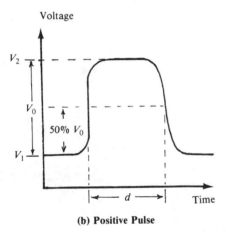

(b) Positive Pulse

FIGURE 7.18. Pulse signals consist of two step signals in succession. The terms that describe pulses are the pulse height V_0 and the pulse width d. Rise and fall times for the leading and trailing edges can also be specified, as for a step signal.

Pulse Signals

Pulse signals are essentially two step signals that occur in succession, one increasing and one decreasing (or vice versa). They can lead to a negative pulse, as shown in Figure 7.18a, or to a positive pulse, as shown in Figure 7.18b.

Because both edges of a pulse signal are step signals, the terms that describe step signals

apply equally well to pulses. The only difference is that the names change slightly.

Pulse Height: The amplitude of the steps is called the **pulse height**, and the rise and fall times must be specified, respectively, to either the **leading edge** or the **trailing edge** of the pulse. Note in Figure 7.18 that the time for the leading edge may be different from that of the trailing edge.

Pulse Width: Another term is required to specify the time between the two steps: the **pulse width** d. As for the rise and fall times, an approximate rule of thumb is required to specify the pulse width because the shape of the leading and trailing edges is variable. Generally, the pulse width is taken as the time between the 50% amplitude points on the leading and trailing edges (Figure 7.18).

EXAMPLE 2:   What is the pulse width of the signal shown in Figure 7.19?

Solution: The first task is to find the pulse height V_0 using Equation 3. The initial and final voltages are approximated by the dashed lines in the figure. They are:

$$V_1 \approx 5.4 \text{ V}$$
$$V_2 \approx 1.8 \text{ V}$$

The pulse height, then, using Equation 3, is:

$$V_0 = V_2 - V_1 = 1.8 \text{ V} - 5.4 \text{ V}$$
$$= -3.6 \text{ V}$$

Fifty percent of the pulse height is:

$$0.50(-3.6 \text{ V}) = -1.8 \text{ V}$$

Thus, the value on the voltage axis at half-height is:

$$V(50\%) = V_1 - 0.50V_0$$
$$= 5.4 \text{ V} - 1.8 \text{ V} = 3.6 \text{ V}$$

EXAMPLE 2

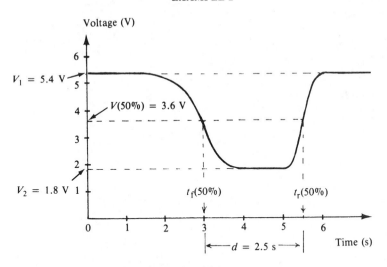

FIGURE 7.19. The pulse width d is the time between the 50% or half-height values of the leading and trailing edges and can be calculated or estimated.

EXAMPLE 3

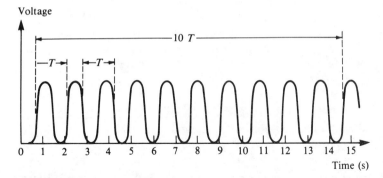

FIGURE 7.20. A sequence of pulses occurring in succession is called a pulse train. The time between pulses is called the period T and can be measured between the same point on two successive pulses. A more accurate method is to measure the time for a number of pulses and divide by the number of pulses.

The times corresponding to 50% of V_0 on the leading and trailing edges are found on the time axis to be:

leading edge: $t_r = 3.0$ s

trailing edge: $t_f = 5.5$ s

Thus, the pulse width is:

$$d = t_f - t_r = 5.5 \text{ s} - 3.0 \text{ s} = \textbf{2.5 s}$$

As for the rise and fall times of a step signal,

the pulse width can often be approximated simply by looking at the pulse and estimating the width at half-height. Be sure, however, to use the half-height of the *pulse only*, and do not include any offset of the baseline.

Period: A train of pulses requires three other descriptive terms. The first is the **period** T. As noted earlier, the period is simply the time between pulses. It can be measured from any point on the pulse — for example, the leading or trailing edge — but the same point must be used for successive pulses (Figure 7.20).

Generally, measuring the period between

only two successive pulses is not very accurate. This accuracy can be considerably increased by measuring the time for a number of pulses to occur and dividing by that number of pulses. This method is particularly useful when the pulses are not exactly the same and it is difficult to find the "same point" on different pulses or when the pulses are not spaced evenly. In the latter case, this method must be used to determine an *average* pulse period.

EXAMPLE 3: What is the period of the pulse train of Figure 7.20?

Solution: Because the pulse period varies slightly from one pulse to the next, an average value must be obtained. Using the leading edge of the pulse as a reference point, the time for 10 pulses is:

$$10T = 14 \text{ s}$$

Thus, the *average* pulse period is:

$$T = \frac{14 \text{ s}}{10} = 1.4 \text{ s}$$

Duty Cycle: Another term that describes a pulse train is the duty cycle. The **duty cycle** is a measure of the percentage of the total time that a pulse is on—that is, while the voltage is high. It is simply the pulse width d divided by the period T expressed as a percentage:

$$\text{duty cycle} = \frac{d}{T} \times 100\% \qquad (4)$$

EXAMPLE 4: What is the duty cycle of Figure 7.21?

Solution: From the figure, the pulse width d and the pulse period T can be estimated as:

$$d = 0.05 \text{ s}$$
$$T = 0.5 \text{ s}$$

By using Equation 4, the duty cycle is:

$$\text{duty cycle} = \frac{d}{T} \times 100\%$$

$$= \frac{0.05 \text{ s}}{0.5 \text{ s}} \times 100\% = 10\%$$

Thus, the voltage is on 10% of the time.

Because the fraction of on time is often of primary interest in a pulse train, the pulse width of a pulse train is sometimes called the **on time** and expressed as t_{on}. Note also that because both the pulse width and the pulse period are measured in the same units, a ruler measurement can be used instead of the time axis.

EXAMPLE 5: Estimate the duty cycle for the pulse train in Figure 7.22.

Solution: A ruler measurement of the period T gives 4.0 cm. The pulse width d measures 3.0 cm. Therefore, the duty cycle is, by Equation 4:

$$\text{duty cycle} = \frac{3.0}{4.0} \times 100\% = 75\%$$

Average Voltage: Frequently, a piece of electrical equipment cannot respond fast enough to keep up with a rapidly varying signal. Rather, it responds to the *average* value of the signal just as if it were a constant dc voltage. Hence, the average voltage is often designated V_{dc} to indicate that it is equivalent to a dc voltage of that value.

EXAMPLES 4 AND 6

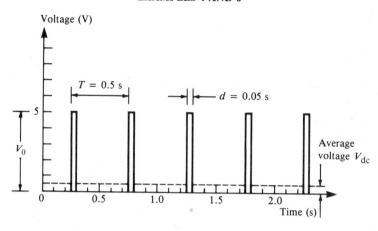

FIGURE 7.21. The duty cycle is the percentage of the time that the pulse is on. Thus, it is the pulse width divided by the period expressed as a percentage.

EXAMPLES 5 AND 7

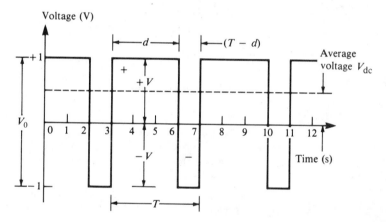

FIGURE 7.22. The average voltage of a pulse train V_{dc} is the value measured by a dc voltmeter. It is the area of the pulse divided by the period, where pulses above the time axis represent positive areas and pulses below the time axis represent negative areas.

The **average voltage** V_{dc} of a pulse train is the area of the signal voltage divided by the period T:

$$V_{dc} = \frac{\text{pulse area}}{T} \qquad (5)$$

In determining pulse areas, *we must take into account the sign of the voltage*. Thus, negative pulses represent a negative area and positive pulses represent a positive area.

EXAMPLE 6: What is the average voltage of the pulse train in Figure 7.21?

Solution: The area of the pulses in Figure 7.21 is the pulse width d times the pulse height V_0:

$$\text{pulse area} = dV_0 = 0.05 \text{ s} \times 5 \text{ V}$$
$$= 0.25 \text{ V-s}$$

Therefore, by Equation 5, the average voltage V_{dc} is:

$$V_{dc} = \frac{\text{pulse area}}{T} = \frac{0.25 \text{ V-s}}{0.5 \text{ s}} = \textbf{0.5 V}$$

This value is indicated by a dotted line in Figure 7.21. If the pulse train in Figure 7.21 is measured by a dc voltmeter, the voltmeter will read 0.5 V.

EXAMPLE 7:   What is the average voltage of the pulse train in Figure 7.22?

Solution: There are two pulse areas to calculate. The first is the area of the positive voltage pulse, for which $+V = +1$ V and $d = 3.0$ s:

$$\begin{aligned}(+\text{pulse area}) &= +V \times d \\ &= 1 \text{ V} \times 3.0 \text{ s} = 3.0 \text{ V-s}\end{aligned}$$

The second is the area of the negative voltage, for which $-V = -1$ V and the time is $(T - d) = 1.0$ s:

$$\begin{aligned}(-\text{pulse area}) &= -V \times (T - d) \text{ s} \\ &= -1 \text{ V} \times 1.0 \text{ s} \\ &= -1.0 \text{ V-s}\end{aligned}$$

Therefore, the total pulse area is:

$$\begin{aligned}\text{total pulse area} \\ &= (+\text{pulse area}) + (-\text{pulse area}) \\ &= (3.0 - 1.0) \text{ V-s} \\ &= 2.0 \text{ V-s}\end{aligned}$$

The period is 5 s, so the average voltage is:

$$V_{dc} = \frac{\text{pulse area}}{T} = \frac{2.0 \text{ V-s}}{4.0 \text{ s}} = \textbf{0.50 V}$$

Again, this value is shown with a dotted line on the graph and is the value that would be read by a dc voltmeter.

PULSE DESCRIPTIONS

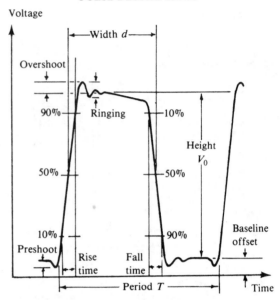

FIGURE 7.23.   This figure summarizes the terms used to describe pulse characteristics and illustrates deviations from the ideal behavior.

Note that the pulse height V_0 is the total height of the pulse, regardless of whether it is $+$ or $-$. Note also that if the area of negative voltage equals the area of positive voltage—as it does for the square, sine, triangle, and sawtooth pulses in Figures 7.10, 7.11, 7.12, and 7.14—then the average dc voltage is zero. A dc meter connected to any of these waveforms would read 0 V.

Other Pulse Characteristics: An ideal pulse is never achieved experimentally. Changes in voltage levels cannot change instantaneously, and the top of an actual pulse may not be flat. The periodic pulse shown in Figure 7.23 summarizes the terms that describe pulses and displays most of the possible deviations from ideal behavior.

Note that the baseline is not at zero volts and that the amplitude—or *pulse height*—is an approximate quantity. The *pulse width* is measured between two half-height points, and the

SINE WAVE VOLTAGE

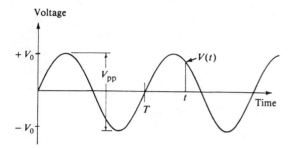

FIGURE 7.24. The most important periodic signal is the sine wave. It can be completely described in terms of two quantities: its peak amplitude V_0 and its period T. Two equivalent terms are its peak-to-peak amplitude $V_{pp} = 2V_0$ and its frequency $f = 1/T$.

times needed for the pulse to rise and fall—the *rise time* and the *fall time*—are measured between the 10% and 90% pulse height points. Other nonideal features, such as *overshoot*, *ringing*, and *preshoot*, are also indicated.

Sinusoidal Signals

Of all the periodic signals, the sine wave is the most common, but its description requires several new terms. A generalized sine wave is shown in Figure 7.24. The sine wave is completely expressed mathematically by Equation 2:

$$V(t) = V_0 \sin \left[2\pi \left(\frac{t}{T} \right) \right]$$

This equation gives the exact voltage value $V(t)$ at any time t after the sine wave has begun.

Evaluating $V(t)$, however, is rarely required, so this procedure is not discussed in detail. On the other hand, the general characteristics and terms that describe sinusoidal waves are extremely important and are used to describe essentially all periodic signals.

Peak Amplitude: The sine wave can be thought of as a train of positive and negative rounded pulses. The pulse height of each pulse is called the **peak amplitude** V_0 (Figure 7.24). In Equation 2, the peak amplitude V_0 precedes the mathematical sine wave expression and determines how high the sine wave will be. The greater the value of V_0, the taller the sine wave.

Peak-to-Peak Amplitude: Determining the peak amplitude requires knowing the midpoint of the wave at 0 V. This point is sometimes difficult to establish exactly, particularly in measurement.

An easier technique is to measure the total voltage excursion of the signal from $-V_0$ to $+V_0$, called the **peak-to-peak amplitude** V_{pp}. For a sine wave, V_{pp} is just twice the peak amplitude V_0:

$$V_{pp} = 2V_0 \qquad (6)$$

The peak-to-peak amplitude is particularly useful for signals that are not sinusoidal and for which a specific amplitude may not be well defined. For example, for the periodic signal in Figure 7.9, there is no quantity that corresponds to V_0 for a sine wave. However, a peak-to-peak amplitude, defined as the maximum voltage change of the signal, is a useful quantity for describing this wave.

Period: The **period** T for any periodic signal is the time required for the waveform to repeat itself. The shape of the waveform does not matter, just as long as the complete shape is repeated.

As for a pulse train, the period can be measured between any two points on successive cycles of the wave provided that the same point is taken. In Figure 7.24, T is measured between successive zero crossings when the voltage is changing from negative to positive, but any other point could be used.

In Equation 2, T appears in the denominator of t/T. Thus, it establishes how frequently the sine wave repeats itself in time. When the time t on the time axis increases by an amount T, the sine wave will have gone through one complete cycle, and the voltage $V(t)$ will be back to its initial value.

Another way of indicating this periodicity mathematically is to note that when t increases by an amount T, $2T$, $3T$,..., t/T increases by a whole number 1, 2, 3,.... Thus, the argument of the sine function $(2\pi t/T)$ increases by 2π, 4π, 6π,.... A characteristic of the sine function is that whenever the argument increases by 2π, a complete cycle takes place.

In summary, *a sinusoidal signal is completely described by two quantities: the peak amplitude V_0 and the period T.* Thus, any additional quantities used to describe a sine wave, such as Equation 6, must be expressed in terms of V_0 and T.

Frequency: A term equivalent to *period* that expresses how often a periodic signal repeats itself is *frequency*. The **frequency** f is simply the reciprocal of the period T:

$$f = \frac{1}{T}$$

Rather than describing the time between successive cycles, the frequency tells how many complete cycles occur in a unit of time. If the period is measured in seconds, then the frequency is measured in cycles per second, or **hertz** (Hz). Thus, a frequency of 10 Hz means that 10 complete waves occur in 1 s.

EXAMPLE 8: Calculate the periods that correspond to the following frequencies: 1 Hz, 10 Hz, 0.2 Hz, and 1000 Hz.

Solution: Since f (Hz) = $1/T$ (s),

$$T = \frac{1}{f}$$

If $f = 1$ Hz, $T = \dfrac{1}{1} = \mathbf{1\ s}$

If $f = 10$ Hz, $T = \dfrac{1}{10} = \mathbf{0.1\ s}$

If $f = 0.2$ Hz, $T = \dfrac{1}{0.2} = \mathbf{5\ s}$

If $f = 1000$ Hz, $T = \dfrac{1}{1000} = 0.001\ \text{s}$

$\qquad = \mathbf{1\ ms}$

Equation 1 can be substituted into Equation 2 to give an alternative form for the sine wave voltage:

$$V(t) = V_0 \sin(2\pi ft) \tag{7}$$

Average Voltage: The average voltage of a sine wave, as shown in Figure 7.25a, is zero because the wave is positive as much as it is negative. However, consider the wave shown in Figure 7.25b. This wave is called a **half-wave rectified sine wave** — one for which the negative values are missing. The average voltage for this wave is not zero. Like the average voltage for pulses, it is the area of the half–sine pulse divided by the period. Calculation of the area of a half–sine wave is somewhat complicated, but gives the simple result:

$$V_{\text{dc(half-wave)}} = \frac{1}{\pi}\, V_0 = 0.318 V_0 \tag{8}$$

Rather than eliminating the negative part of the sine wave, we can invert it and make it positive, as shown in Figure 7.25c. This wave is a **full-wave rectified sine wave**, and it has twice

AVERAGE VALUE

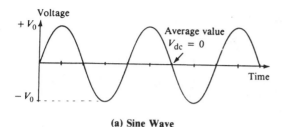

(a) Sine Wave

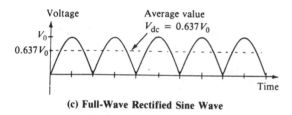

(b) Half-Wave Rectified Sine Wave

(c) Full-Wave Rectified Sine Wave

FIGURE 7.25. The average, or equivalent dc, voltage of a sine wave is 0 because it is positive as much as it is negative. For a half-wave rectified sine wave, in which the positive half-cycle is eliminated, $V_{dc} = 0.318V_0$. For a full-wave rectified sine wave, in which the negative half-cycle is inverted, $V_{dc} = 0.637V_0$.

the area of the half-wave signal. Therefore, its average value is:

$$V_{dc(full-wave)} = \frac{2}{\pi} V_0 = 0.637V_0 \qquad (9)$$

Both of these rectified sine waves are very important in designing sources of dc voltages. They are treated in much more detail later in Chapter 14.

Any wave that has a nonzero average is said to have a *dc component*, or a *dc offset*. Practically speaking, the dc offset is the value that would be read by a dc meter that cannot respond to the rapid variations of most time-varying signals.

rms Voltage: There is one additional term describing sine waves that is used to calculate the electrical power they produce. The electrical power is the product of the electrical current and the electrical voltage. Thus, we need an expression for the electrical current that results from a sine wave electrical voltage. This expression can be obtained from Ohm's law that states that if the voltage $V(t)$ varies with time, so will the current $I(t)$:

$$I(t) = \frac{V(t)}{R} \qquad (10)$$

Substituting in the expression for $V(t)$, Equation 7, gives:

$$I(t) = \frac{V_0}{R} \sin (2\pi ft)$$

$$I(t) = I_0 \sin (2\pi ft) \qquad (11)$$
where:
$$I_0 = \frac{V_0}{R} \qquad (12)$$

as the maximum amplitude of the current. Figure 7.26 shows $V(t)$ and $I(t)$ for a given value of R. Each point on the $I(t)$ curve can be obtained from the corresponding point on the $V(t)$ curve by dividing by R. Note also that the frequency of $I(t)$ is the same as that of $V(t)$.

The electrical power $P(t)$ is the product of $V(t)$, Equation 7, and $I(t)$, Equation 11:

$$\begin{aligned} P(t) &= V(t)I(t) \\ &= V_0 \sin (2\pi ft)I_0 \sin (2\pi ft) \\ &= V_0I_0 [\sin (2\pi ft)]^2 \qquad (13) \end{aligned}$$

SINE WAVE CURRENT

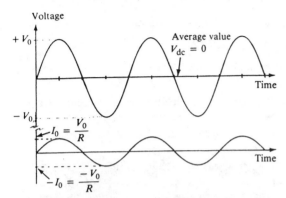

FIGURE 7.26. The current produced by a sinusoidal voltage is also sinusoidal at the same frequency as the voltage. Its peak amplitude I_0 is determined by the circuit resistance R according to Ohm's law, $V_0 = I_0 R$.

Using Equation 10, we can alternatively express Equation 13 as:

$$P(t) = \frac{V(t)^2}{R} = \frac{V_0^2}{R} [\sin (2\pi ft)]^2 \qquad (14)$$

or

$$P(t) = I(t)^2 R = \frac{I_0^2}{R} \sin (2\pi ft)^2 \qquad (15)$$

just as for the equivalent dc expressions.

According to Equations 13, 14, and 15, the electrical power also varies with time, but by the *square* of a sine wave. As with $V(t)$, evaluating $P(t)$ is rarely required. More important is the average or equivalent dc power P_{dc} supplied to a load of resistance R. Using Equation 14, for example, we must evaluate:

$$P_{dc} = [P(t)]_{dc} = \frac{1}{R} [V(t)^2]_{dc} \qquad (16)$$

Thus, we need the average value of the following curve:

$$V(t)^2 = V_0^2 [\sin (2\pi ft)]^2 \qquad (17)$$

Equation 17 is graphed in Figure 7.27.

SINE SQUARED VOLTAGE

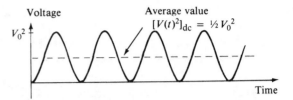

FIGURE 7.27. The electrical power produced by a sinusoidal voltage varies as the square of the time-varying voltage. The average power is thus proportional to the average value of $V(t)^2$. As shown, $[V(t)^2]_{dc} = \frac{1}{2} V_0^2$.

Note that the sine-squared expression is always positive and that it is at twice the frequency of $V(t)$ and $I(t)$ (Figure 7.26).

Calculation of the average value of the sine-squared curve in Figure 7.27 is somewhat complicated, but it gives the simple result:

$$[V(t)^2]_{dc} = \frac{1}{2} V_0^2 \qquad (18)$$

This result can be seen from Figure 7.27 to be just half the amplitude of the wave, as is expected because the sine-squared wave is symmetric about the half-amplitude.

The quantity $[V(t)^2]_{dc}$ is called the **mean square amplitude** because it is the mean, or average value, of the squared amplitude. More common, however, is the square root of this value, called the **root mean square amplitude** (rms). In equation form, using Equation 18:

$$V_{rms} = ([V(t)^2]_{dc})^{1/2} \qquad (19)$$

Substituting in Equation 19 gives the simple result:

$$V_{rms} = \left(\frac{1}{2} V_0^2 \right)^{1/2} = \frac{1}{\sqrt{2}} V_0$$

$$\boxed{V_{rms} = 0.707 V_0 \qquad (20)}$$

In a similar manner, we can find:

EXAMPLE 9

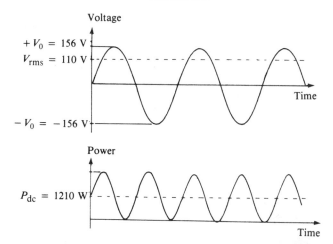

FIGURE 7.28. The equations for the electrical power produced by a sine wave are the same as for dc power if root mean square (rms) values of voltage and current are used. They are simply 0.707 times the peak value. The line voltage specification of 110 V is an rms value to make power calculations easier. The peak value is actually 156 V.

$$I_{\text{rms}} = 0.707 I_0 \qquad (21)$$

In terms of V_{rms} and I_{rms}, the average electrical power P_{dc} is given by any of the following expressions:

$$P_{\text{dc}} = V_{\text{rms}} I_{\text{rms}} = \frac{1}{2} V_0 I_0 \qquad (22)$$

$$P_{\text{dc}} = \frac{V_{\text{rms}}^2}{R} = \frac{1}{2} \frac{V_0^2}{R} \qquad (23)$$

$$P_{\text{dc}} = I_{\text{rms}}^2 R = \frac{1}{2} I_0^2 R \qquad (24)$$

If I_0 is expressed in amperes, V_0 in volts, and R in ohms, then P_{dc} is in watts.

Many sinusoidal signals, particularly the ac voltages that come from the power company, are expressed by their rms value rather than their peak amplitude V_0. For example, the 110 V line voltage is an rms value. This is because the electrical power delivered to an electrical device is of primary importance in electricity.

To indicate that a voltage or current is an rms value, the letters *rms* should be added—for example, 6 V rms or 3 A rms.

EXAMPLE 9: What are the peak and peak-to-peak values of a 110 V rms sine wave voltage? What is the electrical power delivered to an electric heater with a resistance of 10 Ω?

Solution: The rms voltage is:

$$V_{\text{rms}} = 110 \text{ V rms}$$

Using Equation 20 to determine the peak value V_0:

$$V_0 = \frac{V_{\text{rms}}}{0.707} = \frac{110 \text{ V rms}}{0.707} = \textbf{156 V}$$

The peak-to-peak voltage from Equation 6, is just twice this value:

$$V_{\text{pp}} = 2V_0 = 2 \times 156 \text{ V} = \textbf{312 V}$$

The peak-to-peak value of a 110 V line voltage is 312 V. The power delivered to a 10 Ω heater is given by Equation 23:

$$P_{\text{dc}} = \frac{V_{\text{rms}}^2}{R} = \frac{(110)^2}{10} = \textbf{1210 W}$$

Figure 7.28 shows both the voltage and the power for Example 9.

SLEW RATE

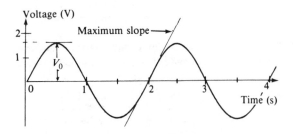

FIGURE 7.29. The final term used to describe time-varying signals is the slew rate. This is the rate at which the signal is changing at a particular instant in time in V/s. The slew rate at a given time is the slope of the signal at that time, and for a sine wave, it continually changes.

As a final note, it must be emphasized that the equations for the root mean square values *apply only to sine waves*. While rms values can be determined for other types of waves, the results will be different from Equations 20 to 24. Also, meters used to measure rms values are generally calibrated for sine waves and can only be used for sine waves.

Slew Rate: One final term used to describe time-varying signals is particularly important when considering an instrument to measure a given signal. The **slew rate** is the rate at which a signal changes. If the time-varying signal is a voltage, then the slew rate is measured in **volts per second** (V/s).

Graphically, the slew rate is the slope of the time-varying signal at a particular time. The slope is a line drawn tangent to the signal at a given point. As shown in Figure 7.29, the slope of the signal may change as the signal changes, so it is not a single value. The slew rate must be specified for a given point on the wave.

Generally, we want to know the maximum slew rate because this rate determines what instrument we must use to measure it. As shown in Figure 7.30, the slope can vary from zero (horizontal) to infinity (vertical) and can be

CALCULATING SLEW RATE

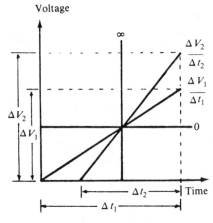

(a) Positive Slopes

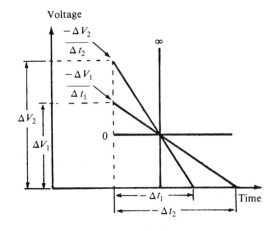

(b) Negative Slopes

FIGURE 7.30. The slew rate at a given point on a curve can be calculated from the slope at that point as the difference in voltage ΔV divided by the corresponding difference in time Δt. The slew rate can vary from zero (horizontal slope) to infinity (vertical slope) and be either positive or negative.

either positive or negative. For a perfect square wave, the slope (and, therefore, the slew rate) is only zero or infinity. No real signal, however, can have an infinite slew rate, so finite values can always be determined.

EXAMPLE 10

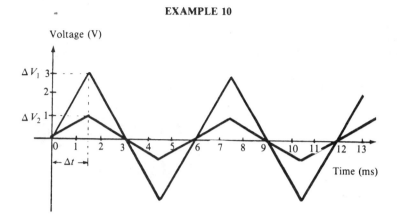

FIGURE 7.31. The triangle wave has a constant slope that alternates between positive and negative values. It can be simply calculated from the leading (or trailing) edges. At a given frequency, the slew rate of a triangle wave increases as the amplitude increases.

The simplest curves for determining slew rate are triangle waves because they have only two equal but opposite slopes (Figure 7.31). The calculation of slew rate from the slope is simply the change in amplitude in volts ΔV divided by the corresponding time in seconds required to produce the change Δt:

$$\text{slew rate} = \frac{\Delta V}{\Delta t} \qquad (25)$$

EXAMPLE 10:   What are the slew rates for the two triangular waves in Figure 7.31?

Solution: The leading edge of the taller triangular wave increases by $\Delta V_1 = 3$ V in a time of $\Delta t = 1.5$ s. Thus, the slew rate, according to Equation 25, is:

$$\text{slew rate} = \frac{\Delta V_1}{\Delta t} = \frac{3.0 \text{ V}}{1.5 \text{ s}} = 2 \text{ V/s}$$

The leading edge of the shorter triangular wave increases by $\Delta V_2 = 1$ V during the same time of $\Delta t = 1.5$ s. Thus, its slew rate, by Equation 25, is:

$$\text{slew rate} = \frac{1 \text{ V}}{1.5 \text{ s}} = 0.67 \text{ V/s}$$

The slew rates for the falling edges of each cycle are, of course, the negatives of these values. Note also that even though the frequency of the two waves is the same, the slew rates are quite different.

As indicated previously, it is generally the maximum slew rate that is of concern. For triangular waves that have only one slope, that value is also the maximum value. For sine waves, however, the value continually varies (Figure 7.32). Its maximum value occurs when it crosses the 0 V axis, where it is given by the equation:

$$\text{slew rate (sine wave max)} = 2\pi V_0 f \qquad (26)$$

Note that the slew rate depends on, and only on, both V_0 and f. The greater the peak amplitude, the greater the maximum slew rate. Also, the higher the frequency, the greater the maximum slew rate. This dependency can easily be seen in Figure 7.32.

FIGURE 7.32. When selecting a measuring instrument for a time-varying signal, it is generally the maximum slew rate that is important. It determines whether the measuring instrument can keep up with the changing voltage. For a sine wave, the maximum slew rate occurs when the signal crosses 0 V and is given by $2\pi V_0 f$.

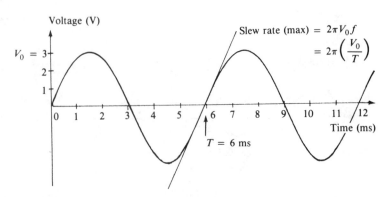

EXAMPLE 11

EXAMPLE 11: What is the maximum slew rate of the sine wave in Figure 7.32?

Solution: The peak amplitude of the sine wave in Figure 7.32 is seen to be:

$$V_0 = 3.0 \text{ V}$$

The period of the sine wave is:

$$T = 6.0 \text{ ms}$$

Thus, the frequency is:

$$f = \frac{1}{T} = \frac{1}{0.006 \text{ s}} = 167 \text{ Hz}$$

By Equation 26, the maximum slew rate is:

slew rate (sine wave max)
$$= 2\pi V_0 f = 2\pi (3.0 \times 167 \text{ Hz})$$
$$= 3150 \text{ V/s} = \textbf{3.15 V/ms}$$

This value can also be determined graphically using the method in Example 10.

7.4 GENERATING TIME-VARYING SIGNALS

Physical Sources

Time-varying signals can be produced in many different ways. In most measurement applications, the signals come from a physical source, such as the temperature of an object. A transducer is connected to the physical source to convert a physical change, such as temperature, into an electrical change (voltage, current, or resistance) that appears as a time-varying signal.

It is beyond the purpose of this book to describe all of the ways in which physical changes can be made to produce time-varying signals. Several examples were given in Figures 7.3, 7.5, 7.6, and 7.7, and throughout this book, other examples will be described.

Function Generator

This section describes another way of generating electrical time-varying signals using a very versatile piece of laboratory equipment called a **function generator**. Basically, the job of a function generator is to produce time-varying signals of different shapes over a wide range of frequencies. These signals are used in the laboratory to test the performance of electrical circuits before they are used in measurement. They are basic electronic test gear and are used constantly in the remainder of the book.

A typical function generator is shown in Figure 7.33, along with its electronic symbol. There are four basic controls: waveform selection, amplitude, frequency, and dc offset, in

FUNCTION GENERATOR AND SYMBOL

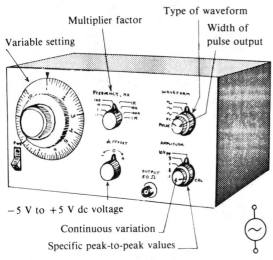

−5 V to +5 V dc voltage

Continuous variation

Specific peak-to-peak values

FIGURE 7.33. A function generator is a basic laboratory test instrument that can produce a variety of time-varying signals for testing circuit performance. The figure shows a typical function generator and its basic controls: WAVEFORM selection, peak-to-peak AMPLITUDE setting, FREQUENCY selection, and DC OFFSET capability.

addition to the on/off power switch and the output terminals.

Waveform Selection: Located in the upper right-hand corner is the WAVEFORM control consisting of two knobs—an inner one and an outer one. The outer knob selects the type of waveform output. The model shown provides five options: square wave, sine wave, triangular wave, dc, and pulses.

In the DC switch position, no time-varying signal is generated. A variable dc voltage can be produced using the OFFSET control to be described. In the PULSE position, the generator provides pulses whose width can be varied by turning the inner knob of the waveform control.

Amplitude: The AMPLITUDE control also consists of an inner and an outer knob. The outer knob selects specific values of output amplitude

ranging from $0.1V_{pp}$ to $10.0V_{pp}$. Note that they use *peak-to-peak* values.

Intermediate values of peak-to-peak amplitude can be selected by turning the inner knob. The full clockwise (CW) position is a switch lock that selects calibration (CAL). In this position, the peak-to-peak output voltage is internally calibrated to give the values set by the outer knob. However, turning this knob counter clockwise (CCW) releases the lock and permits continuous variation from the selected amplitude down to $0V_{pp}$.

Frequency: The FREQUENCY control consists of two separate knobs: a continuously variable dial and a multiplier switch. The variable dial permits very accurate setting of the frequency from 1 to 10. The multiplier switch sets the range of the frequency, indicating the factor of 10 to be multiplied by the variable setting. Eight frequency multipliers are possible from 0.1 to 1,000,000 (1M). As one example, the frequency for the setting shown in Figure 7.33 is:

$$\begin{array}{ccc} \text{dial} & & \\ \text{setting} & \text{multiplier} & \\ 1.0 & \times \quad 100 & = 100 \text{ Hz} \end{array}$$

Direct Current Offset: The DC OFFSET control permits the addition of a constant dc voltage to the output waveform. In the model shown, this voltage is continually adjustable from −5 V to +5 V. The dc offset essentially shifts the baseline from ground to positive or negative dc levels. With the WAVEFORM control set to dc, the OFFSET control permits the function generator to be used as a variable ±5 V dc power supply.

Output Terminals: The output of the function generator is via the OUTPUT terminal, shown at the lower right in Figure 7.33. The connection shown is the BNC type, in which the center jack carries the signal and the outer metal sleeve is ground. BNC connectors are a common and simple connector used by test instruments.

Note also that the OUTPUT terminal is marked 50 Ω. This value refers to the **output re-**

OUTPUT RESISTANCE

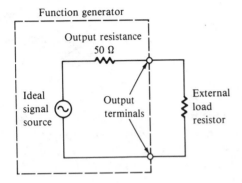

FIGURE 7.34. Connected to a test circuit, the function generator can be modeled as an ideal signal source in series with an internal 50 Ω output resistor. If the external load resistance is large, the 50 Ω output resistance can be neglected, but if it is on the order of 50 Ω, then the circuit must be considered.

sistance of the function generator. The output resistance is the internal resistance of the function generator's output circuit. It can be thought of as a 50 Ω resistor in series with an ideal signal source (Figure 7.34).

The setting of the AMPLITUDE control determines the amplitude of the ideal signal source that is supplied both to the 50 Ω resistor and to a load connected in series at the output.

If the resistance of the load is much larger than 50 Ω, then the function generator's output resistance can be ignored and essentially all of the output voltage will appear across the load. If the load resistor is small (on the order of 50 Ω), however, the output resistance is significant and must be considered. The circuit model in Figure 7.34 can be used to determine the actual output voltage that will appear across the load in this case.

The function generator controls discussed here are found in one form or another on essentially all modern instruments. The names vary, however. For example, "AMPLITUDE" may be called "GAIN," and the ranges of possible waveforms, amplitudes, and frequencies may vary.

The controls may also have additional capabilities, such as frequency sweep and voltage-controlled oscillation. **Frequency sweep** means that the output frequency can be made to change continuously (or sweep) from one value to another. **Voltage-controlled oscillation** (VCO) means that an externally applied voltage can be used to modulate the output frequency. These capabilities are beyond the present discussion, however, and will be described when they are needed.

7.5 MEASURING TIME-VARYING SIGNALS

In the discussion of time-varying signals thus far, it was taken for granted that there are ways of observing them. Indeed, several classes of instruments perform the function. The question is: For a particular signal, what instruments should be selected?

Type of Signal

The basic problem is to choose an instrument that is well matched to the characteristics of the incoming signal. While the incoming signal can represent changes in voltage, current, or resistance, in most cases, the signals are voltages and the majority of instruments are designed to measure them (Figure 7.35). Technically, this means that the input circuitry of the instrument responds to voltage changes and that its input resistance is very large, thus adding a negligible load to the signal source. The examples in this section deal exclusively with voltage-measuring equipment.

When the signal is a current, special current-sensitive circuits can be used; or, more commonly, the current can be converted to an equivalent voltage. Ohm's law is always valid

TYPES OF TIME-VARYING SIGNALS

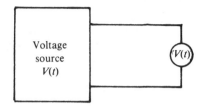

(a) Variable Voltage: $V(t) = V(t)$

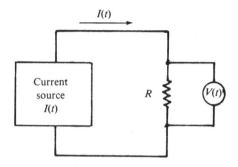

(b) Variable Current: $I(t) = \dfrac{V(t)}{R}$

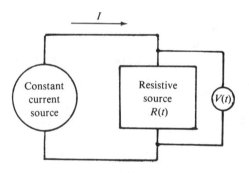

(c) Variable Resistance: $R(t) = \dfrac{V(t)}{I}$

FIGURE 7.35. A signal source can produce a time-varying voltage, current, or resistance. Generally, a voltage-sensitive measuring instrument can be used to measure all three. For a time-varying current $I(t)$, the voltmeter measures the voltage across a fixed resistor. For a time-varying resistance $R(t)$, an external constant-current source is applied and the voltage measured.

for resistors, so any current I can be passed through a known resistor R to produce a voltage V. The voltage is then measured, and the current can be calculated from $I = V/R$.

Similarly, if changes in resistance must be measured, the current through it can be set by a constant-current source. A variable resistance will then lead to a variable voltage across it that can be measured with a voltage-sensitive instrument.

Assuming that the measuring instrument is voltage sensitive, two primary signal factors—amplitude and frequency—determine which of several instruments should be used to measure it. A signal with a peak-to-peak amplitude of 500 V requires a different instrument from one designed to measure amplitudes of a few millivolts. Similarly, signals that change within a fraction of a second are measured with instruments that may be unsuitable for signals that change over periods of hours or days.

Amplitude and Range

When discussing signals and how they are measured, we must keep in mind two sets of characteristics: those that describe the signal and those that describe the measuring instrument. For example, the size of a signal is described by its **amplitude**. Correspondingly, the largest amplitude signal that an instrument can measure is called its **range**, or its full-scale sensitivity. Thus, a signal with an 8 V amplitude requires an instrument with a full-scale range slightly larger than 8 V. A range of 10 V is suitable, but ranges of 1 V or 100 V are not.

Most measuring instruments are equipped with a switch to select the proper range for a specific signal (Figure 7.36). Values of full-scale sensitivity from 10 mV to 10 V are common in most voltmeters. More expensive instruments have ranges above and below these values. However, equipment that is designed to operate at either high or low voltage cannot usually be used in the other range.

FIGURE 7.36. When selecting a voltmeter to measure the voltage from a signal source, make certain that its range is slightly greater than the maximum amplitude of the signal source. The full-scale ranges of most common voltmeters vary from about 0.01 V to 100 V.

SIGNAL AMPLITUDE-INSTRUMENT RANGE

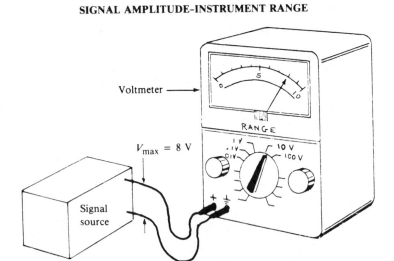

Slew Rate and Response Time

For a signal that varies with time, the way a signal changes can be just as important as its amplitude. In order to measure a time-varying signal, the instrument must be able to respond to the changes as they occur.

The speed with which a signal changes is determined by its **slew rate**. On the other hand, the ability of an instrument to respond to voltage changes is determined by an important characteristic called its **response time**. If a perfect step signal (slew rate = infinity) is applied to the input of an instrument, the instrument cannot instantaneously respond to the change in voltage. The response time is a measure of how long it actually takes the instrument to react to the incoming signal.

By convention, the response time is defined as the time required for an instrument to reach 90% of its final full-scale value when supplied with a full-scale step signal. For example, if a 1 V step signal is applied to an instrument with a full-scale sensitivity of 1 V, the response time is the time required for the instrument to reach 0.9 V (Figure 7.37).

The response time for an instrument determines the maximum slew rate that it can accu-

rately measure. Comparison of values, however, requires some care. Note that instrument response time is in seconds, while signal slew rate is in V/s. An approximate comparison can be made by converting the instrument response time into an equivalent "instrument slew rate." Generally, only an order of magnitude comparison is necessary, so an approximate value for instrument slew rate is sufficient.

An approximate value of **instrument slew rate** is the full-scale sensitivity divided by the response time:

$$\text{instrument slew rate} = \frac{\text{full-scale sensitivity (V)}}{\text{response time (s)}} \quad (27)$$

As a general rule of thumb,

The slew rate of a measuring instrument should be at least two times greater than the maximum slew rate of the signal to be certain that the signal is being accurately recorded.

For example, if the maximum slew rate of a sig-

SIGNAL SLEW RATE–INSTRUMENT RESPONSE TIME

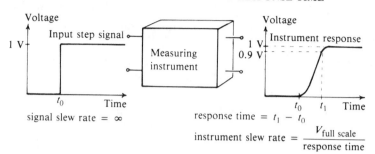

FIGURE 7.37. A second factor to consider when selecting a voltmeter is its response time. Instrument response time is the time required to reach 90% of its full-scale deflection when provided with a full-scale step input. An equivalent instrument slew rate can be calculated as the full-scale deflection divided by the response time. For an accurate measurement, the instrument slew rate should be two times the signal slew rate.

nal is 1 V/s, then the instrument slew rate must be at least 2 V/s.

EXAMPLE 12: What is the maximum slew rate that can be measured by a voltmeter with a response time of 0.5 s on the 1 V scale? On the 10 V scale?

Solution: On the 1 V scale, the instrument slew rate can be calculated by using Equation 27:

$$\text{instrument slew rate} = \frac{1 \text{ V}}{0.5 \text{ s}} = 2 \text{ V/s}$$

The maximum signal slew rate is one-half this value:

$$\text{signal slew rate (max)} = \textbf{1 V/s}$$

On the 10 V scale, the instrument slew rate, according to Equation 27, is:

$$\text{instrument slew rate} = \frac{10 \text{ V}}{0.5 \text{ s}} = 20 \text{ V/s}$$

Thus the maximum signal slew rate is:

$$\text{signal slew rate (max)} = \textbf{10 V/s}$$

Note that a meter's response time is independent of the scale setting, so the instrument slew rate increases with the range setting. Therefore, if a signal cannot be measured on one range setting, it may be measurable on a higher setting, provided, of course, that the signal amplitude can be increased sufficiently to be measured on the higher range.

The two-times-greater instrument slew rate can sometimes be reduced, depending on what signal information is important. For example, in the conveyor belt example (Figure 7.6), only the existence of a pulse is important. Hence, an instrument that only partially responds to the full amplitude of the pulse signal may be adequate to activate a counter or other recording instrument. On the other hand, for the camera example of Figure 7.4, for which an accurate measurement of amplitude is required, the measuring system must be able to respond fast enough to record the full signal amplitude.

Another term that is often used to describe the response time of a measuring system is its frequency response. The **frequency response** is a measure of the maximum frequency sine wave that the instrument can record. This term is used because sine waves are so common in electronics that they provide a reference standard for other types of time-varying signals.

The frequency response of an instrument can be expressed in terms of an equivalent instrument slew rate using the maximum slew rate expression for a sine wave (Equation 26). The sine wave frequency f is replaced by the instrument's frequency response, and V_0 is replaced by the instrument's full-scale voltage:

instrument slew rate

$$= 2\pi V_{\text{full scale}}$$
$$\times \text{ frequency response} \qquad (28)$$

EXAMPLE 13: If a voltmeter has a frequency response of 90 Hz, what is the maximum slew rate signal it can measure on the 100 mV scale?

Solution: The instrument slew rate can be calculated using Equation 28:

instrument slew rate

$$= 2\pi \times 0.1 \text{ V} \times 90 \text{ Hz} = 57 \text{ V/s}$$

The maximum signal slew rate is one-half this value, or:

signal slew rate (max) = **30 V/s**

7.6 TYPES OF MEASURING INSTRUMENTS

The output of almost all electronic measuring circuits is a voltage. The device that records this voltage, so that the system operator can determine the measured quantities, is generally the most expensive part of the system. For voltages that are constant dc values, a voltmeter—either the D'Arsonval or digital type—is adequate. But even those devices are expensive relative to the cost of electronic circuit components and can often dominate the cost of the overall system.

For measuring time-varying signals, a meter is quite limited because its response time is relatively slow. Instruments of considerably greater response time are required for most applications. The most common instrument used for measuring time-varying signals is the cathode-ray oscilloscope (CRO). The CRO, the multimeter, and function generator are the minimum essential instruments for studying electronic and instrumentation systems.

The following paragraphs describe the operation of a CRO, along with other common but more specialized devices, such as the chart recorder. The choice of an instrument to make a given measurement will often be determined by its response time compared to the maximum slew rate of the signal.

Figure 7.38 shows typical values of response time for various instruments. Models with response times faster than the values given are available, but they are considerably more expensive.

In addition to slew rate and response time, there is another major difference between dc measurements and the measurement of time-varying signals. When measuring a dc voltage, there is only a single output value. This value may change depending on some input change, but the correlation between input signal and output voltage can generally be recorded in a simple table of values. These data may be graphed for analysis, but a graph is not always essential.

For time-varying signals, however, a graph of voltage versus time is almost always required. The instruments that measure time-varying signals generally are designed to provide such graphs directly. However, dc meters can also be used in some limited applications.

Meters

We generally think of a dc voltmeter as an instrument for measuring steady values of volt-

**BASIC MEASURING INSTRUMENTS
AND THEIR RESPONSE TIMES**

**(a) Meter: D'Arsonval (1 s)
Digital (1 s)**

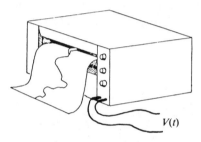

**(b) Chart Recorder: D'Arsonval (0.01 s)
Servo (0.5 s)**

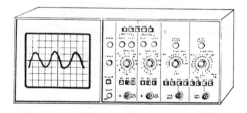

(c) Cathode-Ray Oscilloscope: CRO (0.0000001 s)

FIGURE 7.38. The three basic instruments for measuring time-varying signals are the meter, the chart recorder, and the cathode-ray oscilloscope (CRO). Each instrument is limited by its response time to signals whose slew rates are about a factor of two slower. Thus, meters can be used only for very slowly varying signals, chart recorders for somewhat faster signals, and CROs for essentially all others.

D'ARSONVAL VOLTMETER

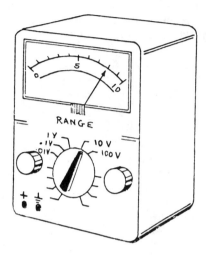

response time ≈ 1 s

For 10 V full-scale deflection:
    instrument slew rate ≈ 10 V/s
    signal slew rate (max) ≈ 5 V/s

FIGURE 7.39. The response time of a D'Arsonval voltmeter is typically 1s. On the 10 V scale, the instrument slew rate is about 10 V/s. Thus, it can measure signals whose slew rates are about 5 V/s or less.

age. However, for slowly varying signals, a meter and a clock can be used to measure a voltage as it changes with time. There is little difference between a dc signal and a time-varying signal that changes very slowly.

The response time of a D'Arsonval voltmeter with a 100 μA meter movement is typically about 1 s (Figure 7.39). Thus, on the 10 V scale, it has an instrument slew rate given by Equation 27:

$$\text{instrument slew rate} = 10 \text{ V/s}$$

On the 10 V scale, then, the meter can accurately measure time-varying voltages whose slew rates are about half this, or about:

$$\text{signal slew rate (max)} = 5 \text{ V/s}$$

FIGURE 7.40. Measuring a time-varying signal with a voltmeter requires a timer, such as a stop watch. When the voltage changes too fast to take the data point by point, you can sometimes make several trials, measuring the time in each trial to reach a different voltage.

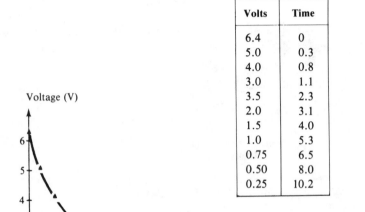

VOLTMETER MEASUREMENT

Volts	Time
6.4	0
5.0	0.3
4.0	0.8
3.0	1.1
3.5	2.3
2.0	3.1
1.5	4.0
1.0	5.3
0.75	6.5
0.50	8.0
0.25	10.2

If the measured voltage changes much slower than this—for example, 1 V in 10 s—it is a simple matter to read the meter at 10 s intervals and record the data. However, if the signal changes at its maximum permissible rate of 5 V in 1 s, it is difficult to read and record the data that rapidly. In this case, however, the experiment can sometimes be altered so that the measurement can be made.

For example, suppose we need a graph of the voltage versus time for some process after a switch is closed, and the entire process lasts for only a few seconds. One method of making this measurement is to repeat the experiment several times, taking a single datum point for each trial. That is, we measure the time to reach a specific voltage during each trial, gradually changing the voltage measured. Many trials provide sufficient data to make a graph of volt-

age versus time. The data plotted in Figure 7.40 were obtained in this way.

When a signal varies too rapidly for a meter or must be studied in more detail than the method illustrated in Figure 7.35 can provide, a simple meter is not equal to the task. Two other ways to record the signal use a chart recorder or a cathode-ray oscilloscope.

Chart Recorders

A **chart recorder** is a dc voltmeter with an added feature: The meter needle deflections are recorded on paper. There are two common types of chart recorders: galvanometric recorders and servo recorders.

A **galvanometric recorder** is a heavy-duty D'Arsonval meter movement with a writing pen attached to the end of the needle (Figure 7.41).

GALVANOMETRIC CHART RECORDER

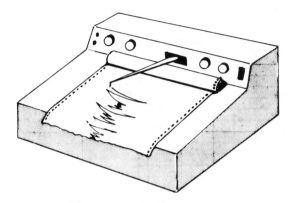

frequency response ≈ 60 Hz to 90 Hz

For 10 V full-scale deflection:
instrument slew rate ≈ 4000 V/s
signal slew rate (max) ≈ 2000 V/s

FIGURE 7.41. When more signal detail is required than a meter can provide, a chart recorder can be used. A galvanometric recorder is essentially a heavy-duty D'Arsonval meter movement with a pen attached. Its frequency response is about 60 Hz to 100 Hz. On the 10 V scale, it can measure signal slew rates up to 2000 V/s.

Chart paper from a roll is pulled past the pen tip at a constant rate that determines the scale of the time axis. With no input signal fed to the chart recorder, the pen makes a continuous line along the base of the paper at 0 V. When a signal is applied to the recorder, the needle deflects and the pen accurately records the voltage as it changes with time.

The response time of a galvanometric recorder is generally given as a frequency, typically 60 to 90 Hz. For a frequency response of 60 Hz and a full-scale deflection of 10 V, the instrument slew rate is given by Equation 28:

instrument slew rate
$$= 2\pi \times 10 \text{ V} \times 60 \text{ Hz} = 4000 \text{ V/s}$$

Thus it can measure voltages whose slew rates are half this, or about:

SERVO CHART RECORDER

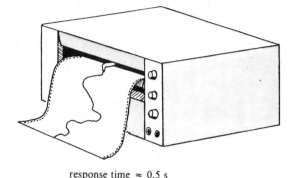

response time ≈ 0.5 s

For a 10 V full-scale deflection:
instrument slew rate = 20 V/s
signal slew rate (max) = 10 V/s

FIGURE 7.42. Another type of chart recorder uses a servomotor. The response time of a servo recorder is comparable to that of a meter (≈ 0.5 s), but it gives more accurate graphs than the galvanometric recorder.

signal slew rate (max) = 2000 V/s

This response time is about 400 times faster than for a meter.

One problem with this type of recording is that the recording is curvilinear, meaning that the pen's movement is a curved arc, while the graph paper lines are linear. Thus, some error is introduced into the recording. For small deflections, however, this error can be neglected.

Galvanometric recorders are typically used where detailed data are not as important as relative changes or peak values. For example, electrocardiogram (ECG) signals in hospitals, such as the ones shown in Figure 7.5b, are made with galvanometric recorders.

A second type of chart recorder is the **servo recorder** that is not curvilinear (Figure 7.42). The circuitry in this instrument converts an input voltage into a pen deflection by means of a servomotor. The design of these instruments makes them more versatile and mechanically

FIGURE 7.43. For time-varying signals above about 100 Hz, a cathode-ray oscilloscope must be used. The model shown can display two time-varying signals simultaneously.

CATHODE-RAY OSCILLOSCOPE

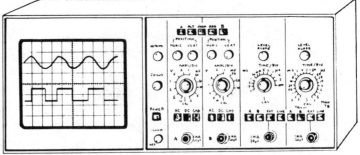

sturdy than the galvanometric type, though their response time is generally not as fast.

Pen deflections, however, are linear, and it is a simple matter to provide a wide range of input voltage sensitivities. Typically, the full-scale sensitivity can be varied from 1 mV to 100 V. Chart speeds may range from 1 cm/h to 100 cm/min. In addition, the 0 V position of the pen can be adjusted to any point on the scale to enable the recorder to measure signals that have both positive and negative values.

The response time of a servo recorder is about the same as that of a meter, typically 0.5 s. Thus, on the 10 V scale, its slew rate is:

$$\text{instrument slew rate} = 20 \text{ V/s}$$

If a shorter response time is essential, a factor of about 20 in signal slew rate can be gained by using a galvanometric recorder. However, if a response time of 0.5 s is not a severe limitation, the versatility and linearity of a servo recorder make it the usual choice for a general laboratory instrument.

Cathode-Ray Oscilloscope

The factor that limits the response time of a meter or a chart recorder is the mass of its moving parts. The pen, which may have a mass of a gram or so, must move quickly across the chart. The smaller the mass, the faster the pen can re-

spond. In a **cathode-ray oscilloscope** (CRO), the "pen" is a beam of electrons, each with a mass of 9.1×10^{-28} g. Clearly, the performance of a CRO is not limited by the electron's mass.

A typical CRO is shown in Figure 7.43. The model shown is a **dual-trace** oscilloscope, which means that two time-varying signals can be viewed simultaneously. Most modern CROs are dual trace because the ability to see two signals—for example, an input and an output voltage—at the same time can be very important.

In contrast to a chart recorder, in which the pen moves back and forth and the time scale is established by pulling the chart paper at a constant speed, only the beam of electrons moves in an oscilloscope. The voltage signal at the input causes the beam to move vertically, and the time axis is produced by moving the beam at a constant speed from left to right. Together, the two motions trace the input voltage as a function of time.

Cathode-Ray Tube: The heart of an oscilloscope is a long evacuated glass tube in which electrons produced at one end cause an image to appear on a glass screen at the other. Electrons are produced, accelerated, and focused in an assembly called an **electron gun.** A schematic drawing of a typical tube is shown in Figure 7.44.

CATHODE-RAY TUBE

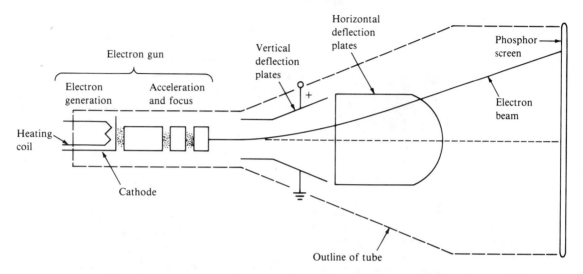

FIGURE 7.44. The basic component of a CRO is the cathode-ray tube (CRT) that produces the signal display. Electrons are produced at the cathode, focused into a beam, and directed to a phosphor screen that emits light when the beam strikes it. The beam can be moved vertically and horizontally by voltages applied to the deflection plates.

When a piece of metal called a **cathode** is heated, it emits electrons. Applying positive voltages between the cathode and the other metal electrodes in the gun assembly accelerates the electrons and focuses them into a narrow beam that is directed toward a screen at the opposite end of the tube. Because the electron beam is produced at the cathode, these glass tubes are appropriately named cathode-ray tubes (CRT).

The inside surface of the end of the CRT is coated with a phosphorous material that emits light when struck by high-energy electrons. Therefore, when the accelerated electron beam strikes the phosphor screen, it forms a small visible spot on the outside surface. The brightness, or intensity, of the spot depends on the number of electrons in the beam. This number is controlled by the INTENSITY control knob on the front panel that varies one of the accelerating voltages (Figure 7.45).

ELECTRON BEAM CONTROLS

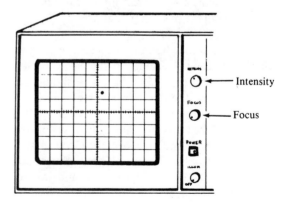

FIGURE 7.45. Two external knobs control the characteristics of the electron beam. The INTENSITY knob adjusts the number of electrons in the beam and thus the spot intensity. The intensity should be just bright enough to make the spot visible. The FOCUS knob determines the sharpness of the spot.

CRT SCREEN AND BEAM DEFLECTION PLATES

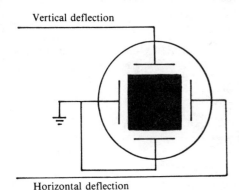

Vertical deflection

Horizontal deflection

FIGURE 7.46. An end view of a CRT shows the deflection plates and the beam spot centered on the oscilloscope screen. A positive voltage applied to the vertical deflection plate will move the spot up, a negative voltage, down. Similarly, a positive voltage on the horizontal deflection plates moves the spot toward the right, and a negative voltage moves it toward the left.

The beam intensity should be set at the lowest level that still gives a good visible image. Because 90% of the electron's energy goes into heat, not light, a too bright image can literally burn through the phosphor coating.

The sharpness of the spot is determined by the FOCUS control knob that adjusts the voltage for the focusing electrodes in the electron gun (Figures 7.44 and 7.45). The focus should be adjusted to give the sharpest possible image.

The position of the electron beam on the phosphor screen is controlled by two sets of deflection plates located between the electron gun and the phosphor screen. These plates are shown in side view in Figure 7.44 and in end view in Figure 7.46.

The first set of plates deflects the beam vertically. A positive voltage applied across these plates moves the beam upward, and a negative voltage moves it downward. The amount the beam moves is directly proportional to the voltage applied to the plates.

The second set of plates deflects the electron beam in the horizontal direction. A negative voltage deflects the beam to the left. A positive voltage deflects it to the right. Again, the deflections are linear.

Without a voltage on either set of plates, the beam will form a small visible spot in the exact center of the screen. By adjusting two separate controls called VERTICAL POSITION and HORIZONTAL POSITION, the beam position can be moved from the center position. The settings of these two knobs control the internal dc voltages that are applied to the vertical and horizontal deflection plates and, hence, determine the beam's position on the screen (see Figure 7.43).

When the electron beam appears as a single spot, the CRO can function as a dc voltmeter. A dc voltage applied to the vertical input deflects the beam vertically. The amount of the deflection is a measure of the input voltage. Adjusting the settings of the POSITION controls simply changes the point from which the spot moves.

If a time-varying signal, such as a sine wave, is applied to the vertical deflection plates, as shown in Figure 7.47, then the spot will move up and down with the frequency of the sine wave. If the frequency is fast enough, the spot moves so rapidly that the eye is unable to follow it, and it appears as a straight line (Figure 7.47). This rapid motion, of course, is not suitable for measurement except for the peak-to-peak value.

In order for an oscilloscope to measure such rapid time-varying signals, the horizontal axis must become a time axis. If the sawtooth voltage shown in Figure 7.48 is applied to the horizontal deflection plates, with the vertical plates grounded, the beam will begin at the left side of the screen when the voltage is negative and move to the right side as the voltage increases to its maximum positive value.

Because the sawtooth voltage increases linearly with time, the beam will move with constant speed from left to right. The horizontal

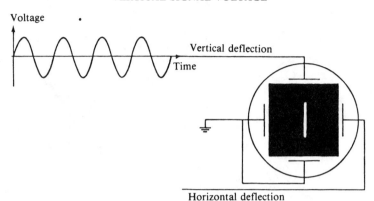

VERTICAL SIGNAL VOLTAGE

FIGURE 7.47. If a sine wave is applied to the vertical deflection plates, the spot will move up and down following the voltage. At frequencies that are sufficiently high (above about 20 Hz), the spot will move so fast that it will appear as a blurred line.

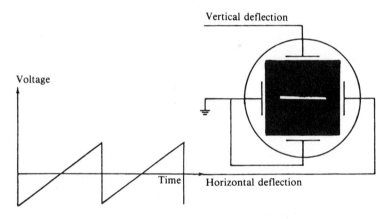

HORIZONTAL SWEEP VOLTAGE

FIGURE 7.48. A sawtooth wave applied to the horizontal deflection plates will cause the beam to move from left to right at a constant speed. The sawtooth wave is called the sweep voltage. It provides a linear time axis for a signal applied to the vertical plates.

axis thus becomes a time axis, with the time determined by the slew rates (V/s) of the sawtooth wave.

The sawtooth voltage in Figure 7.48 is called the **sweep voltage** and is produced by an internal circuit called the **time-base generator**. The time needed for the beam to cross the screen is called **sweep time**—the shorter the sweep time, the faster the speed of the beam. Note that as soon as the beam reaches the right-hand side of the screen, the voltage returns quickly to its maximum negative value so that the beam can begin another sweep.

To get the sinusoidal signal shown in Figure 7.47, the sine wave signal is simultaneously applied to the vertical plates. The beam thus sweeps from left to right, tracing out the sine wave as it passes. At any time, the vertical position of the beam is determined by the sine wave voltage, while the horizontal position is fixed by the sweep voltage. Together, the two motions of the beam trace out the sine wave on the screen of the CRO. In Figure 7.49, these motions lead to four complete cycles on the oscilloscope screen.

If the period of the sweep voltage is halved

FIGURE 7.49. If a time-varying signal is applied to the vertical plates while a sawtooth sweep voltage is applied to the horizontal plates, the combined action reproduces the signal on the CRT screen.

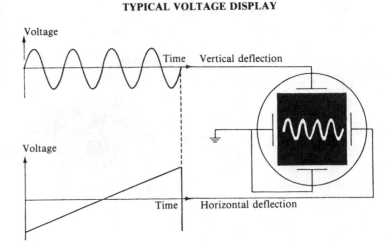

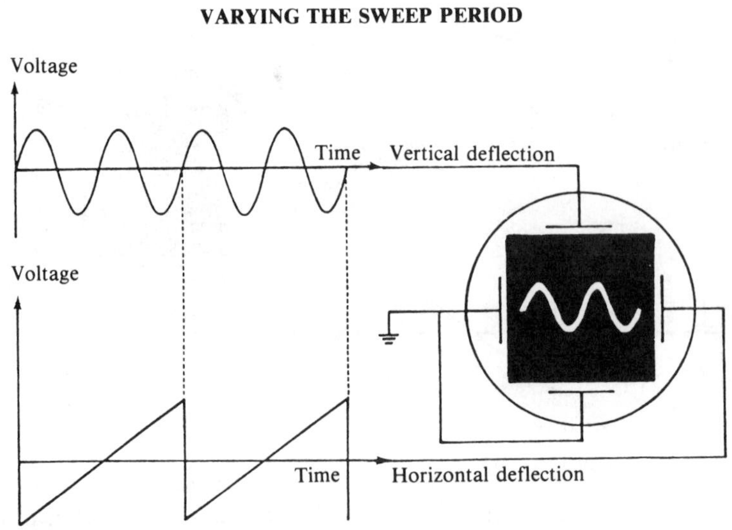

TYPICAL VOLTAGE DISPLAY

FIGURE 7.50. The length of the vertical signal that will be displayed — that is, the scale of the time axis — is determined by the period (frequency) of the sweep voltage. If the period is halved (frequency doubled), only half as much signal voltage will appear.

VARYING THE SWEEP PERIOD

(frequency doubled), as in Figure 7.50, fewer cycles of the sine wave will be traced out on the screen. On the other hand, if the period is increased (frequency decreased), then more cycles will be seen. Thus, the frequency of the sweep voltage determines the scale of the time axis.

Figures 7.46 to 7.50 illustrate the general principles of an oscilloscope. However, to accurately display a range of time-varying signals re-

quires numerous electronic circuits that process the input signals, sweep the electron beam, and synchronize the two to give a steady visible image. The details of these other circuits and their front panel controls are presented next.

Vertical Amplifier: Several hundred volts are required to deflect the electron beam from the bottom of the oscilloscope screen to the top.

VERTICAL AMPLIFIER AND INPUT SELECTION

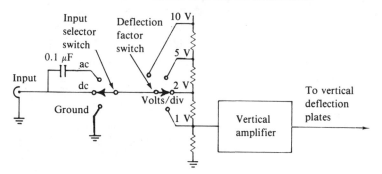

FIGURE 7.51. Plate voltages of several hundred volts are required to deflect the electron beam. Therefore, a high-gain vertical amplifier is required to increase the signal voltage. A variable voltage divider preceding the amplifier attenuates incoming signals so that signals of various amplitudes can be viewed.

Voltages in electronic circuits, however, are typically a few volts or less. Therefore, a high-gain **vertical amplifier** is built into the input circuitry to increase the size of input signals so they can be clearly displayed on the screen.

Because an oscilloscope must measure both large and small signals, it is necessary to vary the size of the display. This is accomplished by an attenuator switch that is placed between the input terminals of the CRO and the input of the vertical amplifier (Figure 7.51). Each switch position selects a different **deflection factor** that is defined as the input voltages needed to produce one unit of deflection on the scope face.

For convenience, the oscilloscope screen is subdivided into a measurement scale by a grid called a **graticule.** The graticule is typically eight units high by ten units wide; each unit is generally a centimeter (Figure 7.52). The deflection factor control knob thus determines the number of volts that correspond to each vertical division.

Note also that the graticule can often be illuminated to highlight the grid markings. On the control panel in Figure 7.52, the illumination control (ILLUM) is combined with the ON/OFF power switch.

A typical deflection factor control knob is shown in Figure 7.52. Its name, AMPL/DIV, indicates that the voltage amplitude for each setting will produce one division of deflection on the

oscilloscope screen. Note also that it consists of an outer and an inner knob. The outer knob sets the deflection factor given on the circular scale.

The scale numbers are only valid if the inner knob is in the locked position for calibration (CAL). Rotating the inner knob clockwise releases the lock and permits continuous (uncalibrated) variation of the amplitude from the scale value down to zero.

EXAMPLE 14:   If the peak-to-peak amplitude of a sine wave measured on the oscilloscope in Figure 7.52 is 4.2 divisions, what is the peak-to-peak amplitude in volts?

Solution: The numbers on the control knob are valid only if the center knob is set to CAL, or locked. This is the case in Figure 7.52, so the deflection factor can be read from the dial as:

deflection factor = 2 V/div

Thus, 4.2 divisions on the oscilloscope screen correspond to a peak-to-peak amplitude of:

$$V_{pp} = 2 \text{ V/div} \times 4.2 \text{ div} = \textbf{8.4 V}$$

EXAMPLE 14

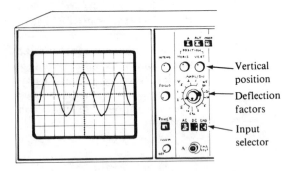

                Vertical
                position

                Deflection
                factors

                Input
                selector

FIGURE 7.52. The external AMPL/DIV knob selects the proper resistor of the voltage divider string to give various deflection factors. The deflection factor in V/div is the value of the signal voltage that will produce a beam deflection of one division on the graticule.

If an input signal is too large and deflects the beam off the screen, increasing the deflection factor will bring it back into view. For too small an image, a smaller deflection factor will enlarge it. Thus, *decreasing* the deflection factor is the same as *increasing* the sensitivity; 1 V/div is ten times more sensitive than 10 V/div.

Vertical Inputs: The input to the vertical amplifier can be connected in one of three ways: dc, ac, and ground (see Figures 7.51 and 7.52). In Figure 7.52, the INPUT SELECTOR is the switch below the rotary switch and can select AC, DC, or ground (GND).

    The schematic diagram of the three switch positions is shown in Figure 7.51. In the ac position, the signal must first pass through a capacitor that permits the ac portion of the signal to be observed, but blocks any dc component. The use of a blocking capacitor will be further described in Chapter 8.

    With the switch in the ground position, the amplifier input is connected directly to ground. The trace on the scope face is then a horizontal line — or spot — that represents an input of 0 V. The ground setting permits the 0 V baseline to

be adjusted to any desired position using the VERTICAL POSITION control knob.

    As an example of the use of the INPUT SELECTOR switch, consider the measurement of the voltage output of a laboratory power supply. After the voltage output of the supply has been connected to the input of the vertical amplifier, the INPUT SELECTOR switch is set to GND. Using the VERTICAL POSITION control, the scope trace is adjusted to some convenient reference position. For example, if the supply voltage were positive, the baseline would be set near the bottom of the screen. For a negative supply voltage, it would be set near the top.

    With the INPUT SELECTOR switch in the dc position, the supply voltage will cause the trace to move up or down depending on the polarity of the signal. The displacement between this trace and the baseline is a measure of the dc voltage. In Figure 7.53, the displacement of the trace is 2.5 divisions from the center baseline, and the deflection factor is 1 V/div, which gives a measured output voltage of 2.5 V for the power supply.

    With a deflection factor of 1 V/div, any small ac variation in the power supply output (on the order of a few millivolts) cannot be observed. An increase in sensitivity, however, will only cause the beam to deflect off scale.

    The solution is to turn the INPUT SELECTOR switch to the ac position. The 2.5 V dc component will be blocked by the capacitor bringing the trace back to 0 V. The sensitivity can now be increased to view any ac signal that might be superimposed on the dc signal.

    In Figure 7.53, with the deflection factor reset to 2 mV/div, the ac variation in the signal — sometimes called **ripple** — is seen to have a peak-to-peak amplitude of approximately 1 mV. It is interesting to note that had we not had the benefit of the blocking capacitor, we would have needed an oscilloscope screen 20 m high in order to observe this same ac waveform.

Input Resistance: With the INPUT SELECTOR switch set to DC or AC, the input resistance of

**INPUT SELECTION
FOR DC POWER SUPPLY VOLTAGE**

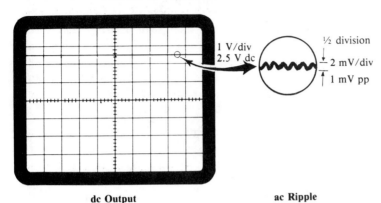

dc Output   ac Ripple

FIGURE 7.53. The input selector switch determines how the input signal is connected to the amplifier. It can ground the input in order to adjust the 0 V baseline, be directly coupled so that dc voltage can pass, or be coupled through a capacitor so that only ac signals can be viewed. Alternating current coupling is useful for observing small ac signals superimposed on a dc voltage.

the vertical amplifier is 1 MΩ between the input terminal and ground. An input resistance of 1 MΩ usually has a negligible effect on the circuit being tested. However, for cases in which the circuit resistances are comparable to 1 MΩ, a special, high-resistance, 10× probe can be added to the input of the oscilloscope (Figure 7.54).

This probe increases the input resistance by a factor of 10 to 10 MΩ, but at the expense of the signal amplitude. That is, the probe also decreases the signal amplitude by a factor of 10. Thus, the signal at the input of the vertical amplifier is only *one-tenth* the signal at the probe, and the reading on the AMPL/DIV control must be increased by a factor of 10. Forgetting this fact is a common error when using the 10× probe.

The probe circuit also contains an adjustable capacitor. Its purpose is to compensate for the input capacitance when the ac input is selected. The adjustment of the compensating capacitor is described in the oscilloscope instruction manual.

INPUT RESISTANCE

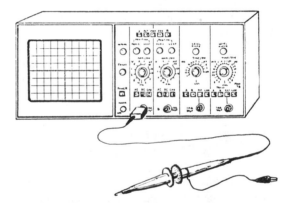

FIGURE 7.54. The resistance of the signal input terminal is typically 1 MΩ. It can be increased to 10 MΩ by using a 10× probe, as shown. However, the 10× probe also reduces the signal amplitude by a factor of 10, so the AMPL/DIV reading should be correspondingly increased.

V/div, what is the voltage amplitude of the signal?

Solution: With a 10× probe, a deflection factor of 10 V/div is equivalent to 100 V/div. Therefore, 6.3 divisions correspond to a voltage of **630 V**.

EXAMPLE 15: If a signal measures 6.3 divisions on a CRO screen using a 10× probe, and the deflection factor is set at 10

DUAL TRACE

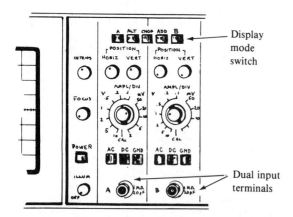

FIGURE 7.55. Most modern CROs have dual-trace capability. Thus, two sets of input terminals (A and B) and controls are available. Separate display mode switches permit viewing A and B separately, alternating between A and B on successive sweeps, chopping back and forth between A and B continually, or adding A and B to give a single trace, which is their sum.

Dual-Trace Oscilloscope: The oscilloscope described in this chapter has dual-trace capability: Two input signals can be traced simultaneously. Hence, it has two input channels, A and B, each of which has a separate input terminal, IN-PUT SELECTOR switch, vertical amplifier with AMPL/DIV control, and HORIZONTAL and VERTICAL POSITION controls (Figure 7.55).

Most dual-trace oscilloscopes use only a single electron gun. Internal circuitry alternates using the single electron beam to trace out the A and B channel input signals. Controls for this purpose are shown in Figure 7.55.

The A and B buttons cause the input signal from either channel A or B to be displayed separately as a single trace. The alternate (ALT) button causes the beam to alternate between the two channels, first tracing a complete A signal, then a complete B signal, then back to an A signal, and so on. The CHOP button causes the beam to "chop" back and forth between the A and B channels many times during each trace, showing a small segment of A, a small segment of B, back to a small segment of A, and so on. In both the ALT and CHOP settings, both signals appear complete, and the alternations and choppings are rarely noticed. The ADD button results in a single trace that is the algebraic sum of the A and B signals at each point in time.

Time-Base Generator: We saw in Figures 7.49 and 7.50 how sweep voltages of different periods cause different lengths of the input signal to appear on the CRO screen. In Figure 7.49, four cycles of the sine wave were displayed. By reducing the sweep period by one-half, only two cycles appeared in Figure 7.50. Thus, by choosing other values for the sweep period, more or fewer cycles can be examined.

For most oscilloscopes, 20 or more sweep period values are available. These values are selected by means of the time-per-division (TIME/DIV) knob on the front panel that controls the frequency of the sawtooth waves produced by the time-base generator (Figure 7.56). The calibrated outer knob can select times that range from 0.2 μs/div to 0.5 s/div, while the inner knob permits continuous variation so that intermediate values can be obtained. But remember, the variable control must be set in the CAL (locked) position before quantitative measurements can be made.

EXAMPLE 16: What are the pulse width, period, and duty cycle for the square wave in Figure 7.57a? The time per division is 0.1 ms/div.

Solution: In Figure 7.57a, the horizontal graticule has 10 divisions, and each large division is subdivided further into 5 divisions. For convenience of measurement, the display can be moved either to the right or to the

TIME/DIV CONTROL

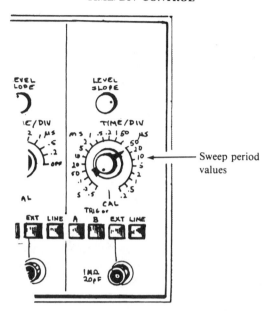

Sweep period values

FIGURE 7.56. The scale of the time axis is controlled by the TIME/DIV knob. This sets the period of the sawtooth wave and, thus, the time for a single sweep. The outer knob selects one of 20 calibrated sweep times, and the inner knob permits continuous variation among calibrated values.

left. In Figure 7.57a, the leading edge of the pulse has been set at a principal graticule line.

The pulse width in figure 7.57a is 1.4 divisions. For a sweep time of 0.1 ms/div, the pulse width d is, therefore:

$$d = 1.4 \text{ div} \times 0.1 \text{ ms/div} = \textbf{0.14 ms}$$

The period of the pulse is seen to be exactly 4 divisions. Thus, the period for the square wave is:

$$T = 4 \text{ div} \times 0.1 \text{ ms/div} = \textbf{0.4 ms}$$

and the duty cycle is:

$$\frac{d}{T} = \frac{0.14 \text{ ms}}{0.4 \text{ ms}} = \textbf{0.35}$$

EXAMPLE 17: What is the frequency of the sine wave in Figure 7.57b? The TIME/DIV switch is set at 0.1 ms/div.

Solution: For a sinusoid, the maximum slew rate occurs when the signal goes through the axis at 0 V. To determine 0 V, the INPUT SELECTOR switch is set to 0 V, and the baseline is lined up on a horizontal graticule line using the VERTICAL POSITION control.

The INPUT SELECTOR switch is then set to ac to eliminate any dc components and the sine wave zero crossing is moved horizontally with the HORIZONTAL POSITION control to line up with a vertical graticule line (Figure 7.57b).

The period is then measured between zero-crossing points. In Figure 7.57b, the horizontal position has been adjusted so that the zero crossing occurs at a principal graticule marking. The next zero crossing *with the same slope* occurs 2.5 divisions later. Therefore, if the time-per-division switch is set at 0.1 ms/div, the period is:

$$T = 2.5 \text{ div} \times 0.1 \text{ ms/div} = 0.25 \text{ ms}$$

The frequency of the sine wave can be calculated by taking the reciprocal of the period:

$$f = \frac{1}{T} = \frac{1}{0.25 \text{ ms}} = \frac{1}{0.25} \times 10^3$$
$$= \textbf{4000 Hz}$$

Note that for a more accurate measurement of the period, the time between three zero crossings could be found and divided by three.

EXAMPLES 16 AND 17

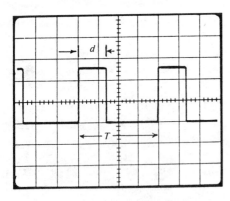

(a) Square Wave

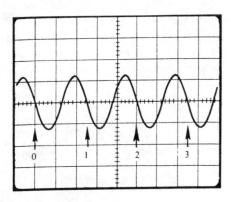

(b) Sine Wave

FIGURE 7.57. The calibrated TIME/DIV settings can be used to determine pulse widths and periods. For ease of measurement, the 0 V baseline is adjusted to line up on a horizontal graticule line, and the point of maximum slew rate is lined up on a vertical graticule line before taking the data.

Triggering and Synchronization: To display a steady waveform, the electron beam traces the same path over and over again. Two conditions must be met: The signal being displayed must be periodic, and the electron beam must always

TRIGGERING AND SYNCHRONIZATION

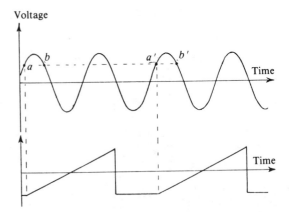

FIGURE 7.58. In order for the displayed wave form to appear stationary on the screen, the sweep voltage must be synchronized with the signal voltage to begin at the same point on each trace.

begin its trace of the waveform at exactly the same point on the wave.

Another way of stating the latter condition is that the start of the sawtooth voltage must be synchronized with a specific point on the signal. Synchronizing the sweep of the electron beam with the input signal is the job of the scope's triggering circuits.

Figure 7.58 shows both an input sine wave signal and the sweep voltage produced by the internal time-base generator of a CRO. The signal that would be displayed on the screen of a CRO is shown in Figure 7.59. Note that the sawtooth wave starts its sweep at point *a* of the sine wave. Once it has completed a single trace, it drops back to its initial voltage and then waits for the sine wave to come to the same voltage *a'* before beginning its next trace.

Because the signal is periodic and because the time base is synchronized to display exactly the same portion of the signal on each sweep, the repeated image appears steady to the eye. If the signal voltage were to change between sweeps or if the sweep began at random points

TRIGGERING ON POSITIVE OR NEGATIVE SLOPE

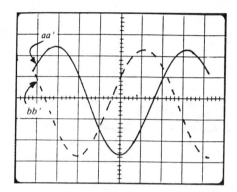

FIGURE 7.59. The sweep voltage can trigger the same voltage on either a positive or a negative slope. On a positive slope (*aa'* of Figure 7.58), the solid line would be traced out; on a negative slope (*bb'* of Figure 7.58), the dashed line would be traced out.

on the signal voltage, the image on the face of the scope would be different on each trace and would not be readable.

It is like writing your signature over and over again, one on top of another. If each signature does not begin at exactly the same point and if they do not have exactly the same shape, the final result will not be legible.

Trigger Level and Slope: There are two controls that let us select the point on the signal voltage that the sweep begins. The TRIGGER LEVEL control sets the voltage level at which the sweep begins, and the SLOPE control determines whether that voltage occurs on a positive slope of the signal or on a negative slope.

In Figure 7.60, these controls are combined on one knob. Turning the knob sets the voltage level, either positive or negative. Pushing in the knob causes the sweep voltage to start when that voltage has been reached on an increasing, or positive, slope (points *a* and *a'* in Figure 7.58). Pulling out the knob causes the sweep voltage to start when it has reached a decreasing, or negative, slope (points *b* and *b'* in Figure

TRIGGER LEVEL AND SLOPE CONTROLS

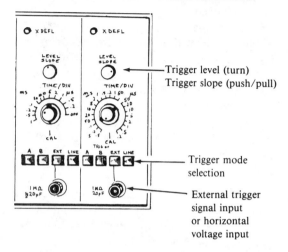

Trigger level (turn)
Trigger slope (push/pull)

Trigger mode selection

External trigger signal input or horizontal voltage input

FIGURE 7.60. Several trigger modes are possible. Trace A can be triggered to begin when B begins (or vice versa) or either signal can be triggered by an external signal or by the line voltage. An external input jack is provided both for external trigger signals and for use of the horizontal axis to display a voltage.

7.58). The negative slope trace is shown dashed in Figure 7.59.

Trigger Signal: There are several ways of triggering the start of the sweep voltage other than at a particular point on the input signal. These alternatives are controlled by a series of TRIGGER MODE switches below the TIME/DIV knob.

For example, when viewing two traces simultaneously, we may want them to begin at the same point in time in order to see what each is doing at the same time. In this case, the triggering of trace A would be determined by its LEVEL/SLOPE control with the A button pushed for channel A. Pressing the A switch on channel B would cause trace B to begin at exactly the same time that A begins.

If the external (EXT) button is pushed, traces A and/or B can be triggered by an external signal applied to the EXT INPUT jack below the TIME/DIV control.

If the line button (LIN) is selected, the sweep is synchronized to the frequency of the line voltage that is either 60 Hz (United States) or 50 Hz (European). The line position is particularly useful when troubleshooting a circuit that has some unwanted signals. If these extra signals—called "noise"—appear stationary when the sweep is synchronized to the line frequency, we know that the noise is probably pickup from the line voltage.

Horizontal Input: Although an oscilloscope is designed primarily to display voltage signals as a function of time, it can also be used to display one signal as a function of another—that is, to make a normal x-y graph of two quantities, neither of which is time. A common example is the measurement of the V-I characteristic of a diode or a transistor in which the voltage is displayed on the horizontal x-axis and the current is displayed on the vertical y-axis.

To make the x-axis a voltage rather than time, an internal horizontal amplifier is available. To select the horizontal amplifier rather than the time-base generator on the CRO of Figure 7.60, a button at the top marked X DEFL (for x deflection) is pushed along with the INPUT button at the bottom. Note, however, that the horizontal amplifier has no deflection factor control, so any amplitude adjustment must be made externally.

To measure the V-I characteristics, the voltage applied to the device is connected to the

x-y DISPLAY

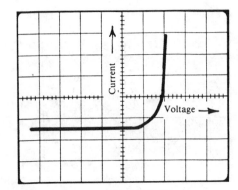

FIGURE 7.61. Using the horizontal axis as a voltage rather than time axis permits the CRO to make x-y graphs, such as the V-I characteristic of a diode shown. The horizontal voltage amplifier has no deflection factor control, so amplitude adjustments must be made externally.

EXT INPUT of the scope, and a signal proportional to the current is connected to the vertical input. As the voltage is varied, the current changes, and the relation between the two is traced on the screen. By repeatedly varying the voltage, an image of the V-I characteristic is displayed (Figure 7.61).

While the discussion here of the CRO has been lengthy, a real understanding of the operation and versatility of the CRO can be obtained only by using one. The foregoing description can be used as a reference when learning to use the CRO controls.

7.7 QUESTIONS AND PROBLEMS

1. Identify the type of each of the time-varying signals in Figures 7.62 and 7.63. Label each of the signals.

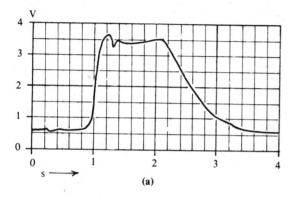

(a)

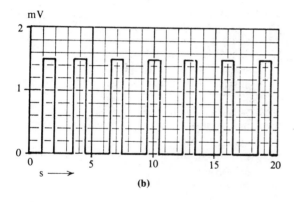

(b)

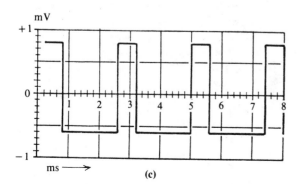

(c)

FIGURE 7.62.

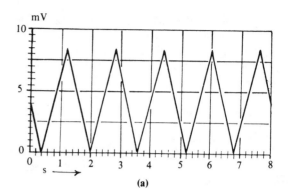

(a)

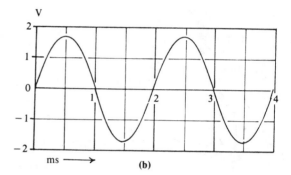

(b)

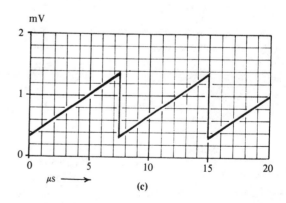

(c)

FIGURE 7.63.

2. Determine the following quantities for the signal of Figure 7.62a:
   (a) Pulse height V_0
   (b) Pulse width d
   (c) Rise time t_r
   (d) Fall time t_f

3. Determine the following quantities for the signals of Figures 7.62b and 7.62c and show these quantities on the figure:
   (a) Pulse height V_0
   (b) Pulse width d
   (c) Period T
   (d) Frequency f
   (e) Duty cycle D
   (f) Average voltage V_{dc}
4. Derive expressions for the slew rate and average voltage of the waveform in Figure 7.63a in terms of its peak-to-peak amplitude and its frequency.
5. Derive expressions for the slew rate and average voltage of the waveform in Figure 7.63c in terms of its peak-to-peak amplitude and its frequency.
6. Determine the following quantities for the signals in Figures 7.63a to 7.63c and show these quantities on the figure:
   (a) Peak amplitude V_0
   (b) Peak-to-peak amplitude V_{pp}
   (c) Period T
   (d) Frequency f
   (e) Average voltage V_{dc}
   (f) rms voltage (sine) V_{rms}
   (g) Maximum slew rate
7. What electrical power would be dissipated by the signal of Figure 7.63b in stock resistors of the following values: 27 Ω, 390 Ω, 22 kΩ, and 1.5 MΩ?
8. Determine the following quantities of the waveform that would be produced by the function generator of Figure 7.64 and sketch the waveform.
   (a) Waveform
   (b) Peak-to-peak amplitude V_{pp}
   (c) Frequency f
   (d) Period T
   (e) dc offset V_b
9. If the output of the function generator in Figure 7.64 is applied across the following load resistances (R), what voltage will appear across the load: 2.2 kΩ, 50 Ω, 180 Ω, and 20 Ω?
10. What is the minimum instrument slew rate that is necessary to measure each of the signals in Figure 7.63?
11. What is the maximum slew rate that can be measured by a meter with a response time of 0.8 s on the following full-scale voltage scales: 0.3 V, 1.0 V, 50 V, and 250 V?

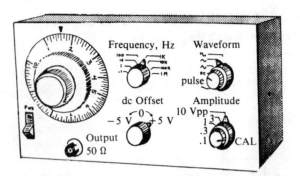

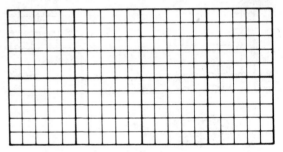

FIGURE 7.64.

12. What is the maximum slew rate that can be measured by a galvanometric recorder with a frequency response of 30 Hz for the following full-scale deflections: 1 mV, 50 mV, 0.5 V, and 2.0 V?
13. Determine the following quantities for the CRO signal in Figure 7.65a:
    (a) Pulse height V_0
    (b) Pulse width d
    (c) Period T
    (d) Frequency f
    (e) Duty cycle D
14. Determine the following quantities for the CRO signal in Figure 7.65b:
    (a) Peak-to-peak amplitude V_{pp}
    (b) Period T
    (c) Frequency f
    (d) Slew rate
15. Determine the following quantities for the CRO signal in Figure 7.65c:
    (a) Peak-to-peak amplitude V_{pp}
    (b) Peak amplitude V_0
    (c) Period T
    (d) Frequency f
    (e) Maximum slew rate

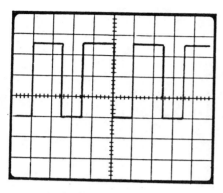

(a) AMPL/DIV = 5 V/div
TIME/DIV = 50 ms/div

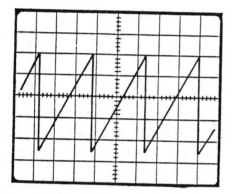

(b) AMPL/DIV = 50 mV/div
TIME/DIV = 5 μs/div

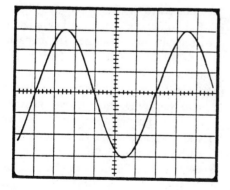

(c) AMPL/DIV = 0.2 V/div
TIME/DIV = 3 ms/div

FIGURE 7.65.

16. What would the peak-to-peak values of the signals in Figure 7.65 be if a high-impedance 10× probe were used?

Chapter

8

CAPACITORS

8.1 OBJECTIVES

Following the completion of Chapter 8, you should be able to:

1. Describe the basic design of a capacitor and explain the principles of its operation.
2. State the relationship among capacitance C, capacitor voltage V, and charge stored Q and explain how changes in C or V affect Q.
3. State the relationship among capacitance C, plate area A, distance of separation d, and dielectric strength ϵ for a parallel plate capacitor and explain the design considerations in selecting A, d, and ϵ to achieve large values of C in a small package.
4. Recognize the units of electrical capacitance and convert between decimal subunits.
5. Determine the equivalent capacitance of capacitors connected in parallel.
6. State the difference between electrostatic and electrolytic capacitors and identify the type of a given capacitor by its component materials.

7. Explain the principal capacitor rating characteristics (tolerance, maximum working voltage, insulation resistance, and frequency range) and properly select a capacitor for a given application based on the circuit requirements.
8. For a capacitor charged by a source of constant current:
   (a) Explain and give the mathematical relationship that describes the capacitor's voltage-time behavior and charging (slew) rate.
   (b) Calculate the voltage-time behavior and charging rate, given the capacitance and charging current.
   (c) Calculate the capacitance and charging current required to achieve a given voltage-time behavior or charging rate.
9. For a capacitor charged through a series resistor by a source of constant voltage:
   (a) Explain and give a graphical representation of the change in capacitor voltage, resistor voltage, and circuit current with time.

(b) Give a mathematical relationship for the time constant of the circuit and explain how it expresses the capacitor's voltage-time behavior.

(c) Determine the time required for the capacitor voltage to reach a specific value, given the source voltage, series resistance, capacitance, and a generalized curve of capacitor charge behavior.

(d) Determine the value of capacitance (or resistance) of the circuit from the capacitor-charging curve, given the circuit resistance (or capacitance) and a generalized curve of capacitor charge behavior.

10. For a capacitor discharging through a series resistor:

(a) Explain and give a graphical representation of the change in capacitor voltage, resistor voltage, and circuit current with time.

(b) Give a mathematical relationship for the time constant of the circuit and explain how it expresses the capacitor's voltage-time behavior.

(c) Determine the time required for the capacitor (or resistor) voltage to reach a specific value, given the initial voltage, series resistance, capacitance, and a generalized curve of capacitor discharge behavior.

(d) Determine the value of capacitance (or resistance) of the circuit from the capacitor's discharge curve, given the circuit resistance (or capacitance) and a generalized curve of capacitor discharge behavior.

11. Explain the general principles underlying the use of a capacitor and a source of constant voltage or current to generate a sawtooth wave.

12. Explain the general principles of operation of a unijunction transistor as a voltage-sensitive switch.

13. Design a sawtooth generator with a specific output frequency using a source of con-stant voltage (or current) and a unijunction transistor.

14. Identify each pin of a 555 timer and explain its purpose in controlling the timer's behavior.

15. Explain the operation of a 555 timer connected as a timed switch, a one-shot multivibrator, a square wave generator, and a sawtooth generator.

16. Design a timed switch and a one-shot multivibrator with a specific output pulse height and width using a 555 timer.

17. Design a square wave generator with a specific output amplitude, frequency, and duty cycle using a 555 timer.

18. Design a sawtooth generator with a specific output amplitude and frequency using a 555 timer.

8.2 INTRODUCTION

The time-varying signals described in Chapter 7 came either from a changing physical source, such as sound, or from an electronic source, such as a function generator. For a physical source to produce an electrical signal, a transducer is required to convert the changing physical quantity into a changing electrical quantity—that is, current, voltage, or resistance.

Figure 8.1 shows a typical example of a time-varying electrical signal produced by a transducer. A photoresistor is connected in series with a resistor and a battery. The resistance of the photoresistor changes when the incident light level changes. By Ohm's law, this change produces a corresponding change in voltage V_o across the photoresistor. If the photoresistor is illuminated by a time-varying source of light, then the voltage across it will vary with time in the same way.

If the photoresistor is replaced by a thermistor, a time-varying temperature will similarly produce a time-varying electrical signal. In fact, a wide variety of resistive transducers can

TRANSDUCER-PRODUCED SIGNALS

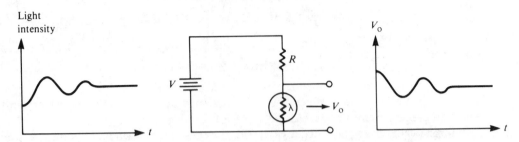

FIGURE 8.1. Time-varying signals can be produced by a changing physical source or electronically. A time-varying light level causes the resistance of a photoresistor to change that in turn produces a time-varying voltage. The electronic generation of time-varying signals depends largely on the characteristics of the capacitor.

be used in place of the photoresistor so that practically any time-varying phenomenon can be converted to a corresponding time-varying electrical signal.

Notice, however, that if the value of resistance is constant, the current and voltage in the circuit of Figure 8.1 will also be constant. Only when the resistance of the transducer changes will the voltage also change. A resistive circuit by itself cannot cause voltages or currents to vary with time.

The time-varying signals of a function generator, on the other hand, are produced electronically. Thus, there are circuit elements that can cause the voltages and currents in a circuit to change with time. The most common of these is the capacitor that is the principal subject of this chapter.

8.3 PRINCIPLES OF OPERATION

A **capacitor** is essentially two sheets of metal, called *plates*, that are separated by a very small distance. There is no electrical connection between the plates, so it is an open circuit. The electrical symbol for a capacitor shows the two

BASIC CAPACITOR DESIGN AND SYMBOL

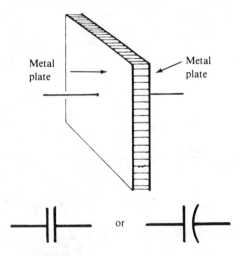

FIGURE 8.2. The basic design of a capacitor is two metal plates separated by a very small distance. Since the plates don't touch, a capacitor is an open circuit, as illustrated by its electrical symbol.

separated plates with an electrical lead attached to each (Figure 8.2).

When a capacitor is connected to a battery, as shown in Figure 8.3a, the plate attached to the positive battery terminal takes on a positive

BASIC CAPACITOR BEHAVIOR

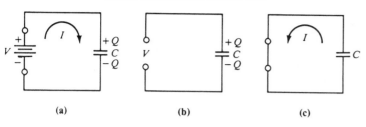

(a)   (b)   (c)

FIGURE 8.3. When a voltage is applied to a capacitor, a current flows briefly to establish a positive charge on one plate and an equal negative charge on the other. When the charges are established, no current flows. If the battery is removed, the charges remain, and the capacitor voltage stays at the battery voltage. If the capacitor terminals are shorted, the charge quickly dissipates. This charge storage capacity of a capacitor is central to its electrical behavior.

charge $+Q$, and the opposite plate takes on an equal negative charge $-Q$. Electrical current flows very briefly in the circuit to charge the capacitor. However, when the charges $+Q$ and $-Q$ are established, no current flows because the circuit is open (Figure 8.3b).

If the battery is removed (Figure 8.3c), the capacitor will remain charged because there is no path through which the charge can flow. The voltage across the capacitor will likewise remain equal to the battery voltage. Thus, a capacitor is able to "store" electrical charge and, hence, voltage. Because a capacitor can store charge, it is often referred to as a *storage circuit element*.

As long as there is no electrical connection between the two plates, there will be a positive charge $+Q$ on one terminal, a negative charge $-Q$ on the other, and a voltage V across the two terminals. However, if the capacitor terminals are shorted together, the capacitor will quickly discharge, and its voltage will drop to zero.

The charge storage capability of a capacitor has a number of important applications. A common example is an electronic flash unit of a camera. The flash of a camera unit is provided by a gas discharge tube that requires a high voltage and a large current. The power source, however, is a low-voltage battery. Thus, an electronic circuit is provided to convert the low bat-

tery voltage into a higher voltage to power the flash tube.

The current output of a high-voltage converter is not great enough to produce a bright flash, so it is used instead to charge a capacitor. When the capacitor is sufficiently charged, a light comes on to indicate that adequate current is available to trigger the flash tube. The time it takes to charge the capacitor is generally several seconds, but the discharge time through the flash tube is a millisecond or less.

8.4 UNITS OF CAPACITANCE

If different voltages are connected to a capacitor, the charge on the capacitor is directly proportional to the applied voltage. Doubling the voltage on a capacitor doubles the charge on each of the plates. Tripling the voltage triples the charge, and so on. In symbols:

$$Q \propto V$$

or

$$\frac{Q}{V} = \text{constant} \tag{1}$$

If several different capacitors are all charged by the same battery, will they all have the same charge? The answer is no; each capac-

FIGURE 8.4. The charge stored on a capacitor is directly proportional to the voltage applied to it. When the voltage is doubled, the charge also doubles. The constant of proportionality is called the capacitance C and is measured in farads (F). The larger the capacitance, the greater the charge stored for a given voltage.

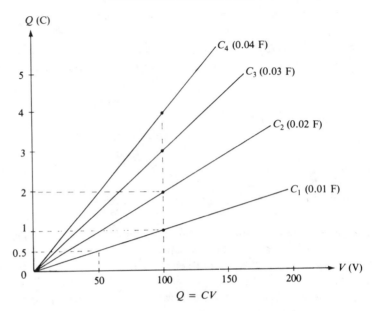

CAPACITOR CHARGE STORAGE

$Q = CV$

itor will take on a different charge that is determined by the way the capacitor is constructed. For a given capacitor, the charge and voltage are always proportional, but the constant of proportionality differs for different capacitors.

This constant of proportionality is called the **capacitance** C. Thus, Equation 1 can be written:

$$\frac{Q}{V} = C$$

$$Q = CV \qquad (2)$$

Figure 8.4 is a graph of charge and voltage for several different capacitors. The larger the capacitance, the larger the amount of stored charge for a fixed value of voltage. When voltage is measured in volts and charge in coulombs, the capacitance is in units of **farads** (F):

1 farad = 1 coulomb/volt

The value of capacitance for a given capacitor depends on several factors: the surface area of the plates A, their distance of separation d, and the material between them (Figure 8.5). The capacitance increases in direct proportion to the plate area, but it is inversely proportional to their separation. Thus, in a large-value capacitor, the plates have a large surface area and are placed very close together.

The mathematical expression for the capacitance of a parallel-plate capacitor is:

$$C = \frac{\epsilon A}{d} \qquad (3)$$

where A is the area of one of the plates, d is the distance between them, and ϵ is a property of the material placed between the plates. The material between the plates is called a **dielectric** and has the property of being an extremely good insulator. A good dielectric material allows the plates to be very close together and large charges to be placed on each plate without the possibility of leakage from one plate to the other. A typical value for ϵ is 2×10^{-11} F/m. Thus, if A is in square meters and d is in meters, C is in farads.

BASIC CAPACITOR DESIGN

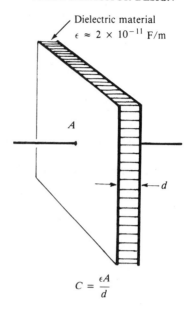

Dielectric material
$\epsilon \approx 2 \times 10^{-11}$ F/m

A

d

$$C = \frac{\epsilon A}{d}$$

EXAMPLE 1: How large must two capacitor plates be if they are separated by 0.01 cm and should have a capacitance of 1 F?

Solution: Rewriting Equation 3 for the area gives:

$$A = \frac{dC}{\epsilon}$$

Substituting $d = 0.01$ cm $= 10^{-4}$ m, $C =$ 1 F, and $\epsilon = 2 \times 10^{-11}$ F/m gives:

$$A = \frac{10^{-4} \text{ m} \times 1 \text{ F}}{2 \times 10^{-11} \text{ F/m}} = 5 \times 10^6 \text{ m}^2$$

FIGURE 8.5. The capacitance of a parallel-plate capacitor depends on the surface area of the plates A, their distance of separation d, and the insulating property ϵ of the material between them. The larger the plate area, the larger the capacitance; and the closer the plates, the larger the capacitance (see Equation 3). Calculation of capacitance is rarely required, however.

Five million square meters is an enormous area—about the size of a thousand soccer fields. Clearly, 1 F is a large amount of capacitance. Generally, capacitors come in much smaller sizes and are expressed in correspondingly smaller units, such as the **microfarad** (10^6 μF $= 1$ F) or **picofarad** (10^{12} pF $= 1$ F) (Figure 8.6).

The calculation of capacitance is rarely required, but it is important to remember how capacitance depends on surface area and plate separation—the larger the surface area, the larger the capacitance, and the closer the plates, the larger the capacitance. Thus, the trick to making large-value capacitors is to package large-area metal plates very close together and use a good dielectric insulating material between them.

UNITS OF CAPACITANCE

millifarad: 1 mF $= 10^{-3}$ F
microfarad: 1 μF $= 10^{-6}$ F
nanofarad: 1 nF $= 10^{-9}$ F
picofarad: 1 pF $= 10^{-12}$ F

FIGURE 8.6. One farad of capacitance is a very large quantity requiring an enormous plate area. Most capacitors are much smaller, typically microfarads (10^{-6} F), but sometimes picofarads (10^{-12} F).

CAPACITORS IN PARALLEL

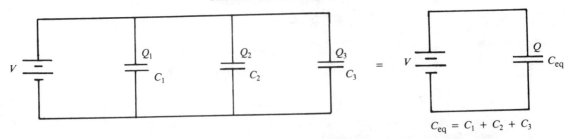

$$C_{eq} = C_1 + C_2 + C_3$$

FIGURE 8.7. Capacitors are often connected in parallel, but rarely in series. The equivalent capacitance of capacitors in parallel is simply their sum (Equation 4). Placing capacitors in parallel increases the area available for storing charge.

8.5 PARALLEL AND SERIES CONNECTIONS

Like resistors, capacitors can be connected in parallel or in series. Figure 8.7 shows three capacitors connected in parallel. All are connected to a battery with voltage V. Because the three are in parallel, the voltage across each capacitor is V. By Equation 2, the charge on each capacitor is:

$$Q_1 = C_1 V$$
$$Q_2 = C_2 V$$
$$Q_3 = C_3 V$$

The total charge Q stored by the circuit is simply the sum of these three values:

$$
\begin{aligned}
Q &= Q_1 + Q_2 + Q_3 \\
  &= C_1 V + C_2 V + C_3 V \\
  &= (C_1 + C_2 + C_3)V
\end{aligned}
$$

Thus, the three capacitors in parallel act like a single equivalent capacitor C_{eq} whose value is the sum of the three individual units (Figure 8.7):

$$Q = C_{eq} V$$

where

> **in parallel:** $C_{eq} = C_1 + C_2 + C_3$ **(4)**

This argument can be extended to any number of capacitors in parallel.

Note that capacitors add in *parallel* much as resistors add in series. Thus, to increase the value of a capacitor in a circuit, we can simply add another capacitor in parallel with the first. This result is also obvious from Equation 3. Two capacitors connected in parallel increase the area A available for storing charge.

EXAMPLE 2: If two 10 μF capacitors are connected in parallel with a 30 μF capacitor, what is the equivalent capacitance of the combination?

Solution: In parallel, capacitance adds. Therefore, by Equation 4:

$$
\begin{aligned}
C_{eq} &= C_1 + C_2 + C_3 \\
       &= 10\ \mu F + 10\ \mu F + 30\ \mu F = \mathbf{50\ \mu F}
\end{aligned}
$$

Capacitors are frequently connected in parallel, but they are rarely connected in series. For completeness, the equation for capacitors in series is given, but you are not likely to encounter it:

$$\text{in series:} \quad \frac{1}{C_{eq}} = \frac{1}{C_1} + \frac{1}{C_2} + \frac{1}{C_3} + \cdots \quad (5)$$

This equation, of course, is identical to the way in which resistors add in parallel.

8.6 TYPES OF CAPACITORS

Figure 8.8 illustrates the numerous shapes and sizes in which capacitors are available. Which type of capacitor should you choose for a particular application? If you look in an electronic parts catalog for information, you will find page after page of capacitors—film, paper, electrolytic, aluminum, tantalum, mica, ceramic, glass, and so on. What do these terms mean? Why do the capacitors in Figure 8.8 differ so in size? How do you choose? This section answers some of these questions.

The different names for capacitors—tantalum, glass, ceramic—refer to the materials used in their manufacture. Each material has certain advantages and disadvantages, so specific types of capacitors serve specific applications.

In selecting a capacitor for a particular application, it is helpful to know that capacitors are grouped in several broad categories. The two primary groups are fixed and variable capacitors. **Fixed-value capacitors** are by far the more common, and they fall into two important subgroups: electrolytic and electrostatic.

Electrolytic capacitors have a specific polarity that must be taken into account when connecting them in a circuit. They are generally named for the material used to make their plates—aluminum, tantalum, or niobium. Because they require a specific polarity, they can-

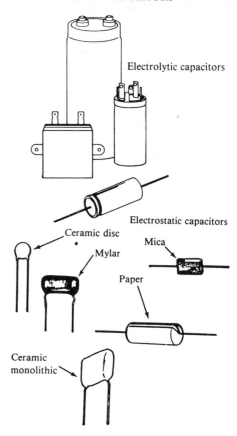

(a) Fixed-Value Capacitors

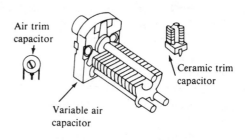

(b) Variable Capacitors

FIGURE 8.8. Capacitors come in a wide variety of packages depending on the material used and the method of construction. Each type of capacitor has characteristics that are better suited for some applications than for others.

FIGURE 8.9.   Capacitors are either fixed-value or variable. Fixed-value capacitors can be electrolytic or electrostatic. Electrolytic capacitors have a specific polarity that must be observed when connecting them in a circuit. They are generally named by the material used for their plates. Electrostatic capacitors have no polarity requirements and are generally named for their dielectric material.

SUMMARY OF CAPACITOR TYPES

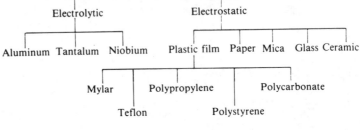

not be used with time-varying signals that change sign.

Caution: An electrolytic capacitor must always have its positive (+) terminal connected to the higher voltage. The + terminal is always clearly marked.

Electrostatic capacitors have no polarity and, like resistors, can be connected either way in a circuit. The term *electrostatic* is rarely used, however, because this subgroup is generally named for the type of dielectric material it contains—for example, ceramic, mica, glass, mylar, and paper. This group has the most stable values of capacitance and ranges in size from tiny beads to cylinders with dimensions of inches.

Variable capacitors permit adjusting the value of capacitance. Their dielectric material can be any of a wide variety of materials—ceramic, air, glass, mica, and plastic. Figure 8.9 summarizes the classification of fixed-value and variable capacitors and the various materials used in their construction.

8.7 CAPACITOR CONSTRUCTION

Because capacitors require large-area plates, they tend to be the largest of the electronic components. Thus, a primary consideration in capacitor design is size—how to store the maximum charge in the minimum amount of space. The maximum charge that can be stored Q_{max} is equal to the product of the capacitance C and the maximum voltage that can be applied V_{max}, or $Q_{max} = CV_{max}$. Increasing either C or V_{max} increases Q_{max} and generally leads to a larger unit.

The dependence of capacitor size on capacitance C is obvious from Equation 3. For large values of C, large-area plates spaced very close together are needed.

The dependence of size on maximum applied voltage V_{max} is determined by both the dielectric material and its thickness. The greater the thickness d, the greater the voltage that can be applied without charge leakage. Thus, given a particular dielectric material, to increase the maximum applied voltage, it is necessary to increase the separation between the plates. However, to get the same value of capacitance, it is necessary to increase the plate area by a comparable amount. Increasing the plate area leads to a larger unit. Thus, both capacitance and maximum applied voltage determine the size of a capacitor.

Other factors, such as weight, cost, and reliability under adverse conditions, are sometimes as important as size. These considerations account for the wide variety of materials and manufacturing techniques. The more common ones are described below.

Aluminum electrolytic capacitors are designed to provide a very large value of capacitance. An aluminum electrolytic capacitor is a sandwich of two aluminum foil ribbons and a porous separator, which contains a fluid electrolyte, that is rolled into a cylinder (Figure 8.10). During manufacture, an aluminum-oxide coating is produced at the boundary between one aluminum ribbon (the anode) and the electrolyte. (The other ribbon makes electrical contact with the electrolyte.) Because of the properties of the electrolyte, the second ribbon must always have a negative voltage with respect to the aluminum anode. If the polarity is reversed, the unit becomes a short circuit and may be destroyed.

Tantalum electrolytic capacitors are manufactured by the same techniques used to make aluminum capacitors except that tantalum film is used instead of aluminum. Another type uses a solid piece of tantalum on which a dielectric layer of tantalum oxide is formed electrochemically. The electrolyte is a solid semiconductor.

Plastic film capacitors are electrostatic capacitors made by depositing vaporized layers of metal on plastic film insulators and then rolling and encapsulating them. The plastic film may be mylar, teflon, polypropylene, polystyrene, polycarbonate, or polyester (Figure 8.10). In place of the vaporized metal, a separate metal foil is sometimes used.

Paper capacitors consist of alternate layers of metal and paper insulation (usually saturated with oil) rolled together into a cylinder. The roller assembly is placed inside a sealed, oil-filled metal case.

Mica and glass capacitors use mica and glass as the insulating dielectric. They are similar in that the insulator is coated with a conductor (often silver), stacked in layers, and then pressed into a compact unit.

Ceramic capacitors consist of layers of conducting material separated by a ceramic dielectric material. Because of its versatility, this type of capacitor is now the most commonly used. Its capacitance, which is very stable, can be produced in a wide range of values.

8.8 CAPACITOR RATINGS

Each type of capacitor comes in a specific range of values determined by its component material and method of construction. The ranges for the more common types are given in Figure 8.11. Note that large-value capacitors are generally aluminum electrolytics. In addition to capacitance, a number of other capacitor characteristics may be important when choosing one for a specific application. These characteristics are described below. Typical values are listed in Figure 8.11.

Tolerance

The capacitance for a given capacitor is usually stated as a specific value with a tolerance of plus or minus some percent. The actual value may be anywhere within the tolerance range. Due to the many variables involved in capacitor construction, the tolerance of capacitors is significantly larger than is resistor tolerance. A tolerance of 100% is not uncommon, particularly for electrolytics. The common mica and ceramic capacitors are typically $\pm 10\%$ to $\pm 20\%$, though units with narrower tolerance are available at somewhat higher cost. In calculating design values for capacitors, we must

CAPACITOR CONSTRUCTION

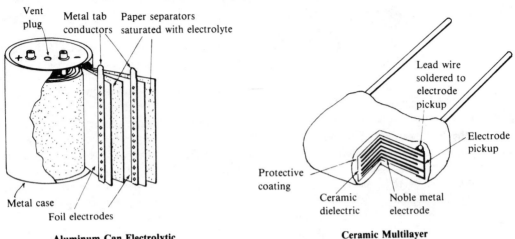

Aluminum Can Electrolytic

Ceramic Multilayer

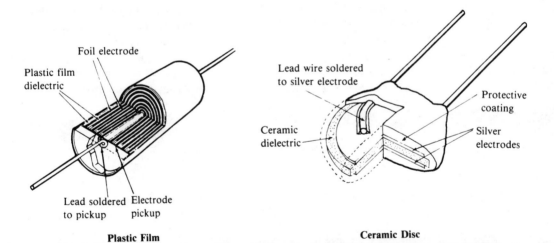

Plastic Film

Ceramic Disc

FIGURE 8.10. Because they require larger-area plates, capacitors are one of the largest electronic components. The figure shows the construction of the more common types of capacitors and illustrates the use of different materials to make them smaller and more compact.

CAPACITOR RATINGS

Type and Material	Available Capacitance Values	Tolerances (%)	Leakage Resistance (MΩ)	Maximum Voltage Ranges (WVDC)	Useful Frequency Ranges (Hz)
Fixed					
Electrolytic					
Aluminum	1 μF–0.5 F	+100 to −20	1	up to 500 V	10–10^4
Tantalum	0.1 μF–2000 μF	+100 to −20	100	up to 300 V	0–10^3
Electrostatic					
Polystyrene	500 pF–20 μF	±0.5	10,000	up to 1000 V	0–10^{10}
Mylar	5000 pF–20μF	±20	10,000	100 V to 600 V	10^2–10^6
Paper (oil-soaked)	1000 pF–50 μF	±10 to ±20	100	100 V to 100 kV	10^2–10^6
Mica (silvered)	1 pF–0.1 μF	±1 to ±20	1000	500 to 75 kV	10^3–10^{10}
Ceramic (low-loss)	1 pF–0.001 μF	±5 to ±20	1000	6000 V	10^3–10^{10}
Variable					
Air	10 pF (unmeshed) to 500 pF (meshed)	±0.1		500 V	

FIGURE 8.11. Each type of capacitor comes in specific ranges of values and has certain performance characteristics. The table summarizes the typical range of values for different capacitor types.

take into account their wide tolerance. Often the right capacitor must be determined by trial and error.

Maximum Working Voltage

The **maximum working voltage** (WVDC) is the maximum dc voltage that can be applied to the capacitor without damaging it or causing significant charge leakage through the dielectric. It is generally expressed in dc volts — for example, 50 WVDC — but may also be expressed in ac volts for ac applications.

Insulation Resistance

Insulation resistance (IR), also called **leakage resistance,** is the effective resistance of the discharge path *through* the dielectric and measures a capacitor's ability to retain its charge. It is ex-

pressed either in megohms or as a time constant ($R \times C$) that is a product of the dielectric's resistance and the value of capacitance (megohms × microfarads). The larger the value of insulation resistance, the smaller the leakage current within the capacitor and the longer it will store a given charge.

In some cases, the maximum dc leakage current (in microamperes) will be given rather than the insulation resistance. The leakage current is the maximum dc current that will flow through the dielectric at the maximum working voltage.

Frequency Range

Every capacitor is designed to operate within a specific range of frequencies. Above the maximum operating frequency, its behavior may depart significantly from that of a simple capacitor.

FIGURE 8.12. Capacitors have a wide range of electronic applications. The application described here in Part III is timing, in which a capacitor in series with a resistor sets a specific time delay in a circuit. Other applications are described in other chapters.

CAPACITOR APPLICATIONS

Application	Part Where Treated
Filtering	VI Power supplies
Bypassing	
Coupling	VII Alternating current, voltage, and impedance
Blocking	
Tuning	
Trimming	VII Alternating current, voltage, and impedance
Timing	III Time-varying signals

8.9 TYPICAL APPLICATIONS

Capacitors are used in filtering, bypassing, coupling, tuning, timing, trimming, and energy storage. These applications are briefly described in the following sections to show what requirements they impose on the choice of capacitor for a job. Figure 8.12 indicates the parts of text where the applications are described in depth.

Filtering

Filtering is required to remove any ac voltages that are superimposed on the dc output in dc power supplies. High-voltage, large-capacitance units are needed, and electrolytics are normally used. The details of selecting a capacitor for dc power supplies are described in Part VI.

Bypassing and Coupling

Bypassing and coupling serve opposite functions. In **bypassing** applications, the capacitor is used to bypass, or short, an ac signal of a certain frequency to ground. In **coupling**, the capacitor is used to pass, or couple, a particular ac signal from one stage of a circuit to the next. The capacitor in the ac input of an oscilloscope is an example of a coupling (or blocking) capacitor (see Figure 7.51). Both applications are treated in more detail in Part VII. Depending on the frequency involved, either electrolytic or electrostatic units may be used in bypassing and coupling.

Tuning and Trimming

Tuning and trimming are special applications that call for variable capacitors. These applications are used in circuits that are designed to be sensitive to only one frequency. Tuning and trimming, too, are treated in Part VII.

Timing

Timing is the capacitor application discussed here in Part III. In timing applications, the capacitor is used in conjunction with a series resistor to produce a specific time delay in the circuit. This time delay is called the **time constant** of the circuit and is determined by the product of the capacitance and the series resistance (RC). This application requires a wide range of stable capacitance values.

Charge leakage is a very important consideration in timing applications because the timing cycle is set by the time for charge or discharge through the series resistor. Any leakage through the capacitor dielectric will change the timing. For small time-constant applications, high–leakage resistance electrostatic units are recommended. For large time constants, high–leakage resistance tantalum capacitors are generally used.

Energy Storage

Energy storage is a special application that requires capacitors to store large amounts of charge and results in high stress on the capacitor's components. Capacitors for this application are rated specifically for this purpose.

8.10 CHARGING A CAPACITOR WITH A CONSTANT CURRENT

To investigate how the voltage across a capacitor changes with time, we next examine a circuit in which a capacitor is charged by a current that is constant. Figure 8.13 shows a capacitor connected directly to a constant-current source. The current source supplies positive charge to the upper capacitor plate at a constant rate while removing an equal amount of positive charge from the lower plate. The rate at which the charge moves (the current) is maintained at a constant value by the constant-current source.

Charge-Current Relation

Because the current has a constant value, the charge Q (in coulombs) stored on either plate after the circuit has been operating for a length of time t is simply the current I in amperes (coulomb/s) multiplied by the time in seconds:

CAPACITOR CHARGED BY CONSTANT CURRENT

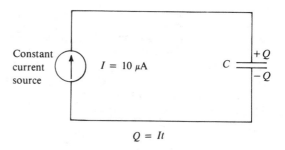

$$Q = It$$

FIGURE 8.13. The simplest analysis of capacitor behavior is in a circuit in which it is charged by a constant current. In this case, the charge on the capacitor is directly proportional to the charging current I and the length of time it is on t.

$$Q = It \qquad (6)$$

EXAMPLE 3: If the charging current is 10 μA for 10 s, how much charge is stored in the capacitor?

Solution: A current of 10 μA is 10×10^{-6} A. The charge after 10 s is:

$$Q = It = (10 \times 10^{-6} \text{ A})10 \text{ s}$$
$$= 10^{-4} \text{ coulombs}$$

Thus, there is a positive charge of 10^{-4} coulombs on one plate and a negative charge of the same value on the other.

Voltage-Time Relation

Of somewhat more interest is the voltage across the capacitor after a time t. As the plates become charged, the voltage drop across the capacitor increases. Remember from Equation 2 that the charge and voltage are related by:

$$Q = CV$$

FIGURE 8.14. When a capacitor is charged by a constant current, the capacitor voltage increases linearly with time. The charging rate in V/s is inversely proportional to the capacitance. The larger the capacitance, the smaller the charging rate and the longer it takes to get to a given voltage.

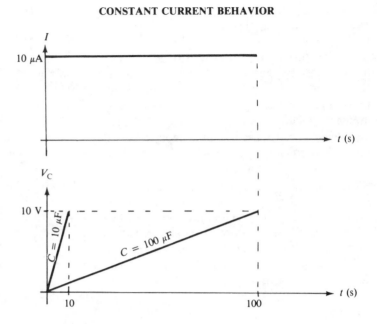

CONSTANT CURRENT BEHAVIOR

For a constant charging current, the charge Q increases with time according to Equation 6:

$$Q = It$$

Putting these two equations together gives:

$$It = CV$$

Rewriting this relationship in terms of voltage gives:

$$V = \frac{It}{C} \qquad (7)$$

When the charging begins at $t = 0$, the voltage across the capacitor is also zero. The voltage then increases linearly with time as long as there is current in the circuit. The graphs in Figure 8.14 show the behavior of current and voltage as a function of time for the circuit in Figure 8.13.

EXAMPLE 4: In Figure 8.13, the charging current is constant at 10 μA and the capacitor is required to charge to 10 V in 100 s. What value of capacitance should be used?

Solution: Equation 7 gives the relationship between voltage and capacitance for various values of charging current I and time t. Rearranging terms for C gives:

$$C = \frac{It}{V}$$

Solving for C gives:

$$C = \frac{(10 \times 10^{-6}\ \text{A})100\ \text{s}}{10\ \text{V}} = 10^{-4}\ \text{F}$$
$$= 100\ \mu\text{F}$$

EXAMPLE 5: If the 100 μF capacitor is

replaced by a 10 μF capacitor, how long will it take the capacitor voltage to reach 10 V?

Solution: Rewriting Equation 7 in terms of the charging time t gives:

$$t = \frac{CV}{I}$$

Solving for t:

$$t = \frac{(10 \times 10^{-6} \text{ F}) \, 10 \text{ V}}{10 \times 10^{-6} \text{ A}} = \textbf{10 s}$$

Thus, the 10 μF capacitor will charge to 10 V ten times faster than the 100 μF capacitor. The charging curves for both capacitors are shown in Figure 8.14. The slope for the 10 μF capacitor is ten times greater than the slope for the 100 μF capacitor.

Note that the slope of the charging curve is the charging rate of the circuit in V/s. This slope is obtained from Equation 7 as the ratio of V/t:

$$\text{charging rate} = \frac{V}{t} = \frac{I}{C}$$

$$\boxed{\text{charging rate} = \frac{I}{C} \qquad (8)}$$

The charging rate is also the slew rate of the capacitor circuit. If the current remains constant, the charging rate is inversely proportional to the capacitance; increase C by a certain factor and the charging rate will decrease by the same factor.

EXAMPLE 6: What are the charging rates for Examples 4 and 5?

Solution: The charging rate is given by Equation 8 in units of V/s:

$$\text{charging rate} = \frac{I}{C}$$

For Example 4, this expression gives:

$$\text{charging rate} = \frac{10 \times 10^{-6} \text{ A}}{100 \times 10^{-6} \text{ F}}$$
$$= \textbf{0.1 V/s}$$

For Example 5, it gives:

$$\text{charging rate} = \frac{10 \times 10^{-6} \text{ A}}{10 \times 10^{-6} \text{ F}}$$
$$= \textbf{1.0 V/s}$$

Note that as the capacitance increases, the charging rate decreases and it takes longer to charge to a given voltage (see Figure 8.14).

A capacitor cannot be charged indefinitely, of course. We have assumed that while the capacitor is charging, the value of the current remains constant. As long as this condition is met, the voltage will continue to increase with time. However, the circuit supplying the constant current cannot do so indefinitely. Generally, a constant-current source can supply a constant current only up to some value of output voltage. After that, the current drops and finally becomes zero. Also, the maximum voltage of the capacitor must not be exceeded. If it is, the dielectric may break down and allow substantial current leakage through the capacitor.

8.11 CHARGING A CAPACITOR WITH A CONSTANT VOLTAGE

Equally common in an electronic circuit is the charging of a capacitor by a constant voltage.

CAPACITOR CHARGED BY CONSTANT VOLTAGE

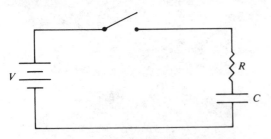

FIGURE 8.15. A capacitor can also be charged by a constant voltage. In this case, a resistor is added in series with the capacitor to limit the charging current. The resistance determines the charging rate and hence the capacitor's voltage-time behavior.

The voltage-time relation for this situation is somewhat more complicated than for a constant current, but it is equally important to learn because it has many electronic applications.

Voltage-Time Relation

Figure 8.15 shows a capacitor being charged by a constant *voltage* rather than a constant current. In this circuit, the capacitor is in series with a battery and a resistor. The resistor has been added to limit the value of the charging current and thus determine how long it will take for the capacitor to charge. The value of the resistance is critical to the capacitor's voltage-time behavior.

If the capacitor is initially uncharged, the voltage across it is zero. When the switch is closed, the voltage across the capacitor is still zero because it has not yet charged. Thus, the initial voltage drop across the *resistor* must be the battery voltage V. By Ohm's law, the initial charging current, then, is $I_0 = V/R$, which is the current that begins to charge the capacitor.

Let us now consider the condition of the capacitor a long time after the switch has been closed. Charge will have flowed from the battery to the capacitor, and the capacitor voltage will have increased. But, the maximum possible value of the voltage across the capacitor is the battery voltage V. When the capacitor voltage equals V, no more charge will flow and the current will be zero. If the current is zero, then the voltage across the resistor will also be zero.

Thus, from the time the switch is closed ($t = 0$) until a long time afterward, the capacitor voltage goes from 0 to V, the resistor voltage goes from V to 0, and the current goes from V/R to 0. The actual charging characteristics of the circuit are shown in Figure 8.16. The change in voltage across the capacitor V_C is shown in Figure 8.16a. Note that it begins at 0 V and gradually approaches the battery voltage V.

Figure 8.16b shows the change in voltage across the resistor V_R. This behavior is simply a mirror image of Figure 8.16a. It can be understood by Kirchhoff's voltage law that requires the sum of V_C and V_R to be constant and equal to the battery voltage:

$$V = V_C + V_R \tag{9}$$

Thus, the sum of V_C and V_R at any instant in time on the curves of Figures 8.16a and 8.16b will equal V.

The charging current I is shown in Figure 8.16c. By Ohm's law, this is the voltage across the resistor at any instant in time divided by its resistance:

$$I = \frac{V_R}{R} \tag{10}$$

Initially (when $t = 0$), the value of V_R is the battery voltage V because the capacitor voltage is zero. Thus, the initial charging current I_0 is:

$$I_0 = \frac{V}{R} \tag{11}$$

The current gradually decreases to zero as V_R

CONSTANT VOLTAGE BEHAVIOR

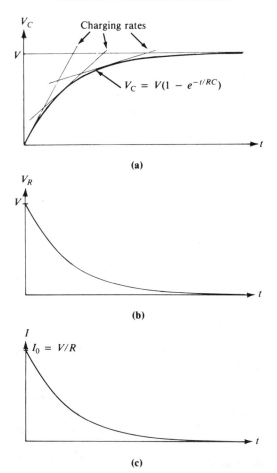

(a)

(b)

(c)

FIGURE 8.16. The capacitor voltage follows exponential behavior going from 0 V to the battery voltage V. By Kirchhoff's law, the voltage across the resistor must simultaneously decrease by an inverted exponential. The charging current similarly follows a decreasing exponential, approaching zero as the capacitor gets fully charged. The slope of the $V_C = t$ curve is the charging rate that decreases to 0 as V_C approaches V.

approaches zero and the capacitor becomes fully charged. Because the charging current is directly proportional to the resistor voltage, the curve of its behavior is identical to the curve for

V_R. The value of I_0 in Figure 8.16c at any time t can be found by dividing the value of V_R at the same time t in Figure 8.16b by the value of R.

Of particular interest is the manner in which the rate of capacitor charging changes. The **charging rate** for the capacitor voltage is given by Equation 8 in units of V/s:

$$\text{charging rate} = \frac{I}{C}$$

Substituting in the initial charge current, Equation 11, gives:

$$\text{initial charging rate} = \frac{I_0}{C} = \frac{V}{RC} \qquad (12)$$

This is the rate at which the capacitor first begins to charge and is thus the initial slope of Figure 8.16a. Of course, this expression is valid only for a very short time when the capacitor voltage is still zero.

At later times, the capacitor current is given by Equation 10, so the capacitor charging rate, Equation 8, becomes:

$$\text{charging rate} = \frac{V_R}{RC}$$

Substituting the value of V_R from Equation 9 gives:

$$\text{charging rate} = \frac{V - V_C}{RC} \qquad (13)$$

Since V_C and R are constant, the charging rate depends solely on V_C. As V_C increases, the charging current I decreases and the charging rate becomes smaller and smaller. Finally, when the capacitor is fully charged, $V_C = V$, the charging current is zero, and the charging rate also is zero (Figure 8.16a).

The curves in Figure 8.16 illustrate a specific and common type of physical behavior. The equation that describes this behavior is called an **exponential function**. For the curve in

Figure 8.16a, it is written:

$$V_C = V(1 - e^{-t/RC}) \tag{14}$$

The term e is a constant that is the base of the natural logarithms.

Equation 14 enables us to calculate the value of capacitor voltage V_C for any time t after the switch has been closed. Calculating this expression is seldom required, however, so the details are not discussed here. It is a subject that is covered in most math texts, and many pocket calculators will calculate an exponential function at the push of a single key. Nevertheless, it is useful to know some of the properties of the exponential function because they are used to describe the charging characteristics of a capacitor and many other electronic phenomena.

Time Constant

One question that arises during the examination of an exponential curve like Figure 8.16a is, How long does it take for the capacitor to become fully charged? Figure 8.16a shows that it is difficult to give a precise answer because the closer the capacitor voltage gets to V, the slower the charging rate. Equation 13 says that as V_C approaches V, the charging rate approaches zero. In fact, according to Equation 14, the capacitor voltage mathematically *never* reaches V; it only approaches closer and closer. This conclusion is not very helpful, however, and we need a way to describe this unusual behavior.

The exponential behavior of voltage versus time shown in Figure 8.16a is the same for any linear RC combination supplied by a constant voltage source. The only difference from one circuit to the next is the final capacitor voltage and the scale of the time axis. The final capacitor voltage, of course, is determined by the value of the voltage source V. The time scale, on the other hand, is determined by the *rate* at

which the capacitor charges (Equation 13):

$$\text{charging rate} = \frac{V - V_C}{RC}$$

This rate (in units of V/s) depends on three circuit quantities: the charging voltage V, the resistance R, and the capacitance C. If the charging voltage is fixed, the rate at which the capacitor charges depends on the constant factor RC.

From Equation 7, we see that the combination RC has the dimension of time:

$$V = \frac{It}{C}$$

or

$$\frac{V}{I}C = t$$

By Ohm's law, V/I is a resistance R, so:

$$RC = t \tag{15}$$

Note that this result is confirmed by Equation 14. The exponent of e—that is, t/RC—must be a unitless quantity. Thus, RC and t must have the same units. If t is measured in seconds, then RC must be in seconds.

The product RC is called the **time constant** for this circuit and is generally represented by the Greek letter *tau* (τ):

$$\tau = RC \tag{16}$$

When R is in ohms and C is in farads, τ will be in seconds.

The time constant determines how rapidly the capacitor will charge and can be used to describe the charging behavior. According to Equation 14, the capacitor voltage V_C is equal to the applied V multiplied by a time-varying factor $(1 - e^{-t/\tau})$, where $\tau = RC$. This factor is dominated by the exponential function $e^{-t/\tau}$. As t goes from zero to infinity, $e^{-t/\tau}$ goes from one to zero—that is, when $t = 0$, $e^{-t/\tau} = 1$, and when $t = \infty$, $e^{-t/\tau} = 0$.

The rate at which $e^{-t/\tau}$ goes to zero as t goes to infinity is determined by the time constant τ (and hence RC). This is shown in Figure 8.17a. The table gives values of $e^{-t/\tau}$ for integer values of $t = \tau$. For example:

For $t = 1\tau$: $e^{-t/\tau} = e^{-1} = 0.368$

For $t = 2\tau$: $e^{-t/\tau} = e^{-2} = 0.135$

For $t = 3\tau$: $e^{-t/\tau} = e^{-3} = 0.050$

and so on.

According to Equation 14, however, we are not interested in $e^{-t/\tau}$ but in the quantity $1 - e^{-t/\tau}$. This is also given in Figure 8.17a. Note that as $e^{-t/\tau}$ goes from one to zero, $(1 - e^{-t/\tau})$ goes from zero to one. The value of $(1 - e^{-t/\tau})$ at any time t multiplied by the charging voltage V will give the capacitor voltage V_C at that time. Essentially, the quantity $(1 - e^{-t/RC})$ represents the fraction of V that the capacitor voltage has reached:

$$V_C = (\text{fraction of charge})V \qquad (17)$$

where $1 - e^{-t/\tau} = $ fraction of charge.

Because the rate at which $(1 - e^{-t/\tau})$ goes to zero is determined by τ (which, in turn, depends on the values of R and C), it is useful to make a graph of $(1 - e^{-t/\tau})$ as a function of τ, as shown in Figure 8.17b. This graph gives values of the *fraction of charge* number for Equation 17. For example, after one time constant τ, the capacitor voltage V_C will be 0.632 of its final value V, or $V_C = 0.632$ V; after two time constants (2τ), $V_C = 0.865$ V; after 3τ, $V_c = 0.95$ V; and so on.

Note that the initial slope of the charging curve (the initial charging rate) intersects the final voltage at $t = \tau$, for which $V_C = 0.632$ V. This fact, as well as the universal data given in Figure 8.17, can be used to solve any series RC circuits charged by a constant voltage.

EXAMPLE 7: In Figure 8.15, if $R = 1$ MΩ and $C = 1$ μF, when would the capacitor be fully charged?

**UNIVERSAL
EXPONENTIAL CHARGE BEHAVIOR**

t	$(1 - e^{-t/\tau})$	$e^{-t/\tau}$
0	0	1.000
τ	0.632	0.368
2τ	0.865	0.135
3τ	0.950	0.050
4τ	0.982	0.018
5τ	0.993	0.007
6τ	0.998	0.002
7τ	0.999	0.001

(a) Exponential Voltage Increase

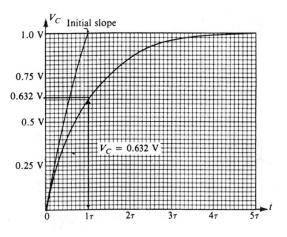

(b) Charge Curve

FIGURE 8.17. The charging curve of a series RC circuit connected to a constant voltage V follows an exponential equation. This behavior is described by the time constant of the circuit τ that is the product of R and C. The table and curve give values of capacitor voltage V_C at various time constants after charging has begun. These values can be used to solve most practical problems.

Solution: Mathematically, the capacitor is *never* fully charged, so we must specify some fraction of full charge that will be acceptable. Let us assume that when the capacitor is 99% charged — $V_C = 0.99$ V — that is close enough for our purposes. Figure 8.17(a) tells us that this condition will occur at a time equal to about five time constants:

$$t = 5\tau$$

Substituting values of R and C for τ gives:

$$t = 5RC = (5 \times 10^6 \, \Omega)10^{-6} \, \text{F} = \textbf{5 s}$$

EXAMPLE 8:   In Example 7, if the battery voltage were 15 V, when would the capacitor voltage reach 12 V?

Solution (a):   To solve this problem, we must first determine what fraction of the final voltage of 15 V is 12 V. Equation 17 gives:

$$\text{fraction of charge} = \frac{V_C}{V} = \frac{12 \text{ V}}{15 \text{ V}} = 0.8$$

$$V_C = 0.8 \text{ V}$$

According to the curve in Figure 8.16b, $V_C = 0.8$ V after a time of about:

$$t = 1.6\tau$$

Substituting values gives:

$$t = 1.6\tau = 1.6RC$$
$$= (1.6 \times 10^6 \, \Omega)10^{-6} \, \mu\text{F} = \textbf{1.6 s}$$

Note that electronic calculators that calculate natural logarithms will produce more accurate results and eliminate the need for Figure 8.17. To determine the time for the capacitor voltage to reach a given value, we simply rewrite Equation 14 for t:

$$t = -\left[\ln\left(1 - \frac{V_C}{V}\right)\right]\tau \qquad \textbf{(18)}$$

Solution (b):   Substituting values gives:

$$t = -[\ln(1 - 0.8)]\tau = -[\ln 0.2]\tau$$
$$= -[-1.6]\tau = 1.6\tau = 1.6RC = \textbf{1.6 s}$$

EXAMPLE 9:   An oscilloscope measures the charging curve shown in Figure 8.18. What is the time constant? If the circuit resistance is known to be 4.7 kΩ, what is the capacitance?

Solution (a):   In Figure 8.18 we see that $V = 7$ V. According to Figure 8.17a, after one time constant, $V_C = 0.63$ V. Thus, $t = \tau$ when the capacitor voltage is:

$$V_C = 0.63(7 \text{ V}) = 4.4 \text{ V}$$

Figure 8.18 shows that $V_C = 4.4$ V at about:

$$t = 8.0 \text{ ms}$$

Thus, the time constant for the circuit is:

$$\tau = 8.0 \text{ ms}$$

If the circuit resistance is 4.7 kΩ, then the circuit capacitance can be calculated from Equation 16:

$$\tau = RC$$
or
$$C = \frac{\tau}{R} = \frac{8.0 \text{ ms} \times 10^{-3} \text{ s/ms}}{4.7 \times 10^3 \, \Omega}$$
$$= 1.7 \times 10^{-6} \, \text{F} = \textbf{1.7 } \mu\textbf{F}$$

Solution (b):   The time constant can be graphically approximated by using the fact that the initial slope of the voltage curve will intersect the final voltage value at a time equal to the time constant. This relationship is shown in Figure 8.18 where the initial slope intersects $V = 7$ V about 8.0 ms after $t = 0$.

EXAMPLE 9

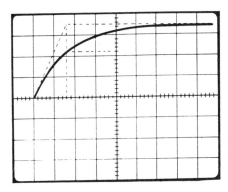

Vertical deflection: 2 V/cm
Sweep speed: 5 ms/cm

FIGURE 8.18. **The value of a capacitor can be determined experimentally by placing it in series with a known resistor and measuring the charging curve. The time constant can be determined from the curve and used to calculate C.**

This method of determining the time constant is not very accurate because it is difficult to accurately determine the initial slope. If only an approximate measure of τ is required, however, it may be suitable, and it has the advantage that no calculations are necessary.

8.12 DISCHARGING A CAPACITOR

The discharge behavior of a capacitor is essentially the same as its charge behavior. After a capacitor has been charged, it can be discharged quickly by putting a short circuit across its terminals; or, it can be discharged slowly by routing the discharge current through a resistor. Discharge through a resistor produces an exponential discharge curve that is simply an inversion of the charge curve.

Figure 8.19 shows a circuit in which a two-position switch can be used to charge and discharge a capacitor C through a resistor R. In switch position 1, the capacitor will charge to a final voltage V with a time constant of $\tau = RC$. Changing the switch to position 2 provides a closed circuit for the capacitor to discharge through the same resistor R. The slope of the discharge curve is identical to the charge curve except that it is inverted (Figure 8.19).

If the capacitor charges through one resistor, but discharges through a different resistor, as shown in Figure 8.20, then there will be a different curve for each part of the cycle. If the charging resistor R_1 is much less than the discharging resistor R_2, the capacitor will require much less time to charge than to discharge.

The shape of the discharge curve, like that of the charge curve, is an exponential function, but it is described by a slightly different expression to account for the fact that the initial voltage is V, not zero:

$$V_C = Ve^{-t/RC} \qquad (19)$$

The time constant for a discharge curve is the same as for a charge curve:

$$\tau = RC \qquad (20)$$

where the resistor R is the resistance of the *discharge path* R_2 in Figure 8.20.

Much like the charging behavior, the capacitor voltage at any time t during discharge is simply the value of an exponential function (in this case $e^{-t/\tau}$) times the initial voltage V:

$$V_C = (\text{fraction of discharge})V \qquad (21)$$

where $e^{-t/\tau}$ = fraction of discharge.

When $t = 0$, $e^{-t/\tau} = 1.00$, and the capaci-

FIGURE 8.19. The discharging behavior of a capacitor through a resistor is similar to the charging behavior. In switch position 1, the capacitor charges through a resistor R to the battery voltage following an increasing exponential with a time constant $\tau = RC$. In position 2, it discharges through the same resistor following a decreasing exponential with the same time constant.

CAPACITOR DISCHARGE

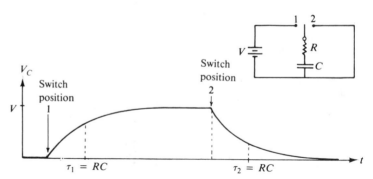

DIFFERENT CHARGE-DISCHARGE BEHAVIOR

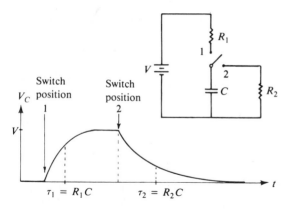

FIGURE 8.20. If the discharge resistor is different from the charge resistor, the discharge curve will be exponential, but the time constant will be different.

tor voltage is the initial voltage V. As time passes, $e^{-t/\tau}$ gradually decays to zero at a rate determined by the time constant τ. Figure 8.21 illustrates the rate at which $e^{-t/\tau}$, and thus the capacitor voltage, decays to zero for various values of τ. Note also that the initial slope of the discharge curve intersects $V = 0$ at the time $t = \tau$, for which $V_C = 0.368$ V.

EXAMPLE 10: In Figure 8.19, if $R = 1$ MΩ and $C = 1$ μF, when would the capacitor be fully discharged?

Solution: As in Example 7, we select 0.99 as the fraction representing "full" discharge. This state occurs after five time constants, or:

$$t = 5\tau = 5RC = 5 \text{ s}$$

EXAMPLE 11: If the battery voltage in Example 10 were 15 V, when would the capacitor voltage reach 12 V?

Solution (a): First, we determine what fraction of the discharge has occurred. The fraction of discharge can be obtained from Equation 19:

$$\text{fraction of discharge} = \frac{V_C}{V} = \frac{12 \text{ V}}{15 \text{ V}}$$
$$= 0.8$$

$$V_C = 0.8 \text{ V}$$

Figure 8.21 shows that $V_C = 0.8$ V after a time of about 0.2τ. Thus:

$$t = 0.2\tau = 0.2RC$$
$$= (0.2 \times 10^6 \ \Omega)10^{-6} \ F = \mathbf{0.2 \ s}$$

Solution (b):   As in Example 8, we can obtain the same result with a calculator by using the equation:

$$t = -\left[\ln\left(\frac{V_c}{V}\right)\right]\tau = -\left[\ln\left(\frac{12}{15}\right)\right]\tau \qquad \textbf{(22)}$$

$$= -[\ln 0.8]\tau = 0.2\tau = \mathbf{0.2 \ s}$$

EXAMPLE 12:   An oscilloscope measures the capacitor discharge curve of Figure 8.22. What is the time constant?

Solution (a):   According to Figure 8.21a, after one time constant, the fraction of discharge is:

$$V_C = 0.37 \ V$$

Figure 8.22 shows that $V \cong 15 \ V$. Thus, $t = \tau$ when the capacitor voltage is:

$$V_C = 0.37 \times 15 \ V = 5.6 \ V$$

Figure 8.22 shows that this occurs at about:

$$t = 2.6 \ ms$$

so that the time constant is:

$$\tau = \mathbf{2.6 \ ms}$$

Solution (b):   We can also solve the problem graphically by extending the initial slope of the discharge curve to cross the line $V_C = 0$. Using Figure 8.22, this method gives the same result as Solution (a), but again, it is only an approximation.

**UNIVERSAL
EXPONENTIAL DISCHARGE BEHAVIOR**

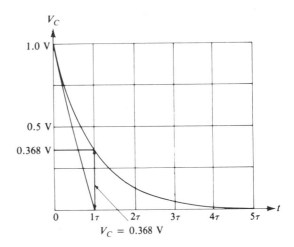

t	$e^{-t/\tau}$
0	1.000
τ	0.368
2τ	0.135
3τ	0.050
4τ	0.018
5τ	0.007
6τ	0.002
7τ	0.001

(a) Exponential Voltage Decrease

(b) Discharge Curve

FIGURE 8.21.   The universal discharge curve is a simple inversion of the charge curve. Because the initial voltage at $t = 0$ is V and not zero, the fraction of discharge data is simply the value of the exponential function $e^{-t/\tau}$. Compare with Figure 8.17.

EXAMPLE 12

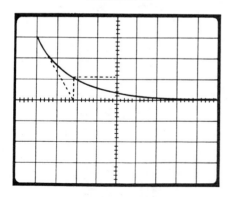

Vertical deflection: 5 V/cm
Sweep speed: 2 ms/cm

FIGURE 8.22. The discharge curve of a capacitor through a known resistor can also be used to experimentally determine its value.

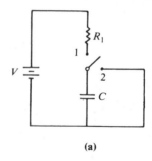

(a)

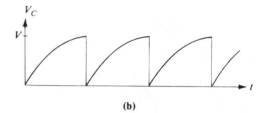

(b)

8.13 SAWTOOTH GENERATOR

Principles of Operation

A common application of the charging and discharging behavior of a capacitor is the sawtooth generator that is used, for example, in the horizontal sweep circuit of an oscilloscope. Consider the RC circuit shown in Figure 8.23a, where the discharge resistor from Figure 8.20 has been reduced to zero. If we rapidly changed the switch in Figure 8.23a from position 1 to position 2, the capacitor would alternately charge through resistor R_1 and quickly be discharged through the short circuit. The resulting voltage across the capacitor is shown in Figure 8.23b.

If the switching rate is increased so that the capacitor is discharged before it is fully charged, as shown in Figure 8.23c, the waveform begins to look like a sawtooth wave, but

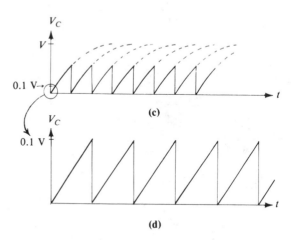

(c)

(d)

FIGURE 8.23. A capacitor can be used to generate a sawtooth wave by charging it through a series resistor and then rapidly discharging it through a short circuit. In circuit (a), this effect is achieved by a mechanical switch. To reduce the curvature of the exponential curve (b), the switch rate is increased (c) to the point that only the initial, nearly linear segment of the charging exponential is used (d).

UNIJUNCTION TRANSISTOR AND SYMBOL

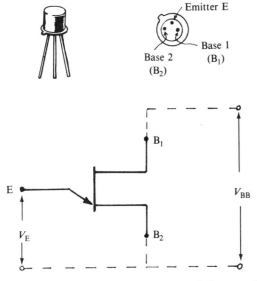

FIGURE 8.24. The rapid switching of the capacitor voltage can be done electronically with a unijunction transistor (UJT). The figure shows a typical unit and its symbol. The switch is between the emitter E and base B_2 junction.

there is a slight curvature due to the exponential nature of the charging curve. In the extreme case, when the switching rate is so fast that the capacitor is discharged just as it begins to charge, the curvature is much less severe. The resulting waveform looks like Figure 8.23d, in which the sawteeth are greatly magnified segments of the initial portions of the charge curve (Figure 8.23b). In practice, the amplitude of this sawtooth wave could be increased to any desired value with a suitable voltage amplifier.

Apparently, then, if we had a way to rapidly charge and discharge a capacitor, we could use a constant-voltage charging circuit, like the one in Figure 8.23, to produce a sawtooth wave. What we need is an electronic, voltage-sensitive switch that will open to let the capacitor charge through a resistor, close when the capacitor reaches some predetermined value to quickly discharge it, and then reopen to permit the capacitor to recharge. Fortunately, such a switching device exists; it is called a unijunction transistor.

Unijunction Transistor

A **unijunction transistor** (UJT) is a three-terminal device with two base terminals, marked B_1 and B_2, and a third emitter terminal, marked E (Figure 8.24). The "switch" of the UJT is between the emitter E and the base B_2. When the voltage of the emitter V_E relative to base B_2 is zero, the EB_2 junction acts like an open switch and conducts no current.

As V_E increases, the EB_2 junction remains an open circuit until V_E reaches some **peak voltage** V_p, determined both by the voltage across the two bases V_{BB} and by a characteristic of the UJT according to:

$$V_p = \eta V_{BB} \tag{23}$$

The proportionality constant η (Greek letter *eta*) depends on the particular UJT and has a typical value in the range of 0.6 to 0.8.

When V_E reaches V_p, the EB_2 junction quickly closes to produce a short circuit across EB_2. This switch will remain closed until the emitter voltage drops to some minimum value, called the **valley voltage** V_v. The valley voltage depends on the base voltage V_{BB}, but it is typically 1 V to 3 V. When $V_E = V_v$, the EB_2 switch abruptly opens again. Hence, the UJT acts like a voltage-sensitive switch that has the following properties (Figure 8.25):

EB_2 switch opened: $\quad V_E < V_p \quad$ **(24)**
EB_2 switch closed: $\quad V_E > V_p$
EB_2 switch reopened: $\quad V_E < V_v$

The UJT can be used to charge and discharge a capacitor in the circuit shown in Figure 8.26. The capacitor C charges exponentially

FIGURE 8.25. The operation of the UJT as a switch can be understood in terms of the equivalent circuit shown. For V_E less than the peak voltage V_p, switch EB$_2$ is open. When V_E exceeds V_p, switch EB$_2$ closes. When V_E drops to the valley voltage V_v, switch EB$_2$ reopens.

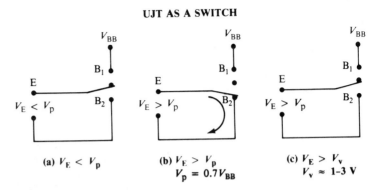

UJT AS A SWITCH

(a) $V_E < V_p$

(b) $V_E > V_p$
$V_p = 0.7V_{BB}$

(c) $V_E > V_v$
$V_v \approx 1\text{--}3$ V

FIGURE 8.26. In the sawtooth wave generator using a UJT switch, the capacitor voltage oscillates continually between V_v and V_p. Two disadvantages of this circuit are the curvature of the sawtooth wave and the fact that V_C does not go to zero.

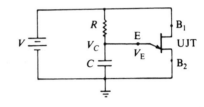

CONSTANT VOLTAGE SAWTOOTH GENERATOR

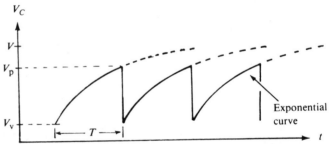

through resistor R according to the time constant RC. The capacitor voltage is connected to the UJT emitter so that $V_C = V_E$. The switching voltage V_p is set by the voltage across the two bases and is simply the battery voltage V. According to Equation 23:

$$V_p = \eta V \approx 0.7 \text{ V}$$

Thus, when the capacitor voltage reaches about 0.7 V, the EB$_2$ switch closes and discharges the capacitor. When the capacitor voltage reaches the valley voltage, the EB$_2$ switch reopens and

the capacitor charges again. The resulting waveform is shown in Figure 8.26. Note that the circuit oscillates continually between V_v and V_p, which is somewhat less than the full range of 0 to V. Note also that even though the voltage source is connected to the capacitor during the discharge, the discharge is so much faster than the charging time constant that we can ignore this fact.

The period of oscillation T is determined largely by the time constant of the circuit RC, but is somewhat influenced also by the value of η, according to:

$$T \approx RC \ln\left(\frac{1}{1 - \eta}\right) \text{ s} \qquad (25)$$

For the typical range of values of η (0.6 to 0.8), this relation gives:

$$T \approx 1.2RC$$

EXAMPLE 13: Design a sawtooth generator with a frequency of 1000 Hz.

Solution: The basic circuit for the sawtooth generator is shown in Figure 8.26. It uses a constant voltage source V to charge an RC combination and a UJT switch. Because the period of the sawtooth generator, Equation 23, does not depend on the value of the voltage source, we can select any convenient value for V, provided that it is somewhat greater than the UJT valley voltage of about 3 V. For example:

$$V = +15 \text{ V}$$

The period of oscillation is simply the inverse of the frequency, Equation 1:

$$T = \frac{1}{f} = \frac{1}{1000 \text{ Hz}} = 0.001 \text{ s}$$

Equation 25 can now be used to calculate R and C. We will assume that the value of η for the UJT is 0.7, in which case Equation 25 becomes:

$$T = 1.2RC = 0.001 \text{ s}$$

or

$$RC = 0.00083 \text{ s}$$

We are free to select any values of R and C whose product is 0.00083. For convenience, we will use:

$$R = 5 \text{ k}\Omega$$

and then calculate a value for C:

$$5000C = 0.00083$$
$$C = \frac{0.83 \times 10^{-3} \text{ s}}{5 \times 10^{3} \ \Omega}$$
$$= 0.17 \times 10^{-6} \text{ F} = 0.17 \ \mu\text{F}$$

To get a frequency of exactly 1000 Hz, we select a stock value capacitor of **0.2 μF** and use a variable resistor of **10 kΩ**. We would then carefully adjust the variable resistor to give the desired frequency.

While the circuit shown in Figure 8.26 will produce some semblance of a sawtooth wave, it has two disadvantages. First, the voltage across the capacitor is not linear, but displays the rounded characteristic of an exponential curve. Second, the capacitor voltage does not go to 0 V, but only to the valley voltage V_v, typically 1 V to 3 V. Both of these problems can be eliminated by using a more complicated circuit that is beyond the scope of this book.

It may have occurred to you that the first problem can be solved simply by charging the capacitor with a constant current rather than a constant voltage. This approach is shown in Figure 8.27, along with the resulting waveform. The period of oscillation for this circuit depends on the time it takes the constant-current source to charge the capacitor from V_v to V_p. According to Equation 23, the value of V_p is determined by the value of V_{BB} that in Figure 8.26 is the voltage V:

$$V_p = \eta V$$

The time it takes to charge the capacitor by an amount $V_p - V_v$ is given by Equation 7:

$$t = \frac{CV}{I} = \frac{C}{I}(V_p - V_v)$$

FIGURE 8.27. The curvature in the sawtooth wave produced by a constant voltage source can be eliminated by charging the capacitor with a constant-current source.

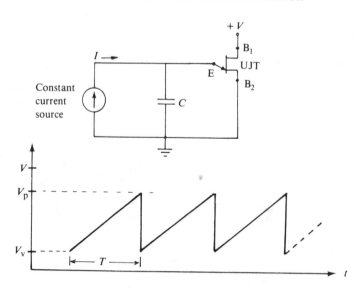

CONSTANT CURRENT SAWTOOTH GENERATOR

So, the period of oscillation is:

$$T = \frac{C}{I}(V_p - V_v) \qquad (26)$$

In this case, the period depends on the capacitance C, the charging current I, and the values of V_p and V_C that are determined by the UJT.

8.14 555 TIMER

Early in the chapter, it was stated that the capacitor is the key component in circuits that electronically generate and process time-varying signals. The capacitor characteristic that permits these applications is the time delay that occurs as it charges. By combining this time delay with an electronic switch, such as the UJT, we can construct a variety of signal-generating circuits. In fact, there is a large class of electronic circuits whose performance relies on these two basic elements—a voltage that increases (or de-

creases) with time, and a circuit or circuit element that changes its behavior at a specific voltage level. These circuits are called **voltage level detectors**, or voltage threshold detectors.

The circuits in Figures 8.26 and 8.27 are basic forms of the voltage level detector circuit in which a capacitor and a UJT are used to produce a sawtooth wave. The following paragraphs describe a versatile integrated circuit called the **555 timer** that can act as a voltage level detector and perform many other tasks. While the internal operation of the 555 is relatively complicated, controlling its performance is done quite simply with an external resistor-capacitor combination.

Symbols and Terminals

The 555 integrated circuit is shown in Figure 8.28. It has eight terminals and is available in either a cylindrical TO-99 or a rectangular **dual-in-line package** (DIP). The focus here is on the DIP package.

An integrated circuit functions only when

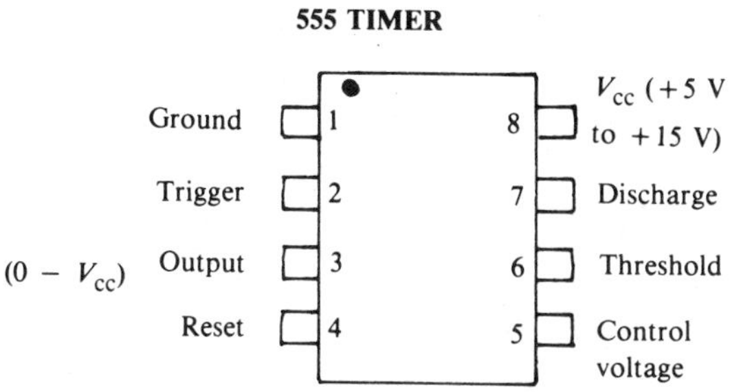

555 TIMER

(a) Top View: DIP (8-Pin) Case

(b) Top View: TO-99 Case

FIGURE 8.28. Another application of an *RC* network is controlling the behavior of an integrated circuit, such as the 555 timer. The 555 is a versatile integrated circuit with a wide range of applications. The applications described here use pin 1 (ground), pin 8 (external power supply), pin 2 (input trigger), pin 3 (output), and pins 6 and 7 for timing.

external power is applied. For the 555 DIP, the external voltage connections are pins 1 and 8. Pin 1 is connected to *ground* and pin 8 is connected to an external *dc supply voltage* V_{cc} that can be any value in the range of +5 V to +15 V.

The *output terminal* is pin 3. The output signal will have one of two values—either 0 V or a positive voltage that is about equal to the external supply voltage V_{cc}.

There are four control terminals. The most important of these are pins 2 and 6, labeled *trigger* and *threshold*. Both can act as voltage level detectors, and both can cause the output signal to switch from one value to the other.

These two pins play a key role in all of the circuits discussed in this section.

Pins 4 and 5, labeled *reset* and *control voltage*, are sometimes used to control the output of the 555, but in most applications they are not needed. Usually pin 5 is connected to ground through a 0.01 μF capacitor, and pin 4 is connected to the supply voltage.

Pin 7, labeled *discharge*, is an internal switch to ground. This switch is closed when the output is low (0 V) and open when the output is high ($\approx V_{cc}$). It is pin 7 that provides the path through which a capacitor can be discharged.

With eight pins, the 555 integrated circuit may appear to be a very complicated device. However, remember that a 555 was designed to be extremely versatile. This section explores only a few of the possible applications using only some of the terminals: pins 2 and 6 for control, pin 7 as a discharge switch, and pin 3 for output.

The principal point of this section is simply to demonstrate how capacitors can be used to control the timing of time-varying signals in a practical way. Hence, the 555 is treated essentially as a "black box," and only its basic functions are described. However, this introduction to the 555 should equip you to understand a more extensive description of its operation and applications as found in a text on integrated circuits.

Applications

The simplest way to explain the operation of the 555 timer is to describe some typical applications. The following paragraphs consider two of the most common applications of the 555: the timed switch and square wave generator.

Timed Switch: A **timed switch** is a switch that will close for a specified amount of time and then reopen. This circuit can be used in photography, for example, to open and close the shut-

TIMED SWITCH

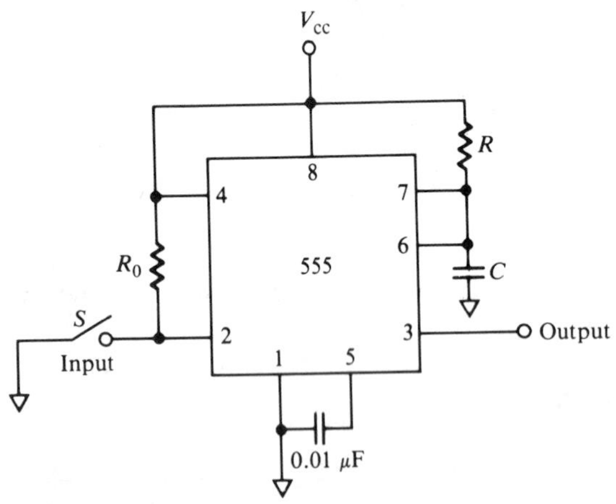

FIGURE 8.29. A basic application of the 555 timer is the timed switch. When switch S is closed, the output voltage goes from low (0 V) to high ($\approx V_{cc}$). It stays high for a period determined by the time constant RC. The high output voltage can be used to turn on a variety of devices – lights, motors, heaters, and so on.

ter of a camera or to turn on and off the light of a photographic enlarger. Another typical application is the **delayed switch**, in which the timer keeps a control switch closed for a specified amount of time after some event has occurred.

Figure 8.29 shows a 555 timer that has been wired to act as a timed switch. The general operation of the circuit is as follows: When the external switch S is open, the voltage at pin 2 is high at a value of V_{cc} that is provided through resistor R_0. If switch S is closed, the voltage at pin 2 goes low, 0 V. This causes the output at pin 3 to jump from low (0 V) to high (a voltage equal to the supply voltage V_{cc}). Meanwhile, capacitor C begins to charge through the resistor R.

The output voltage of pin 3 remains high until the capacitor voltage at pin 6 reaches a preset value. At this value, the output returns to

its low state of 0 V and the internal switch at pin 7 closes to discharge the capacitor. The circuit is then ready to be reactivated by closing the switch. Note that the time during which the output is high is determined by the time constant of the RC combination and not by the time that switch S is closed. This time delay is also variable with R.

This description of the circuit's operation is very general. Let us now look at the circuit in more detail to see what each pin controls and, in particular, how R controls the length of time the output is high.

Figure 8.30 shows the voltage as a function of time at the control pins 2 and 6 and the output pin 3. Closing switch S at $T = 0$ causes pin 2 to go low (0 V), as indicated in Figure 8.30a. The 555 is designed so that when pin 2 first goes low, the output pin 3 will go high (Figure 8.30c), and the internal switch of pin 7 will open to permit the capacitor to charge. The capacitor then charges with a time constant of RC (Figure 8.30b):

$$\tau = RC \qquad (27)$$

Then, if switch S is opened, the voltage at pin 2 will return to V_{cc}, but the rest of the circuit will be unaffected. In fact, to obtain a proper output pulse, switch S must be reopened before the output pulse is over. Hence switch S often is a *touch switch*, closing when it is pushed, but immediately springing back open when the finger is removed.

Pin 6 is the threshold terminal. As soon as the voltage of pin 6 reaches two-thirds of the supply voltage, $V_6 = \frac{2}{3}V_{cc}$, an internal threshold circuit triggers the output voltage at pin 3 to go low. When pin 3 goes low, the capacitor voltage at pin 6 also drops abruptly to zero. This sharp drop in voltage at pin 6 is caused by the triggering action of pin 6 on pin 7. At the same time that a voltage of $\frac{2}{3}V_{cc}$ causes pin 6 to drive pin 3 to ground, it also changes pin 7 from an

555 TIMED SWITCH OPERATION

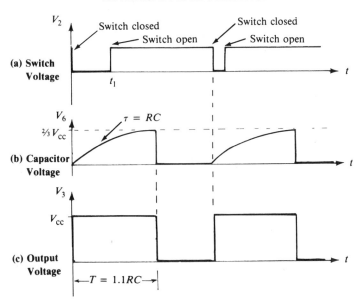

FIGURE 8.30. The 555 timed switch operates as follows: When switch S is closed, the input voltage at pin 2 goes low (a), the output at pin 3 goes high (c), and the internal switch to ground, pin 7, opens to let capacitor C charge through resistor R (b). When the voltage at pin 6 reaches $\frac{2}{3}V_{cc}$, the output goes low (c), and the switch to ground of pin 7 closes, discharging the capacitor. The period of high output voltage is $T = 1.1RC$.

open circuit to a short circuit to ground that quickly discharges the capacitor.

To summarize, grounding pin 2 makes pin 3 go high and opens the internal switch to ground at pin 7. The capacitor voltage increases from $\frac{1}{3}V_{cc}$ to $\frac{2}{3}V_{cc}$, with a time constant set by $\tau = RC$. When pin 6 reaches $\frac{2}{3}V_{cc}$, pin 3 goes low, the internal switch 7 closes, and the capacitor discharges to ground. The time interval T during which the output is high is determined by the capacitor charging time constant $\tau = RC$ according to:

$$T = 1.1\tau$$

$$\boxed{T = 1.1RC \qquad (28)}$$

Monostable Multivibrator: The timed switch shown in Figure 8.29 can be activated electronically by applying a suitable external voltage to pin 2 (Figure 8.31). The only requirement for triggering the timed switch is that the voltage

on pin 2 go from high to low. In Figure 8.29, the high voltage comes from the 555 power supply through resistor R and the low voltage from the closing of switch S to ground.

In Figure 8.30, the high and low voltages are both supplied by external time-varying signals. The requirements for the external trigger signal are that the high level of input voltage be greater than $\frac{2}{3}V_{cc}$ and the low level go below $\frac{1}{3}V_{cc}$ and that the transition back to a voltage greater than $\frac{2}{3}V_{cc}$ occur before the output signal returns to low (Figure 8.31).

The high output voltage of the 555 timer can be used to activate a switch, such as a relay, that can then be used to turn on and off a variety of devices—a light, a heater, a motor, and the like. Depending on the values of R and C, the time interval during which the output is high can vary widely from milliseconds to hours. Therefore, this simple electronically timed switch has many practical applications.

This type of circuit is called a **monostable multivibrator** because its output has only one

ONE-SHOT MULTIVIBRATOR

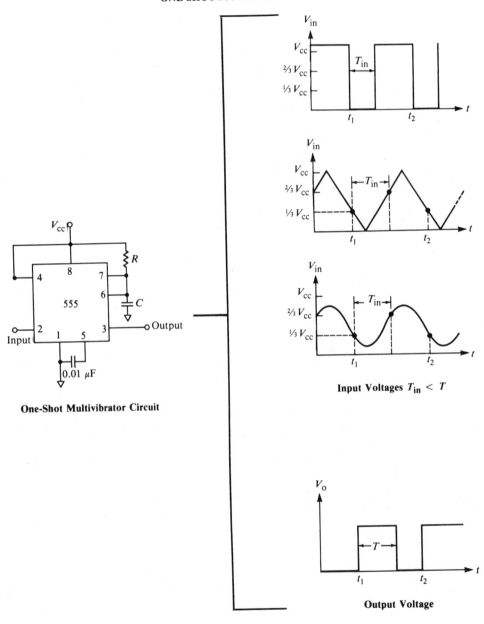

One-Shot Multivibrator Circuit

Input Voltages $T_{in} < T$

Output Voltage

FIGURE 8.31.   The timed switch can also be triggered electronically by applying a suitable input voltage to pin 2. For proper triggering, the input signal must go from high (more than $\frac{2}{3}V_{cc}$) to low (less than $\frac{1}{3}V_{cc}$) and then back to more than $\frac{2}{3}V_{cc}$ in a time less than the output pulse width. In this application, the timed switch is called a one-shot (or monostable) multivibrator.

(mono-) stable value: 0 V. Closing the switch raises the output to +V, but only for a fixed length of time determined by τ. The output then returns to its original, stable value of 0 V. Hence, the output will "vibrate" from 0 V to +V and back again, once for each time the input switch is closed.

This circuit is also called a **one-shot multivibrator** or *Schmidt trigger* because a trigger signal applied to pin 2 will produce one fixed-width output pulse for each closing of the switch. An important application of a one-shot multivibrator is wave shaping. As shown in Figure 8.31, a wide variety of input signals can be fed into the one-shot, but the output will always have the same pulse height and width. Thus, the one-shot often serves as an interface between an input transducer whose output may vary and subsequent circuits that require a standard input pulse for their operation.

EXAMPLE 14: Design a one-shot multivibrator with a pulse width of 3 s.

Solution: The basic circuit for the one-shot is shown in Figure 8.29. It uses a 555 timer. The pulse width for this circuit is given by Equation 28:

$$T = 1.1\,RC = 3\text{ s}$$
or
$$RC = 2.7\text{ s}$$

We can choose any values of R and C whose product is 2.7. As in Example 13, we select:

$$R = 5\text{ k}\Omega$$

and then calculate a value for C:

$$5000C = 2.7$$
$$C = \frac{2.7}{5000\ \Omega} = 5.4 \times 10^{-4}\text{ F} = 540\ \mu\text{F}$$

This is a fairly large value of capacitance, one that will be somewhat more expensive than smaller values. Therefore, let us reduce it by a factor of 10 and correspondingly increase the value of R by a factor of 10. Thus, our design values are:

$$R = 50\text{ k}\Omega$$
$$C = 54\ \mu\text{F}$$

As in Example 13, for a precise setting of $T = 3$ s, we would probably select a stock capacitor of **50 μF** and a variable resistor of **100 kΩ**.

Square Wave Generator: In contrast to a monostable multivibrator that has only one stable state, a square wave generator has no stable value. The output vibrates back and forth continuously between two values—usually 0 V and some positive voltage. This type of circuit is called an **astable multivibrator**. Figure 8.32 shows a 555 integrated circuit wired to make a square wave generator that incorporates a capacitor and two resistors. Note that the capacitor voltage, pin 6, is connected directly to the input trigger, pin 2.

To see how this circuit works, we begin by assuming that the capacitor is uncharged. Pins 6 and 2 are low, and the output (pin 3) is high. When the output is high, pin 7 is an open circuit and the capacitor begins to charge. Both resistors R_1 and R_2 are in the charging circuit, so the time constant for charging is:

$$\tau_1 = (R_1 + R_2)C \tag{29}$$

As for the timed switch, when the capacitor voltage at pin 6 reaches $V_6 = \frac{2}{3}V_{cc}$, an internal threshold circuit changes the output to low and pin 7 becomes a short to ground, discharging

SQUARE WAVE GENERATOR (ASTABLE MULTIVIBRATOR)

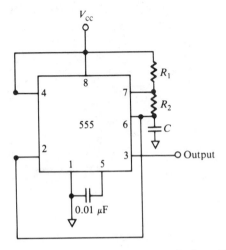

FIGURE 8.32. Another application of the 555 timer is the square wave generator (also called an astable multivibrator). For this application, the capacitor voltage at pin 6 is connected to the input trigger, pin 2, and a second resistor R_2 is placed in the discharge path.

the capacitor. During discharge, however, only resistor R_2 is in the discharge path. Thus, the time constant for discharge is:

$$\tau_2 = R_2C \tag{30}$$

which is less than τ_1.

The capacitor discharges until the voltage at the trigger terminal (pin 2) drops below $\frac{1}{3}V_{cc}$. At that point, an internal circuit switches the output to high and opens the switch at pin 7, and the capacitor begins to charge again. This cycle repeats continuously.

Figure 8.33 shows the output voltage V_o at pin 3 and the voltage across the capacitor V_C (pins 2 and 6). Neither of the output voltages is stable. If it is high, the capacitor is being charged; if it is low, the capacitor is being discharged. Thus, connecting the capacitor volt-

age to the input causes the output to vibrate back and forth between the two states and produces a square wave output.

The width of the output pulse t_1 is determined by the charging resistors of R_1 and R_2 according to:

$$t_1 = 0.7\tau_1$$

$$t_1 = 0.7(R_1 + R_2)C \tag{31}$$

while the time between output pulses t_2 is determined by the single discharge resistor R_2:

$$t_2 = 0.7\tau_2$$

$$t_2 = 0.7R_2C \tag{32}$$

The total period of the oscillation T is simply the sum of t_1 and t_2:

$$T = t_1 + t_2 = 0.7(R_1 + R_2 + R_2)C \tag{33}$$

$$T = 0.7(R_1 + 2R_2)C \tag{34}$$

The frequency of the square wave is, of course, the inverse of the period, or:

$$f = \frac{1}{T}$$

$$f = \frac{1}{0.7(R_1 + 2R_2)C} \tag{35}$$

The duty cycle of the square wave is the time that the voltage is high t_1 divided by the total period T, according to Equation 4 of Chapter 7:

$$\text{duty cycle} = \frac{t_1}{T} \times 100\%$$

$$= \frac{0.7(R_1 + R_2)C}{0.7(R_1 + 2R_2)C} \times 100\%$$

SQUARE WAVE GENERATOR OPERATION

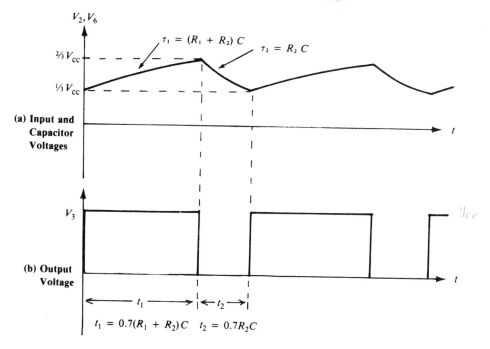

FIGURE 8.33. When the capacitor voltage is used as the input voltage, the output voltage goes high if V_6 drops below $\frac{1}{3}V_{cc}$ and low if V_6 exceeds $\frac{2}{3}V_{cc}$. The output pulse width t_1 is determined by the resistance of the charging path $(R_1 + R_2)$, while the time between pulses t_2 is determined by the resistance of the discharge path R_2. The period of the square wave is $t_1 + t_2$.

$$\text{duty cycle} = \frac{R_1 + R_2}{R_1 + 2R_2} \times 100\% \qquad (36)$$

A simple analysis of Equation 36 shows that the maximum duty cycle is 100% when R_1 is much greater than R_2. As R_1 decreases relative to R_2, the duty cycle decreases. The minimum possible duty cycle is 50%, when $R_1 = 0$. Hence, the duty cycle of the circuit in Figure 8.32 is limited to values between 50% and 100%.

The duty cycle can be decreased below 50% if the capacitor can be charged and discharged through separate resistors. This situation is shown in Figure 8.34, where a diode has been placed across resistor R_2. During the charging cycle, the diode is forward biased and acts as a short circuit around resistor R_2. Thus, the charging time constant is simply R_1C, and the width of the high output is:

$$t_1 = 0.7R_1C \qquad (37)$$

During the discharge cycle, however, the diode will not pass reverse current, so the discharge time constant and the width of the low output remain the same:

$$t_2 = 0.7R_2C \qquad (38)$$

The total period and duty cycle, then, are:

$$T = t_1 + t_2 = 0.7R_1C + 0.7R_2C$$

$$T = 0.7(R_1 + R_2)C \qquad (39)$$

$$\text{duty cycle} = \frac{t_1}{T} \times 100\%$$

$$= \frac{0.7(R_1C)}{0.7(R_1 + R_2)C} \times 100\%$$

$$\text{duty cycle} = \frac{R_1}{R_1 + R_2} \times 100\% \qquad (40)$$

R_1 and R_2 can be varied separately to give nearly a full range of output duty cycles from 0% up to 100%.

SQUARE WAVE GENERATOR
(EXTENDED DUTY CYCLE)

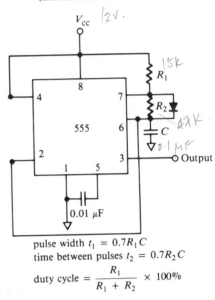

pulse width $t_1 = 0.7R_1C$
time between pulses $t_2 = 0.7R_2C$

$$\text{duty cycle} = \frac{R_1}{R_1 + R_2} \times 100\%$$

FIGURE 8.34. The duty cycle of the square wave generator shown in Figure 8.32 can be extended to 100% by shorting resistor R_2 with a diode during the charge cycle. Then, R_1 sets the pulse width and R_2 sets the time between pulses.

EXAMPLE 15: Design a square wave generator with a frequency of 500 Hz and a duty cycle of 25%.

Solution: Because the duty cycle is less than 50%, we must use the circuit shown in Figure 8.34. Thus, we need values of R_1, R_2, and C. (V_{cc} may be any convenient value between 5 V and 18 V because it was not specified.) The required period can be calculated from Equation 1 of Chapter 2:

$$T = \frac{1}{f} = \frac{1}{500} = 0.002 \text{ s}$$

Given the period, the duty cycle will determine the width of the high output pulse (Equation 3):

$$\text{duty cycle} = \frac{t_1}{T} \times 100\% = 25\%$$

$$t_1 = 0.25T = 0.25 \times 0.002 = 0.0005 \text{ s}$$

This value can be substituted into Equation 37 to get the charging time constant:

$$t_1 = 0.7R_1C = 0.0005 \text{ s}$$

$$R_1C = 0.00071 \text{ s}$$

Any values of R_1 and C whose product is

0.00071 will suffice. Let us select a stock capacitance of:

$$C = 10 \ \mu F$$

and calculate R_1 from the above equation:

$$R_1(10 \times 10^{-6} \text{ F}) = 0.00071$$

$$R_1 = \frac{7.1 \times 10^{-4}}{10^{-5}} = 71 \ \Omega$$

The value of R_2 can be calculated from Equation 39 or from Equation 38 after we have determined t_2 from Equation 33:

$$T = t_1 + t_2$$

or

$$t_2 = T - t_1 = 0.002 - 0.0005$$
$$= 0.0015 \text{ s}$$

Therefore, by Equation 38:

$$t_2 = 0.7R_2C = 0.0015 \text{ s}$$

or

$$R_2 = \frac{0.0015 \text{ s}}{0.7(10 \times 10^{-6} \text{ F})} = 200 \ \Omega$$

As in previous examples, we would select potentiometers of somewhat larger values than those calculated and adjust them carefully to get the desired frequency and duty cycle.

Sawtooth Generator: The 555 timer can also be used to generate a sawtooth wave, as shown in Figure 8.35. As in the square wave generator, the capacitor voltage is connected to the input trigger, pin 2, to produce astable operation. The capacitor, however, is charged by a constant-current source rather than by a constant voltage, and the sawtooth output is the capacitor voltage at pin 6 rather than that at pin 3. Note that because the sawtooth is taken from pin 6, it does not go to ground, but oscillates between $\frac{1}{3}V_{cc}$ and $\frac{2}{3}V_{cc}$. This behavior is a consequence

SAWTOOTH GENERATOR

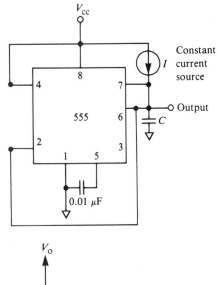

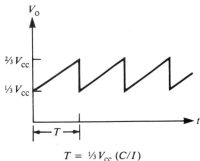

$$T = \frac{1}{3}V_{cc} (C/I)$$

FIGURE 8.35. The 555 timer can be used to generate a sawtooth wave by charging the capacitor with a constant current. In this case, the capacitor voltage at pin 6 is the output. The frequency of the sawtooth is determined by C, the charging current I, and the voltage supply V_{cc}.

of the fact that when the trigger voltage at pin 2 drops below $\frac{1}{3}V_{cc}$, the internal switch to ground at pin 7 opens, reinitiating the charging cycle.

The period of oscillation T can be calculated from Equation 7 for the voltage across a capacitor charged by a constant current:

$$V = \frac{It}{C}$$

The period T is the time it takes for the voltage V to go from $1/3V_{cc}$ to $2/3V_{cc}$, or:

$$\frac{2}{3}V_{cc} - \frac{1}{3}V_{cc} = \frac{IT}{C}$$

$$T = \frac{1}{3}V_{cc}\left(\frac{C}{I}\right) \qquad (41)$$

8.15 QUESTIONS AND PROBLEMS

1. Given the following capacitors, show in Figure 8.36 the configuration that will give circuit capacitances of (a) 76 μF, (b) 21 μF, and (c) 0.0016 μF. Be sure to label each capacitor type.

Al Electrolytic	Mylar	Ceramic
100 μF (1)	10 μF (1)	0.1 μF (4)
50 μF (1)	5 μF (2)	0.01 μF (4)
10 μF (1)	1 μF (2)	0.001 μF (4)
5 μF (1)	0.01 μF (4)	500 pF (2)
1 μF (4)	0.001 μF (4)	100 pF (4)

2. With a constant-current source of 150 μA, what value of capacitance will charge to 5.0 V in (a) 0.30 s, (b) 5.5 ms, and (c) 80 μs? What is the slew rate in each case?

3. With a constant-current source of 70 μA, how long will it take the following capacitors to charge to 0.3 V: (a) 200 μF, (b) 500 pF, and (c) 0.10 μF?

4. Design a circuit that will produce constant slew rates up to 10 V of (a) 1 V/s, (b) 10 V/s, and (c) 100 V/s. Draw the circuit.

5. In Figure 8.37, if $V = 6$ V, how long will it take for the voltage at point B to reach 1.5 V after switch S_1 is closed, given the following values of resistance and capacitance:
   (a) $R_1 = 22$ kΩ and $C = 100$ μF
   (b) $R_1 = 1.5$ MΩ and $C = 0.15$ μF
   (c) $R_1 = 470$ Ω and $C = 100$ pF

6. How long will it take for the voltage V_{AB} across resistor R_1 in Figure 8.37 to become 2.7 V for each of the RC combinations given in Question 5?

7. An oscilloscope measures the charging curve of Figure 8.38. What is the time constant for each of the following oscilloscope settings? If the circuit resistance is 56 kΩ, what is the value of circuit capacitance in each case?
   (a) AMPL/DIV = 2 V/div and TIME/DIV = 0.1 s/div
   (b) AMPL/DIV = 50 mV/div and TIME/DIV = 20 ms/div
   (c) AMPL/DIV = 2 mV/div and TIME/DIV = 5 μs/div

8. If the capacitor in Figure 8.37 is fully charged, how long after the switch S_2 is changed to R_2 will it take for the voltage at point B to reach 1.8 V for the following values of capacitance and resistance?
   (a) $R_2 = 22$ kΩ and $C = 0.20$ μF
   (b) $R_2 = 820$ kΩ and $C = 1500$ pF
   (c) $R_2 = 1.8$ kΩ and $C = 250$ μF

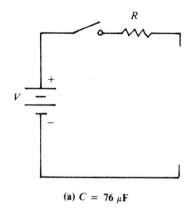

(a) $C = 76 \ \mu F$

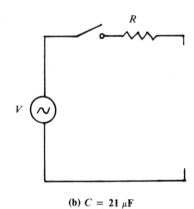

(b) $C = 21 \ \mu F$

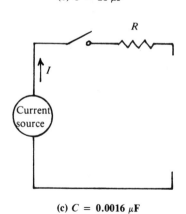

(c) $C = 0.0016 \ \mu F$

FIGURE 8.36.

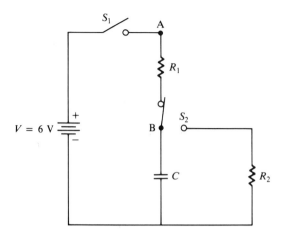

FIGURE 8.37.

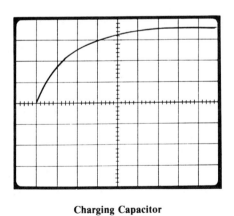

Charging Capacitor

FIGURE 8.38.

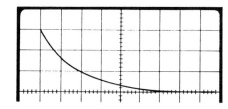

Discharging Capacitor

FIGURE 8.39.

9. An oscilloscope measures the discharge curve shown in Figure 8.39. What is the time constant for each of the following oscilloscope settings? If the circuit capacitance is 25 μF, what is the value of circuit resistance?
   (a) AMPL/DIV = 1 V/div and TIME/DIV = 0.2 s/div
   (b) AMPL/DIV = 20 mV/div and TIME/DIV = 10 μs/div
   (c) AMPL/DIV = 5 mV/div and TIME/DIV = 5 ms/div
10. Design a variable-frequency sawtooth generator with a frequency range of 100 Hz to 100 kHz using a constant voltage source and unijunction transistor. Draw the circuit showing component values and frequency controls.
11. Using a 555 timer, design a one-slot multivibrator that has a variable pulse height of 0 V to 10 V and a variable pulse width of 1 ms to 1.0 s. Draw the circuit showing component values and pulse height and width controls.
12. Using the thermistor shown in Figure 8.40, design an audible thermometer in which a 555 timer serves as a square wave generator. The thermistor resistance should set the square wave frequency to fall in the audible range of about

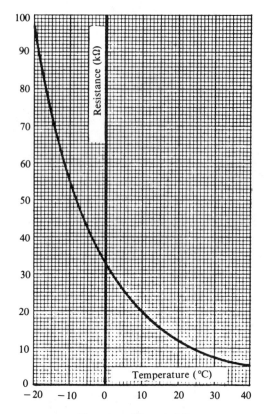

Thermistor Characteristics

FIGURE 8.40.

100 Hz to 5000 Hz over a temperature range of about −20°C to +40°C. Draw the circuit showing component values.
13. Design a variable-frequency sawtooth generator with a frequency range of 100 Hz to 100 kHz using a constant-current source and a 555 timer. Draw the circuit showing component values and frequency controls.

Part

IV

OPERATIONAL AMPLIFIERS

ELECTRONIC
DEVICE
EVOLUTION

Vacuum Tube:
Developed in early 1900s
Primarily used
until mid-1950s

Integrated Circuit:
Developed in late 1950s
Primarily used in
mid-1960s to present

Transistor:
Developed in late 1940s
Primarily used in
mid-1950s to late 1960s

FIGURE 1. Devices that change and control current and voltage in an electronic circuit are the key elements in modern electronics. In the first half of the century, these devices were primarily vacuum tubes, which were replaced by the transistor after its development in the 1950s. Today, circuits are designed primarily around integrated circuits.

INTEGRATED CIRCUITS

The field of electronics has resulted from the development of a variety of electronic devices that can change and control circuit currents and voltages (Figure 1). Until the mid-1950s, these circuit elements were vacuum tubes, which gave rise to bulky systems that required dc power supplies of several hundred volts for their operation.

Following the invention of the transistor in 1948 and its refinement during the 1950s, circuit engineers were able to design circuits around transistors. The newer designs greatly reduced the size, expense, and voltage requirements of earlier models. By the early 1960s, transistor manufacturers could place two or three transistors on a single chip. This capability permitted the circuit designer to reduce greatly the number of circuit parts, again reducing system cost while increasing reliability and performance.

In the mid-to-late 1960s, transistor manufacturers made another major breakthrough that affected circuit cost, complexity, and design. They were able to fabricate thirty or more transistors and resistors on an area the size of a single transistor. Devices manufactured by this method are called **integrated circuits**, or simply ICs.

The development of the IC has led to an era of microelectronics in which new fabrication techniques are placing more and more components in less and less space. More than 250,000 separate components have been successfully fabricated in a single IC unit, and a single IC on a chip about 1/4″ square can contain more electronic elements than the most complex piece of electronic equipment built in 1950.

Integrated circuits have greatly reduced the size, weight, and price of most electronic systems and increased their ruggedness and dependability. Today, ICs are used in a wide variety of applications from home appliances to computers. Also, as sales of ICs have soared, prices have increased so that some ICs can be purchased for less than a dollar, and complete microcomputers can be purchased for as little as ten dollars.

Integrated circuits are categorized by construction and application. They are classed according to construction as either monolithic or hybrid. A **monolithic IC** is one in which the entire circuit, including transistors, resistors, capacitors, and all the interconnections, are fabricated on a single chip of silicon (Figure 2). A **hybrid IC**, on the other hand, is fabricated on an insulating material from both monolithic ICs and discrete components. Connections between the circuit parts are made by external wires or printed circuit conductor patterns.

TYPICAL MONOLITHIC IC

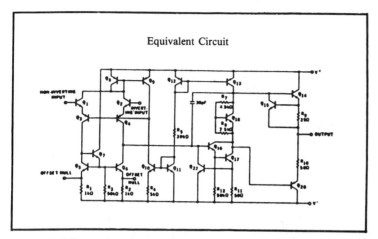

Equivalent Circuit

FIGURE 2. ICs are electronic circuits, including transistors, resistors, capacitors, and their connections, fabricated as a single unit. Monolithic ICs, such as the operational amplifier shown, are complete circuits fabricated on a single chip of silicon. Hybrid ICs are composed of monolithic ICs and discrete components fabricated on an insulating material with wire or printed-circuit connections between components.

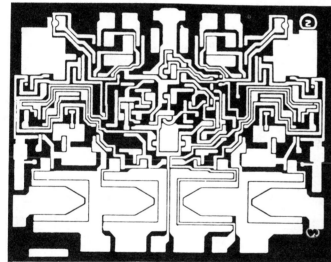

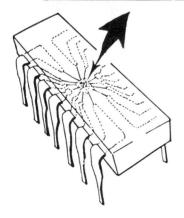

Integrated circuits are manufactured in a sequence of automated steps and in lots of hundreds or thousands. This technique produces ICs in large quantities at low cost and with remarkable reliability. However, if wire connections must be added, as for hybrid ICs, the additional steps increase the cost and reduce reliability because such connections have a relatively high failure rate. Thus, monolithic ICs, which have fewer connections, are becoming increasingly popular.

Integrated circuits are classed by application as either digital or analog. Digital ICs have an output voltage that is either a low voltage (typically 0 V) or a high voltage (typically 5 V) depending on some characteristic of the input voltage. They are used primarily in switching operations and as the basic components of digital computers.

Analog ICs have input and output voltages that are continuously related to one another. For example, the output voltage might always be twice the input voltage: If the input voltage were 2 V, the output voltage would be 4 V, and if the input voltage were 3 V, the output would be 6 V. This particular relationship is termed **linear** because a graph of the output versus the input voltage is a straight line.

Almost all electronic circuits designed today use one or more analog ICs as the heart of the system. Chapters 9, 10, and 11 examine the most versatile and widely used linear IC, the **operational amplifier**, or simply op amp (Figure 3).

FIGURE 3. Part IV focuses on the operational amplifier, or op amp. The op amp is an analog IC whose output can be a continuous function of its input. In contrast, the output of digital ICs can be in only one of two states: high, typically 5 V, or low, typically 0 V.

OPERATIONAL AMPLIFIER IC

μA741C
HIGH PERFORMANCE OPERATIONAL AMPLIFIER
FAIRCHILD LINEAR INTEGRATED CIRCUITS

FEATURES:
- NO FREQUENCY COMPENSATION REQUIRED
- SHORT-CIRCUIT PROTECTION
- OFFSET VOLTAGE NULL CAPABILITY
- LARGE COMMON-MODE AND DIFFERENTIAL VOLTAGE RANGES
- LOW POWER CONSUMPTION
- NO LATCH UP

GENERAL DESCRIPTION — The μA741C is a high performance monolithic operational amplifier constructed on a single silicon chip, using the Fairchild Planar* epitaxial process. It is intended for a wide range of analog applications. High common mode voltage range and absence of "latch-up" tendencies make the μA741C ideal for use as a voltage follower. The high gain and wide range of operating voltages provide superior performance in integrator, summing amplifier, and general feedback applications. The μA741C is short-circuit protected, has the same pin configuration as the popular μA709C operational amplifier, but requires no external components for frequency compensation. The internal 6dB/octave roll-off insures stability in closed loop applications. For full temperature range operation (−55°C to +125°C) see μA741 data sheet.

Many analog devices, however, are **nonlinear**—that is, a graph of output versus input voltage is not a straight line. For example, the output voltage might be the square of the input voltage. Graphs of output versus input voltage for a device are called **voltage transfer characteristics**, and they are critical to understanding the device's behavior in an electronic circuit (Figure 4).

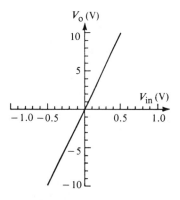

OP AMP VOLTAGE
TRANSFER CHARACTERISTIC

FIGURE 4. The circuit performance of an op amp in a given application is described by its voltage transfer characteristic, a graph of the output voltage for all input voltages. The voltage transfer characteristic shown is for an op amp used as a voltage amplifier. The straight line indicates that the op amp is a linear IC.

Chapter 9

OPERATIONAL AMPLIFIERS AS VOLTAGE COMPARATORS

9.1 OBJECTIVES

Following the completion of Chapter 9, you should be able to:

1. Recognize and draw the symbol for an operational amplifier and correctly identify its terminals.
2. Explain the maximum ratings (supply voltage and input voltage) and basic performance characteristics (open loop gain, output saturation voltages, maximum output current, and input resistance) of an op amp and determine their values from a specification sheet.
3. Determine the output voltage of an op amp that is operated open loop, given the voltage at the two input terminals.
4. Identify an op amp connected as a voltage comparator (zero-crossing detector or positive or negative voltage level detector) and draw its voltage transfer characteristic and output voltage for a given input voltage.
5. Design a voltage comparator circuit that has a given voltage transfer characteristic (positive or negative, inverting or noninverting) and obtain any required reference voltages from the op amp power supply.
6. Design an output circuit for a voltage comparator that will limit the output to either a positive or a negative voltage or light an LED.
7. Identify an op amp connected as a window detector and draw its voltage transfer characteristic and output voltage for a given input voltage.
8. Design a window detector circuit that has a given voltage transfer characteristic and obtain any required reference voltages from the op amp power supply.
9. Design an output circuit for a window detector circuit that will light an LED when the input voltage falls within the voltage window.
10. Explain the meaning and purpose of hysteresis as applied to a voltage level detector.

BASIC AMPLIFIER BEHAVIOR

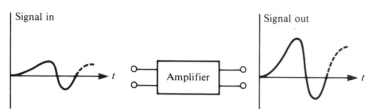

FIGURE 9.1. The primary purpose of an amplifier is to increase the magnitude of an electrical signal, voltage and/or current, while maintaining the basic signal characteristics—for example, frequency.

11. Explain the meaning of feedback, both positive and negative, as applied to an op amp.

12. Identify an op amp connected as a zero or nonzero voltage level detector with hysteresis and draw its voltage transfer characteristic and output voltage for a given input voltage.

13. Design a voltage level detector with hysteresis (inverting or noninverting) that has a specific center voltage and hysteresis width and obtain any required reference voltages from the op amp power supply.

9.2 AMPLIFIERS

An **amplifier** is a circuit that accepts an input signal and produces an enlarged reproduction of that signal at its output, as shown in Figure 9.1. In the process of amplification, only the voltage (or current) is enlarged. The frequency of the signal does not change and is the same frequency at the output as it was at the input.

In a typical amplifier application, the input signal may come from another electronic system (such as an audio oscillator, function generator, or AM/FM tuner) or from a transducer (such as a microphone, phonograph cartridge, thermocouple, and the like). The amplifier's output signal may go to another electronic circuit (such as another amplifier or various signal-processing circuits) or to some output device (such as a voltmeter, oscilloscope, loudspeaker, relay, motor, and so on). A typical ap-

TYPICAL AMPLIFIER APPLICATION

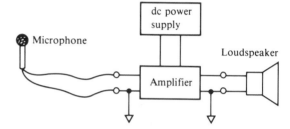

FIGURE 9.2. One typical amplifier application is in a public address system. The amplifier is used to increase the small electrical output of a microphone sufficiently to drive a loudspeaker. The electrical energy required to drive the speaker comes from the external power supply.

plication is shown in Figure 9.2. A microphone serves as the input device and a speaker serves as the output device for a public address system.

An amplifier is necessary because most input signals, such as those from a transducer, are not capable of powering most other circuits, particularly output devices. In Figure 9.2, for example, the output of the microphone is on the order of milliwatts, yet the speaker may require tens of watts. The role of the amplifier is to provide the necessary electrical power while maintaining the important characteristics of the transducer's input signal.

The electrical energy needed for an amplifier to produce the larger output signal comes from a separate dc power supply. Thus, all am-

AMPLIFIER VOLTAGE GAIN

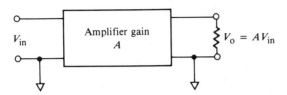

FIGURE 9.3. **The amount by which an amplifier increases a signal is specified by the voltage gain A that is the ratio of the output voltage V_o to the input voltage V_{in} (Equation 2).**

plifiers (or any device that increases power output) require a power supply for their operation.

Voltage Gain

An important specification for any amplifier is its voltage gain A. **Voltage gain** is the number by which the input signal V_{in} is multiplied to give the output voltage V_o (Figure 9.3). In equation form:

$$V_o = AV_{in} \qquad (1)$$

Note that A is simply a number with no units.

Rearranging Equation 1 defines the voltage gain of an amplifier as the ratio of output voltage to input voltage:

$$A = \frac{V_o}{V_{in}} \qquad (2)$$

Gain for various amplifiers ranges from 1 to 100,000 or more depending on the application.

EXAMPLE 1: An amplifier has a voltage gain of 20 and an input signal of 50 mV. What is the output voltage?

Solution: Using Equation 1, we have:

$$V_o = (20)(50 \text{ mV}) = 1000 \text{ mV} = \mathbf{1.0 \ V}$$

If the 50 mV input signal refers to a dc, rms, peak, or peak-to-peak (pp) value, the output voltage is also a dc, rms, peak, or peak-to-peak value, respectively. The gain of an amplifier is only a multiplication factor: It does not change the characteristics of the input voltage.

EXAMPLE 2: An audio amplifier with a gain of 40 has an output voltage of 12 V pp. What is the magnitude of the input signal?

Solution: Rearranging Equation 1 gives:

$$V_{in} = \frac{V_o}{A} = \frac{12 \text{ V}}{40} = \mathbf{0.3 \ V \ pp}$$

Amplifier Coupling

Amplifiers are often designed and built with several different stages, typically two or three. A **three-stage amplifier** consists of an input stage, an isolation stage, and an output stage, as shown in Figure 9.4. The **input stage** is the voltage gain stage, receiving the input signal and amplifying it to a larger output signal. This stage usually has a high input resistance. The **output stage** increases the output current capability, generally without a change in voltage, and provides the amplifier with a low output resistance. To separate these two stages, a third **isolation stage** is normally required. This stage minimizes the effects that the input and output stages have on each other.

Two of the most common techniques used to interconnect amplifier stages are the capacitor-coupled and direct-coupled methods. A **capacitor-coupled amplifier** has a capacitor between each stage, while a **direct-coupled am-**

THREE-STAGE AMPLIFIER

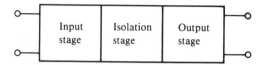

FIGURE 9.4. Most amplifiers are composed of several stages, typically three. An input stage increases the voltage, an output stage increases the current, and an isolation stage minimizes effects between the two. Coupling between stages can be direct, as in an op amp, or pass through a capacitor if only ac signals are to be amplified.

plifier has no capacitor. An amplifier must be direct coupled to amplify dc input signals because capacitors block dc. When ac signals are amplified, however, capacitor coupling is used to block unwanted dc signals that may be superimposed on the ac signals. Using the concepts of voltage gain and amplifier coupling, the characteristics of operational amplifiers can now be described.

9.3 OPERATIONAL AMPLIFIERS

Operational amplifiers are high-gain, direct-coupled amplifiers. (See Figure 4 in the introduction to Part IV.) Their name derives from their initial applications in the 1940s to perform mathematical operations—for example, addition, subtraction, differentiation, and integration—in analog computers. The early op amps used vacuum tubes, but in the late 1950s and early 1960s, op amps were designed using transistors. By the late 1960s, the development of IC technology had led to the introduction of monolithic IC op amps.

Present-day op amps have a wide range of applications, and the original mathematical operations represent only a small fraction of their uses. Some of the systems that use op amps as

OP AMP SYMBOL

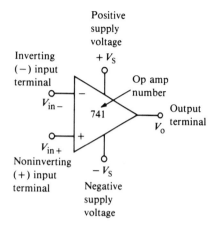

FIGURE 9.5. The operational amplifier was originally designed to perform mathematical operations in computers, but it now has many other uses. Shown are the op amp symbol and basic circuit connections. Note there are two input terminals, minus (−) and plus (+).

the basic building block are amplifiers, measuring instruments, process controls, signal generators, analog-to-digital converters, and communications circuits. In essence, the applications of operational amplifiers are limited only by the circuit designer's imagination.

Symbols and Terminals

The symbol for an op amp is a triangle that points in the direction of signal flow—that is, from input to output—as shown in Figure 9.5. All op amps have five terminals: two input terminals, one output terminal, and two power supply terminals. The input terminals are shown at the base of the triangle and the output terminal is at its apex. The power supply leads (when shown) are drawn on the sides of the triangle. The type of op amp, or manufacturer's part number, is generally printed inside the triangle.

OP AMP POWER SUPPLY CONNECTIONS

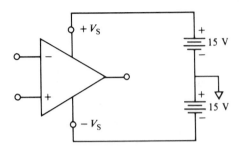

FIGURE 9.6. Op amps require an external power supply, generally of dual polarity. They can operate over a range of supply voltages (± 9 V to ± 18 V), but most applications use ± 15 V with the connections shown here.

One input terminal is called the **inverting**, or minus ($-$), input terminal, and it is shown above the second input terminal, called the **noninverting**, or positive ($+$), input terminal. The two input terminals are termed a **differential input pair** because the output voltage depends on the *difference* between the voltages at the ($+$) and ($-$) inputs. We will see how these terminals are used and the relationship between input and output voltage that their names describe.

Most general-purpose op amps require a dual power supply. On the circuit diagram, the positive and negative supplies are labeled $+V_S$ and $-V_S$, respectively. Although the typical power supply voltage is ± 15 V, the op amp will generally operate over a range of supply voltages from ± 9 V to ± 18 V. Thus, two 9 V batteries can be used to power an op amp, making op amps ideal for portable instrument applications. Figure 9.6 shows how a ± 15 V supply is connected to an op amp.

Caution: Circuit diagrams do not always show the power supply connections. The user must remember that the dc supply voltages must be connected before applying any signal at the input

terminals. Otherwise, the op amp may be damaged.

A resistive load is connected both to the op amp output terminal and to ground by resistor R_L, as illustrated in Figure 9.7. The voltage across R_L is the output voltage V_o. The polarity of V_o with respect to ground depends on the input voltages. Because an op amp has only one output terminal, it is referred to as a **single-ended output**.

Types of Cases

Operational amplifiers are commonly packaged in three types of cases: TO-5 metal package, dual-in-line package, and flat pack (Figure 9.8). Devices in the **TO-5 metal package** have 8, 10, or 12 leads. **Dual-in-line packages** (DIP) are either 8 or 14 pin packages. **Flat packs** have 10 pins. A dual-in-line package is made of either ceramic or plastic. Ceramic packages have better heat-dissipation qualities than plastics but cost considerably more, so plastic packages are more widely used.

Though an op amp has five basic terminals that are used in every application, additional pins are provided. Some pins have special purposes. Others have no connection (NC) and are provided only for reliable insertion of the package into standard sockets. Special-purpose pins will be discussed when they are needed in a particular application.

The op amps that we will examine are designated 741 and 301. These are general-purpose op amps, meaning that they can be used in a variety of applications. These ICs are available in the TO-5 package, both DIP packages, and the flat pack, and they are manufactured by several companies.

Pins used in the different types of packages for a 741 op amp are shown in Figure 9.8. As viewed from the top of the circuit, the pins are numbered counterclockwise. The tab on the TO-5 case locates pin 8. All manufacturers use

OP AMP LOAD CONNECTION

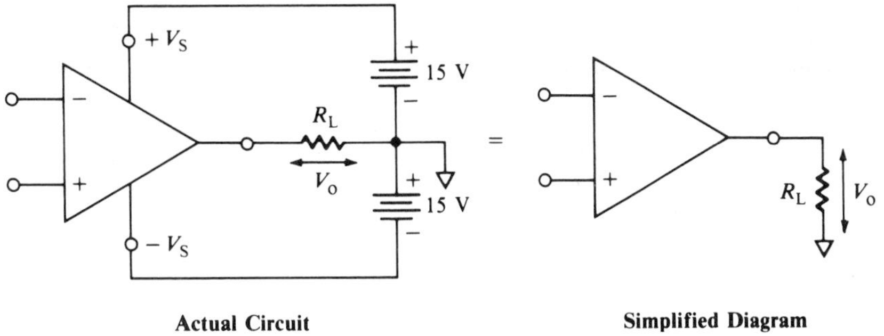

Actual Circuit   Simplified Diagram

FIGURE 9.7. An output device is connected both to the op amp's output terminal and to ground. It is generally represented as a resistive load R_L. In the circuit diagram, the power supply connections and terminals are often omitted, but it is understood that a power supply must be connected.

a notch on dual-in-line packages, and some also include a dot. On a flat pack, pin 1 is identified solely by a dot.

Maximum Ratings

The maximum ratings for the 741 op amp are given in its specification sheet, shown in Figure 9.9. For example, the **maximum dc supply voltage** is ± 18 V (Figure 9.9a). Generally, if this voltage is exceeded, the op amp will be damaged. The dc supply voltage also determines the maximum value of output voltage V_o, as we will see in the following subsections.

The **maximum input voltage** that can be applied to the $+$ and $-$ terminals of an op amp is generally equal to the dc supply voltage. Thus, if the supply voltage is ± 15 V, the maximum voltage that should be applied to either input terminal is also ± 15 V (Figure 9.9b). As with the supply voltage, if the input voltage exceeds this value, the op amp will be damaged.

Caution: Exceeding the input voltage rating is the most common cause of op

amp failure. Extreme care should be exercised when applying an input voltage to an op amp.

For most applications, the output current drawn from the op amp should be limited to 5 mA, but the op amp can supply a **maximum output current** of about 25 mA (Figure 9.9d). The 741 is also **short-circuit proof** because it contains an internal current-limiting circuit that limits the output curent to a maximum value, typically 25 mA. Therefore, even if a low-resistance load (or even a short circuit) is accidentally placed on the output terminals, the op amp will not be damaged.

Open Loop Gain

The voltage gain of the op amp is called its **open loop gain** A_{OL}, and it is the ratio of output voltage V_o to the voltage *difference* between the two input terminals (Figure 9.10). If the voltage at the noninverting input is V_{in+} and at the inverting input is V_{in-}, their **difference voltage** is V_d. In equation form:

OP AMP CASE STYLES

Pin 8 identified by tab

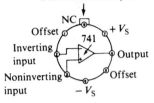

8 Pin TO-5 Case

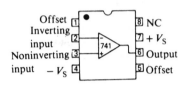

8 Pin Minidip

Pin 1 left of notch or
identified by dot

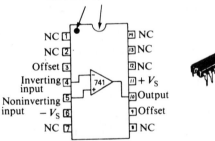

14 Pin Dual-in-Line Package

Pin 1 identified by dot

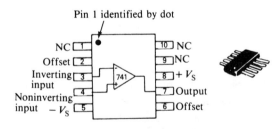

10 Pin Flat Pack

FIGURE 9.8. IC op amps come in several case styles. Pin connections for each case are given.

$$V_d = V_{in+} - V_{in-} \qquad (3)$$

where both V_{in+} and V_{in-} are *measured with respect to ground*.

Using Equation 2, we get the equation for the open loop gain:

$$A_{OL} = \frac{V_o}{V_d} \qquad (4)$$

Rewriting Equation 4 to solve for the output voltage:

$$V_o = A_{OL} V_d \qquad (5)$$

In an ideal op amp, the value of A_{OL} is infinite, and, in general, the higher the value of A_{OL}, the better the op amp's performance. In real op amps, of course, there are practical limits to A_{OL}. A 741 op amp, for example, has a typical open loop gain for dc input voltages of about 200,000. More expensive op amps have open loop gains as high as 10^7.

Note also in Equation 5 that A_{OL} is *always a positive number*. Therefore, the sign of V_o is determined by the sign of V_d. If V_d is positive, then V_o will be positive, and vice versa. Even though both V_{in+} and V_{in-} may be negative voltages, as long as V_d is positive, V_o will be a positive voltage. In summary: If V_d is positive, V_o will be positive; if V_d is negative, V_o will be negative. In equation form:

$$\text{If } V_d > 0, V_o > 0$$
$$\text{If } V_d < 0, V_o < 0 \qquad (6)$$

Saturation Voltages

Since A_{OL} is typically 200,000, we might think that if $V_d = 1$ V, the output voltage would be $V_o = 200,000$ V. This voltage is not possible,

741 SPECIFICATION SHEET

μA741C
HIGH PERFORMANCE OPERATIONAL AMPLIFIER

FAIRCHILD LINEAR INTEGRATED CIRCUITS

ABSOLUTE MAXIMUM RATINGS

(a)

(b)

Supply Voltage	±18 V
Internal Power Dissipation (Note 1)	500 mW
Differential Input Voltage	±30 V
Input Voltage (Note 2)	±15 V
Voltage between Offset Null and V–	±0.5 V
Storage Temperature Range	−65°C to +150°C
Operating Temperature Range	0°C to +70°C
Lead Temperature (Soldering, 60 sec)	300°C
Output Short-Circuit Duration (Note 3)	Indefinite

ELECTRICAL CHARACTERISTICS ($V_S = \pm 15$ V, $T_A = 25°C$ unless otherwise specified)

(c)

(d)

PARAMETERS (see definitions)	CONDITIONS	MIN.	TYP.	MAX.	UNITS
Input Offset Voltage	$R_S \leq 10$ kΩ		2.0	6.0	mV
Input Offset Current			20	200	nA
Input Bias Current			80	500	nA
Input Resistance		0.3	2.0		MΩ
Input Capacitance			1.4		pF
Offset Voltage Adjustment Range			±15		mV
Input Voltage Range		±12	±13		V
Common Mode Rejection Ratio	$R_S \leq 10$ kΩ	70	90		dB
Supply Voltage Rejection Ratio	$R_S \leq 10$ kΩ		30	150	μV/V
Large-Signal Voltage Gain	$R_L \geq 2$ kΩ, $V_{out} = \pm 10$ V	20,000	200,000		
Output Voltage Swing	$R_L \geq 10$ kΩ	±12	±14		V
	$R_L \geq 2$ kΩ	±10	±13		V
Output Resistance			75		Ω
Output Short-Circuit Current			25		mA
Supply Current			1.7	2.8	mA
Power Consumption			50	85	mW
Transient Response (unity gain)	$V_{in} = 20$ mV, $R_L = 2$ kΩ, $C_L \leq 100$ pF				
Risetime			0.3		μs
Overshoot			5.0		%
Slew Rate	$R_L \geq 2$ kΩ		0.5		V/μs
The following specifications apply for $0°C \leq T_A \leq +70°C$:					
Input Offset Voltage				7.5	mV
Input Offset Current				300	nA
Input Bias Current				800	nA
Large-Signal Voltage Gain	$R_L \geq 2$ kΩ, $V_{out} = \pm 10$ V	15,000			
Output Voltage Swing	$R_L \geq 2$ kΩ	±10	±13		V

NOTES:
(1) Rating applies for ambient temperatures to +70°C.
(2) For supply voltages less than ±15 V, the absolute maximum input voltage is equal to the supply voltage.
(3) Short circuit may be to ground or either supply.

FIGURE 9.9.   The specification sheet for the op amp gives its maximum ratings. For the 741 op amp, the supply voltage should not exceed ±18 V and the input voltage should not exceed ±15 V. Exceeding these ratings is the most common cause of IC failure. The maximum 741 output voltage is ±14 V, and the maximum output current is ±25 mA.

OPEN LOOP GAIN

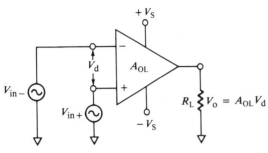

$$V_d = V_{in+} - V_{in-}: \text{If } V_d > 0, V_o > 0$$
$$\text{If } V_d < 0, V_o < 0$$

FIGURE 9.10. An op amp amplifies the difference in voltage between its two input terminals: $V_d = V_{in+} - V_{in-}$. The amplification factor is called the open loop gain A_{OL} that is a positive number. Therefore, if V_d is positive, V_o will be positive, and if V_d is negative, V_o will be negative.

however, because the maximum output voltage is limited to a value slightly less than the dc supply voltages $V_\pm$. For the 741 op amp, the *maximum output voltage* is $V_{o(max)} = \pm 14$ V (see Figure 9.9c). The maximum output voltages for an op amp are called its **saturation voltages.**

The positive saturation voltage $+ V_{sat}$ is the maximum possible positive output voltage with respect to ground, and the negative saturation voltage $- V_{sat}$ is the maximum possible negative output voltage with respect to ground. The op amp is said to be driven into saturation, either positive or negative, whenever the input voltage exceeds that necessary to make V_o equal to the supply voltage. In this case, V_o will remain at the saturation voltage as long as V_d remains too high (Figure 9.11). The following examples illustrate the very small value of V_d that will cause the op amp's output to saturate.

EXAMPLE 3: What value of V_d will make $V_o = + V_{sat} = 14$ V? Assume that $A_{OL} = 200,000$.

OUTPUT SATURATION VOLTAGES

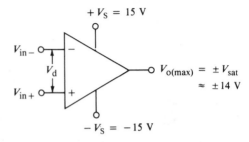

FIGURE 9.11. The maximum output voltage $V_{o(max)}$ of an op amp is slightly less than the supply voltage $\pm V_S$. If V_d is too large, the output will go to this maximum value and stay there until V_d is reduced. When V_o is at $V_{o(max)}$, the op amp is said to be in saturation.

Solution: By rearranging Equation 4, we have:

$$V_d = \frac{V_o}{A_{OL}} = \frac{14 \text{ V}}{200,000} = 70 \ \mu\text{V}$$

Thus, if the voltage difference between the $+$ input and the $-$ input is 70 μV or more, the output voltage will go into saturation at $+ V_{sat} = 14$ V.

EXAMPLE 4: Calculate the value of V_d that will make $V_o = - V_{sat} = -14$ V if $A_{OL} = 200,000$.

Solution: Again, by using Equation 4, we have:

$$V_d = \frac{V_o}{A_{OL}} = \frac{-14 \text{ V}}{200,000} = -70 \ \mu\text{V}$$

Thus, when the voltage at the $-$ input is 70 μV more than the voltage at the $+$ input, V_o is at $- V_{sat}$.

EFFECT OF HIGH OPEN LOOP GAIN

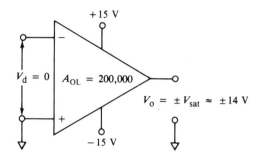

FIGURE 9.12. The gain of the op amp is so large that input voltages as small as $V_d = 70\ \mu V$ will cause it to go to saturation. Even if the inputs are shorted ($V_d = 0$) as shown, random noise will result in output saturation. This feature leads to important op amp applications as voltage comparators.

INPUT RESISTANCE

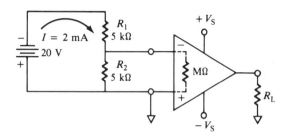

FIGURE 9.13. The input resistance between the + and − input terminals of an op amp is very high (MΩ). Thus, for most applications, the input terminals can be assumed to be an open circuit that draws no current.

Voltages of $\pm 70\ \mu V$ are extremely small—so small that they cannot be measured with ordinary laboratory voltmeters. For example, when the supply voltages V_{in+} and V_{in-} are connected to an op amp and the terminals have no connection, the output voltage will immediately go to $+V_{sat}$ or $-V_{sat}$ due to random electrical noise. Even if both inputs are grounded, as in Figure 9.12, there is sufficient electrical noise and internal voltage fluctuation to cause the output voltage to go to saturation.

This property of the very high gain of the operational amplifier has a variety of important uses. Voltage comparators apply the property of output saturation that results from very high gain and are examined in succeeding pages.

Input Resistance

The high dc voltage gain of an op amp is one of its most attractive properties, but another important feature is its **high input resistance**—that is, the resistance measured between its two input terminals. An ideal amplifier has both

infinite gain and infinite input impedance. Therefore, when it is connected to a voltage source, it draws zero current.

The input resistance of a real op amp, though not infinite, is very large, typically several megohms. Thus, an op amp will draw some current at its input pins, but it will generally be less than 0.1 μA. For most applications, the op amp's input terminals can be considered an open circuit that draws zero current.

For example, the current in the series circuit in Figure 9.13 is 2 mA. If an op amp's input terminals are connected across R_2, the current drawn by the input pin is sufficiently small that the current through R_1 and R_2 can be assumed to remain at 2 mA. Thus, op amps are particularly suited to voltage measurements since they represent no current load to the circuit.

9.4 OPERATIONAL AMPLIFIERS AS VOLTAGE COMPARATORS

A **voltage comparator**, as the name implies, compares two voltages and provides an indication of which one is greater. The op amp is ideally suited for this application because of its differential input and single-ended output. If a reference voltage is applied to one input and an unknown voltage to the other, the value of output saturation voltage indicates whether the unknown voltage is greater than or less than the reference voltage.

In the simplest case, the reference voltage V_{ref} is at ground potential. This type of comparator is called a **zero-crossing detector** because the output voltage changes polarity every time the input voltage passes through 0 V. (It is assumed that the $\pm 70\ \mu V$ required for saturation is so small that it can be neglected.)

The reference voltage may also be set to values other than zero, resulting in both positive and negative voltage level detectors. A **positive voltage level detector** is a comparator circuit that has a positive reference voltage connected to one of the inputs. The output voltage changes polarity whenever the input voltage exceeds the positive reference voltage. A **negative voltage level detector** is one in which the output voltage changes polarity whenever the input exceeds a negative reference level.

Zero-Crossing Detector

Figure 9.14 shows a circuit in which the reference voltage of 0 V (ground potential) is applied to the − input. Thus, $V_{in-} = V_{ref} = 0$. The input signal V_{in} is applied to the + input: $V_{in+} = V_{in}$. Using Equation 3 to solve for V_d, we get:

$$V_d = V_{in+} - V_{in-} = V_{in} - 0 = V_{in}$$

Whenever V_{in} is slightly greater (about 70 μV) than 0 V, V_d will be positive and V_o will go to $+ V_{sat}$. On the other hand, if V_{in} is slightly less than 0 V, V_d will be negative and V_o will go to $- V_{sat}$. If V_{in} is a randomly varying signal, as shown in Figure 9.14, the output will switch from $- V_{sat}$ to $+ V_{sat}$ whenever V_{in} crosses zero

in the positive direction and will switch from $+ V_{sat}$ to $- V_{sat}$ whenever V_{in} crosses zero in the negative direction.

The resulting output voltage is shown superimposed on the input signal in Figure 9.14. Note that the output snaps from one saturation level to the other every time the input crosses zero and that the output voltage is of the same polarity as the input signal; hence, the name **noninverting zero-crossing detector.**

These output voltages can be used to turn various output devices on or off in response to a changing input signal. Thus, the voltage comparator is often used as a voltage-sensitive switch. Remember, however, that the cause of the switching action is the very high gain of the amplifier that causes it to go into saturation from a very small voltage, $V_d \approx 70\ \mu V$, at the comparator's input.

If the ground reference voltage is applied to the + input and V_{in} to the − input, the comparator is still a zero-crossing detector, but the output saturation voltages reverse polarity. In this case, Equation 3 gives:

$$V_d = V_{in+} - V_{in-} = 0 - V_{in} = - V_{in}$$

Whenever V_{in} goes through zero to negative values, V_d becomes positive and V_o will go to $+ V_{sat}$. Similarly, if V_{in} goes through zero in the positive direction, V_d will be negative and V_o will switch from $+ V_{sat}$ to $- V_{sat}$, as shown in Figure 9.15.

This comparator is called an **inverting zero-crossing detector** because it inverts the polarity of the input signal. Therefore, it is used for the opposite purpose of that in Figure 9.14—that is, to turn off a device when V_{in} is greater than 0 V and turn it back on when V_{in} is less than 0 V.

Nonzero-Crossing Detector

Earlier, it was mentioned that the reference voltage need not be zero. Figure 9.16a shows the reference voltage at the − input to be 2 V,

ZERO-CROSSING DETECTOR (NONINVERTING)

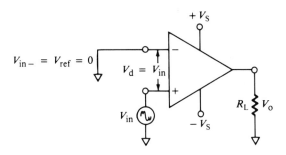

FIGURE 9.14. A zero-crossing detector can be constructed by grounding the – input (V_{in-} = V_{ref} = 0 V) and applying an input signal to the + input, V_{in+} = V_{in}). The difference voltage is then V_d = V_{in}, and the output will change saturation levels every time V_{in} crosses 0 V to the same polarity as V_{in} (noninverting).

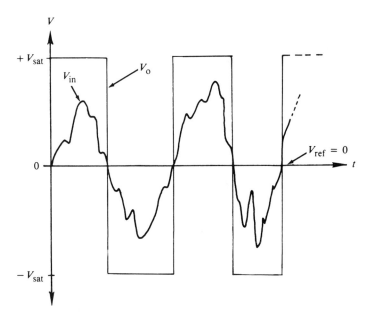

and the same input voltage waveform V_{in} is applied to the + input. For V_o to switch to $+V_{sat}$, V_d must be positive ($V_d > 0$). According to Equation 3, V_o will be at $+V_{sat}$ if:

$$V_d = V_{in+} - V_{in-} > 0$$

Substituting in values gives:

$$V_{in} - 2\ V > 0$$

or

$$V_{in} > 2\ V$$

As long as V_{in} is greater than 2 V, V_d will be positive and V_o will be at $+V_{sat}$. If V_{in} is less than 2 V, however, V_d will be negative and V_o will be at $-V_{sat}$. Thus, Figure 9.16a represents a voltage-sensitive switch that can turn on an output device whenever V_{in} is greater than 2 V and turn it off whenever V_{in} is less than 2 V.

Figure 9.16b shows the behavior of an inverting voltage level detector where the 2 V reference voltage is applied to the + input. Figures 9.16c and 9.16d show the correspond-

FIGURE 9.15. An inverting zero-crossing detector can be constructed by grounding the + input ($V_{in+} = V_{ref} = 0$ V) and applying the input signal to the − input ($V_{in−} = V_{in}$). Then, $V_d = -V_{in}$, and the output will change saturation levels every time V_{in} crosses 0 V, with opposite polarity.

ZERO-CROSSING DETECTOR (INVERTING)

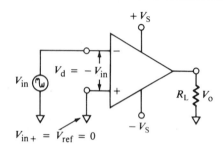

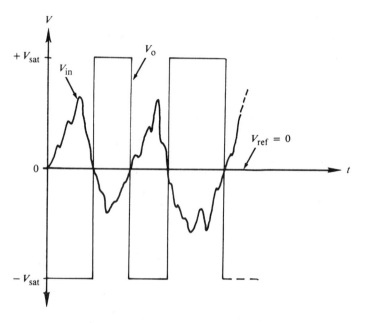

ing behavior of a −2 V reference voltage applied to each input terminal.

Figure 9.17 summarizes voltage comparators by type of construction. **Positive voltage comparators** change output polarity whenever the input exceeds a positive reference voltage. **Negative voltage comparators** change output polarity whenever the input exceeds a negative reference voltage. For each type of comparator, the output change can go from $+V_{sat}$ to $-V_{sat}$ (noninverting) or from $-V_{sat}$ to $+V_{sat}$ (inverting).

EXAMPLE 5: Design a voltage comparator that will go to $+V_{sat}$ whenever V_{in} exceeds −4.5 V.

Solution: According to Figure 9.17, for V_o to go to $+V_{sat}$ whenever $V_{in} > -4.5$ V, a noninverting, negative voltage detector is required. Thus, a negative voltage of $V_{ref} = -4.5$ V must be applied to the − terminal and the input voltage V_{in} to the + terminal.

NON-ZERO CROSSING DETECTORS

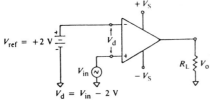

$V_{ref} = +2$ V V_d V_{in} $+V_S$ $-V_S$ R_L V_o

$V_d = V_{in} - 2$ V

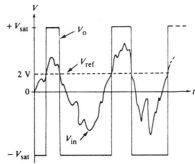

(a) +2 V Voltage Level Detector (Noninverting)

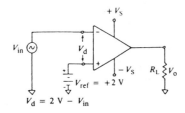

V_{in} V_d $+V_S$ $-V_S$ $V_{ref} = +2$ V R_L V_o

$V_d = 2$ V $- V_{in}$

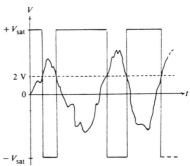

(b) +2 V Voltage Level Detector (Inverting)

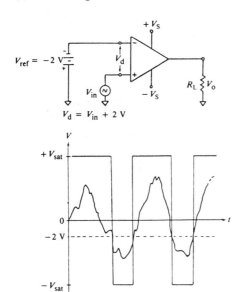

$V_{ref} = -2$ V V_d V_{in} $+V_S$ $-V_S$ R_L V_o

$V_d = V_{in} + 2$ V

(c) −2 V Voltage Level Detector (Noninverting)

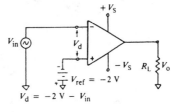

V_{in} V_d $+V_S$ $-V_S$ $V_{ref} = -2$ V R_L V_o

$V_d = -2$ V $- V_{in}$

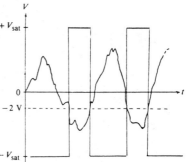

(d) −2 V Voltage Level Detector (Inverting)

FIGURE 9.16.   Nonzero-crossing detectors can be constructed by applying a nonzero reference voltage to one of the input terminals. In (a), the +2 V reference voltage is applied to the − input to give a noninverting, +2 V level detector. Other options are shown in (b), (c), and (d).

FIGURE 9.17. The table gives input terminal conditions for constructing positive and negative voltage level detectors with both inverted and noninverted outputs. The table simplifies circuit design problems. See Example 5.

TERMINAL CONDITIONS FOR VOLTAGE LEVEL DETECTORS

	V_{in}	V_{in+}	V_o
Noninverting	$+V_{ref}$	$V_{in} > +V_{ref}$ $V_{in} < +V_{ref}$	$+V_{sat}$ $-V_{sat}$
Inverting	$V_{in} > +V_{ref}$ $V_{in} < +V_{ref}$	$+V_{ref}$	$-V_{sat}$ $+V_{sat}$

Positive Voltage Detectors

	V_{in}	V_{in+}	V_o
Noninverting	$-V_{ref}$	$V_{in} > -V_{ref}$ $V_{in} < -V_{ref}$	$+V_{sat}$ $-V_{sat}$
Inverting	$V_{in} > -V_{ref}$ $V_{in} < -V_{ref}$	$-V_{ref}$	$-V_{sat}$ $+V_{sat}$

Negative Voltage Detectors

This solution can be confirmed by Equation 3. For V_o to be at $+V_{sat}$, V_d must be positive (> 0). If $V_{in-} = V_{ref} = -4.5$ V and $V_{in+} = V_{in}$, Equation 3 gives:

$$V_d = V_{in+} - V_{in-} > 0$$
$$= V_{in} - (-4.5 \text{ V}) > 0$$

or

$$V_d = V_{in} > -4.5 \text{ V}$$

Thus, for V_{in} greater than -4.5 V, V_d will be positive and $V_o = +V_{sat}$, as required. Similarly, for V_{in} less (more negative) than -4.5 V, V_d will be negative and $V_o = -V_{sat}$.

The reference voltage can be conveniently obtained from the op amp's own power supply by using a simple voltage divider, as shown in Figure 9.18. The value of the reference voltage for a given power supply voltage is derived from the voltage divider equation:

$$V_{ref} = \frac{R_2}{R_1 + R_2} \times \pm V_S \qquad (7)$$

Note in Figure 9.18 that the op amp's input terminals are connected in parallel with the voltage divider resistance R_2. Because the input resistance of the op amp is so large (on the order of several megohms), however, the input terminals draw negligible current (on the order of nanoamperes). Therefore, the voltage divider circuits of Figure 9.18 are not loaded down when connected to the op amp input terminals. This property of high input resistance is one of the reasons that the op amp is such a versatile building block in electronic circuits.

EXAMPLE 6: In the circuit shown in Figure 9.18a, let $R_1 = 15$ kΩ, $R_2 = 1.8$ kΩ, and $V_{in+} = +15$ V. What is the positive reference voltage between point A and ground?

Solution: From Equation 7, we have:

$$V_{ref} = \frac{1.8 \text{ k}\Omega}{15 \text{ k}\Omega + 1.8 \text{ k}\Omega} \times 15 \text{ V}$$

$$\approx 1.6 \text{ V}$$

If this 1.6 V reference voltage at point A is connected to the $-$ input and V_{in} to the $+$ input, $V_o = +V_{sat}$ when V_{in} is greater than 1.6 V, and $V_o = -V_{sat}$ when V_{in} is less than 1.6 V. If the reference voltage at point A is applied to the $+$ input, an inverted output will result.

A voltage divider can also be connected to the V_{in-} supply to obtain a negative reference voltage and, thus, a negative voltage comparator (Figure 9.18b). The reference voltage is the voltage across R_2 from point A to ground. Point A can be connected to either the $+$ or $-$ input terminal, with the input signal connected to the other input terminal to give a negative voltage comparator of either output polarity.

EXAMPLE 7: Design a voltage comparator with an output that switches from $-V_{sat}$ to $+V_{sat}$ whenever V_{in} drops below -6 V. Assume the power supply voltage to be $\pm V_S = \pm 15$ V.

Solution: According to Figure 9.17, for V_o to go to $+V_{sat}$ when V_{in} is less (more negative) than -6 V, an inverting, negative voltage detector is required. A negative voltage of $V_{ref} = -6$ V should be applied to the $+$ terminal and V_{in} applied to the $-$ terminal, as shown in Figure 9.19. For example, with V_{in+} at -6 V, if V_{in-} is at -7 V, then:

REFERENCE VOLTAGE CIRCUITS

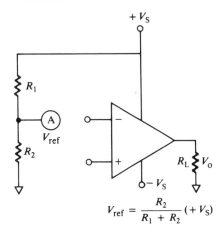

$$V_{ref} = \frac{R_2}{R_1 + R_2}(+V_S)$$

(a) Positive Reference Voltage

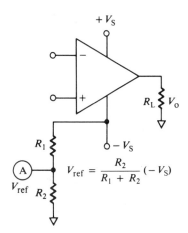

$$V_{ref} = \frac{R_2}{R_1 + R_2}(-V_S)$$

(b) Negative Reference Voltage

FIGURE 9.18. Reference voltages can be obtained from the op amp power supply by using a voltage divider. The figure shows reference voltages obtained from both the positive voltage supply and the negative voltage supply.

$$V_d = V_{in+} - V_{in-} = -6 \text{ V} + 7 \text{ V}$$
$$= 1 \text{ V}$$

and V_o will be $+V_{sat}$, as required.

EXAMPLE 7

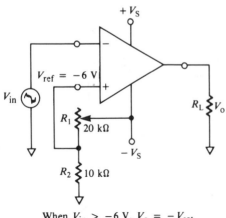

When $V_{in} > -6$ V, $V_o = -V_{sat}$
When $V_{in} < -6$ V, $V_o = +V_{sat}$

FIGURE 9.19. Shown is an inverting, negative voltage detector designed with the aid of Figures 9.17 and 9.18.

A reference voltage of $V_{ref} = -6$ V can be obtained from the negative power supply $-V_S$ with the voltage divider shown in Figure 9.19. To keep the current drawn from the power supply small (on the order of 1 mA), the voltage divider resistors should be on the order of kilohms. For simplicity, we select R_2 to be a standard 10 kΩ resistor and solve for the proper value of R_1 using Equation 7:

$$V_{ref} = \frac{R_2}{R_1 + R_2}(-V_S)$$

Rearranging for R_1 gives:

$$R_1 = R_2\left(\frac{-V_S}{V_{ref}} - 1\right)$$

$$= 10 \text{ k}\Omega\left(\frac{-15 \text{ V}}{-6 \text{ V}} - 1\right)$$

$$= 10 \text{ k}\Omega\ (1.5) = \mathbf{15\ k\Omega}$$

For an accurate setting of $V_{ref} = -6$ V, R_1 should be a 20 kΩ potentiometer (pot) that can be accurately adjusted to the proper resistance (Figure 9.19).

Voltage Transfer Characteristics

A useful method of illustrating the input–output behavior of a voltage comparator is to graph V_o versus V_{in}. This plot is called a **voltage transfer characteristic.**

Figure 9.20 shows the voltage transfer characteristics for the four comparator circuits in Figure 9.16. Figure 9.20b shows that when V_{in} is greater than the 2 V reference voltage, the output of the comparator is at $-V_{sat}$. As V_{in} drops below the 2 V reference level, the output switches to $+V_{sat}$. (The width of the switching region, $\pm 70\ \mu$V, is far too small to be seen on the scale.) The voltage transfer characteristic of Figure 9.20b gives the value of V_o for any value of V_{in} and, therefore, fully describes the input–output behavior of the circuit of Figure 9.16b. Take a moment to see how the other voltage transfer characteristics of Figure 9.20 describe the input–output behavior of the corresponding comparator circuits in Figure 9.16.

Limiting the Output Voltage

Some devices that may be connected to a comparator's output will be damaged if the full output saturation voltage is applied to them. Examples are the reverse-bias voltages of light-emitting diodes (LEDs), the base-to-emitter junctions of transistors, the gate-to-cathode junctions of silicon-controlled rectifiers (SCRs), digital logic, and so on.

To protect a device from a large reverse-bias voltage, a diode can be inserted between the

VOLTAGE TRANSFER CHARACTERISTICS

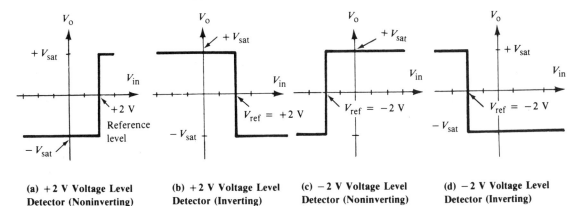

(a) +2 V Voltage Level
Detector (Noninverting)

(b) +2 V Voltage Level
Detector (Inverting)

(c) −2 V Voltage Level
Detector (Noninverting)

(d) −2 V Voltage Level
Detector (Inverting)

FIGURE 9.20.   The performance of a voltage level detector can be described by its voltage transfer characteristic, simply a graph of the output voltage versus input voltage. The figure gives voltage transfer characteristics for the voltage level detectors shown in Figure 9.16.

output terminal and the device. Depending on the diode's direction, one output polarity (either $+V_{sat}$ or $-V_{sat}$) will be dropped across the diode rather than across the device. In Figure 9.21, for example, if V_o is positive (and greater than 0.6 V), the diode will act as a short circuit and $V_L \approx V_o = +V_{sat}$. On the other hand, if V_o is negative, the diode will act as an open circuit and $V_L = 0$. All of the output voltage is dropped across the diode — that is, $V_D = -V_{sat}$. Therefore, V_L switches between $+V_{sat}$ and zero as V_o switches between $+V_{sat}$ and $-V_{sat}$. If the diode is reversed, V_L will switch between zero and $-V_{sat}$ as V_o switches between $+V_{sat}$ and $-V_{sat}$.

If the load resistor R_L in Figure 9.21 were replaced by a light-emitting diode (LED) as in Figure 9.22, the LED would light whenever $V_o = V_{sat}$. This situation will occur whenever V_{in} is greater than the reference voltage. Since the maximum reverse-bias voltage an LED can withstand is approximately 3 V, the diode protects the LED when the output voltage of the op amp is at $-V_{sat}$.

Note that when two forward-biased diodes (the protection diode and the LED) are connected to a 741, they represent a low-resistance load — almost a short circuit. The current drawn from the op amp will go to its maximum limit, approximately 25 mA, which is adequate for lighting most LEDs.

Figure 9.23 shows a zero-crossing detector with two LEDs (one red and one green) connected in parallel at the output. They are arranged so that when V_{in} is positive, V_o tries to go to $+V_{sat}$ and the green LED will light. When V_{in} is negative, V_o tries to go to $-V_{sat}$ and the red LED will light. In this case, no protective diode is required because the forward-bias voltage of the "on" LED is only about 1.5 V. Thus, when the LED conducts, V_o goes only to 1.5 V, even though the full 25 mA of output current is supplied. Since 1.5 V is less than the maximum permissible reverse-bias voltage of the "off" LED (about 3 V), one diode protects the other. Although Figure 9.23 shows a 0 V reference voltage, a positive or negative reference voltage can be used instead.

OUTPUT LOAD PROTECTION

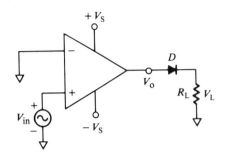

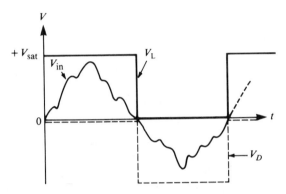

FIGURE 9.21. Inserting a diode in series with a load causes the voltage to have only one polarity, the other polarity being dropped across the diode. This arrangement can be used to protect a load from an excess reverse-bias voltage that might damage it.

LED OUTPUTS

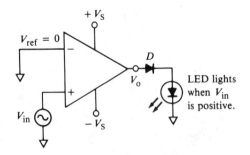

FIGURE 9.22. The maximum reverse-bias voltage for an LED is about 3 V. Inserting the diode in the direction shown causes the LED to light when $V_o = V_{sat}$, but not when $V_o = -V_{sat}$.

DUAL LED OUTPUTS

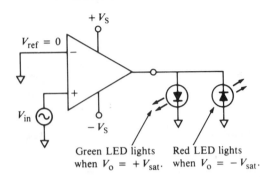

Green LED lights when $V_o = +V_{sat}$.   Red LED lights when $V_o = -V_{sat}$.

FIGURE 9.23. If two LEDs are placed in parallel but opposite directions, one LED will protect the other. When one LED lights, it acts as a short circuit and draws the maximum output current of 25 mA. The voltage drop across the conducting LED is less than the maximum 3 V reverse bias for the other LED.

9.5 COMPARATORS AS WINDOW DETECTORS

A **window detector** is a circuit that uses two comparators to indicate when an input voltage is within a specified voltage range. These circuits are often used in monitoring a signal voltage to determine when it falls below a given low reference voltage or exceeds a higher reference voltage.

For example, the recommended operating voltage for many digital logic circuits is between 4.5 V and 5.5 V. Below 4.5 V, circuit performance is marginal, while above 5.5 V, circuit components may be damaged. The operating voltage can be monitored by applying it to the input of a window detector whose output will sound an alarm or turn the system off if the voltage drops below 4.5 V or goes above 5.5 V.

Another application of a window detector is in quality control systems. For example, it

BASIC WINDOW DETECTOR CIRCUIT

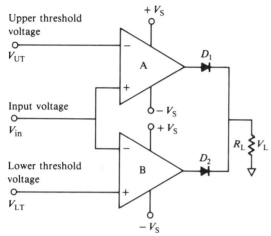

If $V_{LT} < V_{in} < V_{UT}$, the $V_L = 0$
If $V_{in} > V_{UT}$ or $V_{in} < V_{LT}$, then $V_L \approx V_{sat}$

FIGURE 9.24. A window detector is used to indicate when an input voltage falls within a specified voltage range. In the circuit shown, if V_{in} is between a lower threshold voltage V_{LT} and an upper threshold voltage V_{UT}, the load voltage will be $+V_{sat}$ minus the diode drop.

VOLTAGE TRANSFER CHARACTERISTIC

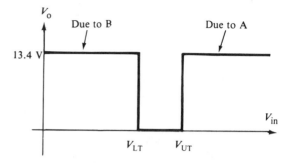

FIGURE 9.25. Shown is the voltage transfer characteristic for the circuit in Figure 9.24. The voltage window between V_{LT} and V_{UT} is set by reference voltages applied to comparators B and A, respectively.

can be used to monitor the optical density of a liquid or the size or weight of an object. Converting the quantity to be controlled to a voltage and requiring that the voltage fall within a specified range permits the window detector to indicate when an object falls outside the range and should be rejected.

Figure 9.24 illustrates the connections for the basic window detector. The input signal V_{in} is applied to both comparators: to the $+$ input of comparator A and the $-$ input of comparator B. The range of the window detector is set by the reference voltages for the two comparators. The reference voltage at the $-$ input of comparator A serves as the upper threshold voltage V_{UT}, while the reference voltage at the $+$ input of comparator B serves as the lower threshold voltage V_{LT}.

When $V_{in} < V_{LT}$, the output voltage of

comparator B is at $+V_{sat}$ and the output of comparator A is at $-V_{sat}$. Diode D_1 blocks current flow from comparator A, so its output can be ignored. The voltage across the load V_L is therefore approximately $+V_{sat} - 0.6$ V, where the 0.6 V is the diode voltage drop across D_2. If $+V_{sat} = 14$ V, then:

$$V_1 = 14 \text{ V} - 0.6 \text{ V} = 13.4 \text{ V}$$

When $V_{in} > V_{UT}$, comparator A is at $+V_{sat}$ and comparator B is at $-V_{sat}$. Diode D_2 blocks current flow from comparator B so that once again, $V_o \approx 13.4$ V.

If V_{in} is *between* V_{LT} and V_{UT}, the output of both comparators is $-V_{sat}$, resulting in both diodes being reverse biased. The current through R_L is zero and $V_L = 0$ V. Figure 9.25 shows the voltage transfer characteristic of the window detector circuit. Note that when $V_{LT} < V_{in} < V_{UT}$, $V_o = 0$ V. For all other values of V_{in}, $V_o \approx 13.4$ V.

Design Example

An application of a window detector is shown in Figure 9.26. This circuit could be used to test resistors to determine whether or not their re-

CIRCUIT FOR TESTING RESISTOR TOLERANCE

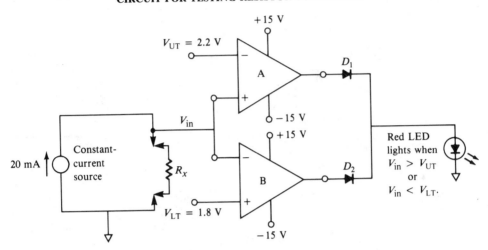

FIGURE 9.26. This window detector circuit can be used to test 100 Ω resistors to a tolerance of ± 10%. If R_x is within ± 10% of 100 Ω, V_{in} will lie between V_{LT} = 1.8 V and V_{UT} = 2.2 V, and the LED will be off. If R_x is greater than ± 10% of 100 Ω, V_{in} will lie outside the voltage window, and the LED will light.

sistance is outside a prescribed tolerance range. The tolerance range is set by the threshold voltages V_{LT} and V_{UT}.

When the input voltage V_{in} due to the resistor is between the threshold voltages, the output voltage is zero and the LED is off. When V_{in} is either less than V_{LT} or greater than V_{UT}, indicating that the resistor is outside its tolerance range, the output voltage is positive and the LED is on. This network can be used for any values of V_{in}, V_{LT}, and V_{UT} that are less than the power supply voltages.

For example, assume that the circuit in Figure 9.26 will be used to test 100 Ω resistors within a tolerance of ± 10%. The unknown resistor under test R_x is supplied by a constant-current source of 20 mA. If the resistor is exactly 100 Ω, the window detector's input voltage V_{in} equals:

$$V_{in} = (0.10 \text{ k}\Omega)(20 \text{ mA}) = 2.0 \text{ V}$$

A resistance tolerance of ± 10% means that

this voltage can vary by ± 10% and the resistor will still be acceptable:

$$V_{in} = 2.0 \text{ V} \pm 10\% = 2.0 \text{ V} \pm 0.2 \text{ V}$$

The threshold voltages, therefore, should be set to reject resistors whose input voltages fall outside the range of 2.0 V ± 2 V:

$$V_{LT} = 2.0 \text{ V} - 0.2 \text{ V} = 1.8 \text{ V}$$

or

$$V_{UT} = 2.0 \text{ V} + 0.2 \text{ V} = 2.2 \text{ V}$$

These threshold voltages for each comparator can be set by the resistor network shown in Figure 9.27. Adjusting R_2 sets V_{LT} and adjusting R_1 sets V_{UT}.

An output circuit can also be designed to light an LED when the unknown resistor falls *within* the tolerance range, as Figure 9.28 shows. When V_{in} is within the voltage window, V_L is at 0 V and the − 15 V supply turns the LED on. The 680 Ω resistor limits the LED cur-

THRESHOLD VOLTAGE CIRCUIT

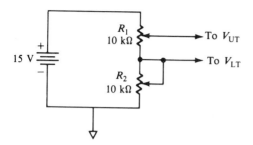

FIGURE 9.27. This circuit can be used to set the upper and lower threshold voltages for the window detector circuit in Figure 9.24. Adjusting R_1 sets V_{UT}, and adjusting R_2 sets V_{LT}. The circuit voltage of 15 V can be obtained from the op amp's power supply.

rent to about 20 mA. When V_{in} falls outside the voltage window, V_L goes to about 13.4 V. Point C is then slightly positive with respect to ground and the LED is off. The 470 Ω resistor protects the LED against large reverse-breakdown voltages.

9.6 COMPARATORS WITH HYSTERESIS

If a comparator is used in an environment of high electrical noise or the input signal has slight voltage fluctuations, the comparator may switch saturation levels, due as often to the electrical noise as to the signal. Figure 9.29 shows a pure sine wave applied to the input of a zero-crossing detector and the desired output. There is only one transition for each zero crossing.

If there is a small amount of electrical noise superimposed on the sine wave, however, as shown in Figure 9.30, instead of a single zero crossing, several crossings may actually occur. A true zero-crossing detector will sense each of these zero crossings, and V_o will switch between

ALTERNATIVE OUTPUT CIRCUIT

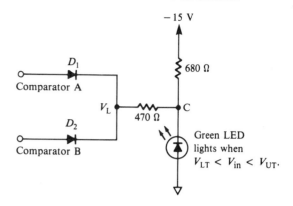

FIGURE 9.28. This circuit will light the LED when V_{in} falls within the voltage window. When V_L is at 0 V, the −15 V supply makes point C negative and the LED is on. When V_L is 13.4 V, point C is positive and the LED is off.

$\pm V_{sat}$ for each one. The resulting output voltage V_o will be as shown in Figure 9.30.

Although it may be the user's intention only to sense once when V_{in} crosses zero, the output voltage waveform in Figure 9.30 shows that each crossing is detected. This effect can be a problem if the comparator is controlling a load that must be turned on or off at a precise time.

A noise voltage can be considered a voltage source in series with the signal source, as shown in the circuit diagram in Figure 9.30. It is often impossible or extremely costly to eliminate, or even reduce, this noise. The alternative is to make the comparator less susceptible to noise. One of the easiest and least-expensive ways to reduce the noise sensitivity of a comparator is to add hysteresis to its voltage transfer characteristic.

Hysteresis

A comparator that changes from $+ V_{sat}$ to $- V_{sat}$ at one voltage (V_{UT}) and back to $+ V_{sat}$ at

**ZERO-CROSSING DETECTOR
(PURE WAVE INPUT)**

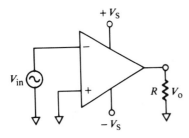

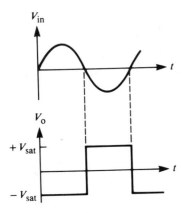

FIGURE 9.29. If a pure sine wave is applied to the input of a zero-crossing detector, its output will switch saturation levels only once each time the sine wave crosses zero volts.

**ZERO-CROSSING DETECTOR
(NOISY INPUT)**

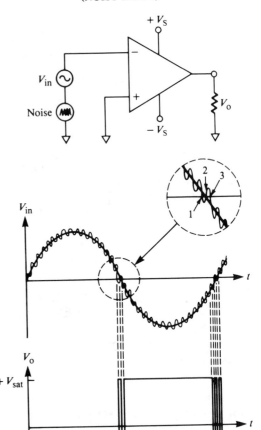

FIGURE 9.30. If there are noise voltage fluctuations superimposed on the sine wave, the input voltage may actually cross zero volts several times for each crossing of the sine wave. The zero-crossing detector output will switch saturation levels for each crossing, giving the output waveform shown above.

a different voltage (V_{LT}) is said to have **hysteresis.** This property is shown by the voltage transfer characteristic in Figure 9.31. As V_{in} increases from negative values (point 1 to point 2), V_o switches from $+V_{sat}$ to $-V_{sat}$ at $V_{in} = 1.0$ V. When V_{in} decreases, however (from point 3 to point 4), V_o does not switch back to $+V_{sat}$ until $V_{in} = -1.0$ V. Note that this voltage transfer characteristic is that of an *inverting* zero-crossing detector.

The width of the input voltage region between the switching thresholds is called the hysteresis width, or **hysteresis voltage** V_H. In

equation form, V_H is the difference between the upper threshold voltage V_{UT} and the lower threshold voltage V_{LT}:

VOLTAGE TRANSFER CHARACTERISTIC OF ZERO-CROSSING DETECTOR WITH HYSTERESIS

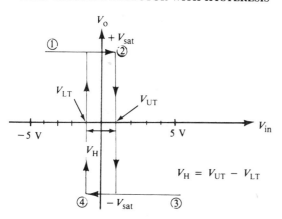

FIGURE 9.31. To eliminate unwanted transitions due to electrical noise on the input signal, hysteresis can be added to the voltage transfer characteristic. As shown, the output switches from $+V_{sat}$ to $-V_{sat}$ at one voltage V_{UT}, but from $-V_{sat}$ to $+V_{sat}$ at a lower voltage V_{LT}. The difference between the voltages is called the hysteresis width V_H.

$$V_H = V_{UT} - V_{LT} \qquad (8)$$

For Figure 9.31, the hysteresis voltage is:

$$V_H = 1.0\,V - (-1.0\,V) = 2.0\,V$$

If the input voltage of Figure 9.30 were applied to a comparator with the voltage transfer characteristic of Figure 9.31, the noise fluctuations would all be within the hysteresis region, and the comparator would not respond to them (Figure 9.32). To eliminate spurious output transitions in a given application, the hysteresis width should be just slightly larger than the peak-to-peak amplitude of the largest noise voltage fluctuation.

INPUT-OUTPUT WAVEFORMS

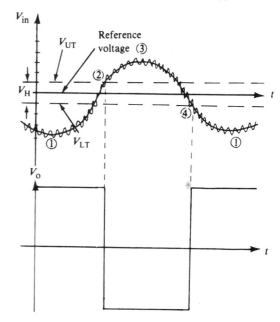

FIGURE 9.32. If the hysteresis width is larger than the amplitude of the noise fluctuations, multiple zero crossings of the input signal due to noise will all lie within the hysteresis region. Thus, the output will change levels only once for each zero crossing of the primary input signal.

Positive Feedback

The easiest way to incorporate hysteresis into a comparator circuit is to feed back a portion of the output signal to one of the inputs. A **feedback circuit** is one in which a connection is made between the output terminal and one of the input terminals. The feedback connection may be direct—for example, a wire—or it may be via a component, such as a resistor or capacitor. Depending on the connection, all or only a fraction of the output voltage may be fed back to one of the input terminals.

When the connection is made between the output and the + input, it is called **positive feedback** (Figure 9.33a). If the connection is

FEEDBACK CIRCUITS

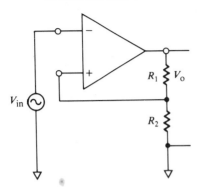

(a) Positive Feedback

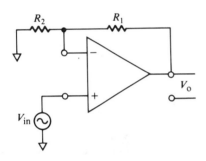

(b) Negative Feedback

FIGURE 9.33. **Hysteresis can be added to a voltage comparator by feeding back a fraction of the output voltage to the positive op amp input. This connection is called positive feedback. Feedback to the negative input is also possible, and its use is examined in Chapter 10.**

made to the − input, it is called **negative feedback** (Figure 9.33b). Applications of op amps as amplifiers use negative feedback because it stabilizes the circuit performance. (Negative feedback is discussed in Chapter 10.) Positive feedback, on the other hand, generally leads to circuit instability. When applied to a comparator circuit, however, it can result in hysteresis.

Zero-Crossing Detector with Hysteresis

To add hysteresis to a zero-crossing detector, a fraction of the output voltage V_o is fed back through a voltage divider to the + input (Figure 9.34). Since V_o is always at either $+V_{sat}$ or $-V_{sat}$, a nonzero reference voltage will always be present at the + input. The value of the reference voltage, however, depends on whether V_o is at $+V_{sat}$ or $-V_{sat}$. Note, also, that because this feedback reference voltage is applied to the + input, the output is inverted (see Figures 9.17 and 9.31).

To begin the analysis, assume $V_o = +V_{sat}$ (Figure 9.34a). Then, the fraction of the voltage that is fed back to the + input is given by the voltage divider ratio:

$$V_{UT} = \frac{R_2}{R_1 + R_2}(+V_{sat}) \qquad \textbf{(9)}$$

This feedback voltage becomes the comparator's reference voltage and is called the **upper threshold voltage** V_{UT}. For the output voltage V_o to switch from $+V_{sat}$ to $-V_{sat}$, the signal voltage V_{in} applied to the − input must be *greater* than V_{UT}.

When $V_o = V_{sat}$, the feedback conditions change to those shown in Figure 9.34b. Now, the voltage fed back to the + input is given by:

$$V_{LT} = \frac{R_2}{R_1 + R_2}(-V_{sat}) \qquad \textbf{(10)}$$

According to Equation 10, the reference voltage at the + input is the negative of Equation 9. This voltage then becomes the **lower threshold voltage** V_{LT}. In order for the output voltage to switch back to $+V_{sat}$, the input signal must be *less* than V_{LT}. Equations 9 and 10 describe the desired voltage transfer characteristic of Figure 9.31.

The hysteresis width is the difference between V_{UT} and V_{LT} (Equation 8):

$$V_H = V_{UT} - V_{LT}$$

Substituting the values of V_{UT} and V_{LT} in Equations 9 and 10 gives:

$$V_H = \frac{R_2}{R_1 + R_2}(+V_{sat})$$
$$- \frac{R_2}{R_1 + R_2}(-V_{sat})$$

Combining terms gives:

$$V_H = \frac{2R_2}{R_1 + R_2}(V_{sat}) \qquad (11)$$

Therefore, the upper and lower threshold values and the hysteresis width are determined by the two feedback resistors, R_1 and R_2. If R_2 is much smaller than R_1, the hysteresis width is small and proportional to R_2, which is the general case when hysteresis is used to eliminate small noise fluctuations. Reducing R_2 to zero, of course, reduces the hysteresis width to zero.

EXAMPLE 8:   Calculate V_{UT}, V_{LT}, and the hysteresis width for $\pm V_{sat} = \pm 14$ V, $R_1 = 50$ kΩ, and $R_2 = 50$ Ω.

Solution: Applying Equation 9 gives:

$$V_{UT} = \frac{R_2}{R_1 + R_2}(V_{sat}) = \frac{0.05 \text{ k}\Omega}{50.05 \text{ k}\Omega}(14 \text{ V})$$
$$= \mathbf{0.014 \text{ V}}$$

From Equation 10, we get:

$$V_{LT} = \frac{0.05 \text{ k}\Omega}{50.05 \text{ k}\Omega}(-14 \text{ V}) = \mathbf{-0.014 \text{ V}}$$

**ZERO-CROSSING DETECTOR
WITH HYSTERESIS**

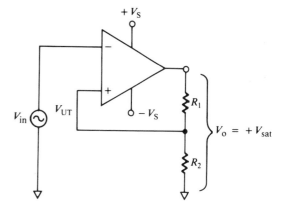

$$V_{UT} = \frac{R_2}{R_1 + R_2}(+V_{sat})$$

(a) Upper Threshold Voltage

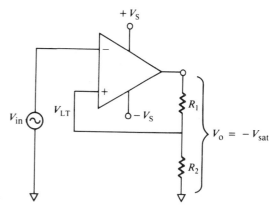

$$V_{LT} = \frac{R_2}{R_1 + R_2}(-V_{sat})$$

(b) Lower Threshold Voltage

FIGURE 9.34.   Shown are the conditions for determining the upper and lower threshold voltages for a comparator with positive feedback. See Examples 8 and 9.

The hysteresis width is found from Equation 8 to be:

$$V_H = V_{UT} - V_{LT}$$
$$= 0.014 \text{ V} - (-0.014 \text{ V}) = \textbf{28 mV}$$

EXAMPLE 9: Design a zero-crossing detector with a hysteresis width of 100 mV. Assume $\pm V_{sat} = \pm 14$ V.

Solution: If $V_H = 100$ mV, then $V_{UT} = +50$ mV and $V_{LT} = -50$ mV. To obtain hysteresis, a fraction of the output voltage must be fed back to the $+$ input via a voltage divider like the one shown in Figure 9.34. For simplicity of calculation, we select a convenient value for R_2—for example, a stock resistor of **200** Ω—and solve for R_1 using Equation 9. Rearranging Equation 9 gives:

$$R_1 = \frac{R_2(V_{sat} - V_{UT})}{V_{UT}}$$
$$= 0.20 \text{ k}\Omega \left(\frac{14 \text{ V} - 0.050 \text{ V}}{0.050 \text{ V}} \right)$$
$$= \textbf{56 k}\Omega$$

The result could also have been obtained using $-V_{sat}$, V_{LT}, and Equation 10.

Nonzero Reference Voltage with Hysteresis

Hysteresis can also be added to nonzero voltage level detectors (Figure 9.35). This addition is accomplished with the feedback circuit shown in Figure 9.36, in which a reference voltage V_{ref}, either positive or negative, is added to the feedback voltage divider. Note again that because the reference voltage is applied to the $+$ input, the output is inverted.

To determine the voltage transfer characteristic of Figure 9.36, we must solve for the voltage at the $+$ terminal V_{in+} and compare it to the voltage at the $-$ terminal V_{in-}, remembering that the output voltage V_o will be at $+V_{sat}$ when $V_{in} < V_{in+}$ and at $-V_{sat}$ when $V_{in} > V_{in+}$.

Voltage V_{in+} can be found by superposition. That is, we first let $V_o = 0$ V and solve for the voltage at V_{in+} due to V_{ref}, which we will call V_A (Figure 9.37a). Then, we let $V_{ref} = 0$ and solve for the voltage at V_{in+} due to V_o, which we will call V_B (Figure 9.37b). The actual voltage at V_{in+} will be:

$$V_{in+} = V_A + V_B \tag{12}$$

Letting $V_o = 0$ V, the portion of V_{in+} due to V_{ref} is given by:

$$V_A = \frac{R_1}{R_1 + R_2} V_{ref} \tag{13}$$

Letting $V_{ref} = 0$, the portion of V_{in+} due to V_o is given by:

$$V_B = \frac{R_2}{R_1 + R_2} V_o \tag{14}$$

The total value of V_{in+} is the sum of V_A and V_B, or:

$$V_{in+} = \frac{R_1}{R_1 + R_2} V_{ref}$$
$$+ \frac{R_2}{R_1 + R_2} V_o \tag{15}$$

where V_{ref} can be either a positive or a negative voltage and V_o is either $+V_{sat}$ or $-V_{sat}$.

The upper threshold voltage V_{UT} occurs when $V_o = +V_{sat}$, while the lower threshold

NONZERO-CROSSING DETECTOR WITH HYSTERESIS

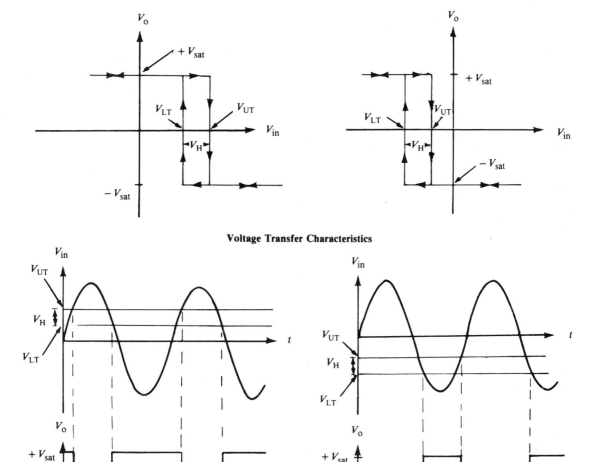

Voltage Transfer Characteristics

Typical Input and Output Waveforms

FIGURE 9.35. Nonzero voltage level detectors with hysteresis can be constructed. The figures show the required voltage transfer characteristics and typical input and output waveforms for both positive and negative versions. Note that the output waveform is inverted.

NONZERO-CROSSING DETECTOR WITH HYSTERESIS

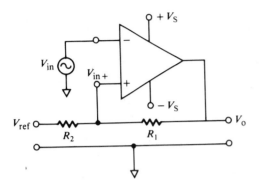

FIGURE 9.36.   A nonzero-crossing detector with hysteresis can be obtained by adding a reference voltage to the feedback voltage divider. The reference voltage can be either positive or negative, to produce either a positive or negative voltage level detector.

voltage V_{LT} occurs when $V_o = -V_{sat}$. In equation form:

$$V_{UT} = \frac{R_1}{R_1 + R_2} V_{ref}$$
$$+ \frac{R_2}{R_1 + R_2} V_{sat} \qquad \textbf{(16)}$$

$$V_{LT} = \frac{R_1}{R_1 + R_2} V_{ref}$$
$$- \frac{R_2}{R_1 + R_2} V_{sat} \qquad \textbf{(17)}$$

Note that when $V_{ref} = 0$ V, Equation 16 reduces to Equation 9, and Equation 17 reduces to Equation 10.

The hysteresis width is:

$$V_H = V_{UT} - V_{LT} = \frac{2R_2}{R_1 + R_2} V_{sat} \qquad \textbf{(18)}$$

which is simply $2V_B$ with $V_o = V_{sat}$. Note also that Equation 18 is identical to that for a zero-crossing detector (Equation 11).

Equations 16, 17, and 18 can be learned and used to solve circuit problems, but it is usually simpler to remember only Equation 15, the general equation that describes the behavior of the comparator circuit in Figure 9.36.

EXAMPLE 10:   Calculate the upper and lower threshold voltages and the hysteresis width for the circuit in Figure 9.36 if $R_1 = 10$ kΩ, $R_2 = 4.0$ kΩ, $V_{ref} = 7.0$ V, and $\pm V_{sat} = \pm 14$ V.

Solution: V_{UT} is found from Equation 15 by setting $V_o = +V_{sat}$.

$$V_{UT} = \frac{10\ k\Omega}{10\ k\Omega + 4\ k\Omega} (7\ V)$$
$$+ \frac{4.0\ k\Omega}{10\ k\Omega + 4\ k\Omega} (14\ V)$$
$$= 5.0\ V + 4.0\ V = \textbf{9.0 V}$$

V_{LT} is found from Equation 15 by setting $V_o = -V_{sat}$:

$$V_{LT} = 5.0\ V - 4.0\ V = \textbf{1.0 V}$$

The hysteresis width is the difference between V_{UT} and V_{LT} (Equation 8):

$$V_H = V_{UT} - V_{LT} = 9.0\ V - 1.0\ V = \textbf{8.0 V}$$

The voltage transfer characteristic for Example 10 is drawn in Figure 9.38. Note that the hysteresis loop is not centered about $V_{ref} = 4.0$ V, as we might expect, but rather about V_A, as given by Equation 13. The **center voltage** V_C of the hysteresis width can be found from Equations 16 and 17, where V_A is the first term in each. That is, the center voltage V_C is simply V_A, or:

CIRCUIT ANALYSIS USING SUPERPOSITION

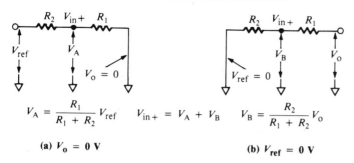

$$V_A = \frac{R_1}{R_1 + R_2} V_{ref} \qquad V_{in+} = V_A + V_B \qquad V_B = \frac{R_2}{R_1 + R_2} V_o$$

(a) $V_o = 0$ V   **(b)** $V_{ref} = 0$ V

FIGURE 9.37. The threshold voltage for the circuit in Figure 9.36 can be obtained by superposition. First, let $V_o = 0$ and solve for V_A due to V_{ref} (a). Then, let $V_{ref} = 0$ and solve for V_B due to V_o (b). The voltage at the + terminal is then $V_{in+} = V_A + V_B$. See Equation 15.

VOLTAGE TRANSFER CHARACTERISTIC
OF EXAMPLE 10

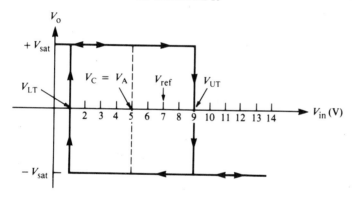

FIGURE 9.38. Shown is the voltage transfer characteristic of Example 10. Note that the center voltage V_C is not V_{ref} but V_A. See Equation 19.

$$V_C = \frac{R_1}{R_1 + R_2} V_{ref} \qquad (19)$$

which, for Example 10, is:

$$V_C = \frac{10 \text{ k}\Omega}{10 \text{ k}\Omega + 4 \text{ k}\Omega} (7 \text{ V}) = \textbf{5.0 V}$$

Note that in Equation 19, the center voltage V_C is directly proportional to V_{ref}, but the hysteresis width (Equation 18) does not depend on V_{ref}. Therefore, by changing V_{ref}, the voltage transfer characteristic of Figure 9.38 can be moved back and forth along V_{in} without a change in shape. Equations 18 and 19 can also be used to design

a voltage level detector centered on a specific voltage and having a specific hysteresis width.

EXAMPLE 11: Design a voltage level detector centered on 9.0 V and having a hysteresis width of 2.0 V (Figure 9.39).

Solution: The basic circuit is the one in Figure 9.36, for which values of reference voltage V_{ref} and feedback resistors R_1 and R_2 must be selected. These values can be determined from the design value of $V_C = 9.0$ V and $V_H = 2.0$ V and from Equations 18 and

FIGURE 9.39. Shown is a voltage level detector with a specific center voltage and hysteresis width.

EXAMPLE 11

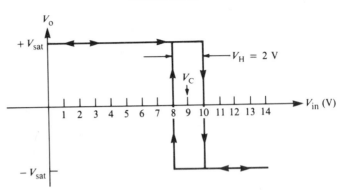

19. Assuming $V_{sat} = \pm 14$ V, Equation 18 gives:

$$V_H = \frac{2R_2}{R_1 + R_2}(14 \text{ V}) = 2.0 \text{ V}$$

or

$$\frac{R_2}{R_1 + R_2} = \frac{1}{14}$$

This equation can be used to obtain values for the feedback resistors. Simply select a value for one of the resistors (R_1 or R_2) and calculate the other from the equation. Since the hysteresis width is fairly small, R_2 must be smaller than R_1. Therefore, we select R_2 = **1.0 kΩ** and solve for R_1:

$$\frac{1 \text{ k}\Omega}{R_1 + 1 \text{ k}\Omega} = \frac{1}{14}$$

$$R_1 = 14 \text{ k}\Omega - 1 \text{ k}\Omega = \textbf{13 k}\Omega$$

The required reference voltage can now be determined from Equation 19:

$$V_C = \frac{R_1}{R_1 + R_2} V_{ref} = 9.0 \text{ V}$$

$$V_{ref} = 9.0 \text{ V} \left(\frac{13 \text{ k}\Omega + 1.0 \text{ k}\Omega}{13 \text{ k}\Omega}\right) = \textbf{9.7 V}$$

The final circuit is shown in Figure 9.40, where a 15 kΩ potentiometer is used for R_1 so that its value can be accurately adjusted. A variable voltage source is used to set V_{ref} = 9.7 V accurately. Again, note that changing V_{ref} simply moves the voltage transfer characteristic of Figure 9.39 along V_{in} without changing its width.

The reference voltage for the voltage level detector with hysteresis can also be obtained from the op amp supply voltage, just as it was for the detector without hysteresis. Figure 9.41 shows a variable reference voltage circuit using both the $+V_S$ and $-V_S$ supply voltages. By changing the potentiometer setting, the reference voltage is continuously adjustable from $+V_S$ to $-V_S$.

The equations for the center voltage and

**CIRCUIT DESIGN FOR
EXAMPLE 11**

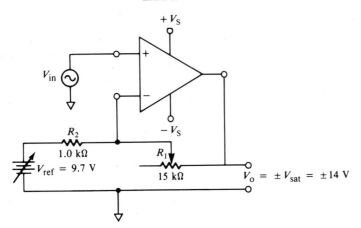

FIGURE 9.40. Shown is the circuit design for a voltage level detector with the voltage transfer characteristic of Figure 9.39.

the hysteresis width for the circuit in Figure 9.41 are not as simple as Equations 18 and 19 because the voltage at the $+V_S$ terminal for the condition of $V_o = \pm V_{sat}$ depends in a complicated way on the setting of the potentiometer R_{ref}. However, provided that R_1 is much larger than R_2 and that R_2 is somewhat larger than R_{ref}, Equations 18 and 19 can be used to obtain approximate values for the center voltage and the hysteresis width. Also, changing V_{ref} will still change the center voltage, though with some change in hysteresis width. Specific values of V_C and V_H will generally have to be determined by trial and error.

It should be noted that the circuit in Figure 9.41 is essentially the same as the one used in the trigger circuit of an oscilloscope. The potentiometer R_{ref} corresponds to the trigger level control on the front panel of an oscilloscope that determines the voltage level of an input signal that will initiate the sweep circuit.

VARIABLE REFERENCE VOLTAGE CIRCUIT

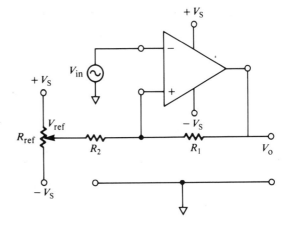

FIGURE 9.41. A variable reference voltage can be obtained from the op amp supply voltage. The equations for V_C and V_H for this circuit are quite complicated, but reduce approximately to Equations 18 and 19 if R_1 is much larger than R_2 and R_2 is somewhat larger than R_{ref}.

**NONINVERTING VOLTAGE LEVEL DETECTOR
WITH HYSTERESIS**

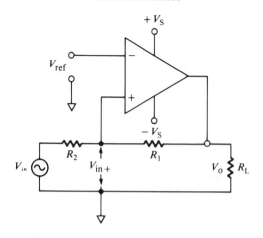

FIGURE 9.42. A noninverting voltage level detector with hysteresis can be constructed by applying the input voltage to feedback resistor R_2 and the reference voltage to the op amp's − input. In this case, however, the input resistance is the sum of R_2 + R_1 + R_L and not the high value of the op amp.

Finally, it is possible to build a voltage level detector with hysteresis for which the output is not inverted. We simply apply the reference voltage to the − op amp terminal and input signal to the end of the feedback voltage divider (Figure 9.42). Equations for the upper and lower threshold voltages can be found by superposition as before, and these equations can be used to get values for the center voltage and the hysteresis width. Deriving these expressions is left as an exercise for the end of the chapter.

Note in Figure 9.42 that the input signal is connected directly across the output through the feedback resistors. Thus, the input resistance of this circuit is not the very high value of the op amp, as it was for the inverting case (see, for example, Figure 9.41). If R_1 and R_2 are very large compared to the resistance of the signal source, this condition may be a problem. However, if a very high input resistance circuit is required, then the inverting voltage level detector must be used.

9.7 QUESTIONS AND PROBLEMS

1. Draw the symbol for a 741 op amp. Label each terminal as to its function, pin numbers (for a 14 pin DIP package), and maximum ratings.
2. What would be the output voltage of the op amp in Problem 1 for the following input voltages?
   (a) $V_{in+} = 6.3$ V and $V_{in-} = 7.2$ V
   (b) $V_{in+} = 0.016$ V and $V_{in-} = 0.008$ V
   (c) $V_{in+} = 0.635$ mV and $V_{in-} = 0.593$ mV
   (d) $V_{in+} = 0$ V and $V_{in-} = 0.135$ mV
3. Draw the voltage transfer characteristic for the circuit shown in Figure 9.43.

4. Design a voltage level detector that will turn on a positive voltage switch whenever the input voltage exceeds 3 V. Obtain reference voltages from the op amp power supply voltage. Draw the circuit and, using Figure 9.44, draw its voltage transfer characteristic and the input and output waveforms that would result from a 10 V pp sine wave at 500 Hz.

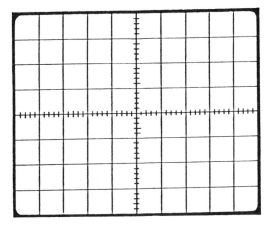

Voltage Transfer Characteristic

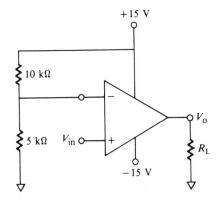

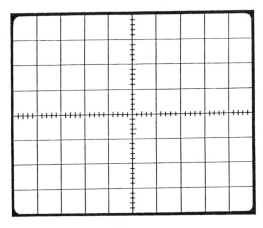

Voltage Transfer Characteristic

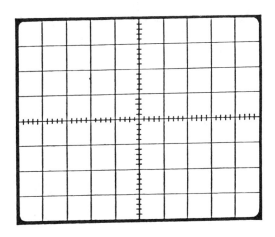

Input and Output Waveforms

FIGURE 9.43.

FIGURE 9.44.

5. Design a voltage level detector with the voltage transfer characteristic shown in Figure 9.45. Obtain reference voltages from the op amp power supply. Draw the circuit.

6. Design a portable battery tester that will signal "ok" if the voltage of a 6 V battery is greater than 6.0 V and "bad" if its voltage is less than 6.0 V. Use two 9 V batteries to power the tester. Draw the circuit diagram.

7. Design a circuit that can be used with a sine wave generator (2 V pp) to produce a variable-amplitude, variable–duty cycle pulse generator. Obtain reference voltages from the op amp power supply. Draw the circuit.

8. Calculate the upper and lower threshold voltages and draw the voltage transfer characteristic for the circuit shown in Figure 9.46.

9. Design a window detector circuit with the voltage transfer characteristic shown in Figure 9.47. Obtain reference voltages from the op amp power supply. Draw the circuit. Also, draw the input and output voltages for a 20 V pp, 250 Hz sine wave input.

10. Calculate the upper and lower threshold voltages and hysteresis width for the circuit shown in Figure 9.48. Draw the voltage transfer characteristic and input and output waveforms for a 20 V pp, 250 Hz sine wave input.

11. Design a zero-crossing detector with a hysteresis width of 20 mV. Draw the circuit.

12. Design an inverting voltage level detector centered on 6.0 V with a hysteresis width of 50 mV. Draw the circuit.

13. Derive expressions for the upper and lower threshold voltages, hysteresis voltage, and center voltages for the noninverting voltage level detector.

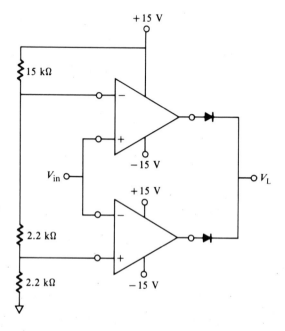

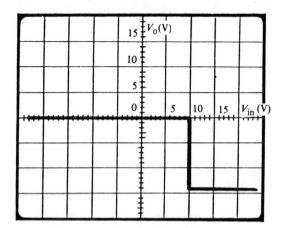

Voltage Transfer Characteristic

FIGURE 9.45.

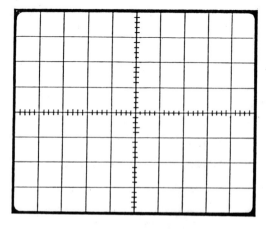

Voltage Transfer Characteristic

FIGURE 9.46.

14. Design a noninverting voltage level detector centered on 6.0 V with a hysteresis width of 50 mV. Maximize the input resistance of the circuit for any size load. Draw the circuit.

15. Design an inverting voltage level detector with variable center voltage and hysteresis width. Draw the circuit. Minimize interaction between the two variable controls.

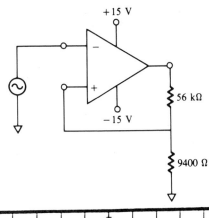

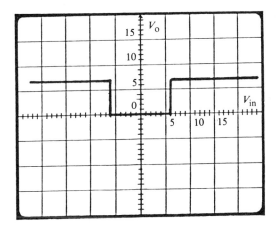

Voltage Transfer Characteristic

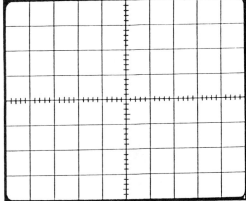

Voltage Transfer Characteristic

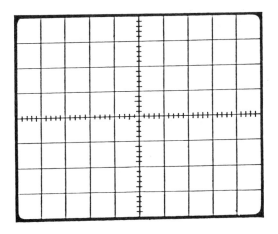

Input and Output Voltages

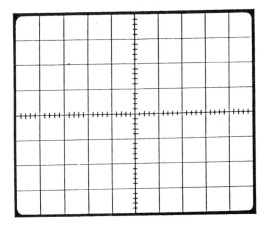

Input and Output Waveforms

FIGURE 9.47.

FIGURE 9.48.

Chapter

10

OPERATIONAL AMPLIFIERS AS VOLTAGE AMPLIFIERS

10.1 OBJECTIVES

Following the completion of Chapter 10, you should be able to:

1. Draw the op amp circuit model and state the conditions of input current and voltage that simplify op amp circuit analysis.
2. Identify and draw an op amp connected as a voltage follower and derive an expression for its voltage transfer function and closed loop gain.
3. State the most important characteristic of a voltage follower and give two typical applications.
4. Identify an op amp connected as a noninverting amplifier and determine its voltage gain, input resistance, and voltage transfer characteristic.
5. Identify an op amp connected as an inverting amplifier and determine its voltage gain, input impedance, and voltage transfer characteristic.

6. Give limitations on values for the input voltage, closed loop gain, output current, load resistance, and feedback resistors for an op amp connected as a voltage follower, inverting amplifier, and noninverting amplifier.
7. Identify and draw circuits for an op amp that will adjust the output offset voltage and boost the output current.
8. Design a noninverting amplifier with a specific value of closed loop gain and draw its voltage transfer characteristic.
9. Explain how to use the noninverting amplifier as a voltage-to-current converter and give two typical applications.
10. Design an inverting amplifier with a specific value of closed loop gain and draw its voltage transfer characteristic.
11. Explain how the inverting amplifier can be used as an inverting adder and give two typical applications.

10.2 OVERVIEW

The comparator circuits discussed in Chapter 9 had only two values of output voltage: $+V_{sat}$ and $-V_{sat}$. Through a variety of schemes, the input voltage that produced an output transistion could be set to any voltage between the two supply voltages ± 15 V. With positive feedback, the comparator could even have hysteresis — that is, change output voltages at one value of input voltage if $V_o = +V_{sat}$ and at a different value if $V_o = -V_{sat}$. However, in none of the cases studied could the output voltage be any value between $\pm V_{sat}$.

In this chapter, we examine circuits for which the output voltage is between $\pm V_{sat}$. For all of these circuits, the output voltage is directly proportional to the input voltage (Figure 10.1). This relationship is described by the simple voltage gain expression of Chapter 9, Equation 1:

$$V_o = AV_{in}$$

where the voltage gain A is the constant of proportionality.

Using a variety of new schemes, we can make A any value from 1 to about 1000 and either positive or negative. Thus, an input signal can be amplified by factors of up to 1000 and have either the same or an inverted polarity. The maximum value of output voltage, of course, will still be limited by the maximum op amp output voltage of $\pm V_{sat}$.

10.3 NEGATIVE FEEDBACK

To reduce the amplifier gain from its very large open loop value of $A_{OL} = 200,000$ to a lower, more stable value, **negative feedback** is used. Negative feedback is simply a connection between the output terminal and the $-$ input terminal. In the simplest case, Figure 10.2a, the connection is a wire from V_o to V_{in}. This con-

VOLTAGE AMPLIFIER
VOLTAGE TRANSFER CHARACTERISTICS

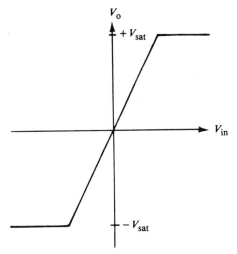

Noninverting: $V_o = AV_{in}$

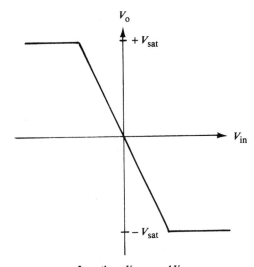

Inverting: $V_o = -AV_{in}$

FIGURE 10.1. An op amp can be made to have outputs between $\pm V_{sat}$ by using negative feedback. For the circuits in this section, the output voltage will be directly proportional to the input voltage. The constant of proportionality A can be any value from 1 to about 1000 and either positive, as in the noninverting op amp, or negative, as in the inverting op amp.

NEGATIVE FEEDBACK CONNECTIONS

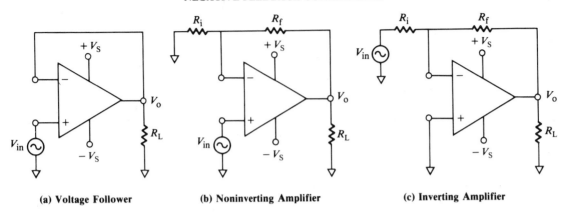

(a) Voltage Follower   (b) Noninverting Amplifier   (c) Inverting Amplifier

FIGURE 10.2. Negative feedback involves a connection from the output terminal to the − input terminal. This connection can be direct to produce a voltage follower with a gain of 1 or a connection through a voltage divider to produce noninverting or inverting amplifiers with gains up to about 1000.

nection makes $A = 1$ and produces a **voltage follower** for which the output voltage is the same magnitude and polarity as the input voltage. This circuit has a number of practical applications that are described in the following paragraphs.

The feedback connection can also be made through a voltage divider, as shown in Figures 10.2b and 10.2c. With these circuits, A can be varied from 1 to about 1000, depending on the voltage divider ratio, and be either positive (Figure 10.2b), to make a **noninverting amplifier**, or negative (Figure 102c), to make an **inverting amplifier**. The remainder of this chapter comprises the operation and application of these two amplifiers.

As these ideas are developed, it is important to recall a fundamental characteristic of op amps discussed in Chapter 9. Examples 4 and 5 of Chapter 9 showed that for the output voltage to go to saturation, the difference in voltage V_d between the op amp's two input terminals need only be greater than $+75\ \mu V$ or less than $-75\ \mu V$.

For the output voltage of an op amp to be between the saturation levels requires that the op amp's input differential voltage be within the range of $V_d = \pm 75\ \mu V$. This range was not possible to achieve through external circuitry due to electrical noise and other problems.

With negative feedback, however, it is possible to force the input differential voltage into this small range and thus produce an output voltage that is between $\pm V_{sat}$. Since the voltage at the + input is only a few microvolts different from the voltage at the − input, we can assume that *the two op amp inputs are at the same voltage*. Thus, we have the first general op amp characteristic:

$$V_{in+} = V_{in-} \tag{1}$$

At this point, an op amp circuit model is useful as an aid for analyzing the various circuits using negative feedback (Figure 10.3). Note that the input circuit is electrically iso-

OP AMP CIRCUIT MODEL

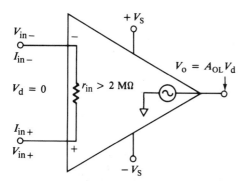

FIGURE 10.3. The circuit model for an op amp is very helpful in analyzing circuits. Note that the input terminals are electrically isolated from the output, the input differential voltage $V_d = 0$, and the input resistance r_{in} is very large. This model provides two useful characteristics for circuit analysis: $V_{in+} = V_{in-}$ and $I_{in+} = I_{in-} = 0$.

VOLTAGE FOLLOWER

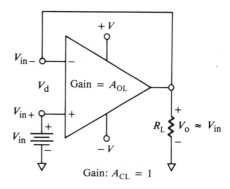

FIGURE 10.4. In the voltage follower circuit, the input voltage is applied to the + input terminal, $V_{in+} = V_{in}$, and the output voltage is connected back to the – input terminal, $V_{in-} = V_o$. Since $V_{in+} = V_{in-}$, $V_o = V_{in}$. Thus, the gain of the voltage follower A_{CL} is 1 and the output voltage is the same value and polarity as the input voltage.

lated from the output circuit. The output voltage is assumed to be supplied by a separate internal voltage source whose value is given by Equation 5 of Chapter 9:

$$V_o = A_{OL}V_d$$

The resistance between the two input terminals of the op amp is called the **input resistance** r_{in}, and it is very large. For the 741 op amp, $r_{in} \approx 2$ MΩ (see Figure 9.13).

According to the circuit model, the op amp can supply no current from its input terminals because they are electrically isolated from the internal voltage source. Even if an external source is connected to one input terminal, essentially no current will flow to the other due to the very large value of r_{in}. Thus, we have the second general op amp characteristic: *No current flows into or out of the input terminals:*

$$I_{in+} = I_{in-} = 0 \tag{2}$$

These two op amp characteristics make analysis of op amp circuits quite simple.

10.4 VOLTAGE FOLLOWER

Figure 10.4 shows the circuit diagram of an op amp connected as a **voltage follower**—that is, a circuit whose output voltage has the same magnitude and polarity as its input voltage. This circuit is also called an isolation amplifier, buffer amplifier, or unity gain amplifier. The reasons for these names will become evident in the following analysis.

Principles of Operation

The input signal V_{in} is applied to the op amp's + input, and the feedback element is a direct connection between the output terminal and the – input. The response of this circuit to an input signal can be understood by applying the gain expression for the op amp (Equation 5, Chapter 9):

$$V_o = A_{OL}V_d$$

where A_{OL} is the high open loop gain of the op amp and V_d is the difference in voltage between the + and the − terminals (Equation 3, Chapter 9):

$$V_d = V_{in+} - V_{in-}$$

To begin the analysis, we assume that V_o is at $-V_{sat}$ and a positive voltage V_{in} is suddenly applied to the + input terminal: $V_{in} = V_{in+}$. According to Equation 3 in Chapter 9, this will make V_d large and positive and, according to that equation, will cause the output voltage to increase to positive values immediately.

As V_o increases, so does the voltage at the − input terminal: $V_{in-} = V_o$. A larger V_{in-}, of course, reduces V_d and leads to a smaller final positive value of V_o. Various other possible situations will lead to the same conclusion: V_o *will always come to the same polarity as* V_{in}.

The final stable value of V_o can be found by applying Kirchhoff's voltage law to Figure 10.4. For the outside loop from input ground, across V_{in}, V_d, and V_o, and back to ground, we have:

$$-V_{in} + V_d + V_o = 0$$

or

$$V_{in} = V_d + V_o \tag{3}$$

Substituting into this equation the value of V_d from Equation 5 in Chapter 9 gives:

$$V_{in} = \frac{V_o}{A_{OL}} + V_o = V_o\left(1 + \frac{1}{A_{OL}}\right)$$

Because A_{OL} is very large ($\approx 200,000$), the factor $1/A_{OL}$ is very small and can be neglected, giving the *voltage transfer function of a voltage follower:*

$$\boxed{V_o = V_{in}} \tag{4}$$

Thus, for all practical purposes, the output voltage is always equal to the input voltage. Since the output voltage will always track, or "follow," the input voltage, this circuit is called a **voltage follower**.

Note that the same result can be obtained by looking at Figure 10.4 and applying the principle that the two input terminals are at the same voltage, $V_{in+} = V_{in-}$. By examination, $V_{in+} = V_{in}$ and $V_{in-} = V_o$, so we must have $V_{in} = V_o$.

In practical terms, with negative feedback, the op amp will always maintain the two input terminals at essentially the same voltage: $V_{in+} = V_{in-}$. If the output voltage is fed back directly to the − terminals, the op amp will force it to be the value applied to the + input terminal. If V_{in} is positive, V_o will be positive, and vice versa.

Voltage Gain

The voltage gain of the *op amp* itself is still the high **open loop gain** A_{OL} that is the ratio of output voltage V_o to the op amp's input differential voltage V_d:

$$A_{OL} = \frac{V_o}{V_d}$$

The voltage gain of the *total circuit*, however, including the feedback loop, is known as the **closed loop gain** A_{CL}. The closed loop gain is the ratio of the output voltage V_o to the input signal voltage V_{in}, as described by Equation 2 from Chapter 9:

$$A_{CL} = \frac{V_o}{V_{in}}$$

Substituting Equation 4 into Equation 5 from Chapter 9 gives the **gain of the voltage follower:**

$$\boxed{A_{CL} \approx 1} \tag{5}$$

The result shows why this circuit is also called a **unity gain amplifier**—that is, a circuit whose voltage gain equals 1. Note also that the result is +1. Thus, the polarity of V_o is the same as that of V_{in}.

Input Resistance

The advantage of using a voltage follower circuit is not its unity gain, but rather its high **input resistance** R_{in}. This is the resistance "seen" by the input voltage source V_{in}, and it determines the current drawn from the source I_{in}.

This input resistance is shown schematically in Figure 10.5a. To determine the value of R_{in}, we will use the op amp circuit model of Figure 10.3 as redrawn in Figure 10.5b. The resistance r_{in} is the resistance of the op amp between its two input terminals.

For the circuit shown in Figure 10.5a, the current drawn from the voltage source is:

$$I_{in} = \frac{V_{in}}{R_{in}} \qquad (6)$$

where R_{in} is the input resistance of the *voltage follower*. For the circuit shown in Figure 10.5b, the same current is given by:

$$I_{in} = \frac{V_d}{r_{in}} \qquad (7)$$

where r_{in} is the input resistance of the *op amp*. Equating these two equations gives:

$$\frac{V_{in}}{R_{in}} = \frac{V_d}{r_{in}}$$

or

$$R_{in} = \frac{V_{in}}{V_d} r_{in} \qquad (8)$$

Substituting into this expression the op amp gain expression $V_d = V_o/A_{OL}$ gives:

$$R_{in} = \frac{V_{in}}{V_o} A_{OL} r_{in} \qquad (9)$$

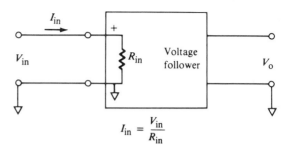

$$I_{in} = \frac{V_{in}}{R_{in}}$$

(a) Input Resistance of Voltage Follower

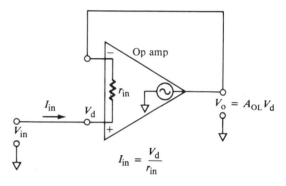

$$I_{in} = \frac{V_d}{r_{in}}$$

(b) Op Amp Circuit Model for Voltage Follower

FIGURE 10.5. The primary advantage of the voltage follower is its very large input resistance R_{in} that is the product of the op amp's large input resistance r_{in} and its large open loop gain A_{OL}. For the 741 op amp, this resistance leads to values of $R_{in} \approx 10^{11}\ \Omega$.

However, according to Equation 5, $V_{in}/V_o = 1$ for a voltage follower, so the *input resistance of a voltage follower* is:

$$R_{in} = A_{OL} r_{in} \qquad (10)$$

Thus, the voltage follower circuit "amplifies" the already high input resistance of the op amp by its similarly high open loop gain. For the 741 op amp, $A_{OL} \approx 200{,}000$ and $r \approx 2\,M\Omega$, giving:

$$R_{in} = (2 \times 10^5)(2 \times 10^6)\,\Omega$$
$$= 4 \times 10^{11}\,\Omega$$

which is an extremely large value.

Because of the very high input resistance, the voltage follower draws negligible current from a voltage source. A voltage follower circuit thus electrically isolates an input signal from the load while allowing the load voltage to equal V_{in}. Hence, the circuit is also referred to as an **isolation**, or **buffer, amplifier**.

Output Current

The **output current** of an op amp should generally be less than about $\pm 5\,mA$:

$$I_{o(max)} = \pm 5\,mA \qquad (11)$$

Though the maximum output current may be specified as ± 25 mA, the voltage regulation at output currents greater than about ± 5 mA may not be maintained. Thus, the voltage gain expression will be true only for output currents less than about ± 5 mA; if this current is exceeded, the output voltage may drop below the expected value.

The output terminal of an op amp can be either a source or a sink of output current. A **current source** means that current flows *out of* the output terminals to a load, as shown in Figure 10.6a. Conversely, an op amp is said to be a **current sink** when current flows *into* the output terminal, as shown in Figure 10.6b.

Also, note in Figure 10.6 the direction of the feedback current I_f that makes a small contribution to I_L. According to the op amp circuit in Figure 10.6b, the feedback current I_f is drawn from the $-$ input as I_{in}:

$$I_f = I_{in} \qquad (12)$$

OUTPUT CURRENT

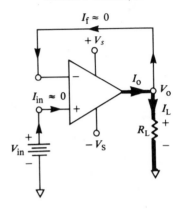

(a) Positive Input Voltage

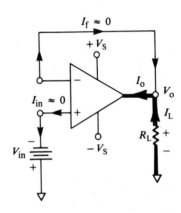

(b) Negative Input Voltage

FIGURE 10.6. The maximum current $I_{o(max)}$ that can be supplied by an op amp is about ± 5 mA, which places a lower limit on the load resistance ($R_{L(min)}$ of about 3 kΩ. When V_{in} is positive, V_o is positive and the output terminal acts as a source of load current. When V_{in} is negative, V_o is negative and the output terminal acts as a sink of load current.

Using Equation 7 for I_{in} gives:

$$I_f = \frac{V_d}{r_{in}} \qquad (13)$$

Using Equation 5 from Chapter 9 for V_d gives:

$$I_f = \frac{V_o}{A_{OL}r_{in}} \qquad (14)$$

and, using Equation 10 for $A_{OL}r_{in}$ gives:

$$I_f = \frac{V_o}{R_{in}} \qquad (15)$$

For a 741 op amp:

$$V_{o(max)} = V_{sat} \approx 14 \text{ V}$$

and

$$R_{in} \approx 4 \times 10^{11} \ \Omega$$

Therefore:

$$I_f \approx \frac{14 \text{ V}}{4 \times 10^{11} \ \Omega} \approx 30 \times 10^{-12} \text{ A}$$

which can clearly be neglected. For all practical purposes, all of the output current goes to the load:

$$I_o = I_L \qquad (16)$$

The real source (or sink) of the output current is, of course, the op amp power supply $+V_S$ and $-V_S$. Figure 10.7 shows the actual paths of these load currents for both positive and negative input voltages. Note that for a positive input voltage, the output current is supplied by $+V_S$ and, for a negative input voltage, by $-V_S$. The negligibly small feedback currents have been omitted in these figures.

Load Resistance

The limitation of $I_o = \pm 5$ mA also limits the resistance of the load that can be attached to the amplifier output. This **minimum load resistance** can be simply calculated from Ohm's law using the output saturation voltages and the maximum output current:

$$R_{L(min)} = \frac{V_{sat}}{I_{o(max)}} = \frac{14 \text{ V}}{5 \text{ mA}}$$

OUTPUT CURRENT SOURCE

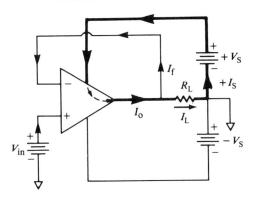

Positive Input Voltage

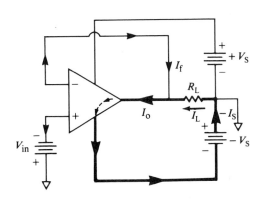

Negative Input Voltage

FIGURE 10.7. The source of output current from an op amp is the external power supply $\pm V_S$. The figure shows the output current paths for both positive and negative input voltages. The feedback current in a voltage follower is negligible, so that essentially all of the output current goes to the load.

which reduces to:

$$R_{L(min)} = 3 \text{ k}\Omega \qquad (17)$$

Smaller load resistances can be used provided that the load current does not exceed $I_{o(max)} =$

FIGURE 10.8. Because the voltage follower's input resistance is very large, it draws negligible current from a voltage source. Thus, it is used to electrically isolate or buffer a voltage source from a load. In the circuit shown, the addition of the voltage follower between a low-sensitivity voltmeter and a circuit to be measured makes a high input resistance voltmeter.

HIGH INPUT RESISTANCE VOLTMETER

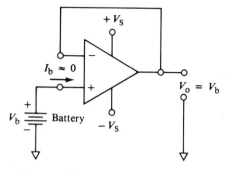

± 5 mA. In general, however, when R_L is less than about 3 kΩ, the amplifier performance should be checked to be certain that it is adequate for the application.

10.5 VOLTAGE FOLLOWER APPLICATIONS

High Input Resistance Voltmeter

The isolation characteristic of a voltage follower that results from its high input impedance has many useful applications. A basic one is the construction of a voltmeter with a very high input resistance. In Part II, we saw that a simple D'Arsonval voltmeter with a low sensitivity (ohms-per-volt rating) can cause the meter to load the circuit whose voltage it is measuring. The result is that the measured value is not the same as it would be without the meter in the circuit and, therefore, is inaccurate.

One solution to this problem is to use a meter with a higher sensitivity. Another solution is to use the same low-sensitivity meter, but with a unity gain buffer amplifier between the meter and the circuit, as shown in Figure 10.8. The high resistance of the voltage follower, $R_m \geq 4$

$\times 10^{11}$ Ω, means that it can be used across circuit resistances up to several megohms with less than 1% error.

Constant-Voltage Source

Another application of a voltage follower is as a stable, constant-voltage source. The voltage of a battery will stay at a constant value pro-

CONSTANT-VOLTAGE SOURCE

FIGURE 10.9. Another application of the voltage follower is to make a constant-voltage source. The voltage across a battery is very constant provided that no current is drawn. Adding a voltage follower makes the stable battery voltage available at the output, $V_o = V_b$, with negligible current drawn from the battery.

vided that no current is drawn from it. Therefore, if a battery is used as the input voltage to a voltage follower, the output of the follower will be a constant voltage (Figure 10.9).

Because the input resistance of the op amp is greater than 4×10^{11} Ω, the current drawn from a 6 V battery is less than:

$$I = \frac{6 \text{ V}}{4 \times 10^{11} \ \Omega} = 15 \times 10^{-12} \text{ A (15 pA)}$$

which is negligible. The current supplied to a load, however, can be up to the maximum op amp value of ± 5 mA. Also, larger currents can be obtained by using an output transistor current boost, as described later in the chapter.

10.6 NONINVERTING AMPLIFIER

The op amp circuit shown in Figure 10.10 is a noninverting amplifier. This circuit is similar to the voltage follower in that the input voltage is applied to the $-$ input terminal, and the output voltage has the same polarity as the input voltage. The circuit does not invert the input signal; hence, the name **noninverting amplifier**.

Principles of Operation

The noninverting amplifier differs from the voltage follower in that only a fraction of the output voltage is fed back to the $-$ input terminal. That fraction is determined by the voltage divider of R_i and R_f, and it leads to voltage gains greater than one. A simple analysis of this circuit can be made using the op amp circuit model and the principle that $V_{in+} = V_{in-}$.

The input voltage V_{in} is applied to the $+$ input terminal so that $V_{in} = V_{in+}$. But, since $V_{in-} = V_{in+}$, we also have $V_{in-} = V_{in}$. This point is important. For the feedback voltage divider, *the voltage across the feedback resistor R_i will always be the input voltage V_{in}.*

NONINVERTING AMPLIFIER WITH
VOLTAGE GAIN

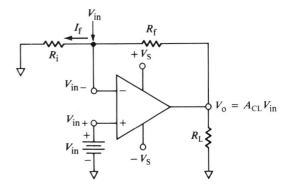

FIGURE 10.10. A noninverting amplifier with a gain greater than 1 can be constructed by placing a voltage divider in the feedback loop. Thus, only a fraction of the output voltage is fed back to V_{in}. The amount of voltage gain A_{CL} is determined by the fraction of the voltage fed back, which is determined by the voltage divider resistors R_i and R_f.

The high input resistance r_{in} of the op amp means that negligible current is drawn from V_{in}. Therefore, the current through R_i and, thus, the voltage at V_{in-} are determined solely by V_o (Figure 10.11). Two conclusions can be drawn from this relationship. First, the values of V_{in} and R_i determine the feedback current I_f drawn from the output terminal V_o:

$$I_f = \frac{V_{in}}{R_i} \tag{18}$$

Second, the feedback current and the feedback resistors R_i and R_f determine the output voltage:

$$V_o = I_f (R_i + R_f) \tag{19}$$

Substituting Equation 18 into Equation 19 gives the **voltage transfer function**—that is, the output voltage in terms of the input voltage and the circuit values:

$$V_o = \frac{V_{in}}{R_i} (R_i + R_f)$$

FIGURE 10.11. The fact that $V_{in-} = V_{in+}$ means that the voltage at V_{in-}, across R_i, is always V_{in}. V_{in} thus determines the feedback current $I_f = V_{in}/R_i$. The output voltage is then related to the input voltage by the voltage divider relation $V_{in} = R_i/(R_f + R_i)V_o$ to give the voltage transfer function and, thus, the closed loop gain A_{CL}.

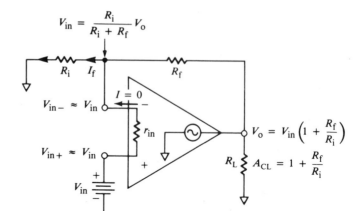

VOLTAGE GAIN

$$V_{in} = \frac{R_i}{R_i + R_f} V_o$$

$$V_o = V_{in}\left(1 + \frac{R_f}{R_i}\right)$$

$$A_{CL} = 1 + \frac{R_f}{R_i}$$

Combining terms gives:

$$V_o = V_{in}\left(1 + \frac{R_f}{R_i}\right) \quad (20)$$

This same result can be obtained by simply noting that the voltage across R_i is V_{in}. This voltage is related to V_o by the voltage divider relation:

$$V_{in} = \frac{R_i}{R_i + R_f} V_o \quad (21)$$

Rearranging terms gives Equation 20 immediately.

Voltage Gain

The **voltage gain** A_{CL} is the term that appears in parentheses in Equation 20:

$$A_{CL} = 1 + \frac{R_f}{R_i} \quad (22)$$

Thus, the greater the ratio of R_f to R_i, the greater the gain of the amplifier. Equation 22

also follows directly from the characteristics of the voltage divider. If the voltage across R_i is always V_{in}, then the greater the ratio of R_f to R_i, the greater the value of V_o for a given V_{in}.

For many cases, $R_f \gg R_i$, so the gain expression reduces approximately to:

$$A_{CL} \approx \frac{R_f}{R_i} \quad (23)$$

which gives the simple proportion:

$$\frac{V_o}{V_{in}} \approx \frac{R_f}{R_i} \quad (24)$$

Note also that the voltage gain of the non-inverting amplifier A_{CL} depends only on the values of the two feedback resistors R_f and R_i. Once the feedback resistors are selected, the amplifier gain is fixed and is independent of the op amp used or the load applied.

Circuit Value Limitations

In the design of noninverting amplifiers, there are some practical limitations on the values of A_{CL}, R_i, and R_f that can be selected. Since V_o is limited to the saturation voltage, A_{CL} must be kept small enough to prevent the output from saturating at any input voltage.

Also, for most general-purpose op amps, A_{CL} should not exceed about 1000 because unstable op amp behavior may result:

$$1 \leq A_{CL} \leq 1000 \qquad (25)$$

Finally, the sum of R_i and R_f should generally be greater than about 1 kΩ but less than about 100 kΩ:

$$1 \text{ k}\Omega \leq (R_i + R_f) \leq 100 \text{ k}\Omega \qquad (26)$$

If R_i and R_f are too small, too much current will be drawn from the output, and if they are too large, they may be comparable to internal op amp resistances, and our simple assumptions would break down. Within these constraints, however, a wide variety of circuits can be designed. These circuit value limitations are summarized in Figure 10.12.

CIRCUIT VALUE LIMITATIONS

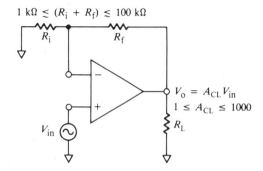

FIGURE 10.12. In the design of noninverting amplifiers, there are some practical limitations on circuit values. The open loop gain should be less than about 1000, and the sum of the feedback resistors should be greater than about 1 kΩ, to limit the feedback current drawn from the output, and less than about 100 kΩ. See Example 1.

EXAMPLE 1: A noninverting amplifier is designed with $R_i = 1$ kΩ and $R_f = 12$ kΩ. What is the value of (a) A_{CL}, (b) the largest input voltage that can be applied before the output saturates, and (c) the maximum current drawn by the feedback resistors?

Solution: (a) The voltage gain is found from Equation 22 and the values of R_i and R_f:

$$A_{CL} = 1 + \frac{R_f}{R_i} = 1 + \frac{12 \text{ k}\Omega}{1 \text{ k}\Omega} = 13$$

(b) The value of input voltage that will produce saturation is found from the gain and the saturation voltage V_{sat}. Assuming $V_{sat} = 14$ V:

$$V_o = A_{CL}V_{in(max)} = V_{sat}$$
$$14 \text{ V} = 13 V_{in(max)}$$
$$V_{in(max)} = \textbf{1.1 V}$$

(c) The feedback current can be found from Equation 15 for $V_{in} = V_{in(max)}$ or from the value of V_{sat} across the sum of R_i and R_f:

$$I_f = \frac{V_{sat}}{R_i + R_f} = \frac{14 \text{ V}}{1 \text{ k}\Omega + 12 \text{ k}\Omega}$$
$$= \textbf{1.1 mA}$$

Note that this current is not insignificant. A gain of 13 could be achieved as well by scaling up the voltage divider by a factor of 5 — for example, $R_i = 5$ kΩ and $R_f = 60$ kΩ — with a corresponding reduction in feedback current drain: $I_f \approx 0.2$ mA.

EXAMPLE 2: Design a noninverting amplifier with a gain of 100.

Solution: First, we select a value of one feedback resistor and then calculate the proper value for the other from the gain equation (Equation 22). To minimize the current drawn from the output, we will select a relatively large stock resistor for R_f (but less than 100 kΩ)—for example, R_f = **56 kΩ**. Using Equation 22 and the required gain of A_{CL} = 100, we have:

$$A_{CL} = 1 + \frac{R_f}{R_i}$$

$$100 = 1 + \frac{56 \text{ k}\Omega}{R_i}$$

$$R_i = \frac{56 \text{ k}\Omega}{99} = 0.57 \text{ k}\Omega$$

The final circuit is shown in Figure 10.13. If the gain must be exactly 100, then both R_i and R_f must be precision resistors. To avoid the cost of precision resistors, one resistor can be the approximate value and the other can be a trimming potentiometer accurately set to give a gain of 100.

Output Offset Voltage

One of the assumptions made in analyzing the noninverting amplifier was that the voltage at the two input terminals was the same: $V_{in+} = V_{in-}$. This assumption is not quite true, however, which leads to another limitation of the op amp.

If the inputs of the op amp are both shorted to ground—that is, $V_{in+} = V_{in-} = 0$ V—the output voltage should be zero. In practice, this zero voltage is often not true. Instead, a small dc voltage, ranging from a small fraction of a volt to possibly several volts, may appear at

EXAMPLE 2

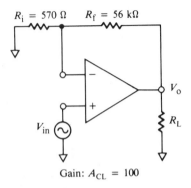

Gain: $A_{CL} = 100$

FIGURE 10.13. The figure shows a typical noninverting amplifier with a gain of about 100. For an accurate setting of A_{CL} = 100, precision feedback resistors should be used or one of the resistors should be a trimming potentiometer.

V_o. This dc voltage is called the **output offset voltage**, and it can lead to an erroneous measurement if not properly accounted for.

The cause of output offset voltages is a slight mismatch of internal op amp components. Normally, an external method of adjusting the amplifier to reduce or eliminate this offset voltage is provided. For example, the op amp may have a set of terminals to connect to an external, zero-balancing potentiometer.

Figure 10.14 shows the pin connections for the 741 in a 14 pin DIP package. In this case, a 10 kΩ potentiometer is required between pins 3 and 9, and the wiper is connected to the $-V_S$ supply voltage, pin 6.

Other case styles may have different pin connections (see Figure 9.8), and other op amps may have different external balancing arrangements. The proper connections will be given in the specification sheet for the particular device used.

With the + input terminal shorted to ground, the 10 kΩ potentiometer is adjusted to set the dc output voltage to zero. Since the output offset voltage generally changes with temperature and time, this zeroing procedure should be repeated from time to time to ensure that no offset value is superimposed on the output.

The magnitude of the output offset voltage, as well as the drift, is determined by the closed loop gain. The greater the gain, the greater the care that must be taken to compensate for the output offset voltage.

The **offset drift** specification for an op amp is usually given as an input specification to account for the dependence on gain. For example, in the specification sheet for the 741 op amp (see Figure 9.9), the input offset voltage is given as ± 6 mV maximum. If the voltage gain of the amplifier is 100, the output offset could be as large as ± 6 mV $\times 100 = \pm 0.6$ V.

The quality and price of an op amp are often determined largely by its offset and drift characteristics. Units with low offset and drift specification generally sell at a higher price than do units with high offset and drift specification.

If the amplifier is used solely for ac amplification, the output offset problem can be largely eliminated by placing a blocking capacitor at the output. Then, it is necessary only to ensure that the output does not drift too close to the output saturation voltage and thus cause the waveform peak to exceed the saturation voltage and be clipped.

Input Resistance

The input resistance of a noninverting amplifier is less than that of a voltage follower. To determine the input resistance, we again use the circuit model (Figure 10.15). Following the same analysis we used for the voltage follower, we arrive at Equation 9:

$$R_{in} = \frac{V_o}{V_{in}} A_{OL} r_{in}$$

OUTPUT OFFSET ADJUSTMENT

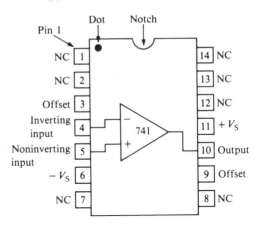

Pin Connections

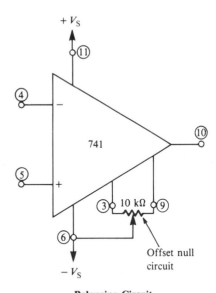

Balancing Circuit

FIGURE 10.14. An additional limitation of the op amp is an output offset voltage that can be significant, particularly at high closed loop gains. Offset voltage can be adjusted to 0 V by adding an external balancing circuit. The figure shows the balancing circuit and pin connections for a 741 op amp in a 14 pin DIP package. Pin 1 is to the left of the notch or identified by a dot.

FIGURE 10.15. The input resistance of a noninverting amplifier is less than that of the voltage follower by a factor equal to the closed loop gain, but it is still very large. For the 741 op amp, it is about 500 MΩ.

INPUT RESISTANCE

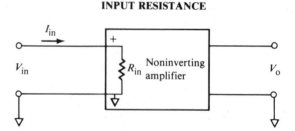

Input Resistance of Noninverting Amplifier

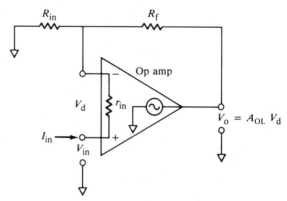

Op Amp Circuit Model for Noninverting Amplifier: $R_{in} = \dfrac{A_{OL}}{A_{CL}} r_{in}$

Since $V_o/V_{in} = 1/A_{CL}$, we have for the **input resistance of a noninverting amplifier**:

$$R_{in} = \frac{A_{OL}}{A_{CL}} r_{in} \tag{27}$$

Thus, the input resistance of the noninverting amplifier is less than that of the voltage follower by a factor equal to the closed loop gain, which can be anywhere from 1 to 1000. However, it is still very large. For the 741 op amp, $A_{OL} = 200,000$, $r_{in} = 2$ MΩ, and, assuming A_{CL} is its maximum of 1000, we have:

$$R_{in} = \frac{200,000}{1000} (2 \text{ M}\Omega) = 400 \text{ M}\Omega$$

which is still very large.

Output Current Boost

The output current of the noninverting amplifier is limited by the maximum output current of the op amp: about ±5 mA. While this current is too small to operate many useful devices, such as a loudspeaker, dc motor, or incandescent light, it can be increased by adding a pair of power transistors to the op amp output inside the feedback loop, as shown in Figure 10.16. This connection consists of a matched pair of NPN and PNP transistors and is called a **complementary emitter-follower amplifier.** Two transistors are required for the amplifier to output both positive and negative current.

The **voltage gain** of the power transistors in the emitter-follower connection is approximately 1, so the load voltage V_L equals the output voltage of the op amp V_o:

POWER AMPLIFIER

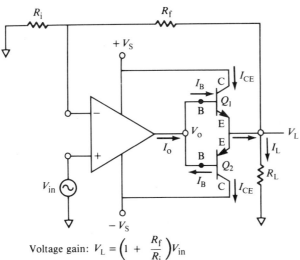

Voltage gain: $V_L = \left(1 + \dfrac{R_f}{R_i}\right)V_{in}$

Current gain: $I_L \approx \beta I_0$ where $50 \le \beta \le 150$

FIGURE 10.16. The current capability of a noninverting amplifier can be significantly increased by adding a complementary pair of NPN and PNP power transistors, as shown. The voltage gain of the transistors is approximately 1, so the voltage transfer function for the circuit is the same as for the noninverting amplifier alone. The current capability of the amplifier, however, is increased by a factor β, the current gain of the transistors, that can be from 50 to 150 depending on the transistors.

$$V_L = V_o \tag{28}$$

Thus, the **voltage transfer function** for the power amplifier circuit is the same as for the noninverting amplifier:

$$V_L = V_{in}\left(1 + \frac{R_f}{R_i}\right) \tag{29}$$

The current capability, however, has been significantly increased. The op amp's output voltage is essentially used to control the current flow through the transistors and, thus, through the load. The higher transistor currents are provided directly by the power supply, $+V_S$ and $-V_S$.

Briefly, the operation of the transistor pair is as follows. For positive values of V_o, Q_1 is on and Q_2 is off, as shown in Figure 10.17a. An off transistor conducts negligible current and, for all practical purposes, is removed from the circuit. For negative values of V_o, Q_1 is off and Q_2 is on, as shown in Figure 10.17b.

When a transistor is turned on, the current in its base is related to the current through its collector-emitter junction by a factor of $1 + \beta$, where β is called the **current gain of the transistor:**

$$I_{CE} = (1 + \beta)I_B$$

Values of β can vary from about 50 to 150 depending on the transistor, reducing this expression approximately to:

$$I_{CE} = \beta I_B \tag{30}$$

The transistor base current I_B is supplied by the output of the op amp I_o, so the **current capability of the amplifier** is given by:

$$I_{CE} = \beta I_o \tag{31}$$

Thus, if the transistors have a β of 100, the op amp's maximum output current of 5 mA in-

FIGURE 10.17. The operation of the power amplifier is shown for both positive and negative input voltages. When V_o is positive, transistor Q_2 is turned off and Q_1 conducts positive current from the $+V_S$ supply to the load. When V_o is negative, Q_1 is turned off and Q_2 conducts negative current from the $-V_S$ supply to the load. The maximum output current is generally determined by the current capability of the power supply $\pm V_S$.

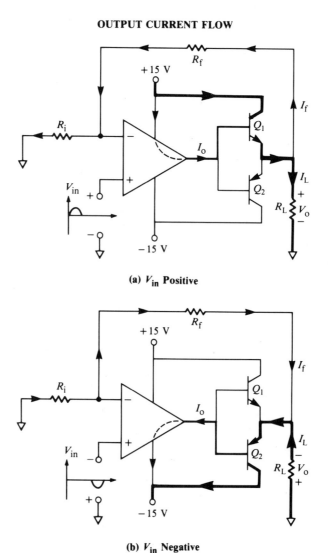

OUTPUT CURRENT FLOW

(a) V_{in} Positive

(b) V_{in} Negative

creases to about 0.5 A, assuming the $\pm V_S$ power supplies can provide this current.

Note in Figure 10.17 that the transistor's collector-emitter current supplies both the feedback current and the load current. However, since the feedback currents are relatively small —a few milliamperes at most— almost all of the increased output current is available for the load.

10.7 NONINVERTING AMPLIFIER APPLICATIONS

Voltage-to-Current Converter

The examples of noninverting amplifiers we have examined thus far are called voltage amplifiers, or **voltage-to-voltage converters**, which means that any input voltage will be amplified (converted) to a higher output voltage by the same constant factor: the closed loop gain. In equation form, the voltage transfer function (Equation 20) is:

$$V_o = V_{in}\left(1 + \frac{R_f}{R_i}\right)$$

The same circuit can also be used as a **voltage-to-current converter.** According to Equation 18, the current in the feedback loop is determined solely by V_{in} and R_i:

$$I_f = V_{in}\left(\frac{1}{R_i}\right)$$

Thus, the current through R_f, or for that matter, any element connected between the output and the − input, is directly proportional to the input voltage by a proportionality constant $1/R_i$. Essentially, $1/R_i$ acts as a "gain" factor in a voltage-to-current "amplifier." As R_i decreases, $1/R_i$ increases, and the feedback current I_f increases proportionally (Figure 10.18).

Note that the "load" for this voltage-to-current converter is connected between V_o and V_{in-} and not to ground. Thus, it can be used only for an electrically "floating" connection — one of this converter's disadvantages. It is possible, however, to design voltage-to-current converters for which the load can be grounded.

Constant-Current Source

The simplest application of this voltage-to-current converter is as a constant-current source. The V_o and V_{in-} terminals can serve as a constant-current source for any circuit attached to them.

For example, if a capacitor is attached in the feedback loop, as shown in Figure 10.19, the circuit will produce a constant charging current. The voltage across the capacitor V_C is related to the charging current I_f by:

VOLTAGE-TO-CURRENT CONVERTER

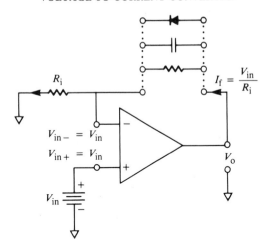

$$I_f = \frac{V_{in}}{R_i}$$

FIGURE 10.18. Because the feedback current I_f is determined only by the input voltage V_{in} and resistor R_i, the current through the feedback element is directly proportional to the input voltage. Thus, the circuit acts as a voltage V_{in}-to-current-I_f converter. Voltage-to-current conversion is true for any feedback element: resistor, capacitor, diode, and so forth.

$$V_C = \frac{I_f}{C}t \qquad (32)$$

If I_f is constant, then the voltage V_C will increase linearly with time. This behavior will produce a linear ramp voltage — for example, for a sawtooth voltage generator (see Part III).

High-Sensitivity, High Input Resistance Voltmeter

In the preceding section, we saw that we can increase the input resistance of a D'Arsonval voltmeter by adding a voltage follower to its input. Since the gain of the voltage follower was 1, however, there was no increase in voltage sensitivity. With the op amp voltage-to-current con-

CONSTANT-CURRENT SOURCE

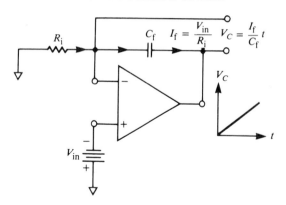

$$I_f = \frac{V_{in}}{R_i} \qquad V_C = \frac{I_f}{C_f} t$$

FIGURE 10.19. One application of the voltage-to-current converter is as a constant-current source. If V_{in} and R_i are constant, I_f will be a constant current, regardless of the feedback element. In the example shown, a constant I_f is used to charge a capacitor to produce a linear ramp voltage.

verter, we can also increase the voltage sensitivity and, at the same time, make the system independent of the resistance of the meter.

Figure 10.20 shows a D'Arsonval meter movement in the feedback loop to record a voltage V_{in} to be measured. The full-scale voltage sensitivity V_{fs} of the meter is the product of its full-scale current sensitivity I_{fs} and its internal resistance R_m:

$$V_{fs} = I_{fs} R_m \qquad (33)$$

The input voltage V_{in} necessary to produce a full-scale current in the feedback loop is given by Equation 18:

$$V_{in_{fs}} = I_{fs} R_i \qquad (34)$$

Thus, the full-scale sensitivity of the meter has been increased by a factor of R_m/R_i.

For example, the meter resistance of a D'Arsonval ammeter with a full-scale sensitiv-

HIGH-SENSITIVITY, HIGH–INPUT RESISTANCE VOLTMETER

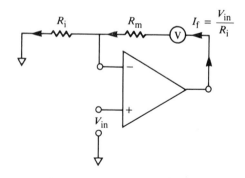

$$I_f = \frac{V_{in}}{R_i}$$

FIGURE 10.20. Another application is to increase the sensitivity and input resistance of a D'Arsonval voltmeter. With the meter in the feedback loop, the meter current is directly proportional to the input voltage. If R_i is less than the meter resistance, R_m, the meter sensitivity will be increased. The meter's input impedance becomes the large value of the noninverting amplifier.

ity of $I_{fs} = 100 \ \mu A$ is typically $R_m = 1000 \ \Omega$. According to Equation 33, its full-scale voltage sensitivity is:

$$V_{fs} = 0.1 \ mA \times 1 \ k\Omega = 0.10 \ V$$

If R_i is 100 Ω, then the input voltage necessary to produce a full-scale reading is, from Equation 34:

$$V_{in_{fs}} = 0.1 \ mA \times 0.1 \ k\Omega = 0.01 \ V$$

Thus, a 0.01 V input signal would produce a full-scale reading on the meter, an increase in sensitivity by a factor of 10. If the meter can be accurately read to 1%, then the voltmeter can measure voltages as small as 0.1 mV, or 100 μV.

Though the input resistance of this circuit is somewhat less than that of the voltage follower, it is still very large. Also, there is the additional advantage of a greater voltage sensitivity.

Note also that the voltage-to-current relation for the meter (Equation 34) is independent

INVERTING AMPLIFIER

$1 \text{ k}\Omega \leq (R_i + R_f) \leq 100 \text{ k}\Omega$

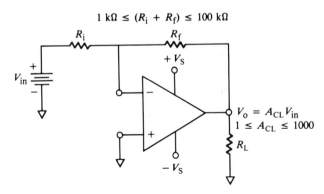

FIGURE 10.21. In an inverting amplifier, the input signal is applied to the – op amp input through the end of the feedback voltage divider. The voltage gain is again determined soley by the values of the feedback resistors. The same limitations on closed loop gain, $1 \leq A_{CL} \leq 1000$, and on feedback resistance values, $1 \text{ k}\Omega \leq (R_i + R_f) \leq 100 \text{ k}\Omega$, also apply.

of the meter resistance. Thus, for a given R_i, any 100 μA meter movement can be used in the feedback loop, and the voltage sensitivity of the system will be the same. Also, changes in the meter resistance—for example, changes due to temperature drifts—will not affect the reading.

10.8 INVERTING AMPLIFIER

An op amp can also be used to make an **inverting amplifier** that has the same range of amplification as the noninverting amplifier: from 1 to about 1000. The basic inverting amplifier circuit is shown in Figure 10.21. For this circuit, the + op amp input is grounded, and the input signal is applied to the inverting – input through the end of the feedback voltage divider. Applying the input signal to the – input inverts the output voltage, making it opposite in polarity from the input voltage.

Principles of Operation

Because the feedback loop is to the – input terminal, the previous assumptions for negative feedback can be used to analyze the circuit (Equations 1 and 2):

$$V_{in+} = V_{in-}$$
$$I_{in+} = I_{in-} = 0$$

In this case, since the + terminal is grounded, we have:

$$V_{in+} = V_{in-} = 0 \qquad (35)$$

Thus, the voltage at the – input terminal, the midpoint of the feedback voltage divider, is at ground, which means that the voltage across R_f is V_o. By Ohm's law, therefore, the input current I_i drawn from the source is:

$$I_i = \frac{V_{in}}{R_i} \qquad (36)$$

The feedback current I_f drawn from the op amp output is:

$$I_f = \frac{V_o}{R_f} \qquad (37)$$

Since no current can flow into or out of the – terminal, these two currents must be equal but opposite in polarity:

$$I_i = -I_f \qquad (38)$$

This fact produces the polarity inversion of the inverting amplifier circuit. For example, if V_{in} is

FIGURE 10.22. The polarity reversal of the inverting amplifier is determined by the condition that the − input is at ground potential, which means that V_{in} and R_i determine the input current I_i that, in turn, determines the feedback current I_f. $I_i = -I_f$ since the op amp input current is 0. Substituting in values gives the voltage transfer characteristic and voltage gain.

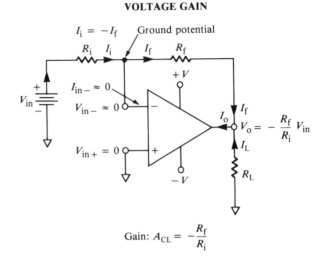

VOLTAGE GAIN

$$\text{Gain: } A_{CL} = -\frac{R_f}{R_i}$$

positive with respect to ground, then I_i flows *toward* the − input terminal, as shown in Figure 10.22. For the net current at the − input terminal to be zero, I_f must flow *away from* the − terminal, as shown in Figure 10.22.

Since V_{in-} is at ground, this feedback current direction makes V_o a negative voltage. Also, because the voltage across the load is negative, the load current I_L flows into the op amp output terminal, as shown. The opposite behavior occurs when V_{in} is negative with respect to ground.

Substituting the values for I_i and I_f into Equation 38 gives the **voltage transfer function** for the inverting amplifier:

$$\frac{V_{in}}{R_i} = -\frac{V_o}{R_f}$$

Transposing terms gives:

$$V_o = -\frac{R_f}{R_i} V_{in} \qquad (39)$$

The minus sign means that V_o and V_{in} are opposite in polarity.

Voltage Gain

The closed loop **voltage gain** of the inverting amplifier is simply the term preceding V_{in}:

$$A_{CL} = -\frac{R_f}{R_i} \qquad (40)$$

The gain of the inverting amplifier is nearly the same as that of the noninverting amplifier (Equation 22). For R_f much greater than R_i, which is often the case, Equation 22 reduces to the same ratio of R_f to R_i, but is opposite in polarity.

The limitations on values for A_{CL}, R_f, and R_i given for the noninverting amplifier (Equations 25 and 26) also apply to the inverting amplifier:

$$1 \le A_{CL} \le 1000$$
$$1 \text{ k}\Omega \le (R_i + R_f) \le 100 \text{ k}\Omega$$

Input Resistance

While the range of possible voltage gains of the inverting amplifier equals that of the noninvert-

INPUT RESISTANCE

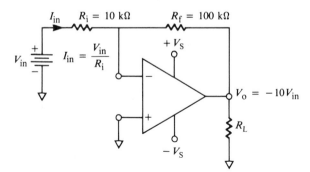

FIGURE 10.23. The input resistance of the inverting amplifier is R_i because it determines the current drawn from a source. The limitation that $R_f + R_i < 100$ kΩ means that R_i will be small compared to that of the noninverting amplifier (> 500 MΩ). The minimum acceptable value of R_i will be determined by the maximum acceptable value of I_{in} from a given voltage source.

ing amplifier, its input resistance is far inferior. The **input resistance** of an inverting amplifier is simply R_i, the smaller resistor of the feedback voltage divider (Figure 10.23):

$$R_{in} = R_i \qquad (41)$$

Since there is an upper limit on R_f of about 100 kΩ, R_i will be a maximum of about 10 kΩ for a gain of 10, which is a relatively small value. Further increases in gain, of course, would lead to further decreases in input resistance.

The minimum acceptable input resistance is determined by the maximum acceptable current drawn from the source and is given by Equation 36:

$$I_i = \frac{V_{in}}{R_i}$$

Thus, the maximum allowable current drain from the source I_i and the maximum value of source voltage V_{in} determine the minimum acceptable value of R_i. This limit on R_i may also place an upper limit on the voltage gain (Equation 40).

EXAMPLE 3: Design an inverting amplifier with a gain of 50 that will draw no more than 0.05 mA from a signal source whose maximum voltage will be about 0.2 V.

Solution: The restriction on the current drawn from the source places a lower limit on the value of R_i. This value can be calculated from Equation 36:

$$R_i = \frac{V_{in}}{I_i} = \frac{0.2 \text{ V}}{0.05 \text{ mA}} = 4 \text{ k}\Omega$$

The value of R_f required to give a voltage gain of $A_{CL} = -50$ can be calculated from Equation 40:

$$R_f = -A_{CL}R_i = -(-50) \times 4 \text{ k}\Omega$$
$$= 50 \times 4 \text{ k}\Omega = 200 \text{ k}\Omega$$

This value is at the upper limit of workable values for R_f. The amplifier should be constructed and its performance tested. If it works, no further design work is needed. If problems develop—for example, excessive noise on the output signal or oscillation of the amplifier—then an alternative circuit will be required.

INCREASED INPUT RESISTANCE

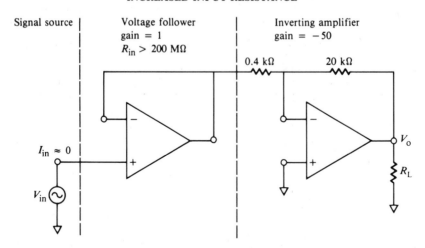

FIGURE 10.24. The input resistance of the inverting amplifier can be increased by placing a voltage follower between it and the signal source. This arrangement will yield the same value for overall voltage gain, but will draw negligible current from the source. See Example 3.

One solution is to add a voltage follower between the inverting amplifier and the circuit (Figure 10.24). The follower's high input resistance will draw negligible current from the source, and it will provide substantially more input current to the noninverting amplifier. As a result, both R_f and R_i will be reduced.

10.9 INVERTING AMPLIFIER APPLICATIONS

A common application of the inverting amplifier is, of course, to invert a signal or to both amplify and invert. When the signal need only be inverted, R_i and R_f are made equal and the circuit operates as an inverting voltage follower: $A_{CL} = -1$.

Inverting Adder

An equally common application of the inverting amplifier is as an adder. An inverting amplifier is used when two or more input signals must be added and an output signal that is proportional to the sum of the input signals must be obtained. The op amp derives its name from this type of application and was initially developed to perform the mathematical "operation" of addition for the early computers.

An inverting amplifier used as an adder is shown in Figure 10.25. If both input resistors R_i are equal to the feedback resistor R_f, then the **output voltage** is given by:

$$V_o = -(V_1 + V_2) \tag{42}$$

Although Figure 10.25 and Equation 42 show only two input voltages, the circuit can be expanded to include many more.

The circuit operation is explained by the facts that for negative feedback op amp circuits, $V_{in+} = V_{in-}$ and $I_{in+} = I_{in-} = 0$. For the inverting connection shown in Figure 10.25, this means that the voltage at the $-$ input is the

INVERTING ADDER

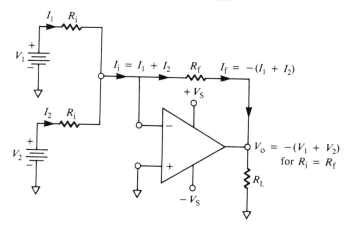

FIGURE 10.25. An important application of the inverting amplifier is as a voltage adder. If two (or more) voltages are applied in parallel to the $-$ input through the same value of resistance R_i (and $R_i = R_f$), then the output voltage will be minus the sum of the input voltages.

CIRCUIT ANALYSIS USING SUPERPOSITION

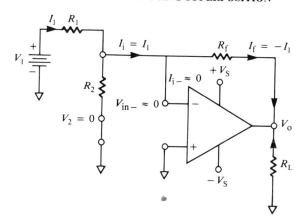

FIGURE 10.26. The operation of the inverting adder can be understood by applying superposition. With $V_2 = 0$, $I_i = I_1 = V_1/R_1$, as shown. Similarly, with $V_1 = 0$, $I_i = I_2 = V_2/R_2$. Superimposing the effects of V_1 and V_2 separately gives $I_i = (V_1/R_1) + (V_2/R_2)$. Using the fact that $I_i = -I_f = -V_o/R_f$ gives the voltage transfer characteristic. For $R_1 = R_2$, this reduces to $V_o = -(V_1 + V_2)$.

same as that at the $+$ input, which is at ground potential, and that the net current flowing into the $-$ input terminal is equal in magnitude but opposite in polarity to the feedback current flowing to the output terminal.

Equation 42 is obtained by applying superposition to the input signals V_1 and V_2. First, consider that voltage source $V_2 = 0$ V, as shown in Figure 10.26. Since the $-$ input terminal is also at 0 V, the voltage across R_2 is zero, and the current through R_2 is zero. The input current I_1 from voltage source V_1, then, is:

$$I_1 = \frac{V_1}{R_1} \qquad (43)$$

The input current produced by signal V_2 is obtained by letting $V_1 = 0$ V. The value of I_2 is:

$$I_2 = \frac{V_2}{R_2} \qquad (44)$$

Superimposing the effects of both input voltages gives a net current of:

$$I_i = I_1 + I_2$$

or

$$I_i = \frac{V_1}{R_1} + \frac{V_2}{R_2} \qquad (45)$$

Since the net current at the − input terminal is zero, the feedback current is equal in magnitude but opposite in polarity to this value (Equation 38):

$$I_f = -I_i$$

or

$$I_f = -\left(\frac{V_1}{R_1} + \frac{V_2}{R_2}\right) \qquad (46)$$

The feedback current, in turn, determines the value of output voltage through Ohm's law:

$$V_o = I_f R_f \qquad (47)$$

Substituting for I_f gives the **voltage transfer characteristic** of the inverting adder:

$$V_o = -R_f\left(\frac{V_1}{R_1} + \frac{V_2}{R_2}\right) \qquad (48)$$

If $R_1 = R_2 = R_f$, then Equation 48 reduces to the inverting adder equation (Equation 42):

$$V_o = -(V_1 + V_2) \qquad (49)$$

where the minus sign indicates that the output is opposite in polarity from the sum of the input voltages. If a noninverted output is required, an inverting follower would be added to change the sign of V_o.

Note that the two input signals are added algebraically *before* multiplying by the minus sign. Thus, if V_1 and V_2 are opposite in polarity, it is still the algebraic sum that is inverted.

Also, regardless of whether the input voltages V_1 and V_2 are positive, negative, or zero, the input currents I_1 and I_2 do not affect one another since they go essentially to ground: $V_{in-} = 0$ V. This property is very important. For example, if the input voltages are signals produced by microphones, there will be no interaction between the two microphones. As a result, this circuit is commonly used to mix audio signals.

Audio Mixer

Figure 10.27 shows how an inverting adder might be used by a band to mix the sounds from four microphones. One input would be used for the singer's microphone, while other inputs would pick up the guitar, piano, and drums. The 100 kΩ potentiometers at the inputs permit adjustment of the volume (voltage) level for each microphone. They would allow for decreasing a drummer's volume level while increasing the singer's volume level.

The transistors Q_1 and Q_2 in Figure 10.27 are used to boost the output current sufficiently to drive an 8 Ω or 4 Ω loudspeaker. The arrangement of Q_1 and Q_2 serves the same purpose as described earlier for the noninverting amplifier. For the inverting amplifier, however, Q_2 will be on and Q_1 will be off when the *sum* of the input voltages is positive. Similarly, when the sum of the input voltages is negative, Q_1 will be on and Q_2 will be off. Finally, when adding audio signals, the inversion of the output voltage has no effect on the sound heard from the loudspeaker.

Direct Current Offset Voltage

Another application of the adder is to add a dc offset to a time-varying signal. This addition is most commonly done in a function generator, where a variable dc voltage V_{dc} is "added" to the time-varying function to produce a dc offset. The dc offset control simply adjusts the magnitude (and polarity) of the added dc voltage (Figure 10.28).

AUDIO MIXER

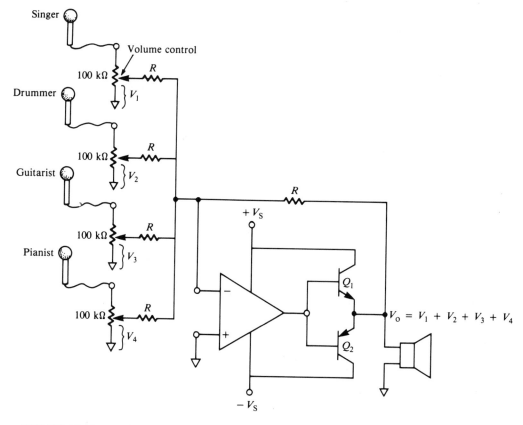

FIGURE 10.27. An important feature of the inverting adder is that the various input voltages do not interact. Consequently, this circuit is commonly used to mix audio signals — for example, from several microphones. Each microphone can also have its own volume control if an attenuating potentiometer is added at the input. If a pair of power transistors is placed at the output, the circuit can drive a loudspeaker.

DIRECT CURRENT OFFSET CONTROL

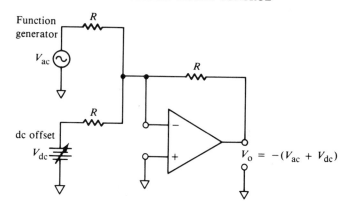

FIGURE 10.28. Another common application of the inverting adder is to add a variable dc offset voltage to the time-varying signal of a function generator.

10.10 QUESTIONS AND PROBLEMS

1. Draw the op amp circuit model. Label each terminal and give the conditions of input current and voltage that simplify most op amp circuit analysis.
2. Draw the voltage follower connection for an op amp and its voltage transfer characteristic (Figure 10.29). Also, give maximum ratings for supply voltage, input voltage, output voltage, output current, typical input resistance, and minimum load resistance.
3. For the noninverting amplifier circuit shown in Figure 10.30, find the following quantities and draw the voltage transfer characteristic:
   (a) Closed loop again
   (b) Maximum input voltage
   (c) Maximum feedback current
   (d) Maximum load current
4. Design a noninverting amplifier with a gain of 250. Draw the circuit and its voltage transfer characteristic (Figure 10.31). Give component values.
5. Design a noninverting amplifier that can accept a microphone's input voltage of 4 mV pp and can drive an 8 Ω loudspeaker at 4 V pp. Include a variable volume control. Draw the circuit.
6. Design a multirange, constant-current source

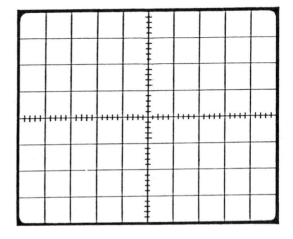

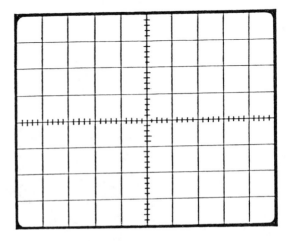

Voltage Transfer Characteristic

FIGURE 10.30.

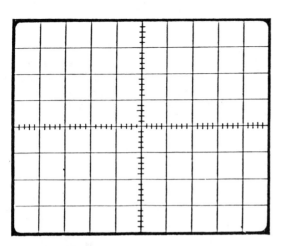

Voltage Transfer Characteristic

FIGURE 10.29.

Voltage Transfer Characteristic

FIGURE 10.31.

that can produce output currents of 10 μA, 100 μA, and 1.0 mA. Draw the circuit.

7. For the inverting amplifier circuit shown in Figure 10.32, find the following quantities and draw the voltage transfer characteristics:
   (a) Closed loop gain
   (b) Maximum input voltage
   (c) Maximum feedback current
   (d) Input resistance

8. Design an inverting amplifier with a gain of -20 and an input resistance of 5 kΩ. Draw the circuit and its voltage transfer characteristic (Figure 10.33). Give component values.

9. The inverting adder can be made into an inverting *averager* with properly selected feedback resistors. According to Equation 47:

$$V_o = -R_f \left(\frac{V_1}{R_1} + \frac{V_2}{R_2} + \frac{V_2}{R_3} + \cdots + \frac{V_n}{R_n} \right)$$

For $R_1 = R_2 = R_3 = \cdots = R_n = R_i$, Equation 48 becomes:

$$V_o = -\frac{R_f}{R_i} (V_1 + V_2 + V_3 + \cdots + V_n)$$

If the ratio R_i/R_f is selected as the number of voltages to be added n, then the equation becomes:

$$V_o = -\frac{V_1 + V_2 + V_3 + \cdots + V_n}{n}$$

which is the equation for an inverting averager.

Design a circuit that can give the negative average of four voltages. Draw the circuit. What will be the output voltage for the following input voltages: $V_1 = 3.6$ V, $V_2 = -4.2$ V, $V_3 = 0.9$ V, and $V_4 = 9.3$ V?

10. The inverting adder can be used also as a current-to-voltage converter. According to Equation 38, the input current I_i is the negative of the feedback current I_f:

$$I_i = -I_f$$

The feedback current is given by Equation 37:

$$I_f = \frac{V_o}{R_f}$$

Combining these equations gives:

$$V_o = -I_i R_f$$

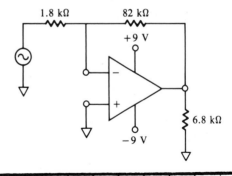

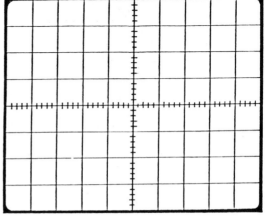

Voltage Transfer Characteristic

FIGURE 10.32.

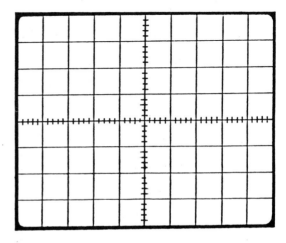

Voltage Transfer Characteristic

FIGURE 10.33.

Thus, the output voltage is directly proportional to the input current with the proportionality constant determined by R_f. Any current source applied to the − input terminal of an inverting amplifier will produce a proportional output voltage, regardless of its resistance.

Design a three-range, digital microammeter (1 μA, 10μA, and 100 μA) using a current-to-voltage converter and a 0.1 V full-scale digital voltmeter. Draw the circuit.

11. When a photodiode is back biased, as shown in Figure 10.34, a small leakage current I_{in} will flow depending on the photodiode's illumination. In total darkness, the leakage current is on the order of nanoamperes. When illuminated, however, the leakage current is directly propor-

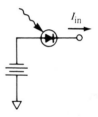

FIGURE 10.34.

tional to the radiant energy striking its surface and can be on the order of 50 μA or more. Use this fact to design a direct-reading solar energy meter that will read 0.50 V full scale for an illumination producing 50 μA leakage current. Draw the circuit.

Chapter 11

OPERATIONAL AMPLIFIERS AS WAVEFORM GENERATORS

11.1 OBJECTIVES

Following the completion of Chapter 11, you should be able to:

1. State the basic behavior and purpose of monostable and astable multivibrators and describe two typical applications.
2. Draw the time behavior of the voltage across a capacitor charged by a source of constant current or voltage.
3. Identify an op amp connected as a square wave generator (symmetrical and nonsymmetrical), determine its peak-to-peak amplitude and frequency of oscillation, and draw the output waveform and the waveforms at its two input terminals.
4. Design a square wave generator with a specific output amplitude and frequency.
5. Design a nonsymmetrical square wave generator and pulse generator with a specific output amplitude, frequency, and duty cycle.
6. Identify an op amp connected as a mono-

stable multivibrator and determine the output amplitude pulse width and the maximum frequency of input trigger signals.
7. Design an input coupling network for a monostable multivibrator that will modify inadequate input signals to trigger the multivibrator.
8. Design a monostable multivibrator circuit that will accept a given input pulse and produce output pulses of a specific amplitude and width.
9. Design an electronic rate meter that will accept a given input pulse and produce a direct reading of the input pulse rate.
10. Identify an op amp connected as a ramp generator (sawtooth or triangle) and determine its output amplitude and frequency.
11. Identify a field-effect transistor (FET) connected as a constant-current source and explain the general principles of its operation.
12. Using an op amp and an FET, design a

sawtooth generator that will have a specific output amplitude, frequency, and baseline voltage.

13. Using an op amp and an FET, design a triangle wave generator that will have a specific output amplitude, frequency, baseline voltage, slew rate for its leading and trailing ramps, and output current capability.

14. Identify an op amp connected as a triangle-to-sine-wave-shaping circuit and explain the general principles of its operation.

11.2 OVERVIEW

The operational amplifier is indeed the workhorse of modern electronic circuit design. Its applications are far-reaching and are limited only by the circuit designer's imagination. This chapter continues the description and analysis of the op amp's basic applications, in this case, as waveform generators. As with the earlier circuits, feedback from the output to the input terminals is the key factor in controlling the op amp's performance.

11.3 MULTIVIBRATORS

In Chapter 9, we saw how use positive feedback from the output to the + input to produce voltage comparators for a variety of switching applications. In Chapter 10, using negative feedback from the output to the − input, voltage amplifiers were produced for a variety of signal-amplifying applications. This chapter focuses on the use of both positive and negative feedback to generate waveforms. The circuits described can be divided into two different categories: monostable multivibrators and astable multivibrators (Figure 11.1).

A **monostable multivibrator** has a single (mono-) stable output voltage, generally 0 V (see Chapter 8, Section 14). When the input voltage exceeds a preset value, however, the output will go temporarily to a different voltage, remain there for a preset length of time, and then return to the stable value. Thus, it is basically a single-pulse generator that can be triggered by an external signal. A monostable multivibrator is used primarily as a wave shaper—that is, to convert irregular input pulses into uniform output pulses for subsequent electronic circuits. Other names for the monostable multivibrator are the *one-shot multivibrator* and the *Schmidt trigger*.

In contrast, **astable multivibrators** (also described in Chapter 8, Section 14) have no stable output state. They continually vibrate back and forth between two voltage levels. The manner in which they change voltage levels can be controlled to produce specific wave shapes: square waves, pulses, sawtooth waves, and triangular waves. Hence, they are basic components of function generator circuits.

In Chapter 8, it was noted that the 555 timer can perform similar multivibrator tasks. The heart of the 555 timer, however, consists essentially of two op amps used as comparators. Now we will see how a single op amp can perform many of the same multivibrator tasks as the 555.

11.4 REVIEW OF CAPACITOR BEHAVIOR

All of the circuits described in this chapter require one or more capacitors to set their timing. The width of the monostable output pulse and the frequency of the astable oscillator depend directly on the time it takes a capacitor to charge or discharge. Therefore, the analysis begins with a brief review of the charge–discharge behavior of capacitors that was described in depth in Chapter 8.

The key property of a capacitor is the time delay it introduces into a circuit when it charges. The characteristics of this time delay

TYPES OF OP AMP WAVE GENERATORS

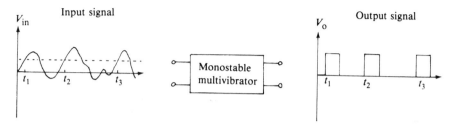

Monostable (One-Shot) Multivibrator

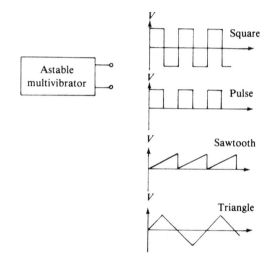

Astable Multivibrator

FIGURE 11.1. Op amps can also be used as either monostable or astable multivibrators to generate waveforms. Monostable multivibrators are used as wave shapers to produce a uniform output pulse whenever the input exceeds a preset voltage level. Astable multivibrators can generate square, pulse, sawtooth, and triangle waves.

differ slightly depending on whether the capacitor is charged by a source of constant current or constant voltage.

Constant Current

Figure 11.2a shows the voltage–time behavior of a capacitor charged by a constant-current source. In this case, the capacitor voltage changes linearly from 0 V to some higher voltage determined by the properties of the current source. The capacitor voltage V_C at any time t after the current-source switch is closed is given by the following equation:

$$V_C = \frac{It}{C} \qquad (1)$$

where I is the charging current in amperes and C is the circuit capacitance in farads. (This

REVIEW OF CAPACITOR BEHAVIOR

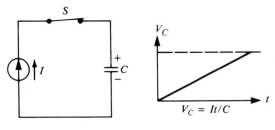

(a) Constant-Current Source

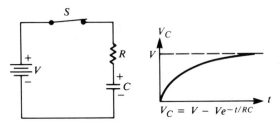

(b) Constant-Voltage Source

FIGURE 11.2. The performance of op amps used as multivibrators depends on a capacitor to set the timing. Depending on the circuit, the capacitor may be charged by a source of constant current or constant voltage. The figure shows typical voltage–time behavior for the two cases.

equation should be familiar from previous chapters; see Chapter 8, Section 10 for its derivation.)

Constant Voltage

Figure 11.2b shows the voltage–time behavior of a capacitor charged by a constant-voltage source, which is the condition of most of the circuits described in this chapter. The addition of a series resistor to set the charging current controls how rapidly the capacitor voltage increases. In this case, the voltage–time curve follows exponential behavior described by the equation:

$$V_C = V(1 - e^{-t/RC}) \tag{2}$$

where V is the applied voltage and e is the base of the natural logarithms (see also Chapter 8, Section 11).

If the initial capacitor voltage is not 0 V but some positive or negative voltage, Equation 2 must be slightly modified as follows:

$$V_C = V - (V - V_{in})e^{-t/RC} \tag{3}$$

where V_{in} is the initial capacitor voltage and V is the applied voltage. Thus $V - V_{in}$ is the total change in capacitor voltage during the charging process, and V is the final capacitor voltage.

This general equation applies to both charging and discharging capacitors. For example, if the final voltage is $V = 0$ V—that is, the capacitor is being discharged—then Equation 2 reduces to the simple capacitor discharge equation:

$$V_C = V_{in}e^{-t/RC} \tag{4}$$

See Figure 11.3.

Equation 3 can be used to calculate the voltage at any time t, but this calculation is rarely required. Of more interest is the quantity RC that is called the charging time constant τ:

$$\tau = RC \tag{5}$$

The time constant specifies the rate at which the capacitor charges (see also Chapter 8, Section 11). Thus, it determines how rapidly the capacitor voltage will reach a specific value for a given applied voltage. This principle is illustrated by Figure 11.3, which gives values of $e^{-t/RC}$ for various time constants. For an applied voltage V, the capacitor voltage will change by $0.369(V - V_{in})$ after one time constant, by $0.135(V - V_{in})$ after two time constants, by $0.050(V - V_{in})$ after three time constants, and so on.

Because the time constant determines the charging rate, it controls the time delay between

GENERAL CONSTANT-VOLTAGE BEHAVIOR:

$$V_C = V - (V - V_{in})\, e^{-t/RC}$$

τ	$e^{-\tau/RC}$
0	1.000
τ	0.368
2τ	0.135
3τ	0.050
4τ	0.018
5τ	0.007
6τ	0.002
7τ	0.001

Values of $e^{-t/RC}$

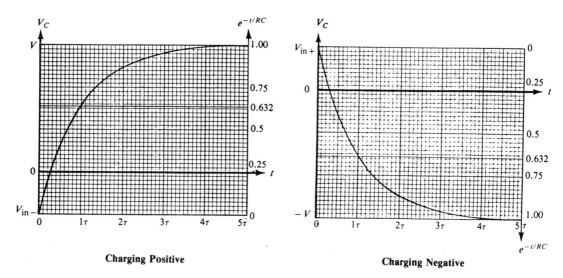

Charging Positive   Charging Negative

FIGURE 11.3. The general behavior of a capacitor charged by a source of constant voltage V can be described by a single expression: $V_C = V - (V - V_{in})e^{-t/RC}$. This equation applies to both positive and negative charging voltages and includes cases in which the initial voltage V_{in} is not zero. The curves show how the quantity $(V - V_{in})$ decays to zero as $e^{-t/RC}$ decays to zero. Values of $e^{-t/RC}$ for various values of $\tau = RC$ are also given.

the application of a voltage V and the subsequent attainment of a given capacitor voltage V_C. This time delay can be varied over a wide range, from nanoseconds (10^{-9} s) to hours, simply by selection of the proper capacitor and resistor.

In most of the applications to be described here, the series RC combination will be placed in the negative feedback loop. Thus, when the output changes voltage levels — for example, from $+V_{sat}$ to $-V_{sat}$ — there will be a delay before this change feeds back to the input terminal. In a monostable multivibrator, the time delay will set the width of the output pulse, and in an astable multivibrator, it will determine the frequency of oscillation (see Figure 11.1).

11.5 ASTABLE MULTIVIBRATORS

Quite often, a periodic wave—square, triangle, sawtooth, sine, and so forth—is required for the operation of an electronic system. Separate function generators are usually too expensive, too impractical, or both. In general, when only a single waveform is required, a simple circuit that generates only the required waveform and employs as few components as possible is preferable. The general term for this type of circuit is an **astable multivibrator**, so called because the circuit has no stable voltage, but continually vibrates back and forth between two voltage levels. By controlling the manner in which the voltage changes between the two levels, we can generate a variety of periodic waveforms (see Figure 11.1). The op amp is ideally suited for astable applications.

11.6 SQUARE WAVE GENERATOR

The first astable multivibrator we will examine is the square wave generator (Figure 11.4). Like other astable multivibrator circuits, it combines feedback from the output to both input terminals. Feedback to the + input terminal is via a simple voltage divider R_1 and R_2, and feedback to the − input terminal is via an RC combination of resistor R_f and capacitor C. Note that the − input voltage V_{in-} is the capacitor voltage V_C that is determined by the output voltage V_o and the time constant $R_f C$.

Principles of Operation

The square wave generator circuit shown in Figure 11.4 can be understood as a zero-crossing detector with hysteresis, in which the input voltage to the − input derives from the op amp's own output, rather than from a separate voltage source (see Figure 9.34).

SQUARE WAVE GENERATOR

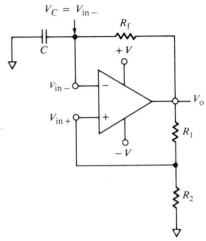

Values of V_{in+}: $V_{UT} = \dfrac{R_2}{R_1 + R_2}(+V_{sat})$

$$V_{LT} = \dfrac{R_2}{R_1 + R_2}(-V_{sat})$$

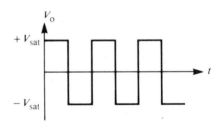

FIGURE 11.4. The simplest astable multivibrator is the square wave generator. Like other multivibrators, it uses both positive and negative feedback. The positive feedback loop is that of a zero-crossing detector with hysteresis and sets upper and lower threshold voltages for the + input. When the capacitor voltage passes the threshold voltage, output transitions occur.

Because of the positive feedback loop, this circuit has only two possible output voltages, $+V_{sat}$ and $-V_{sat}$. The voltage at the − input terminal that will produce transitions between the two levels is given by the upper and lower threshold voltages of the zero-crossing detector.

These threshold voltages are determined by the state of the output voltage $\pm V_{sat}$ and the voltage divider R_1 and R_2. Chapter 9 (Equations 9 and 10, respectively) gives the following for V_{UT} and V_{LT}:

$$V_{UT} = \frac{R_2}{R_1 + R_2}(+V_{sat})$$

$$V_{LT} = \frac{R_2}{R_1 + R_2}(-V_{sat})$$

If the output is at $+V_{sat}$, a transition to $-V_{sat}$ will occur when V_{in-} exceeds V_{UT}. If the output is at $-V_{sat}$, a transition back to $+V_{sat}$ will occur when V_{in-} drops below V_{LT}.

The voltage at the $-$ input terminal that will produce a transition is the voltage across the capacitor V_C. To determine when the capacitor voltage passes the threshold voltages, we must examine the behavior of the $R_f C$ combination for the output conditions of $V_o = \pm V_{sat}$.

To begin the analysis, we assume $V_o = +V_{sat}$ and the capacitor voltage is at 0 V—that is, $V_{in-} = 0$ V. Since $V_o = +V_{sat}$, the voltage at the $+$ terminal is the upper threshold voltage V_{UT}. Hence, we have:

$$V_o = +V_{sat}$$
$$V_{in-} = 0 \text{ V.}$$
$$V_{in+} = V_{UT} = \frac{R_2}{R_1 + R_2}(+V_{sat}) \qquad (6)$$

See Figure 11.5a.

With $V_o = +V_{sat}$, the capacitor will begin to charge toward $+V_{sat}$ through R_f with a time constant $R_f C$. When V_C exceeds V_{UT}, the output will change states to $-V_{sat}$. The condition at this point then changes to:

$$V_o = -V_{sat}$$
$$V_{in-} = V_{UT}$$
$$V_{in+} = V_{LT} = \frac{R_2}{R_1 + R_2}(-V_{sat}) \qquad (7)$$

See Figure 11.5b.

With $V_o = -V_{sat}$, the capacitor will begin to charge toward $-V_{sat}$ with the same time constant $R_f C$. When V_C drops below V_{LT}, the output will change back to $+V_{sat}$. The condition then changes to:

$$V_o = +V_{sat}$$
$$V_{in-} = V_{LT}$$
$$V_{in+} = V_{UT} \qquad (8)$$

With $V_o = +V_{sat}$, the capacitor will again charge toward $+V_{sat}$ as before until V_{in-} again exceeds $V_{in+} = V_{UT}$, at which point the output changes back again to $-V_{sat}$.

This process repeats over and over again, producing the square wave shown in Figure 11.6a. The $+$ input capacitor voltages are also shown in Figure 11.6. Note that the capacitor alternately charges between V_{UT} and V_{LT} with a time constant of $R_f C$.

Period and Frequency

The **period** of the square wave is simply the time it takes for the capacitor to charge from V_{LT} to V_{UT} and then back to V_{LT}. The time to charge from V_{LT} to V_{UT} can be determined with Equation 2, where the initial voltage is the lower threshold $V_{in} = V_{LT}$ and the applied voltage is the positive output saturation voltage $V = +V_{sat}$:

$$V_C = V_{sat} - (V_{sat} - V_{LT})e^{-t/R_f C} \qquad (9)$$

The time it takes for the capacitor voltage to reach the upper threshold $V_C = V_{UT}$ is just half the period: $t = T/2$. Substitution of V_{UT} and $T/2$ for V_C and t, respectively, into Equation 9 gives:

$$V_{UT} = V_{sat} - (V_{sat} - V_{LT})e^{-T/2R_f C} \qquad (10)$$

Rearranging terms and solving for the period T gives:

$$T = 2R_f C \ln \frac{V_{LT} - V_{sat}}{V_{UT} - V_{sat}} \qquad (11)$$

PRINCIPLES OF OPERATION

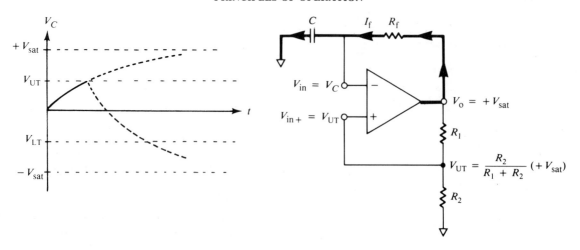

(a) Output = + V_{sat}

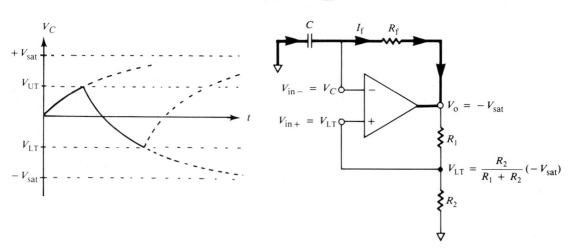

(b) Output = − V_{sat}

FIGURE 11.5.   When $V_o = + V_{\text{sat}}$, the + input voltage is the upper threshold voltage V_{UT}. The negative feedback current I_f charges the capacitor toward $+ V_{\text{sat}}$ with a time constant $R_f C$. When the capacitor voltage exceeds V_{UT}, the output changes to $- V_{\text{sat}}$. The + input voltage is then the lower threshold voltage V_{LT}, and the capacitor charges toward $- V_{\text{sat}}$ with the same time constant. When V_C drops below V_{LT}, the output changes back to $+ V_{\text{sat}}$ and the cycle repeats.

INPUT AND OUTPUT WAVEFORMS

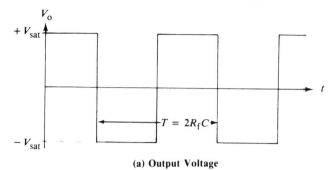

(a) Output Voltage

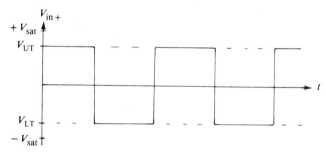

(b) (+) Input Voltage

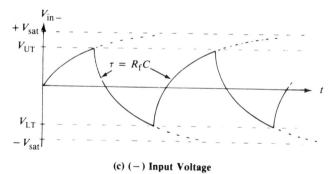

(c) (−) Input Voltage

FIGURE 11.6. Shown are the input and output voltages for the square wave generator. The output oscillates between $+V_{sat}$ and $-V_{sat}$, and the two input voltages oscillate between the upper and lower threshold voltages. The period of the square wave is determined by the time it takes for the capacitor to charge from V_{LT} to V_{UT} and back again. For $R_1 = R_2$, the time is about $T = 2R_fC$.

Substituting Equations 9 and 10 from Chapter 9 for V_{LT} and V_{UT} and rearranging terms gives the period of the square wave:

$$T = 2R_fC \ln \frac{R_1 + 2R_2}{R_1} \qquad (12)$$

In most cases, $R_1 \approx R_2$, in which case, Equation 12 reduces approximately to:

$$T \approx 2R_fC \qquad (13)$$

Thus, the period of the square wave is very

nearly twice the time constant of the negative feedback loop.

The **frequency** of the square wave is, of course, just the inverse of the period, or:

$$f = \frac{1}{T}$$

Substituting in Equation 12 for T gives:

$$f = \frac{1}{2R_f C \ln[(R_1 + 2R_2)/R_1]} \tag{14}$$

which, for $R_1 \approx R_2$, reduces approximately to:

$$f \approx \frac{1}{2R_f C} \tag{15}$$

EXAMPLE 1: Design a square wave generator with a frequency of 500 Hz.

Solution: Equation 15 is approximately correct if we let $R_1 \approx R_2$. Therefore, let us select stock resistors for R_1 and R_2:

$$R_1 = R_2 = 50 \text{ k}\Omega$$

These resistors are relatively large so that the voltage divider will draw negligible current from the output. Approximate values of R_f and C can now be calculated with Equation 15. (Note that only approximate values are required because R_f would be a variable potentiometer if an exact frequency of 500 Hz were required.)

$$f = \frac{1}{2R_f C} = 500 \text{ Hz}$$

$$C = \frac{1}{2R_f \times 500 \text{ Hz}} = \frac{1}{R_f \times 10^3}$$

Let us select a value for R_f of **25 kΩ** with the

expectation that we would use a 50 kΩ variable potentiometer to get a precise frequency setting. Then:

$$C = \frac{1}{R_f \times 10^3} = \frac{1}{(0.25 \times 10^5)10^3}$$
$$= 4 \times 10^{-8} \text{ F} = \textbf{0.04 } \mu\textbf{F}$$

The frequency can be adjusted (or continuously varied) by varying the 50 kΩ potentiometer. A variable output amplitude can be obtained by adding a variable potentiometer to the output. The circuit is shown in Figure 11.7.

Nonsymmetrical Square Waves

We can use the square wave generator to generate nonsymmetrical square waves if the time required for the capacitor to charge from V_{LT} to V_{UT} is different from the time required to charge from V_{UT} to V_{LT}. To achieve this goal, we must use resistors having different values for the charge–discharge paths, as shown in Figure 11.8.

When $V_o = +V_{sat}$ (Figure 11.9a), diode D_- is back biased and conducts no current. Diode D_+, on the other hand, is forward biased so that capacitor C charges to V_{UT} through resistor R_{f+} with a time constant:

$$\tau_+ = R_{f+}C$$

When $V_o = -V_{sat}$ (Figure 11.9b), diode D_+ is back biased and D_- is forward biased. Capacitor C then discharges to V_{LT} through resistor R_{f-} with a time constant:

$$\tau_- = R_{f-}C$$

Typical output and capacitor voltages for R_{f-} larger than R_{f+} are shown in Figure 11.9.

For $R_1 \approx R_2$, the times for the positive- and negative-going pulses are given approximately by:

EXAMPLE 1

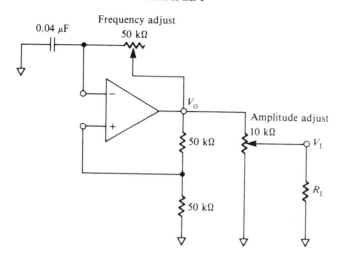

FIGURE 11.7. The square wave generator circuit has a 50 kΩ potentiometer in the negative feedback loop that permits exact adjustment of the frequency to 500 Hz. The 10 kΩ output potentiometer permits variation of the output amplitude.

$$t_+ \approx R_{f+}C$$
$$t_- \approx R_{f-}C \qquad (16)$$

The total **period** of the nonsymmetrical square wave is the sum of t_+ and t_-:

$$T \approx t_+ + t_-$$

Substituting from Equation 16:

$$T \approx (R_{f+} + R_{f-})C \qquad (17)$$

The **frequency** of the nonsymmetrical square wave is, of course, the inverse of Equation 17:

$$f \approx \frac{1}{(R_{f+} + R_{f-})C} \qquad (18)$$

Exact expressions for each of these terms can be obtained from Equation 12 and the values of R_1 and R_2 rather than the approximate Equation 17.

NONSYMMETRICAL SQUARE WAVE GENERATOR

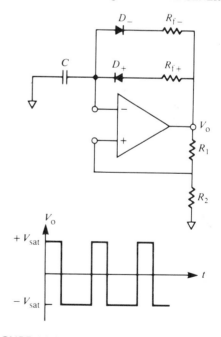

FIGURE 11.8. A nonsymmetrical square wave generator can be constructed by fixing different time constants for the capacitor's positive and negative charging cycles — that is, by having diodes direct the feedback current through different feedback resistors.

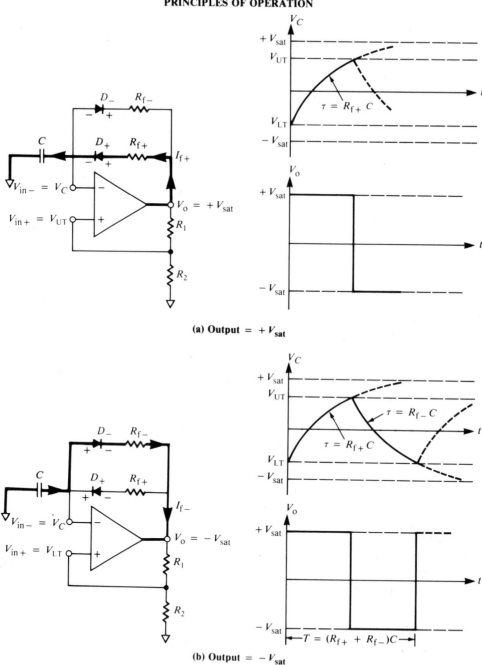

(a) Output = $+V_{\text{sat}}$

(b) Output = $-V_{\text{sat}}$

FIGURE 11.9. When $V_o \approx +V_{\text{sat}}$, diode D_+ is forward biased and conducts positive charging current through resistor R_{f+}. Thus, capacitor C charges toward $+V_{\text{sat}}$ with a time constant $R_{f+}C$. When $V_o = -V_{\text{sat}}$, diode D_- is forward biased and conducts negative charging current through resistor R_{f-}. Capacitor C then charges toward $-V_{\text{sat}}$ with a different time constant $R_{f-}C$. For $R_1 = R_2$, the period of the output wave is given approximately by $T \approx (R_{f+} + R_{f-})C$.

11.7 PULSE GENERATOR

The circuit shown in Figure 11.8 can be converted into a pulse generator by the addition of a diode in series with the load R_L, as shown in Figure 11.10. The op amp's output voltage at point A is the same as that in Figure 11.9, so the behavior of the capacitor and the voltage at the + and − inputs is unchanged. The diode in the direction shown does not conduct load current when $V_o = -V_{sat}$, so the negative portion of the square wave is dropped across the diode rather than across the load R_L. Hence, the load voltage is a series of positive pulses of width t_+. If the diode were reversed, of course, the positive pulses would be blocked and the load voltage would be a series of negative pulses of width t_-.

The **duty cycle** D for the positive pulse circuit shown in Figure 11.10 is the width of the positive pulse t_+ (Equation 16) divided by the total period T (Equation 17), or:

$$\text{duty cycle } D = \frac{t_+}{T} \approx \frac{R_{f+}C}{(R_{f+} + R_{f-})C}$$

which reduces to:

$$\boxed{\text{duty cycle } D \approx \frac{R_{f+}}{R_{f+} + R_{f-}} \qquad (19)}$$

If $R_{f+} = R_{f-}$, the duty cycle is 50% and the output is a square wave. If R_f is less than R_{f-}, the duty cycle is less than 50%; and if R_{f+} is greater than R_{f-}, the duty cycle is greater than 50%.

The pulse width and duty cycle can be varied by using potentiometers for R_{f+} and R_{f-}. Varying R_{f+} would vary the pulse width according to Equation 16, but it would also change the duty cycle according to Equation 19. Varying R_{f-} would change the duty cycle, but would not affect the pulse width.

PULSE GENERATOR

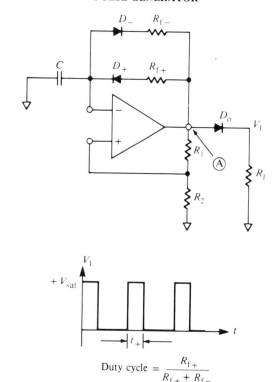

$$\text{Duty cycle} = \frac{R_{f+}}{R_{f+} + R_{f-}}$$

FIGURE 11.10. The nonsymmetrical square wave generator can be converted to a pulse generator by the addition of a diode in series with the load. In the direction shown, the diode blocks the negative portion of V_o to produce only positive load pulses. Reversing the diode would result in negative pulses.

EXAMPLE 2: Design a variable-frequency pulse generator with a pulse width of 20 ms and a minimum frequency of 10 Hz.

Solution: The basic circuit for the pulse generator is shown in Figure 11.10. Let us again select large, equal, stock-value resistors for R_1 and R_2 of 50 kΩ so that Equation 16 will closely approximate the pulse width:

$$R_1 = R_2 = 50 \text{ k}\Omega$$

Values of R_f and C can now be obtained from Equation 16 given the pulse width requirement $t_+ = 20$ ms:

$$t_+ = R_{f+}C = 20 \times 10^{-3}\text{ s}$$

For convienience, we again select a large stock 25 kΩ resistor for R_{f+} and calculate C. (Note that if a precise setting of the pulse width were required, R_{f+} would be a 50 kΩ pot carefully adjusted to give the proper pulse width.)

$$R_{f+} = 25 \text{ k}\Omega$$

$$C = \frac{20 \times 10^{-3}\text{ s}}{R_f} = \frac{20 \times 10^{-3}}{2.5 \times 10^4}$$
$$= 0.8 \times 10^{-6}\text{ F} = 0.8 \text{ }\mu\text{F}$$

The minimum frequency of the pulse generator is determined by resistor R_{f-} according to Equation 18. Using $f = 10$ Hz, we have:

$$f = \frac{1}{(R_{f+} + R_{f-})C}$$

$$R_{f-} = \frac{1}{fC} - R_{f+}$$

$$= \frac{1}{10(8 \times 10^{-7})} - (2.5 \times 10^4)$$

$$= 130 \text{ k}\Omega$$

Thus, R_{f-} would be a 200 kΩ potentiometer. The circuit, shown in Figure 11.11, also includes a 10 kΩ potentiometer attached to the output to allow varying of the output amplitude.

11.8 MONOSTABLE MULTIVIBRATORS

The operation of a wide variety of signal-processing circuits depends on having uni-formly shaped pulses. In electronic measurement, however, the voltage output of a transducer may vary widely depending on the physical situation. As an example, consider the counting system illustrated in Figure 11.12, in which a photocell circuit is triggered by objects passing a light beam. The amplitude and width of the photocell voltage will vary widely depending on the size, shape, and transparency of the object and the speed with which it passes the beam. However, an electronic counting circuit generally requires a uniformly shaped pulse that indicates simply the passing of an object, whatever its physical characteristics. Thus, it is necessary to use an input circuit to "reshape" the various photocell voltages into uniform pulses for the counting circuit. Reshaping is the role of the **monostable, or one-shot, multi-vibrator.**

Ideally, the one-shot produces a rectangular output pulse of constant amplitude and width that is not affected by the shape of the input signal. The characteristics of the rectangular pulse can vary, however, depending on the purpose of the circuit. For example, it can be either positive-going or negative-going, it can have a dc offset between pulses that is at various positive and negative voltages, and it can be triggered by either positive-going or negative-going input signals of various amplitudes.

Principles of Operation

The square wave generator circuit shown in Figure 11.4 can be converted into a monostable multivibrator by the simple addition of a diode across the feedback capacitor (Figure 11.13). The circuit is still essentially a zero-crossing detector with hysteresis, but the addition of the diode limits the values of negative voltage that can occur at the $-$ input. Note that Figure 11.13 provides for an external input trigger signal V_{in} at the $-$ input terminal.

To begin the analysis, we recall that the output voltage can only be $\pm V_{\text{sat}}$, which, in

EXAMPLE 2

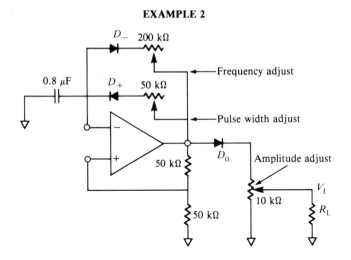

FIGURE 11.11. In this pulse generator circuit, the feedback resistors are potentiometers that permit accurate setting of the pulse width and variation of the frequency. An output potentiometer also permits variation of the output amplitude.

TYPICAL MONOSTABLE MULTIVIBRATOR APPLICATION

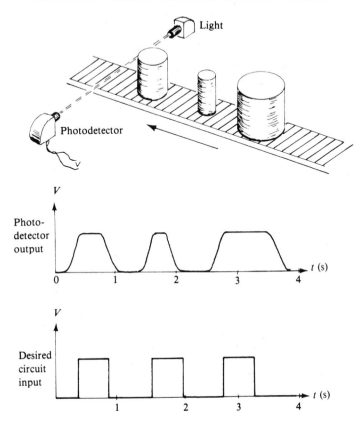

FIGURE 11.12. Monostable multivibrators are used to convert irregularly shaped input pulses into uniform pulses for subsequent circuitry. The shape of the photodetector output pulse shown will vary widely depending on the characteristics of the objects passing the beam. A monostable multivibrator can convert these pulses into identical rectangular waves for processing in a counting circuit.

BASIC MONOSTABLE MULTIVIBRATOR CIRCUIT

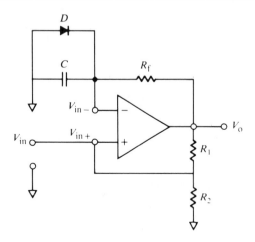

FIGURE 11.13.   The square wave generator can be converted into a monostable multibrator by the addition of a diode across the capacitor in the direction shown. The diode prevents the capacitor voltage from dropping below the diode's forward-bias voltage of about – 0.6 V, thus making the output state $V_o = -V_{sat}$ stable.

turn, determines upper and lower threshold voltages for the + input terminal according to Equations 9 and 10 from Chapter 9:

$$V_{UT} = \frac{R_2}{R_1 + R_2}(+V_{sat})$$

$$V_{LT} = \frac{R_2}{R_1 + R_2}(-V_{sat})$$

When V_{in-} exceeds V_{UT}, V_o goes to $-V_{sat}$, and when V_{in-} drops below V_{LT}, V_o goes back to $+V_{sat}$.

Let us now look at the voltage behavior across the capacitor. If $V_o = -V_{sat}$, diode D is forward biased and it esssentially bypasses capacitor C. There will be a small forward diode voltage of about -0.6 V at the – input, but the rest of $-V_{sat}$ will be dropped across R_f.

The voltage at the + input terminal V_{in+} is given by Equation 10 from Chapter 9, and for reasonable values of R_1 and R_2, it will be

greater than -0.6 V. As a result, the difference voltage $V_d = V_{in+} - V_{in-}$ will be negative, and the output will be at $-V_{sat}$. Hence, the condition $V_o = -V_{sat}$ is stable, and no oscillation will occur (Figure 11.14a).

Now let us consider what happens when an external, positive voltage pulse, sufficiently large to make V_d positive, is suddenly applied to the + terminal. The output switches to $+V_{sat}$. With $V_o = +V_{sat}$, diode D is reverse biased and the circuit is open. Hence, the capacitor C begins to charge through resistor R_f with a time constant $R_f C$ (Figure 11.14b).

As in the case of the square wave generator, charging continues until the voltage at the – input exceeds the upper threshold voltage given by Equation 9 from Chapter 9. When the capacitor voltage exceeds V_{UT}, the output switches back to $-V_{sat}$. We assume that the input trigger pulse is much shorter than $R_f C$ and is no longer present at the + input terminal.

When $V_o = -V_{sat}$, V_C equals V_{UT}, a positive voltage. In order for the cycle to be complete, the capacitor must discharge through resistor R_f back to its initial state of -0.6 V. It does so with the same time constant $R_f C$. When the capacitor voltage drops back to -0.6 V, the one-shot is ready for another trigger pulse. The input and output voltages for the process are shown in Figure 11.15.

Pulse Amplitude and Width

The **peak-to-peak amplitude** of the output pulse is $2V_{sat}$:

$$V_{pp} = 2V_{sat} \tag{20}$$

It can be made into a positive pulse only by the addition of a diode to block the negative voltage from reaching the load.

The width of the output pulse d can be determined in the same manner as for the square

PRINCIPLES OF OPERATION

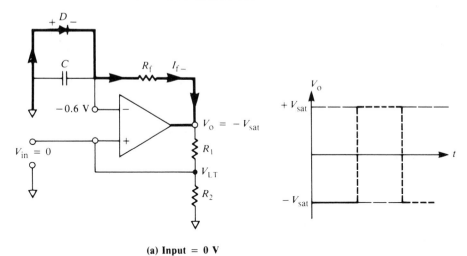

(a) Input = 0 V

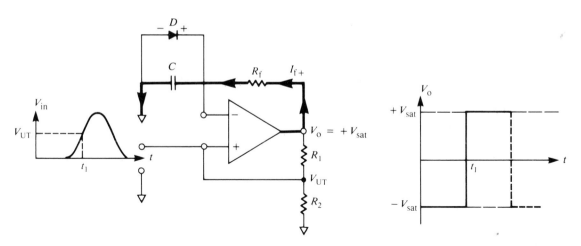

(b) Positive Input Trigger Pulse

FIGURE 11.14.   When $V_o = -V_{sat}$, diode D is forward biased, bypassing the capacitor and putting the $-$ input terminal at -0.6 V. Since this value is less negative than the voltage at the $+$ input, $V_o = -V_{sat}$ is stable. When a positive voltage pulse greater than about V_{UT} is applied to the $+$ input, V_o goes to $+V_{sat}$. Diode D is then back biased and the capacitor charges toward $+V_{sat}$. The output voltage drops back to $-V_{sat}$ when the capacitor voltage exceeds V_{UT}.

INPUT AND OUTPUT WAVEFORMS

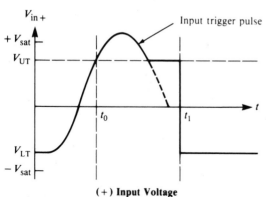

(+) Input Voltage

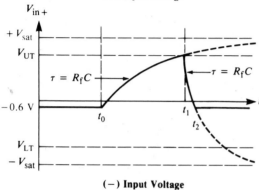

(−) Input Voltage

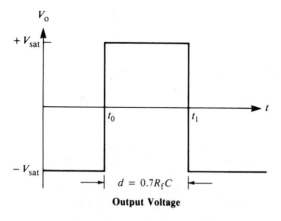

Output Voltage

FIGURE 11.15. Shown are input and output voltages for the monostable multivibrator. An input trigger pulse applied to the + input causes the output to switch to $+V_{sat}$. The width of the output pulse is determined by the time it takes for the capacitor to charge to V_{UT}. For $R_1 \approx R_2$, this time is about $d = 0.7R_fC$.

wave generator except that the capacitor voltage begins charging from the diode voltage of -0.6 V, rather than from the lower threshold voltage (compare Figures 11.15 and 11.9). Thus, Equation 3 for $V_C = V_{UT}$, $V = V_{sat}$, $V_{in} = -0.6$ V, and $t = d$ becomes:

$$V_{UT} = V_{sat} - (V_{sat} + 0.6 \text{ V})e^{-d/R_fC} \quad \textbf{(21)}$$

Rearranging terms and solving for the pulse width d gives:

$$d = R_fC \ln \frac{-0.6 - V_{sat}}{V_{UT} - V_{sat}} \quad \textbf{(22)}$$

Since V_{sat} is much larger than -0.6 V, we can neglect the -0.6 V. Thus:

$$d = R_fC \ln \frac{-V_{sat}}{V_{UT} - V_{sat}} \quad \textbf{(23)}$$

Substituting in Equation 9 of Chapter 9 for V_{UT} and rearranging terms gives the **pulse width** of the monostable multivibrator:

$$d = R_fC \ln \frac{R_1 + R_2}{R_1} \quad \textbf{(24)}$$

For $R_1 = R_2$, $\ln[(R_1 + R_2)/R_1] = 0.7$, so Equation 24 becomes:

$$d = 0.7R_fC \quad \textbf{(25)}$$

Recovery Time

A look at Figure 11.15 shows that after the output has returned to the stable value of $-V_{sat}$, the voltage at the $-$ input does not return to its initial value of -0.6 V until the capacitor has discharged. This process has the same time constant R_fC. Thus, we must wait for a time of about R_fC before V_{in-} is at -0.6 V and the circuit is ready to operate again.

The time required for the circuit to return to its initial state once the output pulse is complete is called the **recovery time** of the circuit.

RECOVERY DIODE

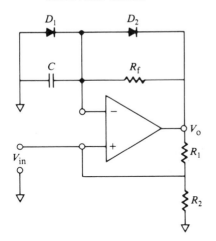

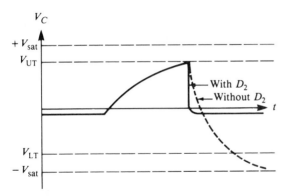

FIGURE 11.16. After the output pulse is complete, the capacitor must still charge back to -0.6 V before another input pulse can be applied. This capacitor recovery time can be significantly reduced by adding a second diode across R_f. When V_o switches back to $-V_{sat}$, D_2 is forward biased and quickly dissipates the capacitor voltage.

When a second trigger pulse is applied during the recovery period, either no output occurs or the output pulse is shorter than T. The reason is that the capacitor begins charging from a voltage higher than -0.6 V. Hence, the recovery time places an upper limit on the frequency of the input trigger pulses.

One way to reduce the recovery time, and thus increase the input trigger pulse frequency,

is to add a diode in parallel with resistor R_f, as shown in Figure 11.16. When V_o switches to $+V_{sat}$, diode D_2 is reverse biased and the capacitor charges through R_f as before. However, when V_o switches back to $-V_{sat}$, the diode is forward biased, thus shorting resistance R_f, and the capacitor can discharge very rapidly to -0.6 V. This situation is shown graphically in Figure 11.16.

Input Trigger Coupling

The ideal trigger signal to the one-shot multivibrator is a narrow positive pulse whose amplitude is sufficient to drive the $+$ input higher than the $-$ input and whose width is much less than the output pulse. The required minimum amplitude is essentially that of the upper threshold voltage V_{UT} given by Equation 9 of Chapter 9. If the amplitude of the signal is not large enough or if the polarity is wrong, it can be made suitable by using either an inverting or a noninverting amplifier.

If the input signal is too wide, however, or contains a dc offset, an input trigger coupling network may be required. This network is shown in Figure 11.17. Input capacitor C_{in} serves to block any unwanted dc signal superimposed on the signal source that might cause the input to trigger falsely. It also blocks the voltage across R_2 from the input signal source.

Diode D_3 couples an input signal to the $+$ input, but immediately decouples it after the output transition has occurred. When the output is at $-V_{sat}$, the diode is forward biased and conducts. Hence, when an input signal arrives, it is immediately passed on to the $+$ input to produce an output transition.

When $V_o = V_{sat}$, however, the diode is back biased and presents an open circuit to the trigger signal. The input capacitor C_{in} then discharges through resistor R_{in} with a time constant $R_{in}C_{in}$.

The constraint on C_{in} and R_{in} is that their time constant must be much shorter than the width of the output pulse d, so that when V_o

INPUT TRIGGER COUPLING

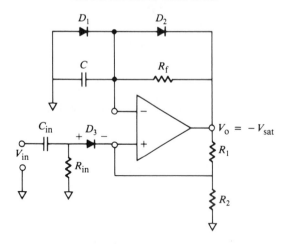

FIGURE 11.17. If the input trigger pulse is too wide or has a dc offset, an input coupling network may be required. Above, capacitor C_{in} blocks any dc offset and diode D_3 couples positive pulses to the + input. When $V_o = +V_{sat}$, however, D_3 is back biased, decoupling the trigger pulse from the + input, and C_{in} discharges through R_{in}. For proper operation, time constant $R_{in}C_{in}$ should be less than $0.1R_fC$.

switches back to $+V_{sat}$ and diode D_3 is forward biased again, the input trigger voltage has fully dissipated. Generally, a time constant of 1/10 the pulse width is sufficient:

$$R_{in}C_{in} < \frac{1}{10d} \qquad (26)$$

EXAMPLE 3: An optical coupler used to measure the rotation rate of a motor puts out voltage pulses having a peak value of about -0.1 V. Design a circuit that will convert these pulses into positive rectangular pulses having an amplitude of 10 V and a width of 10 ms.

Solution: Because the input pulses are small and have negative polarity, an inverting amplifier is required to invert and amplify the input pulses before they can be applied to the one-shot (Figure 11.18). In order that good strong pulses be supplied to the one-shot, a gain of about -100 is desirable. The gain of an inverting amplifier is given by Equation 59 from Chapter 10:

$$A_{CL} = -\frac{R_f}{R_i}$$

If $A_{CL} = -100$ and $R_{in} = 500$ Ω, this equation gives:

$$R_f = 100R_i = 100 \times 500 \ \Omega = 50 \ k\Omega$$

For convenience, let us select equal values for R_1 and R_2 of 50 kΩ:

$$R_1 = R_2 = 50 \ k\Omega$$

Now, the pulse width is given by Equation 25:

$$d = 0.7R_fC$$

For a pulse width of 10 ms, this equation gives a feedback time constant of:

$$0.7R_fC = 10^{-2} \ s$$
$$R_fC = 1.4 \times 10^{-2} \ s$$

Letting $R_f = 25$ kΩ, we find the value of C to be:

$$C = \frac{1.4 \times 10^{-2}}{R_f} = \frac{1.4 \times 10^{-2}}{2.5 \times 10^4}$$
$$= 0.56 \ \mu F$$

The condition of the trigger coupling network is given by Equation 26:

$$R_{in}C_{in} < \frac{1}{10}d = \frac{1}{10} \times 10^{-2} \ s$$

or

EXAMPLES 3 AND 4

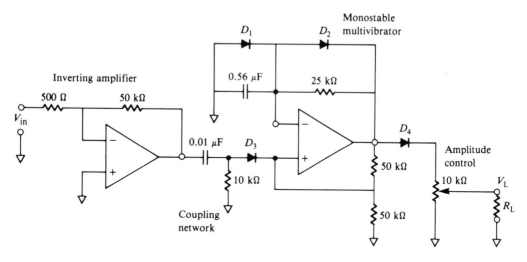

FIGURE 11.18.   A monostable multivibrator circuit designed to shape pulses from an op-tical coupler has an inverting amplifier at the input that both inverts and amplifies the pulses from the coupler so that the pulses will trigger the monostable multivibrator.

$R_{in}C_{in} < 10^{-3}$ s

Letting $R_{in} = 10\text{ k}\Omega$ gives:

$$C_{in} < \frac{10^{-3}}{10 \times 10^3} < 0.10 \ \mu\text{F}$$

Therefore, let us select a value much less than $0.10 \ \mu\text{F}$, say:

$$C_{in} = 0.01 \ \mu\text{F}$$

In order for the output pulses to be positive-going only, a diode should be added to the output to block the negative portion. To adjust the output amplitude to $+10$ V, a 10 kΩ potentiometer should also be added to permit continuous adjustment of the ampli-tudes from $+V_{sat}$ to 0 V. The circuit is shown in Figure 11.18.

Application: Electronic Rate Meter

The basic feature of a one-shot multivibrator is that it produces a uniform output pulse each time it is triggered. This feature can be used to convert the circuit shown in Figure 11.18 into an electronic rate meter that reads directly the rate at which input pulses occur. In Figure 11.12, for example, a pulse will occur each time an object passes the photocell. Thus, the out-put can be made to read the rate at which ob-jects pass on a conveyor.

When a pulse train is applied to the one-shot, the output is also a pulse train whose pulses are all the same height and width. In Part III, we saw that the average value of a pulse train V_{dc} is given by the area of each pulse divided by the period:

$$V_{dc} = \frac{\text{pulse area}}{T} \tag{27}$$

The area of the square output pulses is given by their width d times their peak amplitude V_p, so:

$$V_{dc} = \frac{dV_p}{T} \qquad (28)$$

The pulse frequency f is simply one divided by the period, so the **average value** of dc voltage of a pulse train is:

$$V_{dc} = dV_p f \qquad (29)$$

For the output pulses of the one-shot, d and V_p are constant. Thus, the average dc voltage of the pulse train is directly proportional to the frequency f. When a dc meter is connected to the output of the one-shot, its dc reading is directly proportional to the frequency of the input pulse train.

The values of d and V_p can be adjusted by designing the one-shot so that a frequency of 1000 Hz = 1.000 V. If the meter is a digital meter, then it will read the frequency directly — for example, a voltage of 0.500 V = 500 Hz, 0.750 V = 750 Hz, and so on.

EXAMPLE 4: What design values of d and V_p will make a voltage of 0.100 V represent a rotation rate of 100 rpm in the design circuit shown in Figure 11.18?

Solution: Equation 29 requires that the frequency be expressed in hertz (cycles per second). Thus, we must first convert the rotation rate of 100 rpm into revolutions (cycles) per second:

$$100 \ \frac{rev}{min} \times \frac{1 \ min}{60 \ s} = 1.667 \ Hz$$

Then, using Equation 28 for $V_{dc} = 0.100$ V, we have:

$$V_{dc} = 0.100 \ V = dV_p(1.667 \ Hz)$$

or

$$dV_p = 0.0600$$

If we let $V_p = \mathbf{6.00 \ V}$, then for d we have:

$$d = \frac{0.600}{6.00} = 0.0100 \ s = \mathbf{10.0 \ ms}$$

The circuit shown in Figure 11.18 has a pulse width of 10 ms, so the pulse amplitude need only be adjusted to 6.00 V.

A rate meter that is used to measure rotation rate is called a *tachometer*. Thus, the circuit shown in Figure 11.19 is that of an electronic tachometer.

11.9 SAWTOOTH WAVE GENERATOR

When positive feedback is applied to an op amp, the output can be only $\pm V_{sat}$. This output behavior is ideally suited for generating square waves and rectangular pulses, as the previous applications showed. To generate waveforms that are not square, however, the op amp output of $\pm V_{sat}$ is used in a slightly different way. In these applications, it serves as an alternating voltage source that alternately applies voltages $+V_{sat}$ and $-V_{sat}$ to a constant-current source. If the constant-current source is in the feedback loop and is used to charge a capacitor, a variety of ramp waveforms, including sawtooth and triangular waves, can be generated.

Principles of Operation

To begin the analysis, we recall the voltage–time behavior of a capacitor charged by a constant-current source, as described by Equation 1:

ELECTRONIC RATE METER

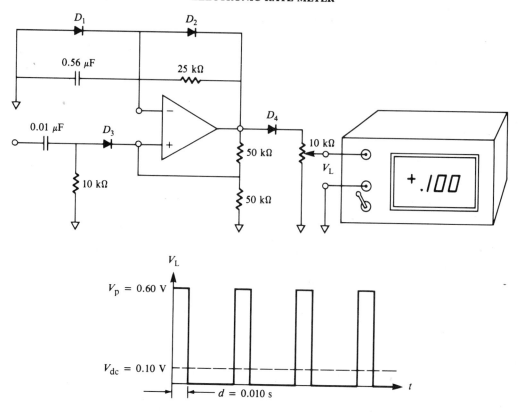

FIGURE 11.19.   The monostable multivibrator becomes an electronic rate meter with the addition of a dc voltmeter to the output. Because the monostable pulses are uniform in amplitude and width, their average dc voltage is directly proportional to their frequency. If the pulse amplitude and width are carefully selected, a digital voltmeter can be made to read rate directly. See Example 4.

$$V_C = \frac{It}{C}$$

For a fixed capacitor C supplied by a source of constant current I, the capacitor voltage V_C increases linearly with time.

In the simplest case, this basic behavior and the circuit design shown in Figure 11.20 can be used to produce a sawtooth wave. When switch S is open, the constant-current source causes the capacitor voltage to increase linearly according to Equation 1. Periodically closing

the switch to discharge the capacitor and then reopening it causes the capacitor voltage to be a sawtooth wave.

In Part III a unijunction transistor (UJT) and a 555 timer were used as electronic switches to produce a sawtooth wave. The same behavior can be achieved with an op amp by placing the current source and capacitor in the feedback loop and powering the current source from the op amp's output voltage, as shown in Figure 11.21.

BASIC SAWTOOTH CIRCUIT

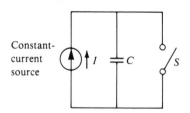

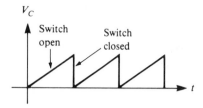

FIGURE 11.20.   The op amp can also be used to generate sawtooth waves. The basic component of a sawtooth generator is a capacitor charged by a constant-current source. If the capacitor is periodically discharged—for example, by closing switch S—the capacitor voltage V_C will produce a sawtooth wave.

OP AMP SAWTOOTH WAVE GENERATOR

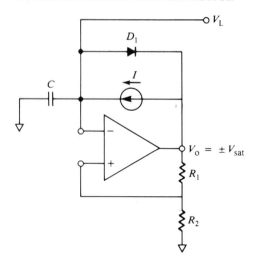

FIGURE 11.21.   When an op amp is used to generate sawtooth waves, the constant-current source and capacitor are placed in the negative feedback loop. The current source is powered by the op amp's output voltage that alternately turns the current source on and off as V_o switches between $\pm V_{sat}$. Note that the output sawtooth voltage V_L is the capacitor voltage.

To begin the analysis, let us assume that the capacitor is initially at the lower threshold voltage $V_C = V_{in} = V_{LT}$, but that $V_o = +V_{sat}$ (Figure 11.22a). Thus, the voltage at the $+$ input terminal is the upper threshold voltage given by Equation 9 of Chapter 9:

$$V_{UT} = \frac{R_2}{R_1 + R_2} V_{sat}$$

When $V_o = +V_{sat}$, the feedback diode D is back biased and acts as an open circuit. The positive voltage on the current source causes it to supply a constant current to the capacitor, and it charges linearly toward $+V_{sat}$. When the capacitor voltage reaches the upper threshold voltage, however, V_{in-} exceeds V_{in+} and the output switches to $-V_{sat}$.

If $V_o = -V_{sat}$ (Figure 11.22b), the feedback diode D is forward biased and quickly discharges the capacitor toward $-V_{sat}$. A negative voltage on the constant-current source renders it inoperative and an open circuit. As before, when the capacitor voltage drops to the lower threshold voltage, the output immediately switches back to $+V_{sat}$ and the cycle repeats. Note that the output waveform of this circuit is the capacitor voltage V_C, also shown in Figure 11.22. Also, the sawtooth alternates between V_{LT} and V_{UT}.

In some applications, a negative portion is acceptable in the sawtooth. If not, the circuit can be made to go between 0 V and V_{UT} by the simple addition of a germanium diode across resistor R_2, as shown in Figure 11.23. Germa-

PRINCIPLES OF OPERATION

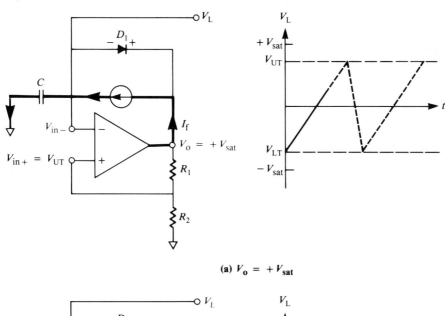

(a) $V_o = +V_{sat}$

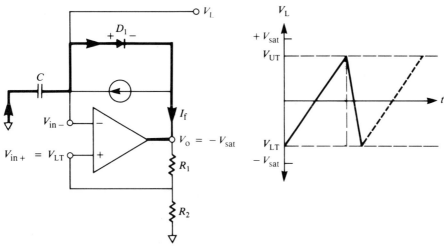

(b) $V_o = -V_o$

FIGURE 11.22. When $V_o = +V_{sat}$, it turns on the current source that supplies current to the capacitor. The capacitor voltage increases linearly toward $+V_{sat}$, producing an output transition to $-V_{sat}$ when it exceeds V_{UT}. When $V_o = -V_{sat}$, the current source is turned off. Diode D_1 is forward biased, however, quickly discharging the capacitor to V_{LT} and producing an output transition back to $+V_{sat}$.

FIGURE 11.23. A sawtooth generator in which there is a diode across resistor R_2 has a baseline of 0 V. When $V_o = -V_{sat}$, diode D_2 is forward biased and bypasses resistor R_2. This design raises the lower threshold voltage to the forward-bias voltage of the diode, about -0.2 V for germanium.

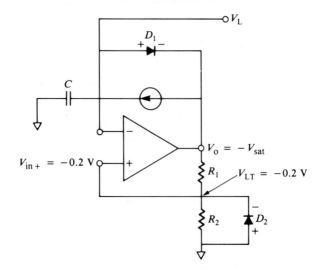

SAWTOOTH WAVE GENERATOR WITH 0 V BASELINE

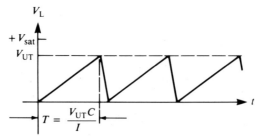

nium diodes are generally preferred due to their low forward-biased voltage of only about -0.2 V. In this case, when $V_o = +V_{sat}$, diode D_2 is back biased and is an open circuit. The voltage at the + terminal is therefore the upper threshold voltage given by Equation 9 of Chapter 9.

When $V_o = -V_{sat}$, however, diode D_2 is forward biased and shorts out resistor R_2. The voltage at the + input is the small forward diode voltage of about -0.2 V, which raises the lower threshold voltage to -0.2 V. The output will then change back to $+V_{sat}$ when the capacitor voltage drops to this value. Adding an appropriate resistor in series with diode D_2, of course, permits adjustment of the baseline to

any value between V_{LT} and -0.2 V. The output waveform of this circuit is shown in Figure 11.23. Note that the vertical portion of the sawtooth has some slope due to the fact that the capacitor does not discharge instantly. This slope can be minimized, however, by using fast-switching diodes. Note, also, that the peak amplitude of the sawtooth is the upper threshold voltage V_{UT}.

Period and Frequency

The period of the sawtooth wave depends on the rate at which the capacitor charges, which is determined by Equation 1, where the sawtooth

CONSTANT-CURRENT SOURCE

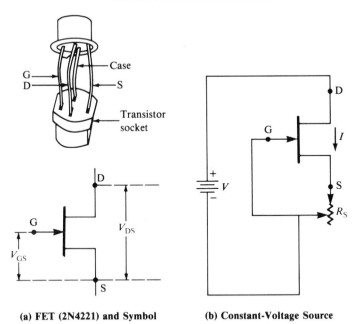

(a) FET (2N4221) and Symbol

(b) Constant-Voltage Source

FIGURE 11.24. The constant-current source is constructed from a field-effect transistor (FET). For a given drain-source voltage V_{DS}, the current from an FET depends on the gate-source voltage V_{GS}. In a circuit, V_{GS} is varied by varying a resistor in series with the source.

period T is the time required for the capacitor to charge from 0 V (for the circuit shown in Figure 11.22) to the upper threshold voltage $V_C = V_{UT}$:

$$V_{UT} = \frac{IT}{C}$$

Rearranging terms gives:

$$T = \frac{V_{UT}C}{I} \qquad (30)$$

If we assume that the discharge time is short compared to T, the sawtooth **frequency** f is the inverse of the period:

$$f = \frac{I}{V_{UT}C} \qquad (31)$$

Thus, the frequency is determined by the charging current I, the capacitor value C, and the upper threshold voltage V_{UT}. Generally, for a given current, values of V_{UT} and C are fixed. Values of I, however, can be varied to change the sawtooth frequency. Hence, we should examine the detailed design of the constant-current source (Figure 11.24).

The constant-current source utilizes the properties of a field-effect transistor (FET). The FET has three terminals labeled gate G, source S, and drain D (Figure 11.24a). The constant current comes from the source S and is determined both by the drain-source voltage V_{DS} and the gate-source voltage V_{GS}. For a given drain voltage V_{DS}, however, we can alter the source current by varying the gate voltage V_{GS}.

A simple method of varying the current is to place a variable resistor in series with the source, as shown in Figure 11.24b. Changing the resistor changes the gate voltage and thus

FIGURE 11.25. The basic circuit of a sawtooth wave generator shows the FET current source in the negative feedback loop. Varying the 100 kΩ potentiometer connected to the FET source terminal varies the sawtooth frequency, and varying the 100 kΩ potentiometer connected to the op amp output varies the sawtooth amplitude.

EXAMPLE 5

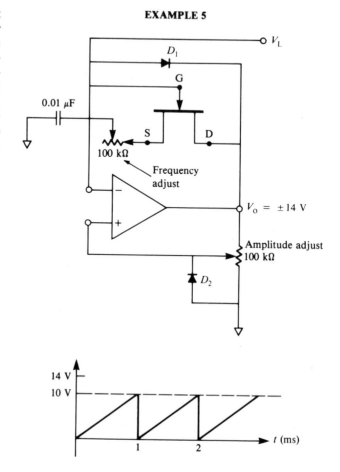

changes the current from the source. The details of this behavior require a more detailed understanding of semiconductor behavior, which is beyond the scope of this discussion.

When the constant-current source in Figure 11.24b is in the feedback loop of an op amp (Figure 11.25), the drain voltage comes from the op amp output voltage $+ V_{sat}$. The op amp can supply a maximum current of only about 5 mA, so there is an upper limit on the current value in Equation 31. For most applications, however, 5 mA is more than adequate.

EXAMPLE 5: Design a sawtooth wave generator with a 10 V peak output and a frequency of 1000 Hz. Assume that the current source can be adjusted to provide a constant current of 0.1 mA.

Solution: The basic circuit is the one shown in Figure 11.25. The 10 V peak value of output voltage sets the value of the upper threshold voltage. In order to make $V_{UT} =$

10 V, we must solve Equation 9 of Chapter 9 to find resistors R_1 and R_2:

$$V_{UT} = \frac{R_2}{R_1 + R_2} V_{sat}$$

We know that $V_{sat} \approx 14$ V, and we select R_2 = **50 kΩ**. Thus:

$$10 \text{ V} = \frac{5 \times 10^4}{R_1 + (5 \times 10^4)} 14 \text{ V}$$

or

$$R_1 = (5 \times 10^4) \frac{14 \text{ V}}{10 \text{ V}} - (5 \times 10^4)$$

$$\approx \textbf{20 kΩ}$$

Thus, we can use a center-tapped 100 kΩ potentiometer to accurately set $V_{UT} = 10$ V.

We assume that the baseline voltage is 0 V so that we can use Equation 31 to determine the value of capacitance C that will give a frequency of 1000 Hz:

$$f = \frac{I}{V_{UT} C}$$

or

$$C = \frac{I}{V_{UT} f} = \frac{10^{-4} \text{ A}}{10 \text{ V} \times 10^3 \text{ Hz}}$$

$$= \textbf{0.01 } \mu\textbf{F}$$

The frequency can be adjusted accurately to 1000 Hz, or continuously varied, by adjusting the current-source resistor.

Negative Sawtooth Wave

A negative-going sawtooth wave can be generated by reversing the direction of the current source and feedback diode D_1, as shown in Figure 11.26. For the circuit to begin at 0 V, diode D_2 across resistor R_2 must also be reversed so that when $V_o = +V_{sat}$, D_2 conducts and lowers the upper threshold voltage to 0.2 V.

NEGATIVE SAWTOOTH GENERATOR

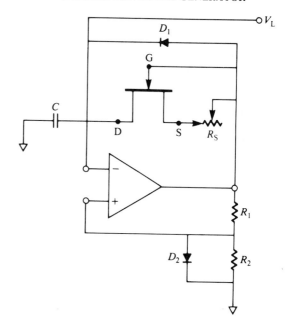

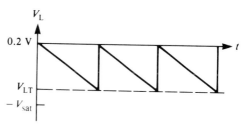

FIGURE 11.26. A negative sawtooth generator can be constructed by reversing the direction of the current source and the feedback diode D_1. Then, the capacitor charges linearly toward the lower threshold voltage and quickly discharges toward the upper threshold voltage. Reversing diode D_2 across R_2 lowers the upper threshold to 0.2 V to provide a baseline near 0 V.

11.10 TRIANGLE WAVE GENERATOR

We can also use the basic principles of the sawtooth generator circuit to produce triangular

FIGURE 11.27. A triangle wave generator can be constructed by placing two current sources in the negative feedback loop to produce both positive and negative charging currents. When $V_o = +V_{sat}$, diode D_{1+} is forward biased and the positive current source charges the capacitor toward V_{UT}. When $V_o = -V_{sat}$, diode D_{1-} is forward biased and the negative current source charges the capacitor toward V_{LT}.

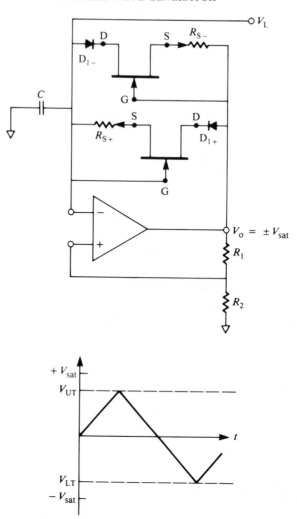

TRIANGLE WAVE GENERATOR

waves. Essentially, a positive current source in the feedback loop generates the increasing ramp of the triangular wave, and a negative current source generates the decreasing ramp.

Principles of Operation

The circuit for the triangular wave generator is shown in Figure 11.27. Two constant-current-source FETs in the negative feedback loop produce currents in opposite directions.

When $V_o = +V_{sat}$, diode D_{1-} is back biased and the negative current source is open circuited. Diode D_{1+}, however, is forward biased and the positive current source causes the capacitor to charge toward $+V_{sat}$. When the capacitor voltage exceeds the upper threshold voltage, V_o goes to $-V_{sat}$.

When $V_o = -V_{sat}$, diode D_{1+} is back biased and the positive current source is open circuited. Diode D_{1-} is forward biased, however, and the negative current source causes the capacitor to charge toward $-V_{sat}$. When the capacitor voltage drops below the lower threshold voltage, V_o again goes to $+V_{sat}$.

Note that the triangle wave oscillates between the upper and lower threshold voltages and that there are no diodes across positive feedback resistor R_2 to shift the 0 V baseline. Adding diodes in the proper direction, however, will cause oscillation between approximately 0 V and V_{UT} or between approximately 0 V and $-V_{LT}$.

One advantage of the circuit shown in Figure 11.27 is that the slew rate of the increasing and decreasing ramps can be made different by varying the positive and negative charging currents. According to Equation 30, the larger the current, the shorter the time needed to reach the required voltage and, therefore, the steeper the slew rate.

Alternative Circuit

When a triangle wave with equal rising and falling ramps is required, we can modify the circuit shown in Figure 11.27 to use only one current source (Figure 11.28). The four diodes around the FET make up a diode bridge that conducts current in one direction or the other depending on the relative voltage at point A.

For example, when $V_o = +V_{sat}$ (Figure 11.28a), point A is positive relative to point B. Thus, diodes D_3 and D_2 are back biased and will not conduct. Diodes D_1 and D_4 are forward biased, however, and conduct positive-source current to the capacitor, causing it to charge toward $+V_{sat}$.

On the other hand, when $V_o = V_{sat}$ (Figure 11.28b), point B is positive relative to point A. Diodes D_1 and D_4 are forward biased, and the current direction through the bridge is reversed. Thus, the capacitor charges toward $-V_{sat}$.

The four-diode bridge, therefore, reduces the number of FETs necessary and assures that the rising and falling ramps of the triangle wave have the same slew rate. As with the other circuits, the frequency of the triangle wave can be changed by varying the charging current—that is, by changing the FET source resistor R_S.

Output Amplitude

Varying the output amplitude requires some consideration. Since the output voltage comes from the capacitor, a load resistance placed across the capacitor to ground would take its current from the capacitor charging current, thus reducing the time for the capacitor to charge to its threshold voltage and changing the frequency of the triangle wave.

The loading effect can be eliminated by inserting a voltage follower between the capacitor and the load, as shown in Figure 11.29. The output of the voltage follower reproduces the triangle wave, but draws negligible input current. This arrangement will not alter the wave frequency. (A noninverting amplifier might also be used in order to obtain some voltage gain.) A potentiometer can then be added to the output of the second op amp to vary the output amplitude (Figure 11.29).

11.11 SINE WAVE GENERATOR

Although sine waves are the most common periodic signal, the design of an electronic sine wave generator is not simple. Sine wave generators utilize one of two basic types of circuits: a sine wave oscillator and a triangle-to-sine-wave-shaping circuit (Figure 11.30).

Many of the function generators available today use the triangle-to-sine method, largely because they also contain a triangle wave generator. The basic strategy is to apply the triangle wave to a circuit that will round the linear

ALTERNATIVE CURRENT SOURCE CIRCUIT

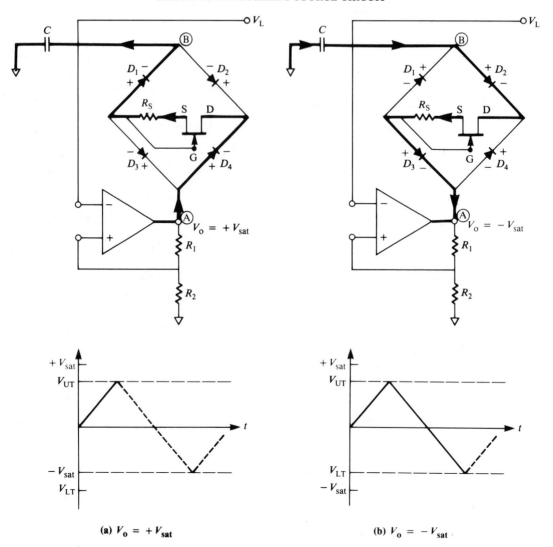

FIGURE 11.28. An alternative current-source circuit uses only one FET and a diode bridge. When $V_o = +V_{sat}$, diodes D_1 and D_4 are forward biased and positive charging current is supplied to the capacitor. When $V_o = -V_{sat}$, diodes D_2 and D_3 are forward biased and negative charging current is supplied to the capacitor. Because the charging current is the same for both cases, the rising and falling ramps of the triangle wave have the same slew rate.

VARIABLE TRIANGLE WAVE GENERATOR

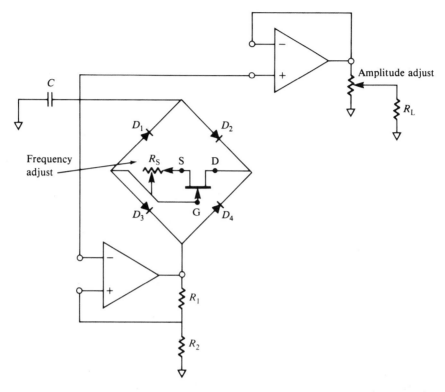

FIGURE 11.29. Because the output voltage of the sawtooth and triangle wave generators is the capacitor voltage, a voltage follower must be added so that a load does not draw any of the capacitor charging current. The voltage follower reproduces the capacitor waveforms across the load, but draws negligible capacitor current.

ramps into a shape that approximates a sine wave.

Principles of Operation

The basic circuit of a sine wave generator, shown in Figure 11.31, uses a noninverting op amp whose gain varies with the amplitude of the input signal. Note that the feedback loop consists of three parallel networks, the upper two of which contain diodes. The feedback diodes may be either germanium or silicon depending on the voltage at which they should conduct. Diodes D_1 and D_2 are typically germanium with a forward conduction voltage of about 0.3 V.

For sufficiently small input voltages, none of the diodes conduct, and the amplifier gain is determined solely by feedback resistors R_i and R_{f_1}. For relatively large output voltages, however, the voltage across diode D_1 exceeds 0.3 V, and it begins to conduct. In this case, the gain depends on the parallel combination of R_{f_1} and R_{f_2} compared to resistor R_i. Since R_{f_1} in parallel with R_{f_2} must be less than R_i alone, the gain for these voltages is smaller.

SINE WAVE GENERATORS

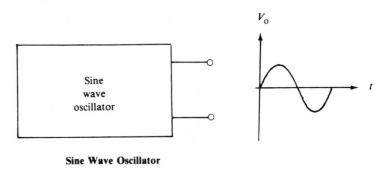

Sine Wave Oscillator

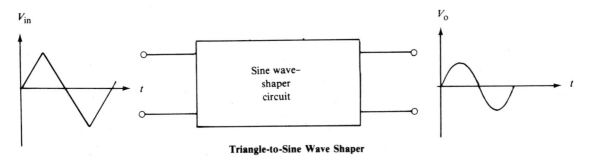

Triangle-to-Sine Wave Shaper

FIGURE 11.30. Sine waves are generated by one of two methods. The first uses a sine wave oscillator to produce sine waves directly. The second begins with a triangle wave and then rounds the peaks to "shape" it into an approximate sine wave. The text describes the second method.

Diodes D_3, D_4, D_5, and D_6 are typically silicon with a forward-conduction voltage of about 0.6 V. When the voltage is sufficiently large, the voltage across series diodes D_3 and D_4 exceeds 1.2 V and they conduct. In this case, the gain is determined by the net resistance of the three parallel networks, which is less than the net resistance of two.

The circuit shown in Figure 11.31, therefore, comprises an amplifier whose gain varies with the input amplitude. As the amplitude increases, the gain decreases. If the input signal is a triangle wave and the component values are carefully chosen, the output will be approxi-

mately a sine wave, at least over some range of frequencies. For example, if $R_i = 2.2 \text{ k}\Omega$, $R_{f_1} = 82 \text{ k}\Omega$, $R_{f_2} = 10 \text{ k}\Omega$, and the input triangle wave has a peak-to-peak amplitude of 2 V, the circuit in Figure 11.31 will produce a reasonable sine wave between about 100 Hz and 5 kHz.

The frequency of the sine wave can be varied by varying the frequency of the triangle wave. To obtain a variable output amplitude, we can add a second noninverting amplifier with voltage gain to the output of the wave-shaping circuit, followed by a variable potentiometer to change the final output amplitude.

TRIANGLE-TO-SINE-WAVE-SHAPING CIRCUIT

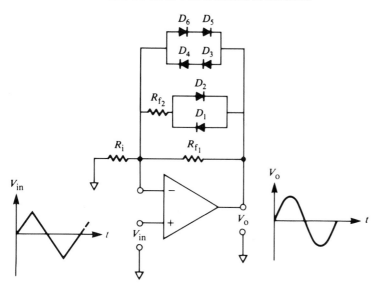

FIGURE 11.31. The basic principle of a triangle-to-sine-wave shaper is a noninverting amplifier whose gain decreases as the input signal amplitude increases. For small V_{in}, all diodes are back biased and the gain is determined by R_1 and R_i. For larger V_{in}, diode D_1 is forward biased and the feedback resistance is reduced to R_{f_2} in parallel with R_{f_1}. For still larger V_{in}, diodes D_3 and D_4 are forward biased to reduce the gain further. The result is an approximate sine wave at the triangle wave frequency.

11.12 QUESTIONS AND PROBLEMS

1. If switch S in Figure 11.32 (capacitor charged by a constant-current source) is closed and then re-opened 10 ms later, draw the resulting capacitor voltage for the following values of constant current I and capacitance C. Be sure to label the axes and each curve.
   - (a) $I = 100\ \mu A$ and $C = 0.05\ \mu F$
   - (b) $I = 1.5\ mA$ and $C = 1.0\ \mu F$
   - (c) $I = 2.5\ mA$ and $C = 2500\ pF$
   - (d) $I = -150\ \mu A$ and $C = 0.1\ \mu F$

2. If switch S in Figure 11.33 (capacitor charged by a constant-voltage source) is closed and then re-opened 8 ms later, sketch the resulting capacitor voltage for the following values of voltage V, resistance R, and capacitance C. Be sure to label the axes and each curve. Also calculate the time constant τ for each set of values.
   - (a) $V = 10\ V$, $R = 20\ k\Omega$, $C = 0.1\ \mu F$
   - (b) $V = 10\ V$, $R = 6.8\ k\Omega$, $C = 0.15\ \mu F$
   - (c) $V = 5\ V$, $R = 250\ \Omega$, $C = 10\ \mu F$
   - (d) $V = -10\ V$, $R = 200\ k\Omega$, $C = 5000\ pF$

3. What are the upper and lower threshold voltages V_{UT} and V_{LT} and frequency of oscillation f of the astable multivibrator circuit shown in Figure 11.34? Draw the output waveform and the waveforms at the two input terminals and label the axes.

4. If the 6.8 kΩ resistor is replaced by a 2.2 kΩ resistor, what are the upper and lower threshold voltages V_{UT} and V_{LT} and frequency of oscillation f of the circuit shown in Figure 11.34?

5. Design a square wave generator with a frequency of exactly 8.0 kHz and an amplitude of 5.0 V. Draw the circuit and give component values.

6. What are the upper and lower threshold voltages V_{UT} and V_{LT}, pulse widths t_+ and t_-, and frequency of oscillation f of the nonsymmetrical

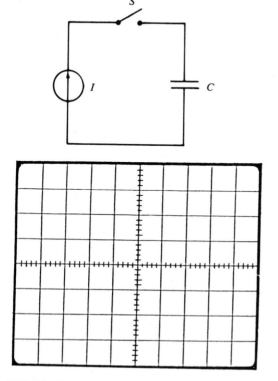

FIGURE 11.32.

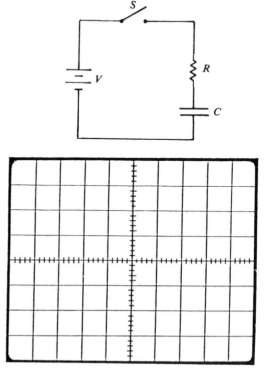

FIGURE 11.33.

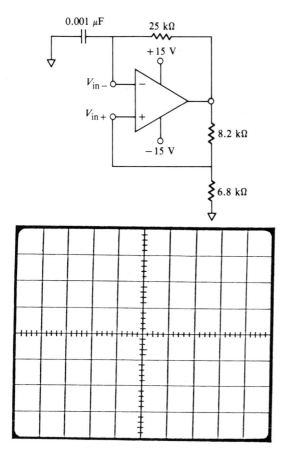

FIGURE 11.34.

Output Waveform

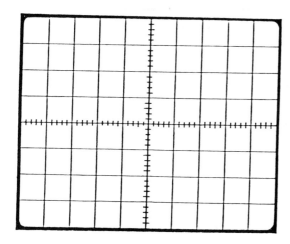

(−) Input Waveform

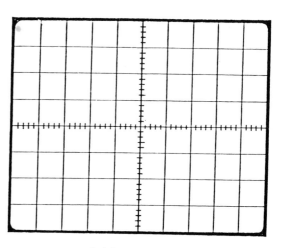

(+) Input Waveform

square wave generator circuit shown in Figure 11.35? Draw the output waveform and label the axes.

7. Design a variable-frequency pulse generator with a pulse width of exactly 0.10 ms and a minimum frequency of 1 kHz. Draw the circuit showing component values.

8. What are the upper and lower threshold voltages V_{UT} and V_{LT} and pulse width d of the circuit shown in Figure 11.36 (monostable multivibrator with input trigger coupling)? Draw the output waveform that would result from an input

pulse signal at a frequency of 500 Hz. Be sure to label the axes.

9. What maximum input frequency f_{max} of trigger signals could be applied to the circuit shown in Figure 11.36? Show how that frequency can be increased and give the approximate new value.

10. Calculate the values of R_{in} and C_{in} in Figure 11.36 and their time constant τ_{in}.

11. Design a monostable multivibrator circuit that will produce output pulses with an amplitude of 5 V and a pulse width of 10 ms. Draw the circuit diagram showing component values.

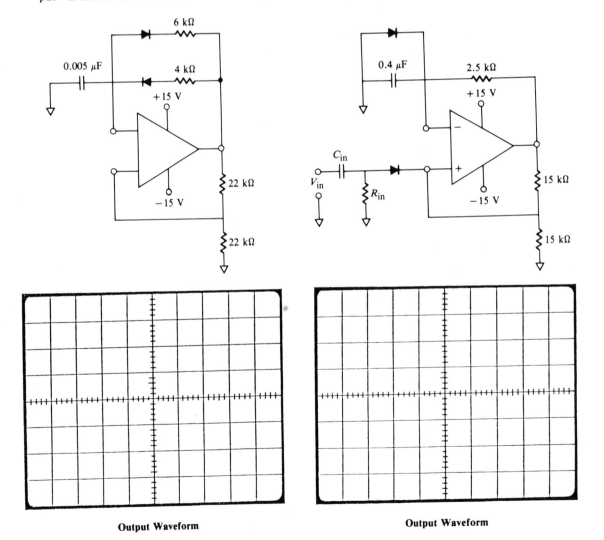

Output Waveform

FIGURE 11.35.

Output Waveform

FIGURE 11.36.

12. Using the circuit diagram drawn for Problem 11, show how the circuit can be made to read the input frequency up to 1000 Hz directly in pulses per second. Use a digital voltmeter with a full-scale sensitivity of 1.000 V as the output meter. What is the required value of output amplitude V_p from the monostable multivibrator?

13. What are the output amplitude V_{pp} and fre-

quency f of the sawtooth wave generator circuit shown in Figure 11.37? Assume that the FET supplies a constant current of 50 μA and that diode D_2 is germanium. Draw the output voltage V_L and label the axes.

14. Derive expressions for the period T and frequency f of the triangle wave generator shown in Figure 11.38.

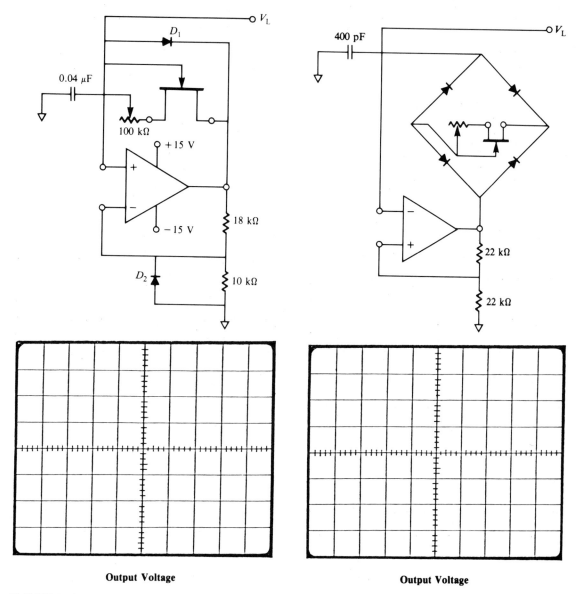

Output Voltage

FIGURE 11.37.

Output Voltage

FIGURE 11.38.

15. What are the output amplitude V_{pp} and frequency f of the triangle wave generator circuit shown in Figure 11.38? Assume the FET supplies a constant current of 100 μA. Draw the output voltage V_L and label the axes.

16. Design a triangle wave generator that could generate triangle waves with a variable amplitude of 0 V to 10 V and frequencies of 100 Hz to 10 kHz. Assume the FET current can be varied from 10 μA to 1 mA. Draw the circuit diagram showing the component values and variable controls.

Part V

DIGITAL ELECTRONIC CIRCUITS

DIGITAL LOGIC IN ELECTRONIC SYSTEMS

The previous parts of this book have developed the basic concepts and tools of what is called **analog** electronic circuits. Analog circuits derive their name from the word *analogy*, meaning *similar* or *like*. They have wide application in electronic systems that measure or reproduce physical quantities, such as temperature, sound, speed, and so on.

Typically, the voltage (or current) in an analog system is directly proportional—that is, analogous, or similar—to the value of a physical quantity. An amplifier is a common example of an analog circuit in which the output voltage may be bigger (or smaller or have more power) than the input signal, but its shape is identical.

Essentially all of electronics was analog until the late 1930s and early 1940s when the power of **digital logic** for electronic computation began to be appreciated. In digital systems, the signal voltages have only two values—usually 0 V and +5 V—in contrast to the continuous range of voltages from zero volts to tens of volts in analog systems.

The two voltage levels represent the two digits 0 and 1. With only these two digits, all of modern computer systems have been developed. The principles of how a two-digit code can be used to perform the many complex tasks in a computer are based on theories of mathematical logic that are introduced here.

The electronic components that permit digital systems to perform their tasks have evolved over several decades of technological development, especially those of semiconductor integrated circuits. In the early 1940s, so-called first generation digital computers were constructed from relays and vacuum tubes. Their capabilities were severely limited by their size, cost, power consumption, and reliability. With the development of the transistor and magnetic core storage elements, second generation computers of the late 1950s and early 1960s had significantly reduced size, weight, and power consumption, as well as improved reliability.

The development of integrated circuits in the mid-1960s led to further improvements in third generation computers, as did the technology of **medium-scale integration** (MSI), **large-scale integration** (LSI), and **very large-scale integration** (VLSI) circuit chips that today's fourth generation computer systems comprise.

Part V introduces some of the basic digital concepts and tools that are used in these computing systems (Figure 1). Only a

DIGITAL COMPUTERS AND PERIPHERAL EQUIPMENT

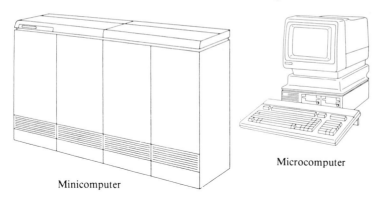

Minicomputer

Microcomputer

Computers

FIGURE 1. The principles and circuits of modern computers are based on mathematical logic in which the digits 0 and 1, represented by 0 V and +5 V, are used to perform computational tasks. The electronic circuits utilize the power of today's very large scale integration (VLSI) ICs to reduce computer size, cost, and power consumption at increased speed and reliability.

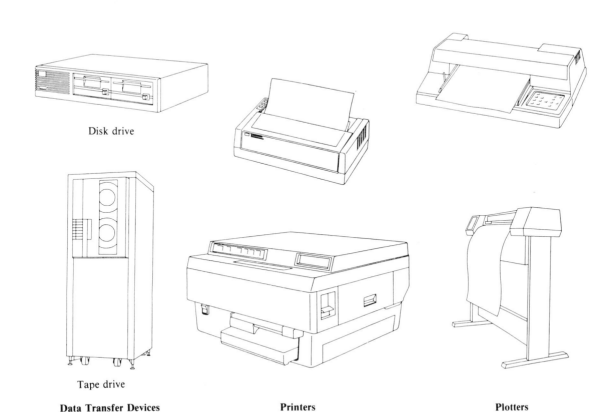

Disk drive

Tape drive

Data Transfer Devices   **Printers**   **Plotters**

very limited view of how electronic circuits are used to construct these modern miracles can be given here. Nonetheless, by the end of Part V, you should have a working language of digital electronic circuits and components, appreciate the mathematical logic on which computing systems depend, and have a knowledge of the basic building blocks from which they are constructed: input/output port, memory, CPU, and system bus.

Chapter 12

DIGITAL ELECTRONIC GATES

12.1 OBJECTIVES

Following the completion of Chapter 12, you should be able to:

1. Identify the five basic gating functions — AND, OR, NOT, NAND, and NOR — draw their circuit symbols, and complete their truth table outputs for all input possibilities.
2. Design a digital logic circuit that solves simple control problems using the three basic gating functions AND, OR, and NOT.
3. Represent the five basic gating functions in Boolean algebraic notation and give their Boolean definition.
4. Determine the truth table for a given Boolean algebraic expression using the Boolean definitions for the gating functions and the laws governing Boolean calculations (commutative, associative, and distributive).
5. State DeMorgan's theorem and show how it yields alternative forms for NAND and NOR gates, including their circuit symbols.

6. Draw the logic flow diagram that is equivalent to a given Boolean algebraic expression.
7. Give the Boolean algebraic expression that represents a given logic flow diagram.
8. Simplify Boolean algebraic expressions and logic flow diagrams using the definitions, laws, and theorems governing their manipulation.
9. Explain the purpose of digital multiplexing and draw a logic flow diagram that shows how this process can be accomplished.
10. Show how the two basic gating functions AND and OR can be achieved using mechanical switches, relays, and resistors.
11. Explain the basic gate performance characteristics of propagation delay time, dc noise margin, and fan-out capability.
12. Describe the behavior of the various semiconductor logic families — RTL, DTL, and TTL — including typical NOT and NOR gate circuit diagrams and voltage transfer

characteristics and explain their relative performance in terms of propagation delay time and dc noise margin.
13. State the various types of MOS logic and give their relative performance in terms of propagation delay time and fabrication capability.

12.2 DIGITAL SIGNAL PROCESSING

If we took a voltmeter and probed the voltages in any computer circuit, we would find only two voltage levels: 0 V and + 5 V. The two voltage levels represent the two mathematical digits 0 and 1. The processing of 0s and 1s by electronic circuitry performs all the complex computational and data processing tasks that computers do (Figure 12.1).

Mathematically, the number of basic operations that can be performed with the digits 0 and 1 is relatively small and can be easily understood. Electronically, these mathematical operations can be produced by a similarly small number of circuit components called **gates**. For these gates, each combination of 0 and 1 inputs yields a specific 0 or 1 output. The performance of all digital systems, including the most sophisticated digital computer, depends on combinations of these simple gating operations.

Furthermore, digital integrated circuits with input/output characteristics that correspond to each of these gates are inexpensive, readily available, and easy to use. Thus, with a surprisingly small amount of knowledge, it is possible to build a variety of digital circuits that perform useful and significant tasks, such as control systems, arithmetic calculations, data processing, storage, and transmission, and interfaces between minicomputers and a variety of peripheral equipment.

DIGITAL COMPUTER OPERATION

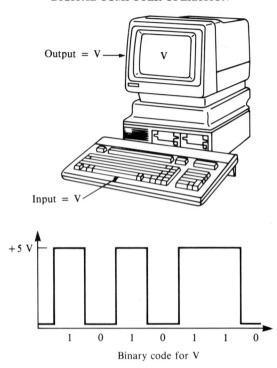

FIGURE 12.1.   The operation of digital computers is based on the electronic processing of the digits 0 and 1 (electronically, 0 V and + 5 V) by circuit components called gates. Chapter 12 introduces the mathematical logic used in digital signal processing and the electronic gates that perform the logic tasks.

Chapter 12 introduces the basic digital electronic gates and the mathematical logic on which their operation depends. The principles are then applied to a simple digital control task — the operation of a vending machine. Finally, the various types and characteristics of electronic components that are available to perform the gating tasks are described. The description includes a brief overview of the modern IC logic families that are used in today's computer circuits.

12.3 BASIC GATING FUNCTIONS

As mentioned previously, all digital electronic circuits utilize only the two voltage levels 0 V and +5 V that in turn represent the two mathematical digits 0 and 1. The two voltages are commonly termed **low** (LO) and **high** (HI). For simplicity in the following discussion, the low voltage is shown as 0 and the high voltage as 1, even though the real circuit voltage of a high condition is +5 V. This convention helps to correlate the mathematical analysis with the electronic circuit behavior.

AND and OR Gates

Like an operational amplifier, the basic electronic gate has two inputs and one output. Each input and output can have only one of two values, 0 or 1. With these two digital possibilities, there are two basic gating operations called **AND** and **OR**. The circuit symbols for these gates are shown in Figure 12.2. The two inputs are usually labeled A and B and the output is labeled X. As with the operational amplifier, there are additional connections to the gate for ground and +5 V power, but, for simplicity, these connections are not generally shown.

The behavior of an AND gate is shown in Figure 12.3a, along with a simple circuit model using switches. Each input, A or B, can be either a 0 (switch open) or a 1 (switch closed). The output X will be a 1 only if both A AND B are 1s (both switches closed). If either A or B is a 0 (one switch open), X will be a 0. The four possible input conditions, along with the corresponding output for each, are shown in the associated table of Figure 12.3 called a **truth table**.

The behavior of an OR gate is shown in Figure 12.3b, along with its switch model and truth table. In this case, the output will be a 1 if

BASIC GATING FUNCTIONS AND AND OR

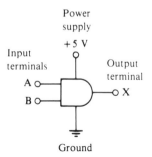

AND Gate

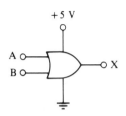

OR Gate

FIGURE 12.2. The two basic gating functions are called AND and OR. These gates have two input terminals, A and B, and an output terminal X. Each combination of 0 and 1 inputs to A and B yields a specific 0 or 1 output at X. The power supply connection of +5 V for the 1 levels and ground are generally not shown.

either A OR B (or both) is a 1. Hence, three out of four possible inputs will yield a 1 for the OR gate, while only one out of four possible inputs will yield a 1 for the AND gate.

NOT Gate

To make things a little more complicated, there is also a one-input gate called a **NOT** gate. The output of a NOT gate is simply the opposite of its input, as shown by its truth table (Figure 12.4).

FIGURE 12.3. The electronic behavior of AND and OR gates can be illustrated by a simple circuit model using switches. The digital behavior is described by a truth table that gives the 0 or 1 output for each possible combination of 0 and 1 inputs.

ELECTRONIC BEHAVIOR OF AND AND OR GATES

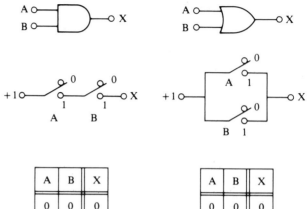

A	B	X
0	0	0
0	1	0
1	0	0
1	1	1

(a) AND Gate Symbol, Circuit Model, and Truth Table

A	B	X
0	0	0
0	1	1
1	0	1
1	1	1

(b) OR Gate Symbol, Circuit Model, and Truth Table

NOT GATE SYMBOL AND TRUTH TABLE

A	X
0	1
1	0

FIGURE 12.4. A third basic gate, called NOT, has only one input. Its output is the opposite of its input.

NAND and NOR Gates

Placing a NOT gate at the outputs of AND and OR gates creates two more basic gates: the **NAND** gate (NOT AND) and the **NOR** gate (NOT OR). These two basic gates, along with

their truth tables, are shown in Figure 12.5. Note that the symbols for the NAND and NOR gates show the small circle of the NOT gate symbol at their output terminal to distinguish them from AND and OR gates.

The importance of the NAND and NOR gates is that they are the simplest to fabricate in integrated circuit form. Futhermore, it is possible to construct all of the computer logic functions using only NOR gates or only NAND gates. Thus, while it may be simpler to design and analyze digital circuits using AND or OR gates, you should also practice using NAND and NOR gates, as they are more likely to be used in actual circuit construction.

So far, gates with only one or two inputs have been discussed. However, more complicated gates can be constructed with more than two inputs. Figure 12.6 shows three-input AND and OR gates, along with their truth tables. Comparable three-input (or more) NAND and

NAND AND NOR GATES

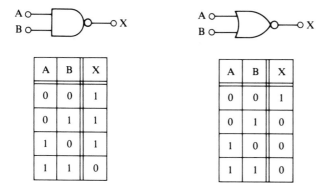

A	B	X
0	0	1
0	1	1
1	0	1
1	1	0

NAND Gate Symbol
and Truth Table

A	B	X
0	0	1
0	1	0
1	0	0
1	1	0

NOR Gate Symbol
and Truth Table

FIGURE 12.5. Combining a NOT gate with AND and OR gates creates two other common gates called NAND and NOR. Using NAND and NOR gates is conceptually more difficult than AND and OR, but they are more commonly used due to their simplicity of fabrication as ICs.

THREE-INPUT GATES

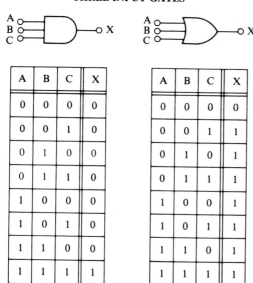

A	B	C	X
0	0	0	0
0	0	1	0
0	1	0	0
0	1	1	0
1	0	0	0
1	0	1	0
1	1	0	0
1	1	1	1

Three-Input AND Gate
Symbol and Truth Table

A	B	C	X
0	0	0	0
0	0	1	1
0	1	0	1
0	1	1	1
1	0	0	1
1	0	1	1
1	1	0	1
1	1	1	1

Three-Input OR Gate
Symbol and Truth Table

FIGURE 12.6. Gates with more than two inputs are also possible. Shown are three-input AND and OR gates as examples of the multiple-input gates – AND, OR, NAND, and NOR – that are used in digital circuit design.

NOR gates can also be constructed. Note from the truth tables that in every case, there is a unique output condition that can be created by only one identical combination of inputs.

12.4 APPLICATION: VENDING MACHINE

We now have enough information about digital principles to design an electronic system that will solve a simple control problem. While this example uses only two gates compared to the millions used in a computer, it is typical of how 0–1 decisions made by gates are used in processing data.

The example used is that of a ten-cent vending machine. Suppose we want to construct a vending machine that will release an item worth ten cents when the proper combination of coins is put in the machine. If we eliminate the use of pennies, then the customer could insert either 2 nickels or 1 dime. These possibilities can be expressed using AND and OR statements as follows:

If (1 nickel AND 1 nickel) OR (1 dime), then (item released).

When a condition is met, we will assign it the digit 1, and, when it is not, a 0. Thus, the two nickel conditions can be inputs to an AND gate whose output will be a 1 if both nickels are inserted and a 0 otherwise. This output can then serve as the input to an OR gate whose other input is the dime. When either 2 nickels OR 1 dime is present, the output of the OR gate will be a 1 that will release the item. The circuit diagram for the design solution is shown in Figure 12.7, along with its truth table.

To implement this design in practice, of course, would require a little more thought. Coin slots would have to be designed that separate nickels from dimes and coin-activated switches would have to be installed that apply a + 5 V signal to the proper input when a coin is present to indicate a 1 condition. The + 5 V output of the OR gate, in turn, would have to activate a mechanism that would release the desired item when its 1 conditions were met. However, these considerations are outside our design responsibilities here.

Note also from the truth table that the last three conditions for which more than ten cents is inserted will also cause the item to be released. As business persons, we would have to decide whether or not we wanted to keep the extra money. If we were honest and valued return business, we would probably want to return the excess money to the customer through the coin return slot. This design problem would read:

If (1 dime) AND (1 nickel OR 1 nickel), then (return nickels).

The design solution for this problem is shown in Figure 12.8, along with its truth table. Note that the last three conditions match the excess money conditions in Figure 12.7 and will cause the excess nickels to be released. To implement the total solution would require the coin input

CONTROLLING A VENDING MACHINE WITH GATES

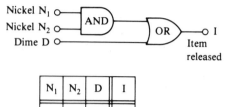

N_1	N_2	D	I
0	0	0	0
0	1	0	0
1	0	0	0
1	1	0	1
0	0	1	1
0	1	1	1
1	0	1	1
1	1	1	1

Item released

FIGURE 12.7.   The figure shows how gates can be used to control a ten-cent vending machine. As the truth table shows, either 2 nickels OR 1 dime will cause a ten-cent item to be released. In practice, the presence of a coin would cause + 5 V to be applied to the proper gate input.

switches to be applied to both circuit inputs so that all desired output conditions are achieved.

This example has shown the power of digital electronic circuits. With only two types of gates, we have solved a useful control problem. Many other comparable examples can be imagined. In addition to more complicated vending machines that include a third coin possibility, the quarter, there are many machine controls that utilize this same digital approach. Some examples are: safety controls that will not permit a machine to be turned on unless certain safety conditions are met—for example, safety belts fastened, lights or alarms activated, safety

SOLVING THE COIN RETURN PROBLEM WITH GATES

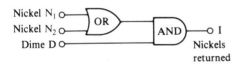

N_1	N_2	D	I
0	0	0	0
0	1	0	0
1	0	0	0
1	1	0	0
0	0	1	0
0	1	1	1
1	0	1	1
1	1	1	1

Nickels returned

FIGURE 12.8.   For the last three conditions of the truth table of Figure 12.7, more than ten cents will also release the item. The gates above will return the nickels when this happens. This is one of many examples of how logic gates can be used to solve a control problem.

guards in place, and so on—or controls that turn something off when certain conditions have been achieved—for example, cooking temperature, pressure, time, and so forth. Some further examples are given as problems at the end of the chapter, but applications in daily life are everywhere.

As design problems get more complicated, and especially when arithmetic calculations and other data processing tasks are considered, the simple if-then, AND-OR, truth table approach we have used thus far becomes unwieldy. This situation can easily be appreciated by considering a thirty-five-cent vending machine for

which there are 6 different coin combinations that will comprise thirty-five cents and many more excess coin combinations for which money should be returned.

To deal with the increasing complexity, digital circuit designers turned to mathematics for shorthand ways of expressing their problems and rules for manipulating the mathematical expressions to find simple solutions. These mathematical techniques are introduced next to give you an appreciation of the ways in which modern digital circuits are designed and the language and basic concepts that underlie their performance. However, it is well beyond the scope of this book to train you as a digital designer, and the information presented thus far is more than adequate for many practical problems you may encounter.

12.5 BOOLEAN ALGEBRA

The mathematical techniques used in digital design problems were developed by the English mathematician George Boole around 1854 to deal with true–false logic statements. His goal was to find a mathematical way of proving logical propositions by posing a series of questions whose answers were either true or false. He showed that the true–false questions could be expressed as independent variables and the true–false answers as dependent variables in algebraic equations that could be "solved" using certain mathematical rules.

In digital circuit design, of course, the true–false questions are the gating operations and the true–false answers are expressed by the digits 0 and 1. The mathematical notation for expressing the problems and the rules for solving them are those developed by Boole and are called **Boolean algebra**. Because this algebra and its notations come directly from mathematical logic, the terms **logic variable** and **logic op-**

eration are generally used instead of *digital variable* and *digital operation*.

Gate Representations

In Boolean algebra, the symbol for the digital AND operation is a dot ($\cdot$). Hence, the statement A AND B is written:

A AND B = A $\cdot$ B

or, simply:

A AND B = AB

The mathematical expression for the AND operation can then be written:

$$A \cdot B = X \qquad (1)$$

which has the mathematical definition:

If A = 1 AND B = 1, then X = 1
If not,   then X = 0   (2)

Thus, we can write the truth table of the AND operation in mathematical notation as:

$$
\begin{aligned}
0 \cdot 0 &= 0 \\
0 \cdot 1 &= 0 \\
1 \cdot 0 &= 0 \\
1 \cdot 1 &= 1 \qquad (3)
\end{aligned}
$$

From this table, we see that the AND operation is a form of multiplication.

Similarly, the symbol for the OR operation is a plus (+) sign. Hence, the statement A OR B is written:

A OR B = A + B

and the mathematical expression for the OR operation is written:

$$A + B = X \qquad (4)$$

with the mathematical definition:

If A = 1 OR B = 1, then X = 1
If not,   then X = 0   (5)

The truth table for the OR operation then becomes:

$$
\begin{aligned}
0 + 0 &= 0 \\
0 + 1 &= 0 \\
1 + 0 &= 0 \\
1 + 1 &= 1 \qquad (6)
\end{aligned}
$$

Thus, the OR operation is a special form of addition with the important difference that 1 + 1 = 1, not 2!

EXAMPLE 1: Express the ten-cent vending machine problem in Boolean notation and evaluate the result.

Solution: For N_1 and N_2 representing the nickels, D representing the dime, and I representing the item being released, the following Boolean algebraic expression represents the problem:

$$\mathbf{I = D + (N_1 \cdot N_2) = D + N_1 N_2}$$

I is evaluated for all possible combinations of N_1, N_2, and D as:

N₁	N₂	(N₁ · N₂)	+ D	= I
0	0	0	0	0
0	1	0	0	0
1	0	0	0	0
1	1	1	0	1
0	0	0	1	1
0	1	0	1	1
1	0	0	1	1
1	1	1	1	1

Compare this truth table with that of Figure 12.7.

The NOT operation produces the opposite value of a variable. In Boolean algebra, the opposite value of a variable is called its **complement** and is indicated by a bar drawn over the variable:

$$\text{NOT } X = \overline{X} \qquad (7)$$

The mathematical definition of the NOT (or inversion) operation is then:

$$\begin{aligned} \text{If } X &= 1, \text{ then } \overline{X} = 0 \\ \text{If } X &= 0, \text{ then } \overline{X} = 1 \qquad (8) \end{aligned}$$

Using the complement notation, the NAND operation can now be defined as:

$$\overline{A \cdot B} = X \qquad (9)$$

with the definition:

$$\begin{aligned} &\text{If } A = 1 \text{ AND } B = 1, \text{ then } X = 0 \\ &\text{If not}, \qquad\qquad\qquad \text{then } X = 1 \quad (10) \end{aligned}$$

and the NOR operation can be defined as:

$$\overline{A + B} = X \qquad (11)$$

with the definition:

$$\begin{aligned} &\text{If } A = 1 \text{ OR } B = 1, \text{ then } X = 0 \\ &\text{If not}, \qquad\qquad\qquad \text{then } X = 1 \quad (12) \end{aligned}$$

Thus, we see that in Boolean algebra there are three basic logic operations: AND, which is identical to normal multiplication; OR, which is similar to addition except that the sum of any two digits can at most be 1; and NOT, which simply changes a digital variable to its opposite value. Figure 12.9 summarizes the Boolean expression for the AND, OR, and NOT gating operations, and Figure 12.10, for the NAND and NOR operations.

Except for the restriction to the two digital possibilities 0 and 1, the other laws of "normal" algebra apply to the evaluation of Boolean algebraic equations:

$$\begin{aligned} &\text{Commutative law:} \\ &A + B = B + A \\ &AB = BA \qquad\qquad\qquad\qquad (13) \\[4pt] &\text{Associative law:} \\ &A + (B + C) = (A + B) + C \\ &A(BC) = (AB)C \qquad\qquad\quad (14) \\[4pt] &\text{Distributive law:} \\ &A(B + C) = AB + AC \qquad\quad (15) \end{aligned}$$

Logic Flow Diagrams

These mathematical expressions can be directly converted into circuit diagrams using the appropriate gates. The circuit diagrams are called **logic flow diagrams** and can be directly constructed as working circuits using the appropri-

BOOLEAN EXPRESSIONS FOR AND, OR, AND NOT

Gating Operation	Boolean Notation	Truth Table
A AND B = X	$A \cdot B = X$	$A \cdot B = X$
		$0 \cdot 0 = 0$
		$0 \cdot 1 = 0$
		$1 \cdot 0 = 0$
		$1 \cdot 1 = 1$
A OR B = X	$A + B = X$	$A + B = X$
		$0 + 0 = 0$
		$0 + 1 = 1$
		$1 + 0 = 1$
		$1 + 1 = 1$
NOT X	$\overline{X}$	$X \quad \overline{X}$
		$0 \quad 1$
		$1 \quad 0$

FIGURE 12.9. The gating operations can be expressed mathematically using Boolean algebra. In this system, AND is a form of multiplication, OR is a form of addition, and NOT is an operation that produces the complement of a variable. Shown are the Boolean notation and truth table results for these operations. Note that for OR, $1 + 1 = 1$, not 2.

BOOLEAN EXPRESSIONS FOR NAND AND NOR

Gating Operation	Boolean Notation	Truth Table
A NAND B	$\overline{A \cdot B} = X$	$\overline{A \cdot B} = X$
		$\overline{0 \cdot 0} = 1$
		$\overline{0 \cdot 1} = 1$
		$\overline{1 \cdot 0} = 1$
		$\overline{1 \cdot 1} = 0$
A NOR B	$\overline{A + B} = X$	$\overline{A + B} = X$
		$\overline{0 + 0} = 1$
		$\overline{0 + 1} = 0$
		$\overline{1 + 0} = 0$
		$\overline{1 + 1} = 0$

FIGURE 12.10. Similar Boolean algebraic expressions exist for the NAND and NOR operations. These expressions are simply the complement of the AND and OR operations.

ate electronic gate components. The correlation between the mathematical expressions and the logic flow diagrams is illustrated in the following examples.

EXAMPLE 2: Write a Boolean algebraic expression for the logic gate connections shown in Figure 12.11.

Solution: The output expressions are:

(a) $\mathbf{X = AC + BC}$

(b) $\mathbf{X = AC + BC + AD + BD}$

(c) $\mathbf{X = \overline{AB} + CDE + \overline{FG}}$

EXAMPLE 3: Draw equivalent logic flow diagrams for Figures 12.11a and 12.11b.

Solution: (a) The logic expression for Figure 12.11a can be rewritten as:

$$X = (A + B)C$$

The logic flow diagram for this expression is shown in Figure 12.12a. Note that by using this equivalent circuit, we have reduced our gate count by 1, an important cost-saving accomplishment for the digital circuit designer. The equivalency of the circuits can be verified for all possible input conditions using a truth table. This verification is left as a problem.

(b) The logic expression for Figure 12.11b can be rewritten as:

$$X = (A + B)C + (A + B)D$$

EXAMPLE 2

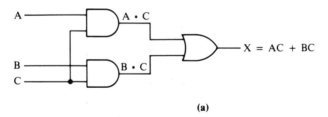

(a)

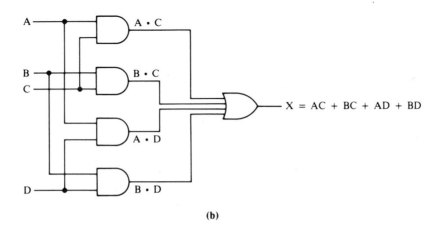

(b)

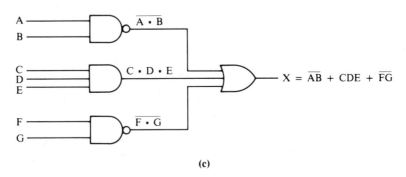

(c)

FIGURE 12.11.   Boolean algebraic expressions can be directly converted into working circuits by substituting the appropriate gate and making the proper input and output connections. Shown are the logic flow diagrams corresponding to the expressions of Example 2.

EXAMPLE 3

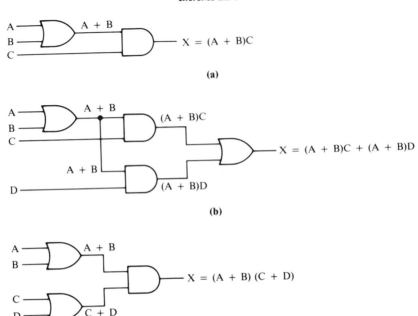

(a)

(b)

(c)

FIGURE 12.12.   By rearranging terms in a Boolean expression, it is often possible to simplify the logic flow diagram and thereby reduce the number of gates. The diagrams above are equivalent simplifications of Figures 12.11a and 12.11b.

or

$$X = (A + B)(C + D)$$

The logic flow diagrams for these two alternatives are shown in Figures 12.12b and 12.12c. Again, note the reductions in component count.

DeMorgan's Theorem

Two important equivalency rules in Boolean algebra are often used to simplify expressions. These rules are called DeMorgan's theorem, which states:

The complement of an OR operation is equal to the AND operation on the complements. In Boolean notation this equivalency is expressed as:

$$\overline{A + B} = \overline{A} \cdot \overline{B} \qquad (16)$$

The complement of an AND operation is equal to the OR operation on the complements:

$$\overline{A \cdot B} = \overline{A} + \overline{B} \qquad (17)$$

Note that when evaluating these expressions,

the operation underneath the complement bar must be evaluated *before* the complement is taken.

Using DeMorgan's theorem, we have the following alternative expressions for the NAND and NOR operations:

$$\text{NAND:} \quad \overline{A \cdot B} = \overline{A} + \overline{B} = X \quad \textbf{(18)}$$
$$\text{NOR:} \quad \overline{A + B} = \overline{A} \cdot \overline{B} = X \quad \textbf{(19)}$$

The DeMorgan equivalent symbols for the NAND and NOR gates are shown in Figure 12.13, along with their truth tables.

DEMORGAN EQUIVALENT FORMS FOR NAND AND NOR

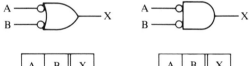

A	B	X
0	0	1
0	1	1
1	0	1
1	1	0

NAND Gate Symbol and Truth Table

A	B	X
0	0	1
0	1	0
1	0	0
1	1	0

NOR Gate Symbol and Truth Table

FIGURE 12.13. DeMorgan's theorem gives two equivalency rules that are often helpful in simplifying Boolean expressions and logic flow diagrams. For example, the theorem shows that the alternative forms for NAND and NOR shown here are equivalent to those of Figure 12.5.

EXAMPLE 4: Verify the equivalent expressions for the NAND operation using Boolean algebra.

Solution: The evaluation of $\overline{A \cdot B}$ for all possible combinations of A and B is:

A	B	A · B	$\overline{A \cdot B}$
0	0	0	1
0	1	0	1
1	0	0	1
1	1	1	0

The evaluation of $(\overline{A} + \overline{B})$ for all possible combinations of A and B is:

A	B	$\overline{A}$	$\overline{B}$	$\overline{A} + \overline{B}$
0	0	1	1	1
0	1	1	0	1
1	0	0	1	1
1	1	0	0	0

Note the identity of the last column of the tables for each expression.

EXAMPLE 5: Design an equivalent logic diagram for the excess nickel return problem of the ten-cent vending machine (a) using only NAND gates and then (b) using only NOR gates.

Solution: The Boolean algebraic expression for the return of excess nickels is:

$$I = (N_1 + N_2) \cdot D$$

This expression says that if one OR two nickels is present AND a dime is present, then return the two nickels.

(a) By DeMorgan's theorem, there are two

FIGURE 12.14. The figure shows how the DeMorgan equivalent NAND and NOR gates can be used in a logic flow diagram for the ten-cent vending machine.

EXAMPLE 5

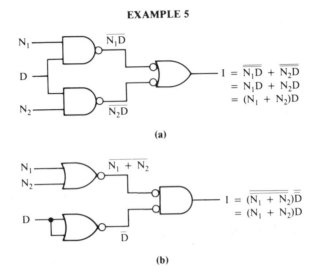

(a)

(b)

alternative but equivalent NAND operations given by Equation 18:

$$X = \overline{A \cdot B}$$
$$X = \overline{A} + \overline{B}$$

Rewriting the top equation for I gives:

$$I = N_1 D + N_2 D$$

An equivalent expression using NAND gates is:

$$I = \overline{\overline{N_1 D} + \overline{N_2 D}}$$

where the upper complement bar inverts the result of $\overline{N_1 D} + \overline{N_2 D}$. The logic flow diagram for this expression using NAND gates is shown in Figure 12.14a. Verification of the result using a truth table is left as Problem 12.

(b) The two alternative NOR operations are given by Equation 19:

$$X = \overline{A + B} = \overline{A} \cdot \overline{B}$$

An expression equivalent to the logic equation for I using NOR operations is:

$$I = \overline{(N_1 + N_2)} \cdot \overline{D}$$

The logic flow diagram for this equation is shown in Figure 12.14b. Note that the inversion of D using a NOR gate is achieved by connecting the two inputs of a NOR gate together. While these circuits have a higher component count than the original circuit, the total cost of construction may be less because of the greater simplicity of fabricating NAND and NOR gates.

12.6 APPLICATION: DIGITAL MULTIPLEXER

A final example of a practical application of digital logic gating is the **digital multiplexer**, a basic circuit found in essentially every data processing system. In a later chapter, you will

MULTIPLEXING

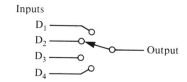

(a) Selector Switch

FIGURE 12.15. A control application of gates in computers is multiplexing. Multiplexing acts like a rotary switch where data on several inputs can be individually selected for a single output. The truth table for the multiplexer shows that for any switch input S = 1, the data on the corresponding data input D will appear at the output.

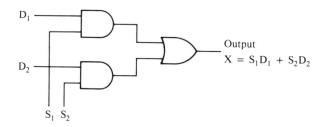

Output
$X = S_1 D_1 + S_2 D_2$

(b) Digital Multiplexer

NAND GATE MULTIPLEXER

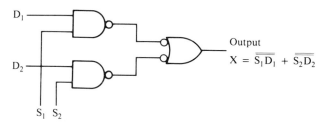

Output
$X = \overline{\overline{S_1 D_1}} + \overline{\overline{S_2 D_2}}$

FIGURE 12.16. Multiplexers that use NAND gates are the most common. Essentially, the NOT inputs of the second NAND gate cancel the NOT outputs of the first to produce the same result as the diagram of Figure 12.15.

see the important role of multiplexing in digital signal processing, but, for the moment, simply think of a multiplexer as a multi-input switch, much like the rotary switch shown in Figure 12.15a. The purpose of a digital multiplexer is to permit digital information coming from several channels to be switched or selected onto a single output channel for further operations.

The logic diagram for a single, two-input multiplexer using AND and OR gates is shown in Figure 12.15b. The data to be selected are applied to the **data input terminals** D_1 and D_2, while digital selection signals are applied to the **select inputs** S_1 and S_2. The combination of a select signal $S_1 = 1$ and a data input signal D_1

$= 1$ will cause the output of the AND gate to be $S \cdot D = 1$. With no other select signals present, all other inputs to the OR gate will be 0, and its output will be $X = S_1 \cdot D_1 + S_2 \cdot D_2 + \cdots = 1 + 0 + 0 + \cdots = 1$.

If $S_1 = 1$ but $D = 0$, then the output of the OR gate will be 0. Thus, a select signal applied to any select input will "read" the digital data on the corresponding data input channel and pass it on to the next stage of the logic system. The same process will work for any number of input data channels. Eight-channel multiplexers are the most common multiplexers available.

The multiplexer circuits shown thus far used AND or OR gates. Figure 12.16 shows an equivalent circuit using NAND gates. Note that the NOT outputs of the input and select gates are cancelled by the NOT inputs of the output gate. As before, the use of NAND gates is preferred due to their lower cost of fabrication. Common applications of multiplexer circuits are reading data "stored" in certain computer circuits or routing data inputs from several sources to a single output line.

The reverse of multiplexing — that is, **demultiplexing** — is an equally important logic circuit task. In demultiplexing, data from a single input channel are switched to one of several output channels for such tasks as storage or routing. The importance of these applications will become clearer in the next chapter as we begin to explore the data processing circuits of digital computers.

While the mathematics of Boolean algebra get increasingly more complex, these examples have illustrated the basic principles and practice. More advanced and complete texts can provide a greater depth and broader range of applications for those interested. The final section of this chapter describes some of the variety of electronic components that have been developed to achieve the gating operations and the basic electronic principles on which their behavior depends.

12.7 ELECTRONIC SWITCHES AS GATES

The rapid expansion of digital electronic systems into our society has depended partly on the power of digital logic to solve data processing and control problems and partly on the rapid advances in technology that brought their cost within the range of affordability. The number of gates required to perform even modest amounts of data processing can quickly get into the millions, so that the cost per gate must be extremely low to permit widescale applications.

As with digital logic, the technology of gate design and fabrication can get very sophisticated, but the discussion here is limited to basic principles and to the language that is used to describe the field.

Mechanical Switch Gates

The principal operating component of all digital gates is a **switch.** In modern gates, these switches are semiconductor devices, but in the earlier systems, they were electromechanical devices, such as relays. In fact, relays are still used as the switching devices in very high power applications.

Figure 12.3 showed the basic way in which two switches in series and in parallel can be used to achieve an AND gate and an OR gate. Figure 12.17 shows how these gating functions can be achieved in a real circuit using mechanical switches and resistors. Note that the logic state 1 is indicated by a $+5$ V voltage level and the logic state 0 by ground.

For the AND gate (Figure 12.17a), if both switches A AND B are open, then voltage appears across R_L. If $R_L \gg R$, the voltage at X will be essentially the full supply voltage of $+5$ V (logic state 1). If either A or B is closed, then the top of R_L is grounded and the voltage at X is 0 V (logic state 0). These possibilities are shown in the accompanying truth table.

Similarly, for the OR gate (Figure 12.17b), if either A OR B is closed, then $+5$ V appears across R_L at X (logic state 1). If both A and B are open, then no voltage appears across R_L and the X voltage is 0 V (logic state 0). Again, these possibilities are shown by the truth table.

Relay Gates

While the circuits of Figure 12.17 show the switches to be mechanical ones, they could equally well be relays. Figure 12.18 shows AND and OR gates constructed with relays. As a commercial unit, the gates would simply be a

AND AND OR GATES USING SWITCHES

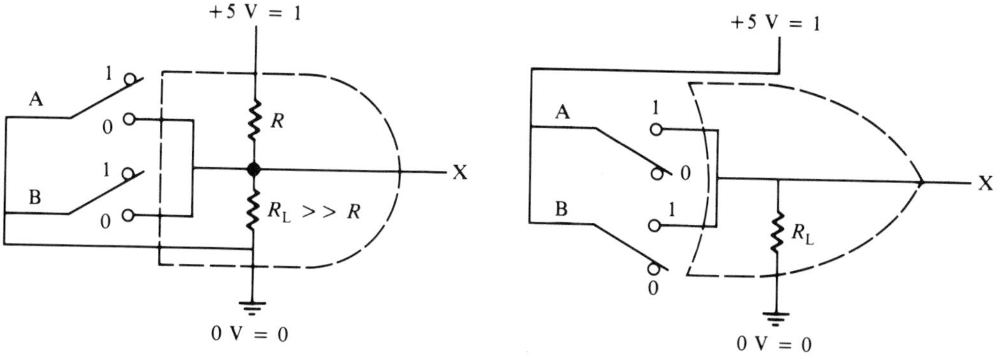

A	B	X
0	0	0
0	1	0
1	0	0
1	1	+5 V

(a) AND

A	B	X
0	0	0
0	1	+5 V
1	0	+5 V
1	1	+5 V

(b) OR

FIGURE 12.17.   The principal operating component of a gate is a 0–1 switch. The figure shows how AND and OR gates can be constructed from ordinary mechanical switches and resistors.

box with relays and resistors for its internal components and with five external connections: input terminals A and B, output terminal X, a power supply connection for +5 V, and ground. These units could be connected together to form the logic flow diagrams described earlier. The relays would be designed to be activated by the +5 V of the power supply, and the power supply would need sufficient power capability to supply current to all of the relay coils.

Gate Performance Characteristics

Early digital computers used relay gates much like those shown in Figure 12.17. In considering

digital computers constructed from thousands of such gates, some of the problems associated with their use become immediately apparent. For example, the gates are physically quite large, are fairly expensive, consume large amounts of electrical power, dissipate considerable heat, and have questionable reliability.

These are some of the factors that must be considered when selecting and using gates. In addition to size, cost, power dissipation, and reliability are the additional performance characteristics of speed and noise immunity. In today's computers, the issue of switching speed is a key performance factor because it determines how rapidly a calculation or data processing step can be made. Since millions of gating steps

AND AND OR GATES USING RELAYS

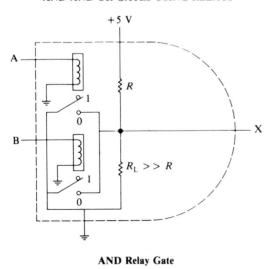

AND Relay Gate

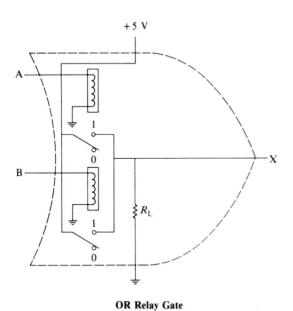

OR Relay Gate

FIGURE 12.18. The mechanical switch design of Figure 12.16 can be duplicated using electromechanical relays to produce relay gates that are still used in high-power control applications. Each gate has five external terminals: inputs A and B, output X, +5 V power supply, and ground.

PROPAGATION DELAY TIME

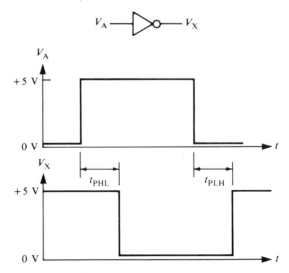

FIGURE 12.19. One of the important performance characteristics of gates is their propagation delay time: the time it takes for the output to change logic levels after a change of input. For relay gates, this time may be 1/100 s, but for semiconductor gates, it can be as fast as 10 ns.

may be required for a given data processing task, each gate transition must be extremely fast so that the total waiting time for the result is not excessive.

For gates made of relays, the switching time is on the order of 1/100 s at best. If 10,000 successive switching steps are required for a given data processing step, it would take about 100 s or two minutes for the result. Imagine how much fun a video game would be if each move took two minutes!

Semiconductor gates used in contemporary circuits have switching times of 10 ns (0.000000010 s) or less, so that 10,000 gating operations would take only 0.0001 s. Clearly, this speed makes digital computers much more practical devices.

To understand better what is meant by the switching speed of a gate, consider the NOT gate of Figure 12.19. For a step voltage input

REAL GATE WAVEFORMS AND
PROPAGATION DELAY TIME

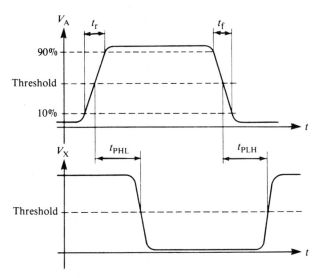

FIGURE 12.20.   In real gates, the waveforms are not as square as those of Figure 12.18. Hence, rise times, fall times, and transition threshold points must be defined as shown. Note that the propagation delay from high to low (t_{PHL}) may be different from low to high (t_{PLH}).

signal from 0 V to +5 V, some finite time later, the output of the NOT gate will switch from +5 V to 0 V. The time between the input transition and the output transition is called the **propagation delay time**, or, more commonly, the **switching speed.**

In a real system, of course, the waveforms are not square, so that arbitrary rise times, fall times, and transition points on each waveform must be defined. In Figure 12.20, for example, the rise and fall times are the times between 10% and 90% of full transition, and the transition, or **threshold point**, is the input voltage level that produces the same output voltage level. Note in Figure 12.20 that the propagation delay from high to low (t_{PHL}) need not be the same as from low to high (t_{PLH}).

Another important factor is the immunity of the gate from noise. While the 0–1 logic levels have nominal voltage values of 0 V and +5 V, respectively, the voltages need not be exactly these values. For example, the 0 V requirement may be met by a range of voltages from 0 V to 1 V, and the +5 V by a similar range from 4 V to

5 V. Thus, an input signal plus noise can be anywhere in that range and not cause a false logic change. The width of the voltage range that defines the logic level—that is, within which a logic transition will not take place—is called the **dc noise margin.** The wider the noise margin, the greater the immunity of the gate from spurious transitions due to noise voltages on the logic voltages.

Noise margins range from a few tenths of a volt to several volts depending on the logic family and logic state. It is this relatively high immunity to noise that makes digital systems preferred over analog systems, especially for long-range data transmission where transmission line noise can be a significant factor.

A final gate characteristic is called **fan-out capability.** As we saw earlier, the output of a gate may be applied to the inputs of several gates. Since each input must draw some power to function properly, there is a limit to the number of inputs a given output can supply. Because all inputs do not require the same amount of power, the input requirements and output ca-

pabilities for each logic family are often specified in terms of a **unit load** that represents the
lowest input requirement in the logic family.
The number of input unit loads that an output
can power is its fan-out capability.

12.8 SEMICONDUCTOR LOGIC FAMILIES

This section introduces the various semiconductor logic families that are available today,
including resistor–transistor logic (RTL), diode–
transistor logic (DTL), transistor–transistor
logic (TTL), and the newer logic families made
from metal oxide semiconductors (MOS). As
these various names suggest, there has been
considerable research and development effort
to produce gates of lower cost, higher speed,
lower power dissipation, and the like.

Because this is just an introduction, only
the basic operating principles are described—
that is, how the logic family produces switching
actions. For a fuller understanding of how the
switching action is used to form all of the gate
alternatives and the detailed performance characteristics of each, a more advanced text should
be consulted.

Resistor–Transistor Logic

One of the earliest of the logic families is **resistor–transistor logic** (RTL). Figure 12.21 shows a
typical RTL inverter circuit or NOT gate. Note
that the supply voltage for this logic family is
normally +3.6 V.

When the input voltage V_A is low (for this
logic family, low means less than about +0.4
V), the transistor does not conduct and acts like
an open circuit. Thus, the output will be high—
nearly +3.6 V. When the input voltage is high
(greater than about +1.1 V), the transistor will
conduct with a saturation voltage of about
+0.3 V across its output terminal. Hence, the

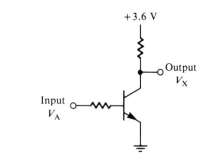

RTL NOT GATE

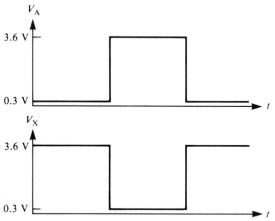

**FIGURE 12.21. An early form of semiconductor
logic used resistors and transistors (RTL). The circuit shows a typical RTL NOT gate. When V_A is
high, the transistor conducts and V_X is low.**

output will be low. Thus, the two logic levels
are:

Input Logic Level	Input Voltage	Output Logic Level	Output Voltage
0	$0 < V_A < 0.4$ V	1	$1.1 < V_X < 3.6$ V
1	$1.1 < V_A < 3.6$ V	0	$0 < V_X < 0.4$ V

This transistor switch behavior is also described by the voltage transfer characteristic for
the RTL inverter shown in Figure 12.22. The

RTL VOLTAGE TRANSFER CHARACTERISTIC

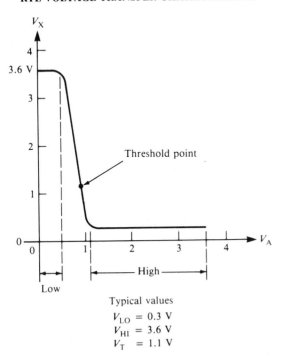

Typical values
$V_{LO} = 0.3$ V
$V_{HI} = 3.6$ V
$V_T = 1.1$ V

FIGURE 12.22. The performance of the RTL inverter is also described by its voltage transfer characteristic. Note that the low logic level has a range of about 0 V to 0.4 V, and the high level, about 1.1 V to 3.6 V. The narrowness of the low range makes for poor immunity from noise.

figure also shows one of the principal disadvantages of the RTL family: its relatively narrow voltage margin for the 0 and 1 logic levels. The low voltage is typically about 0.3 V which is very close to the limit of 0.4 V. Thus, its noise immunity is not as good as that of other logic families.

The result of adding a second transistor to the inverter, as shown in Figure 12.23, is a NOR gate. That is, if either A or B is high, one of the transistors will conduct and the output will be low. If both A and B are low, neither transistor will conduct and the output will be high. These are the truth table requirements of a NOR gate.

RTL NOR GATE

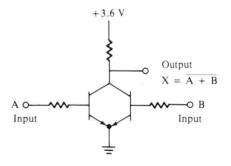

FIGURE 12.23. Adding a second transistor to an RTL inverter makes an RTL NOR gate. If either A or B is high, the output is low. If both A and B are low, the output is high.

Diode–Transistor Logic

A circuit diagram for a NOT gate in the **diode-transistor logic** (DTL) family is shown in Figure 12.24. Note that the supply voltage for DTL is +5 V. Also, recall that for silicon semiconductors, the forward-bias voltage is about +0.6 V — that is, a junction will not conduct until the voltage across it exceeds +0.6 V.

The operation of this circuit is as follows. If V_A is low — that is, less than about 1.2 V — the diode and transistor are back biased and do not conduct. Hence, the output voltage V_X is near the supply voltage of +5 V — that is, high. As V_A exceeds 1.2 V, however, the diode and transistor start to conduct and the output goes low (about +0.3 V). Thus, the two levels are:

Input Logic Level	Input Voltage	Output Logic Level	Output Voltage
0	$0 < V_A < 1.2$ V	1	$1.5 < V_X < 5$ V
1	$1.5 < V_A < 5$ V	0	$0 < V_X < 1.2$ V

The voltage transfer characteristic for the DTL inverter is shown in Figure 12.25. In practice, the low voltage level is usually about +0.3

DTL NOT GATE

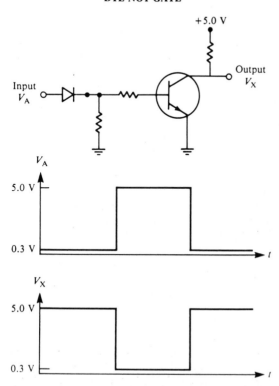

FIGURE 12.24. In DTL logic, gates are con-structed from diodes and transistors. For the NOT gate, when V exceeds about 1.5 V (high), the diode and transistor conduct and the output is low (about 0.3 V). When V is less than 1.2 V (low), nei-ther the diodes nor the transistor conducts, and the output is high.

DTL VOLTAGE TRANSFER CHARACTERISTIC

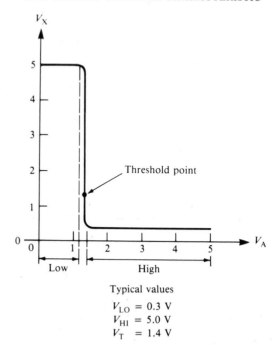

Typical values

$V_{LO} = 0.3$ V
$V_{HI} = 5.0$ V
$V_T = 1.4$ V

FIGURE 12.25. The voltage transfer characteristic of the DTL inverter shows the low range to be about 0 V to 1.2 V and the high range to be about 1.5 V to 5 V. Because the low and high voltages are typically 0.3 V and 5 V, respectively, DTL logic has good noise margins.

V and the high level is near $+5$ V, which gives DTL gates very good dc noise margins.

The propagation delay time for a typical DTL gate is about 30 ns, which is comparable to that of RTL gates. This speed is moderately slow for high-speed computer operation, but more than adequate for such applications as hand-held calculators where the calculation times are well within human reaction time.

When a second diode is added to the DTL inverter input, as shown in Figure 12.26, the result is a NOR gate. That is, if either A or B is high, then one diode and the transistor will conduct and the output will be low. If both A and B are low, neither of the diodes nor the transistor will conduct and the output will be high.

Transistor–Transistor Logic

Transistor–transistor logic (TTL) evolved from DTL, but is unique in that it is fabricated only in integrated circuit form and has no discrete component counterpart. Figure 12.27 shows the basic circuit for a TTL NOT gate. While it is

DTL NOR GATE

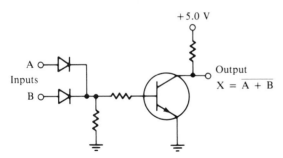

FIGURE 12.26. Adding a second diode to the DTL inverter makes a NOR gate. If either A or B is high, one diode and the transistor will conduct and the output is low. If both A and B are low, neither the diodes nor the transistor will conduct, and the output is high.

TTL NOT GATE

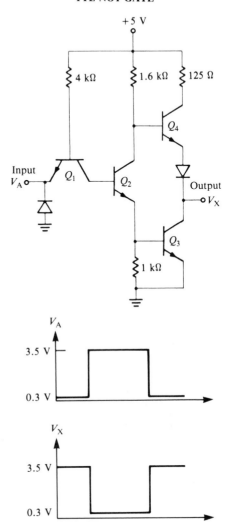

FIGURE 12.27. Gates in transistor–transistor logic (TTL) are somewhat complicated and are fabricated only as integrated circuits. Like all NOT gates, when the input voltage V of the circuit is low, the output voltage V is high and vice versa. Transistors Q_3 and Q_4 are added to increase the gate's switching speed.

not necessary to understand the detailed operation of this circuit, simply note that when the input voltage is low, input transistor Q_1 will conduct, causing Q_2 to shut off (high resistance). As a result, Q_4 conducts and Q_3 shuts off, so that a high voltage of about $V = 3.5$ V appears at the output. The right-hand part of the circuit, involving transistors Q_3 and Q_4, is used to increase the switching speed of the TTL gate.

The voltage transfer characteristic for a typical TTL gate is shown in Figure 12.28. Typical low and high voltage levels are $+0.3$ V and $+2.0$ V, respectively, so that TTL noise margins are not as good as DTL margins. On the other hand, TTL propagation delay times are considerably faster than DTL times, being about $t_{PHL} = 7$ ns and $t_{PLH} = 11$ ns. Thus, TTL is preferred for high-speed data processing applications. Also, when the transistors of TTL conduct, they draw about 1.6 mA from the power source that in turn dissipates heat. This current limits the fan-out capability of TTL to about 10 gates.

TTL VOLTAGE TRANSFER CHARACTERISTIC

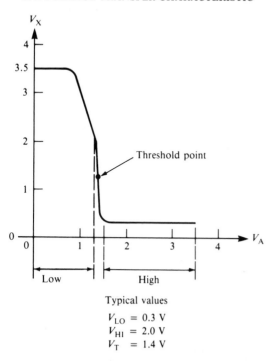

Typical values

$$V_{LO} = 0.3 \text{ V}$$
$$V_{HI} = 2.0 \text{ V}$$
$$V_T = 1.4 \text{ V}$$

FIGURE 12.28.   The voltage transfer characteristic for a TTL gate shows that its noise margins are not as good as those for a DTL. However, it has much higher switching speeds: about 10 ns compared to about 30 ns for DTL.

MOS Logic

The transistors used in typical TTL are bipolar transistors. The newer logic families use an alternative type of transistor called a **metal-oxide semiconductor field-effect transistor** (MOS-FET). The behavior of MOSFETs is beyond the scope of this book. However, note that MOSFETs come in a variety of different forms that make up different logic families: P channel devices make up PMOS logic, N channel devices make up NMOS logic, and, most common, both N channel and P channel devices used to-

TTL–CMOS LOGIC COMPARISON

	TTL	CMOS
Typical number of gates per 40 pin package	10–20	1K–20K
Typical cost of packaged part	$0.20	$3–$15
Typical cost per gate	$0.03	$0.002
dc power per gate	2 mW	< 100 μW
Typical propagation delay time	10 ns	25 ns

FIGURE 12.29.   The newest logic families use metal-oxide semiconductor field-effect transistors (MOSFETs). Switching speeds for MOS logic are also slower than for DTL (about 25 ns), but MOS-FETs are much smaller and consume less power. As many as 200,000 transistors can be fabricated in a single IC. The figure compares TTL and complementary MOS logic (CMOS) characteristics.

gether make up complementary MOS logic, or CMOS.

MOS logic is slower than bipolar transistor logic (typical switching speeds are 25 ns), so it is most useful where speed is not the most important factor. Its advantages are that it requires little power for its operation and, thus, has a very high fan-out capability. In addition, MOS transistors are significantly smaller than bipolar transistors, allowing very dense packing of components in integrated circuit fabrication. Thus, MOS devices permit what is called **large-scale integration** (LSI) and **very large-scale integration** (VLSI) in which more than 200,000 transistors can be fabricated into a single integrated circuit device (Figure 12.29).

12.9 QUESTIONS AND PROBLEMS

1. Correctly identify each gate in Figure 12.30 and determine its truth table.

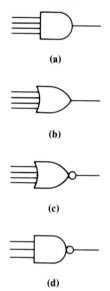

(a)

(b)

(c)

(d)

FIGURE 12.30.

2. Draw the correct symbols for each of the following gates and complete their truth tables:
   (a) four-input OR gate
   (b) five-input AND gate
   (c) three-input NOR gate
   (d) four-input NAND gate
3. Design a digital logic circuit for the safety system of a machine that can be operated with either a foot switch or a hand switch. Make it so that the machine cannot be started until one of the switches is on and a safety guard is in place. Verify the circuit's proper operation with a truth table.
4. Design a digital logic circuit for a fifteen-cent vending machine that will release an item when the correct combination of nickels and a dime is inserted. Verify the circuit's proper operation with a truth table.
5. Design a digital logic circuit for Problem 4 that will release any excess nickels that are present when a dime has been inserted. Verify the circuit's proper operation with a truth table.

6. Design a digital logic circuit for the safety system of a machine so that it will not turn on unless the operator has pressed start buttons with each hand or pressed a start button with one hand and has a hand safety guard in place. Verify the circuit's proper operation with a truth table.
7. Write Boolean algebraic expressions for the solutions to Problems 3 through 6.
8. Verify the logic equivalence of Figures 12.11a and 12.12a and of Figures 12.11b, 12.12b, and 12.12c by comparing their truth tables.
9. Verify the logic equivalence of the basic NAND and NOR gates (Figure 12.5) and their DeMorgan equivalents (Figure 12.12) by comparing their truth tables.
10. Determine the truth table for each of the following Boolean algebraic equations:
    (a) $X = (A + B)(A + C)$
    (b) $X = (A + B)\overline{C}$
    (c) $X = A(B + C) + D(C + B)$
    (d) $X = AC\overline{E} + AD\overline{E} + BC\overline{E} + BD\overline{E}$
11. Draw the logic flow diagram for each of the expressions of Problem 10. Simplify the diagram whenever possible.
12. Using truth tables, verify the equivalence of the DeMorgan expressions for the ten-cent vending machine problem of Example 5a and 5b and the original expressions.
13. Write the Boolean algebraic expression for each of the logic flow diagrams of Figure 12.31.
14. Simplify each of the expressions found in Problem 13 and draw the equivalent logic flow diagrams.
15. Verify that the expressions found in Problem 14 are equivalent to those in Problem 13 by comparing the truth tables for each.
16. Draw the logic flow diagram for an eight-input multiplexer and give the Boolean algebraic expression that represents its performance.
17. Figure 12.32 shows multiple-gate integrated circuits that are available for AND, OR, and NOT applications. Correctly connect (or provide a pin connection table for) these ICs for each of the equations of Problem 10. *Note*: If an input is not used, be sure that it is connected to either the +5 V (logic 1) terminal or ground (logic 0) so that the resulting output will be correct for the other input terminal possibilities.

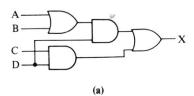

(a)

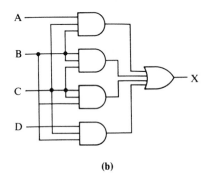

(b)

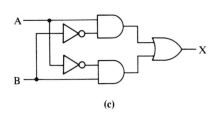

(c)

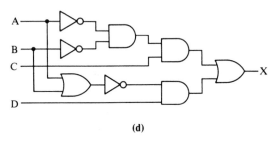

(d)

FIGURE 12.31.

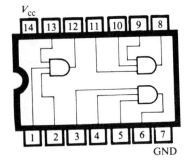

**(a) Fairchild TTL/SSI
Three-Input AND Gates**

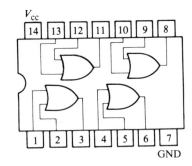

**(b) Fairchild TTL/SSI
Quad Two-Input OR Gates**

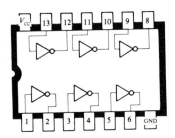

**(c) Sprague 7404
Hex Inverter**

FIGURE 12.32.

Chapter 13

DIGITAL ELECTRONICS AND COMPUTERS

13.1 OBJECTIVES

Following the completion of Chapter 13, you should be able to:

1. Describe the three basic categories of computers—large mainframes, minicomputers, and microcomputers—and give typical applications of each.
2. State the four major components of a microcomputer—I/O port, memory, CPU, and system bus—and explain their respective roles in computer operation.
3. State the difference between volatile and nonvolatile memory and give examples.
4. Explain the basis of the binary number system and convert numbers from binary to decimal and vice versa.
5. State the two basic types of computer language—assembly and programming—give examples, and explain how each is used in computer systems.
6. Explain the binary-coded decimal (BCD) system and convert numbers among BCD, decimal, and binary.

7. Explain the American Standard Code for Information Interchange (ASCII) and convert between the ASCII binary code and its alphanumeric equivalent using a conversion table.
8. Explain the hexadecimal and octal codes and convert between these codes and their alphanumeric, BCD, and ASCII equivalents.
9. Add and subtract binary numbers and verify the results by converting each term to its decimal equivalent.
10. Define the exclusive OR gate in terms of its circuit symbol, Boolean algebraic expression, and truth table.
11. Give the Boolean algebraic expression and logic flow diagram for half-adder and full-adder circuits and show how these circuits are used to perform digital addition.
12. Show how binary subtraction can be accomplished using logic gates.

13.2 INTRODUCTION

In Chapter 13, we will begin to explore how digital electronic circuits are used to construct digital computers (Figure 13.1). Modern computer systems are very complex, of course, so that, as in Chapter 12, we will be able only to scratch the surface of their complexity. Nevertheless, the basic prinicples are fairly straightforward and a good feel for how these marvels of modern technology work will be gained.

First, to place the role of digital electronic circuits in context, an overview of computer technology is given that includes the various categories of computers, the terms that describe these categories, and basic computer architecture. Like building architecture, computer architecture describes the major building block components of computers and how they work together to perform their functions. The remainder of the chapter develops the digital codes or languages that computer components use to communicate among each other and to the user input/output devices. We then see how these codes are used to perform simple data processing tasks, such as addition and subtraction.

13.3 DIGITAL COMPUTERS

Large Mainframes

Digital computers fall into three broad categories distinguished by such factors as size, cost, speed, and application. The fastest and most powerful computers are the large **mainframe computers**. These computers can store huge amounts of data, support many input/output devices, and handle thousands of data processing instructions simultaneously. They are typically used where large databases and many users are required, such as banks, insurance companies, and federal agencies like the Social Security Administration and the IRS. These

HEWLETT PACKARD PERSONAL COMPUTER

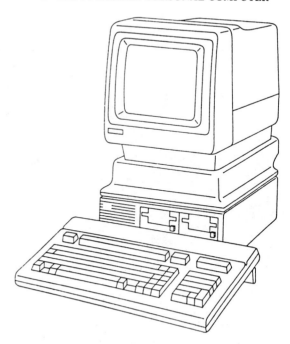

FIGURE 13.1. The operating principles of computers, including their major functional components and the binary codes used for communication and data processing, are the same for all types of computers—large mainframes, minicomputers, and today's desk-top microcomputers like the HP PC shown.

systems cost millions of dollars and take up thousands of square feet of floor space. The early computers, especially those that used tubes and transistors, were all of the large mainframe type.

Minicomputers

With the development of integrated circuits in the mid-1960s, manufacturers introduced smaller and less expensive computers called **minicomputers**. Minicomputers were somewhat slower, could store significantly less data, and handle fewer users simultaneously. However,

MICROCOMPUTER (μC) ARCHITECTURE

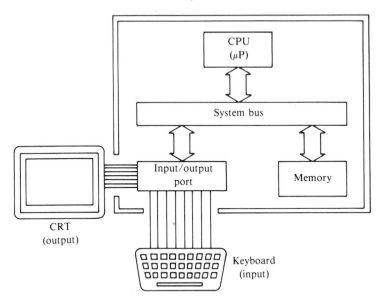

FIGURE 13.2. The architecture of a microcomputer (μC) consists of three basic components – input/output (I/O) port, central processing unit (CPU), and memory – and a system bus that connects these components together.

they were more than adequate for many applications. Originally used for laboratories and industrial applications, they now make up the major portion of computers used by small- and medium-sized businesses. Typical applications are payroll, customer billing, personnel records, inventory management, and a variety of other accounting tasks. Minicomputers cost in the tens of thousands of dollars and take up a few hundred square feet of floor space.

Microcomputers

More recently, with the development of LSI and VLSI components, **microcomputers** (μCs) are now available. Modern personal computers (PCs) costing a few hundred dollars and sized for a desk top are examples of how far computer technology has come. Microcomputers are somewhat slower than minicomputers, but more than adequate for a wide range of applications, including machine and process control,

medical electronics, communication, home appliances, and entertainment devices. Computer games, for example, are essentially microcomputers, using the same basic principles and components.

In the following discussion, the basic computer architecture of these microcomputers is described as an example of the basic design of all digital computers, including the large mainframes.

13.4 MICROCOMPUTER ARCHITECTURE

Microcomputers comprise three basic components: an input/output (I/O) port, a central processing unit (CPU), and memory. Connecting these components together and providing their communication link is the system bus (Figure 13.2). The role of each of these components is briefly described in the following paragraphs.

PERIPHERAL EQUIPMENT

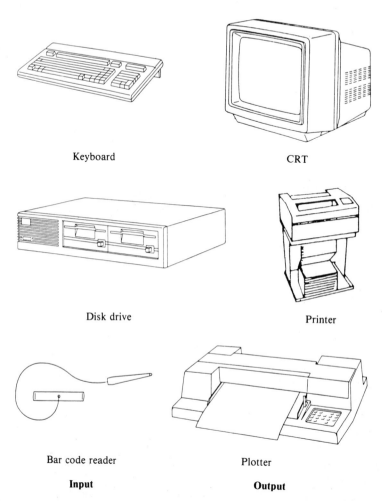

Keyboard

CRT

Disk drive

Printer

Bar code reader

Plotter

Input

Output

FIGURE 13.3. The I/O port is the interface between the μC and input/output peripherals. Input peripherals, such as keyboards, data readers (card, paper tape, magnetic tape,
and magnetic disk), and transducers, supply data to the μC. Output peripherals display
data (CRT and printer), store or transmit data, or perform control tasks.

Input/Output Port

The **input/output** (I/O) **port** is the interface between the microcomputer (μC) and a variety of
peripheral equipment that in turn interfaces
with the outside world. **Peripheral input equipment** supplies data to the computer or instruc-

tions for processing data. Examples of
peripheral input equipment include an operator
keyboard, card or paper tape readers, magnetic
tape readers, magnetic disks, or a process control transducer—thermometer, photocell, pressure gauge, and so forth (Figure 13.3).

Peripheral output equipment displays,

stores, or transmits data or controls some physical system. The common output display devices are the cathode-ray tube (CRT) and printer that are often built into the μC. However, the output port might also be connected to tape or disk storage devices, transmission lines to other computers, or directly to an output control device—heater, light, control motor, and so on.

The I/O port itself consists of one or more LSI chips. Their task is to convert the data from the input device into a form suitable for processing by the CPU, and, conversely, to convert the data from the CPU into a form suitable for use by the appropriate output device.

Memory

Memory is the μC component that stores data or instructions for processing data. These instructions are commonly called the **computer program**, or **software**, while the physical μC components and peripherals are termed the **computer hardware.**

The memory of a computer is also a series of LSI chips that are of two types: volatile and nonvolatile. A **volatile memory** chip will lose the data stored in it when the power is turned off. A **nonvolatile memory** chip, on the other hand, will retain its data whether the power is on or off (Figure 13.4).

Volatile memory chips are typically random access memories (RAMs), while nonvolatile memories can be read only memory (ROM), programmable read only memory (PROM), erasable programmable read only memory (EPROM), or electrical erasable read only memory (EEROM).

RAMs are used for the temporary storage of data—for example, while awaiting further data processing instructions—and can be used over and over again to store and later read back the data they contain. **ROM**s, on the other hand, are preprogrammed with data or data processing steps for use in executing some spe-

MEMORY CHIPS

Ceramic Package

Plastic Package

Frit Seal Package

FIGURE 13.4. Computer memory stores data or the instructions for processing data. Volatile memory chips called RAMs retain data temporarily while the μC is running. Nonvolatile memory chips (ROMs, PROMs, EPROMs, and EEROMs) are preprogrammed with data or instructions and retain these even when the μC is turned off.

cific computer program. The various kinds of ROMs indicate the different ways in which the user accesses (reads) or imprints (writes) data or data processing instructions into the unit.

In a four-function calculator, for example, the numbers that are input via the keyboard and that appear on the display are being tempo-

CENTRAL PROCESSING UNIT CHIPS

Plastic Package

Ceramic Package

FIGURE 13.5. The central processing unit (CPU) is the computer's control center, decoding instructions, transferring data, and performing data processing. In today's microcomputers, this complex component is fabricated as a single IC chip called a microprocessor (μP).

rarily stored in a RAM-type memory. Pressing the CLEAR key or turning off the calculator quickly erases the data. On the other hand, the calculation instructions—for example, $+$, $-$, $\times$, $\div$—are stored in ROM-type memories, and these instructions remain whether or not the calculator is turned off.

Central Processing Unit

The **central processing unit** (CPU) is the heart of the digital computer. It serves as the control center, decoding instructions, transferring data to and from other parts of the system, and performing the data processing instructions. In early computers, the CPU consisted of tens of thousands of vacuum tubes; in today's micro-

computers, these tasks can all be contained in a single integrated circuit chip called a **microprocessor** (μP) whose cost can be on the order of $10 or less (Figure 13.5). In fact, in some cases, such as computer games and household appliance controls, the entire microcomputer—I/O port, CPU, and memory—is contained in a single LSI chip at a cost of a few dollars.

System Bus

The final element of a microcomputer is the **system bus.** A **bus** is an electrical conductor that connects the various circuit components together. In most cases, the bus is an electroplated copper conductor pattern on a printed circuit board that interconnects the leads of the LSI chip components (Figure 13.6).

The system bus is generally made up of three parts: an address bus, a data bus, and a control bus. The **address bus** allows the CPU to locate or address the various memory locations. The **data bus** is used to transmit information (data or instructions) among the I/O port, CPU, and memory. The **control bus** transmits control information among the components to keep the system running in proper sequence.

In order to understand how these computer components perform their tasks, we need one final piece of information—the digital language that they use for communication. The binary number system was briefly introduced in Chapter 6, but we now need a deeper understanding of how the two digital levels of 0 and 1 are used to represent data and instructions in digital computers.

13.5 COMPUTER LANGUAGES

Binary Number System

As described in Chapter 6, the binary number system has only two digits (0 and 1), rather than ten digits (0, 1, 2, 3, . . . , 9) of the decimal

PRINTED CIRCUIT BOARD WITH MICROPROCESSOR

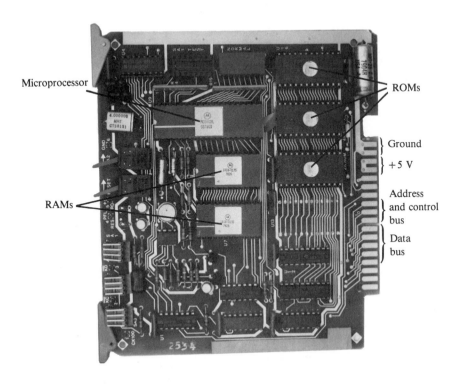

FIGURE 13.6. Connecting the components together is a complex network of electro-plated copper conductors called the system bus. The system bus has three major parts: an address bus to access data from memory, a data bus to route data through the system, and a control bus to transmit control signals.

number system with which we we are most familiar. The sequence of digits (0 or 1) in the binary system gives, from right to left, the number of 1s, 2s, 4s, 8s, 16s, and so on, that make up the number, in contrast with the decimal system in which the sequence of digits (0 to 9) gives the number of 1s, 10s, 100s, 1000s, and so on, that make up the number (Figure 13.7).

Each digit in a binary number is called a **bit**, a shortened form for *binary digit*. Thus, the 8 bit binary sequence 11101101 represents, from right to left, 1 one, 0 twos, 1 four, 1 eight, 0 sixteens, 1 thirty-two, 1 sixty-four, and 1 one

hundred and twenty-eight. The decimal equivalent of this 8 bit number can be obtained by adding up the decimal equivalent components:

$$
\begin{array}{rcrcr}
1 & \times & 1 & = & 1 \\
0 & \times & 2 & = & 0 \\
1 & \times & 4 & = & 4 \\
1 & \times & 8 & = & 8 \\
0 & \times & 16 & = & 0 \\
1 & \times & 32 & = & 32 \\
1 & \times & 64 & = & 64 \\
1 & \times & 128 & = & \underline{128} \\
11101101 & & & = & 237
\end{array}
$$

FIGURE 13.7. The operation of digital computers depends on processing the two binary digits, or bits, 0 and 1. These bits form the basis of the binary number system in which the place digits represent the number of 1s, 2s, 4s, 8s, 16s, 32s, 64s, 128s, and so on, in the number.

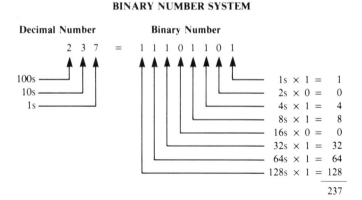

Similarly, in the decimal system, 237 means 7 ones, 3 tens, and 2 hundreds:

$$7 \times 1 = 7$$
$$3 \times 10 = 30$$
$$\underline{2 \times 100 = 200}$$
$$237 = 237$$

Strings of 0s and 1s are difficult to read and write and highly prone to errors. In addition, they represent only numbers and do not include letters or other symbols used in writing, such as periods, question marks, equal signs, and the like, that appear on typewriter keyboards. To meet these needs, several different binary codes have been developed. The codes are used for expressing data and writing instructions in what is called **assembly language**—that is, they are the language codes that convert alphanumeric symbols—decimal numbers, letters, and other symbols—into the proper *machine language* code of 0s and 1s that the internal μC logic gates can understand. Writing programs in assembly language is the most difficult of the programming tasks and is generally left to a design specialist.

The first two assembly language codes to be described—binary-coded decimal (BCD) and the American Standard Code for Information Interchange (ASCII)—are used principally in a μC to communicate with peripheral equip-ment, such as keyboards, printers, plotters, CRT, and so on. The second two codes—octal code and hexadecimal code—are used internally by the μP for data and data processing instructions.

Another group of codes that is probably more familiar is called **programming languages**—for example, BASIC, FORTRAN, COBOL, and PASCAL. These languages are used by most computer programmers to write software for computer applications. Programming languages are considered high-level languages in that a single command can produce thousands of logic steps. The interpretation of these commands and the implementation of the logic steps are all done by assembly language codes like those described above.

The next several paragraphs introduce the different assembly language codes and some of the conversions among them. Remember, however, that the goal is to give you an understanding of how a μC works and not to make you a computer language specialist. Programming languages are beyond the scope of this book and are not discussed further. The various computer codes are summarized in Figure 13.8.

Assembly Languages

Binary-Coded Decimal: As we have seen, translation of a string of 0s and 1s to their deci-

BINARY LANGUAGE CODES

Assembly Languages	Programming Languages
Binary-Coded Decimal (BCD)	BASIC
American Standard Code for	FORTRAN
Information Interchange (ASCII)	COBOL
Hexadecimal (hex)	PASCAL
Octal	

FIGURE 13.8. Communication and data processing in computers is done in a variety of binary language codes: assembly languages for internal data processing and communication with peripherals and programming languages for writing computer application software.

BINARY-CODED DECIMAL SYSTEM

BCD   Decimal
0000 = 0
0001 = 1
0010 = 2
0011 = 3
0100 = 4
0101 = 5
0110 = 6
0111 = 7
1000 = 8
1001 = 9

Example:
Decimal   3   4   9
BCD   0011 0100 1001   ◄──── 4 bit sequence

FIGURE 13.9. The binary-coded decimal (BCD) system is an interface language in which each decimal digit is replaced by its 4 bit binary equivalent.

mal equivalent is not obvious by inspection and usually requires some mental or calculator arithmetic to get the result. A somewhat simpler code for numbers is the **binary-coded decimal** (BCD) system. In the BCD system, groups of binary digits are used to represent the number of 1s, 10s, 100s, and so on, of a decimal number. The following ten 4 bit binary numbers represent the ten decimal digits from 0 to 9:

        0000 = 0
        0001 = 1
        0010 = 2
        0011 = 3
        0100 = 4
        0101 = 5
        0110 = 6
        0111 = 7
        1000 = 8
        1001 = 9

Note that the six remaining 4 bit binary digits — 1010, 1011, 1100, 1101, 1110, and 1111 — are not used in the BCD system.

A decimal number is represented in BCD by simply placing the appropriate 4 bit binary number in place of its decimal equivalent (Fig-

ure 13.9). For example, the three-digit decimal number 349 would be written in BCD as:

Decimal:	3	4	9
BCD:	0011	0100	1001

In straight binary, the number 349 would be written:

    Decimal: 349
    Binary: 101011101

Note that the BCD system requires twelve digits to represent 349, while the binary system requires only nine. Hence, the BCD system requires more memory space than binary, but it is much simpler to translate into its decimal equivalent.

American Standard Code for Information Interchange: A similar code that represents all of the keyboard symbols, as well as some internal machine commands, is the **American Standard Code for Information Interchange** (ASCII).

ASCII CODE

Bit Numbers					0 0 0	0 0 1	0 1 0	0 1 1	1 0 0	1 0 1	1 1 0	1 1 1
	0	0	0	0	NUL	DLE	SP	0	@	P	\	p
	0	0	0	1	SOH	DC1	!	1	A	Q	a	q
	0	0	1	0	STX	DC2	"	2	B	R	b	r
	0	0	1	1	ETX	DC3	#	3	C	S	c	s
	0	1	0	0	EOT	DC4	$	4	D	T	d	t
	0	1	0	1	ENQ	NAK	%	5	E	U	e	u
	0	1	1	0	ACK	SYN	&	6	F	V	f	v
	0	1	1	1	BEL	ETB	'	7	G	W	g	w
	1	0	0	0	BS	CAN	(	8	H	X	h	x
	1	0	0	1	HT	EM	)	9	I	Y	i	y
	1	0	1	0	LF	SUB	•	:	J	Z	j	z
	1	0	1	1	VT	ESC	·	;	K	[	k	{
	1	1	0	0	FF	FS	,	<	L	\	l		
	1	1	0	1	CR	GS	-	=	M	]	m	}
	1	1	1	0	SO	RS	.	>	N	∧	n	~
	1	1	1	1	SI	US	/	?	O	—	o	DEL

Machine
Commands

FIGURE 13.10. The American Standard Code for Information Interchange (ASCII) is a 7 bit binary code that represents 128 different alphanumeric characters and machine commands. In the matrix above, the first 4 bits are in the left-hand row and the final 3 bits are in the top column.

The ASCII system is a 7 bit binary code that permits 128 different characters to be represented (Figure 13.10). Note that Figure 13.10 is a matrix in which the first 4 bits of the ASCII word are along the left-hand column, and the final 3 bits are along the top row. Hence, the letter *G* is given in ASCII by:

Top row
G = $\overbrace{1000}\underbrace{111}$
Left column

EXAMPLE 1: Write Ohm's law, $V = IR$, in ASCII.

Solution: Using Figure 13.10, the characters V, =, I, and R are found to be:

V	1010110
=	0111101
I	1001001
R	1010010

HEXADECIMAL CODE

Decimal	Binary	Hexadecimal
0	0000	0
1	0001	1
2	0010	2
3	0011	3
4	0100	4
5	0101	5
6	0110	6
7	0111	7
8	1000	8
9	1001	9
10	1010	A
11	1011	B
12	1100	C
13	1101	D
14	1110	E
15	1111	F

Example:

Hex   4   B
Binary   0100   1011 ←——— 4 bit
                              sequence
              Byte

FIGURE 13.11.   The hexadecimal code consists of an 8 bit word, called a byte, that is broken into two 4 bit sequences. Each 4 bit sequence represents one of 16 different alphanumeric characters (0 to 9 plus A to F).

In a typical μC system, pressing a key on the keyboard causes the 7 bit binary ASCII code corresponding to that key to be generated and carried to the input port via a 7 line input bus — one line for each of the seven possible bits (see Figure 13.2). At the input port, the 7 bit ASCII code is converted into the machine language code appropriate for the particular μP chip being used. Similarly, the output port converts the internal machine language code into its 7 bit ASCII equivalent and transmits it via a 7 line output bus to an output device — for example, a CRT or printer — that displays the corresponding character. Note also that the binary data are really 0 V and +5 V pulses that ultimately go to logic gates for interpretation and processing.

The two most common internal machine language codes are the hexadecimal code and the octal code. Like the BCD and ASCII codes, the strings of binary bits are broken into groups called **words.** In BCD, a word consists of 4 bits; in ASCII, 7 bits. However, inside a μC, the word length is generally 8 bits. An 8 bit binary pattern is called a **byte.** While longer word lengths are used by larger computers, the 8 bit byte is the most frequently used in μCs.

Hexadecimal Code: The most common internal code is the **hexadecimal code** (hex), where hexadecimal means 16. In this code, the 8 bit byte is broken into two groups of four binary digits. Four binary digits can represent 16 different characters. Thus, in the hexadecimal code, 6 letters are added to the 10 numerical digits to give 16 alphanumeric characters represented by the 4 bit binary patterns shown in Figure 13.11. Note that the hex code is similar to the BCD except that the six unused 4 bit words are assigned characters and used.

EXAMPLE 2:   Write the following binary patterns in their hex equivalent:

(a) 01001011
(b) 10011101
(c) 01100101
(d) 11010111

Solution: Dividing each byte into two groups of 4 bits and referring to the table of Figure 13.11 gives:

(a) 0100   1011 = **4B**
(b) 1001   1101 = **9D**
(c) 0110   0101 = **65**
(d) 1101   0111 = **D7**

FIGURE 13.12. The 8 bit hex byte can represent up to 256 different characters that include the 128 characters of the ASCII code. The matrix gives the two-digit alphanumeric hex equivalent for each ASCII character. Translation between hex and ASCII can be done by comparing Figures 13.10 and 13.12.

HEX TO ASCII TRANSLATION

Column → Row ↓	0	1	2	3	4	5	6	7	← Left digit
0	NUL	DLE	SP	0	@	P	\	p	
1	SOH	DC1	!	1	A	Q	a	q	
2	STX	DC2	"	2	B	R	b	r	
3	ETX	DC3	#	3	C	S	c	s	
4	EOT	DC4	$	4	D	T	d	t	
5	ENQ	NAK	%	5	E	U	e	u	
6	ACK	SYN	&	6	F	V	f	v	
7	BEL	ETB	'	7	G	W	g	w	
8	BS	CAN	(	8	H	X	h	x	
9	HT	EM	)	9	I	Y	i	y	
A	LF	SUB	•	:	J	Z	j	z	
B	VT	ESC	.	;	K	[	k	{	
C	FF	FS	,	<	L	\	l	l	
D	CR	GS	-	=	M	]	m	}	
E	SO	RS	.	>	N	∧	n	~	
F	SI	US	/	?	O	—	o	DEL	

↑ Right digit   Machine Commands

To write a program in hex, the programmer is given a 4 × 4 keyboard with the 16 hex characters. Pressing a key automatically transmits the proper sequence of 0s and 1s that defines the character.

Note also that an 8 bit byte can have up to 256 different combinations. Thus, it can include all of the 128 characters of ASCII shown in Figure 13.10 without using one of the bit locations. This translation is shown in Figure 13.12. Because the maximum character of the first four bits is a 7 (top row of Figure 13.12), represented as 0111, the leading bit of the hex–ASCII translation is always a 0. Translation between hex and ASCII can be done by finding the appropriate character in Figures 13.10 and 13.12 and determining the corresponding ASCII and hex equivalents.

EXAMPLE 3: Write Ohm's law, $V = IR$, in hex and compare with the ASCII equivalent.

Solution: Using Figures 13.10 and 13.12, characters V, =, I, and R are found to be:

	Hex	ASCII	
V	56	0101	0110
=	3D	0011	1101
I	49	0100	1001
R	52	0101	0010

Before leaving the hex code, a brief note about counting in hex should be made because it can be somewhat confusing. Because there

DECIMAL–BINARY CODE COMPARISON

Decimal	Hex	Octal	BCD		ASCII	
0	0	0		0000		0110000
1	1	1		0001		0110001
2	2	2		0010		0110010
3	3	3		0011		0110011
4	4	4		0100		0110100
5	5	5		0101		0110101
6	6	6		0110		0110110
7	7	7		0111		0110111
8	8	10		1000		0111000
9	9	11		1001		0111001
10	A	12	0001	0000	0110001	0110000
11	B	13	0001	0001	0110001	0110001
12	C	14	0001	0010	0110001	0110010
13	D	15	0001	0011	0110001	0110011
14	E	16	0001	0100	0110001	0110100
15	F	17	0001	0101	0110001	0110101
16	10	20	0001	0110	0110001	0110110
17	11	21	0001	0111	0110001	0110111

FIGURE 13.13. Another machine language code is the octal code, a 3 bit binary code that represents the eight decimal digits 0 to 7. The place digits indicate the number of 1s, 8s, 64s, 512s, and so on, in the number. The table compares the different codes up to decimal 17.

are 16 characters, the digit locations represent the number of 1s, 16s, 256s, 4096s, and so on, in the number. Thus, the number 15 is written F, but the number 16 is written 10, and the number 17 is written 11. Because these symbols have decimal meanings also, the symbols H or 16 are often written as a subscript to indicate that it is a hex number—for example, 11_H or 11_{16}. Getting used to this new notation is not easy.

Conversion between hex and decimal is similar to that between binary and decimal, as shown in the following example.

EXAMPLE 4: Convert the number E39B to decimal.

Solution: The conversion is made by converting the hex characters to their decimal equivalent, multiplying each by its place representation, and adding:

Hex	Decimal	$\times$	Place	=	Decimal
B	11	$\times$	1	=	1
9	9	$\times$	16	=	144
3	3	$\times$	256	=	768
E	14	$\times$	4096	=	57,334
			E39B		58,267

Octal Code: A final code that is sometimes used in machine language is the **octal code**. In the octal code, there are only eight digits: 0, 1, 2, 3, 4, 5, 6, and 7. Hence, it has half the number of characters and can be represented by a 3 bit word, as shown in Figure 13.13 where all of the codes are compared up to decimal 17.

Conversion between octal and the other codes is similar to that for hex and is left for the Questions and Problems at the end of the chapter. Note, however, that the place digits for the octal code represent the number of 1s, 8s, 64s, 512s, and so on, and that a number in octal will have a subscript 8 after it to indicate that it is octal — for example, 432_8.

13.6 DIGITAL ARITHMETIC

With the overview of μC architecture and the languages that convert alphanumeric symbols into sequences of 0s and 1s for the μC to process, we are now ready to investigate how the μC circuits react to the 0 V and 5 V logic levels to perform data processing tasks. The two tasks we examine are addition and subtraction. These tasks are the most common functional elements of a μC and are involved in one way or another in essentially every data processing step.

Binary Addition

To see how a logic circuit can perform addition, we must first see how binary numbers add. As shown in Figure 13.14, adding binary numbers is similar to adding decimal numbers except that there are many fewer digits to worry about — two instead of ten. As with decimal addition, when two binary numbers add, their sum can be a single digit or a digit plus a carry to the next column. For example, the two binary digits 0 and 1 have four possible addition combinations, only one of which has a carry (Figure 13.14).

Note also in Figure 13.14 the last binary sum: $1 + 1 = 0$ with a carry of 1 to the next column. In decimal, $1 + 1$ is 2, which is 10 in binary, as our result gave. This simple procedure can be used for any number of binary digits as the following examples illustrate.

EXAMPLE 5: Add the binary numbers 110 and 011 and compare the result with their decimal equivalent.

BINARY ADDITION

A	0	1	0	1$\overline{1}$	← Carry digit
+ B	+ 0	+ 0	+ 1	+ 1	
S	0	1	1	10	← Sum digits

FIGURE 13.14. One of the most common data processing tasks in a computer is the addition of binary numbers. Like decimal addition, adding two binary numbers can result in a simple sum digit or a sum digit plus a carry digit. See Examples 5 and 6.

Solution: The sum of 110 and 011 using the addition rules above is:

$$11 \leftarrow \text{Carry digits}$$
$$110$$
$$+011$$
$$\overline{1001} \leftarrow \text{Sum digits}$$

In decimal notation, these numbers are:

$$\begin{array}{rrr} 110 = & 4 + 2 + 0 = & 6 \\ +011 = & 0 + 2 + 1 = & +3 \\ \hline 1001 = & 8 + 0 + 0 + 1 = & 9 \end{array}$$

EXAMPLE 6: Add the binary digits 110110 and 011101 and compare the result with their decimal equivalent.

Solution: The sum is:

$$1111 \leftarrow \text{Carry digits}$$
$$110110$$
$$+011101$$
$$\overline{1010011}$$

Note in the fifth column from the right that three 1s must be added. This is done two digits at a time in succession — that is, $1 + 1 = 0$ carry 1 and $0 + 1 = 1$. Hence, the result is 1 carry 1.

BINARY SUBTRACTION

				0 2	← Borrow digits
A	0	1	1	1̸ 0̸	
− B	− 0	− 1	− 0	− 0 1	
D	0	0	1	0 1	← Difference digits

FIGURE 13.15. Another common data processing task is the subtraction of binary numbers. In subtraction, we can have a simple difference digit in each column (first four columns) or, when the lower number is larger than the upper number, the requirement of borrowing from the next column (column 5). See Examples 7 and 8.

The decimal equivalent is:

$$
\begin{array}{rlr}
& & \text{Carry digit} \\
& & \downarrow \\
& & 1 \\
110110 = & 32 + 16 + 0 + 4 + 2 + 0 = & 54 \\
+011101 = & 0 + 16 + 8 + 4 + 0 + 1 = & +29 \\
\hline
1010011 = & 64 + 0 + 16 + 0 + 0 + 2 + 1 = & \mathbf{83} \\
& & \uparrow \\
& & \text{Sum digits}
\end{array}
$$

Binary Subtraction

Binary subtraction is like decimal subtraction in that the result of each column's subtraction is a single digit and, when the lower number is larger than the upper number, a borrow from the next column is required. For the two binary digits 0 and 1, only one of the four possible combinations requires a borrow: 0 − 1 (Figure 13.15). In the last column of Figure 13.15, 1 cannot be subtracted without a borrow from the next column, which, in this case, does not have a digit. However, for the following two digit binary number, the borrow procedure is as follows:

$$
\begin{array}{rl}
02 & \leftarrow \text{Borrow digits} \\
1̸ 0̸ = & 02 \\
-01 = & -01 \\
\hline
& 01 \leftarrow \text{Difference digits}
\end{array}
$$

Note that the borrow digit 1 becomes a 0

and puts a 2 in the next column. This step is simply a recognition that binary 10 is 2 in decimal. Writing a decimal 2 in the borrowing column is a convenient device for remembering what is being done. In some cases, the borrowing must be done from several columns over, as shown in the following example.

EXAMPLE 7: Subtract the binary number 0001 from binary 1000 and compare the result with their decimal equivalent.

Solution: The difference between the binary digits 1000 and 0001 using the above rules for *binary subtraction* is:

$$
\begin{array}{l}
0112 \leftarrow \text{Borrow digits} \\
1̸ 0̸ 0̸ 0̸ \\
+0001 \\
\hline
\mathbf{111}
\end{array}
$$

Note that the borrow column is the fourth column from the right and that each successive borrow changes the 0 to 1 except for the right-most column, which becomes a 2. This is equivalent to the *decimal subtraction* of 1000 − 1, which is:

$$
\begin{array}{l}
099(10) \leftarrow \text{Borrow digits} \\
1̸ 0̸ 0̸ 0̸ \\
+0001 \\
\hline
999
\end{array}
$$

FIGURE 13.16. Binary subtraction can be simplified and made into an addition task by using the three-step subtraction-by-complement method. See Example 9. In real computer logic, similar but slightly different subtraction-by-addition methods are used.

SUBTRACTION BY COMPLEMENT

Subtraction Problem	Solution Steps				
	1. Complement larger number:		2. Add the result:		3. Complement the sum:
110010	$\overline{110010}$	=	001101		
− 101101			+ 101101		
000101			111010	=	000101

The binary subtraction is verified in decimal as follows:

$$1000 = 8 + 0 + 0 + 0 = 8$$
$$+0001 = 0 + 0 + 0 + 1 = -1$$
$$111 = 0 + 4 + 2 + 1 = 7$$

EXAMPLE 8: Subtract binary 101101 from binary 110010 and compare the result with their binary equivalent.

Solution: The binary difference is:

```
  01202 ← Borrow digits
  1̶1̶0̶0̶1̶0̶
− 101101
  000101
```

The decimal equivalent is:

$$\begin{array}{r} \text{Borrow digits} \\ \downarrow \\ 4(10) \end{array}$$

$$\begin{array}{rcl} 110010 = & 32 + 16 + 0 + 0 + 2 + 0 = & \cancel{50} \\ -101101 = & -(32 + 0 + 8 + 4 + 0 + 1) = & -45 \\ \hline 000101 = & 0 + 0 + 0 + 4 + 0 + 1 = & 5 \end{array}$$

Subtraction by Complement

Clearly, subtraction is more complicated than addition. Fortunately, there is a method called **subtraction by complement** that converts a number to its complement and then uses addi-

tion. Essentially, the subtraction is performed by converting the larger number to its complement (changing all 0s to 1s and all 1s to 0s) and then adding the complemented number to the number to be subtracted. The complement of the result is the same as the subtraction process described above (Figure 13.16). This procedure is shown in the following example.

EXAMPLE 9: Repeat Example 8 using the method of subtraction by complement.

Solution: The larger binary number of Example 8 is 110010. Its complement is:

$$\overline{110010} = 001101$$

Recall that the bar over a number means its complement. Adding this complemented number to 101101 gives:

```
   11 1
   001101
 + 101101
   111010
```

Complementing the result gives:

$$\overline{111010} = \mathbf{000101}$$

which is the same result as in Example 8.

The advantage of this procedure, of course, is that only a complementing circuit and an addition circuit are needed to perform binary addition and subtraction. The complementing circuit, you will remember, is the NOT gate that converts 0s to 1s and vice versa. The addition circuit is described in the following paragraphs.

One additional note is that the subtraction-by-complement method described above is only one of several possible methods of subtracting by adding. The subtraction-by-complement method has the disadvantage that we must know which of the two numbers to be subtracted is the larger. Determining this fact with a computer is often more complicated than the subtraction itself. Hence, other methods that do not require this procedure are generally used. These methods can be found in most books on digital electronics.

13.7 DIGITAL ARITHMETIC CIRCUITS

Half-Adder Circuit

To develop a digital logic circuit that adds, we begin by determining a Boolean algebraic expression that gives the results of binary addition and then reproducing this Boolean expression with the appropriate logic gates. Recall that binary addition of two digits gives a sum digit S and a carry digit C. These results can be expressed by what is called a **half-adder circuit**. The circuit is called a half-adder because it can add only two digits and, therefore, has limited application. The truth table for the half-adder is:

Inputs		Outputs	
A	**B**	**S**	**C**
0	0	0	0
0	1	1	0
1	0	1	0
1	1	0	1

The next step is to find Boolean algebraic expressions for S and C in terms of A and B. The Boolean expression for C is fairly simple. It is the AND operation:

$$C = A \cdot B \tag{1}$$

Hence, the logic gate for the carry digit is an AND gate.

The expression for S, however, is somewhat more complicated. Note that it is almost, but not quite, the OR operation. While the first three operations are identical to OR, the OR operation on $1 + 1$ equals 1 rather than the 0 required above. Instead, the truth table for S can be obtained by the following somewhat complicated Boolean expression:

$$S = A \cdot \overline{B} + \overline{A} \cdot B \tag{2}$$

The truth table and Boolean expressions for the half-adder circuit are summarized in Figure 13.17.

EXAMPLE 10: Verify the Boolean expression $S = A \cdot \overline{B} + \overline{A} \cdot B$ for the half-adder circuit.

Solution: Substituting in the possible values for A and B gives:

A	B	$A \cdot \overline{B} + \overline{A} \cdot B = S$
0	0	$0 \cdot 1 + 1 \cdot 0 = 0$
0	1	$0 \cdot 0 + 1 \cdot 1 = 1$
1	0	$1 \cdot 1 + 0 \cdot 0 = 1$
1	1	$1 \cdot 0 + 0 \cdot 1 = 0$

which agrees with the half-adder truth table.

This expression can be reproduced using NOT, AND, and OR gates as shown in Figure

BINARY ADDITION

$$
\begin{array}{r}
C \\
A \\
+ \; B \\
\hline
S
\end{array}
$$

Binary Addition

Inputs		Outputs	
A	B	S	C
0	0	0	0
0	1	1	0
1	0	1	0
1	1	0	1

Truth Table

$$C = A \cdot B$$
$$S = A \cdot \bar{B} + \bar{A} \cdot B$$

Boolean Expression

FIGURE 13.17. The logic flow diagram for adding two binary digits requires finding Boolean algebraic expressions for the sum digit S and a carry digit C. The carry-digit expression is the AND operation. The sum-digit expression is more complicated, but yields the proper truth table result.

13.18a. Other Boolean expressions using NAND and NOR gates can also yield the truth table for the half-adder.

The half-adder truth table is so common that it is given a special name and gate symbol called the **exclusive OR** (X-OR) gate (Figure 13.18b). The name simply expresses the truth table result that only when A OR B is exclusively 1 is the output 1; otherwise, it is 0. Combining the exclusive OR gate that yields the Boolean expression for S and the AND gate for

C gives the half-adder circuit shown in Figure 13.19.

Full-Adder Circuit

When adding two multidigit binary numbers, the first column requires the addition of only two digits: $A + B$. However, if this addition yields a carry digit C, then the next column will require the addition of three digits — $A_2 + B_2 + C_1$:

$$
\begin{array}{llll}
C_3 & C_2 & C_1 & \leftarrow \text{Carry digits} \\
A_4 & A_3 & A_2 & A_1 \\
+ B_4 & B_3 & B_2 & B_1 \\
\hline
S_4 & S_3 & S_2 & S_1 & \leftarrow \text{Sum digits}
\end{array}
$$

The half-adder circuit will be adequate for the first column, giving S_1 and C_1, but successive columns may require the addition of three digits — for example, $A_2 + B_2 + C_1$. Hence, a three-digit adder circuit must be developed for all successive columns. A three-digit adder circuit is called a **full adder**. The truth table for a full adder is as follows, where A_2 and B_2 are the two digits in column 2, and C_1 is a possible carry digit from the previous column 1:

A_2	B_2	C_1	S_2	C_2
0	0	0	0	0
0	0	1	1	0
0	1	0	1	0
0	1	1	0	1
1	0	0	1	0
1	0	1	0	1
1	1	0	0	1
1	1	1	1	1

Again, the task is to find Boolean expressions for S_2 and C_2 in terms of A_2, B_2, and C_1. As we would expect, these expressions are significantly more complicated than for the half-

EXCLUSIVE OR GATE

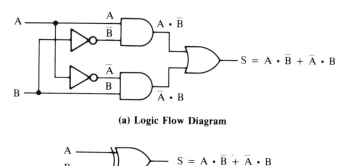

(a) Logic Flow Diagram

$$S = A \cdot \bar{B} + \bar{A} \cdot B$$

(b) Logic Symbol

FIGURE 13.18. The Boolean expression for the sum digit in binary addition can be achieved using the logic flow diagram shown. This circuit is so common that it is given its own special name – exclusive OR (X-OR) – and symbol.

adder and can take several equivalent forms. One set for S_2 and C_2 is:

$$S_2 = P_2 \cdot \overline{C_1} + \overline{P_2} \cdot C_1 \qquad (3)$$

where P_2 is the partial sum:

$$P_2 = A_2 \cdot \overline{B}_2 + \overline{A}_2 \cdot B_2 \qquad (4)$$

and

$$C_2 = A_2 \cdot B_2 + B_2 \cdot C_1 + A_2 \cdot C_1 \qquad (5)$$

Verification of these expressions is also left as Problem 21. Figure 13.20 summarizes the truth table and Boolean expressions for the full-adder circuit.

Note that the expressions for S_2 and P_2 are those for exclusive OR gates. Hence, S_2 can be formed from two exclusive OR gates, as shown in Figure 13.21a. The logic circuit for carry digit C_2 is formed from 3 two-input AND gates and a three-input OR gate, as shown in Figure 13.21b. As before, equivalent expressions and logic circuits using NAND gates, NOR gates, and exclusive NOR (X-NOR) gates can also be found, and, in fact, are more common in actual practice.

HALF-ADDER CIRCUIT

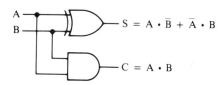

FIGURE 13.19. Combining the exclusive OR gate for the sum digit and the AND gate for the carry digit gives the logic flow diagram for the addition of two binary digits. This diagram is called the half-adder circuit.

When adding multibit digits, each column would have its own full-adder circuit with a carry digit input coming from the previous column's circuit. The general design is shown in Figure 13.22 for a 4 bit addition. Note that each full adder requires about 14 gates (5 for each exclusive OR gate), so that as many as 56 gates are required overall. Thus, the gate count for even the simple task of addition is getting fairly large.

Subtraction Circuit

The simplest way to accomplish subtraction is to use the subtraction-by-complement method described earlier. This method simply requires

MULTIBIT BINARY ADDITION

Inputs			Outputs	
A_2	B_2	C_1	S_2	C_2
0	0	0	0	0
0	0	1	1	0
0	1	0	1	0
0	1	1	0	1
1	0	0	1	0
1	0	1	0	1
1	1	0	0	1
1	1	1	1	1

$$C_2\ C_1$$
$$A_3\ A_2\ A_1$$
$$+\ B_3\ B_2\ B_1$$
$$\overline{\ \ \ \ \ \ \ \ \ \ \ }$$
$$S_3\ S_2\ S_1$$

$$S_2 = P_2 \cdot \overline{C_1} + \overline{P_2} \cdot C_1$$

where $P_2 = A_2 \cdot \overline{B_2} + \overline{A_2} \cdot B_2$

$$C_2 = A_2 \cdot B_2 + B_2 \cdot C_1 + A_2 \cdot C_1$$

Binary Addition   **Truth Table**   **Boolean Expression**

FIGURE 13.20.   The addition of two multibit binary numbers requires the addition of three binary digits: the two digits to be added for each column and a possible carry digit. Shown are the truth table and Boolean expressions for the sum digit and carry digit for this addition.

FULL-ADDER CIRCUIT

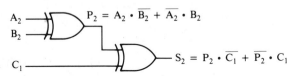

(a) Sum-Digit Diagram

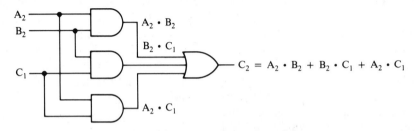

(b) Carry-Digit Diagram

FIGURE 13.21.   The circuit for adding three binary numbers is called the full adder. It consists of logic flow diagrams for the sum digit and carry digit that reproduce the Boolean expression of Figure 13.20.

4 BIT BINARY ADDER

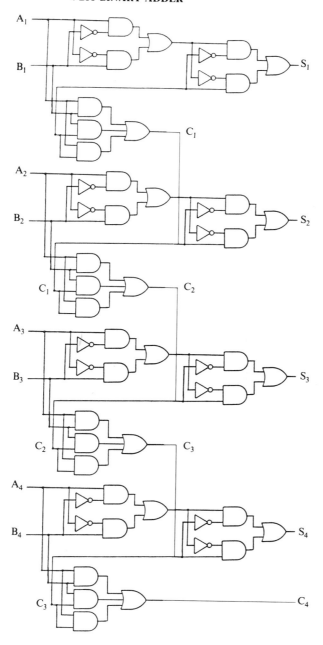

FIGURE 13.22. Shown is the logic flow diagram for a 4 bit binary addition circuit. Since each full adder requires 14 gates, the total number of gates required for this addition is 56.

putting NOT gates on each input of the larger number to produce its complement and NOT gates on each full-adder output to obtain the complement of the sum. The general scheme is shown in Figure 13.23. This circuit requires 8 more gates than the 4 bit addition circuit of Figure 13.21, for a total of 64 gates.

There are more complicated ways to

FIGURE 13.23. The logic flow diagram for binary subtraction shown here uses the subtraction-by-complement method. Each input of the larger number is complemented using NOT gates. The result is then added to the smaller number, and the resulting sum digits are complemented. This diagram requires a total of 64 gates.

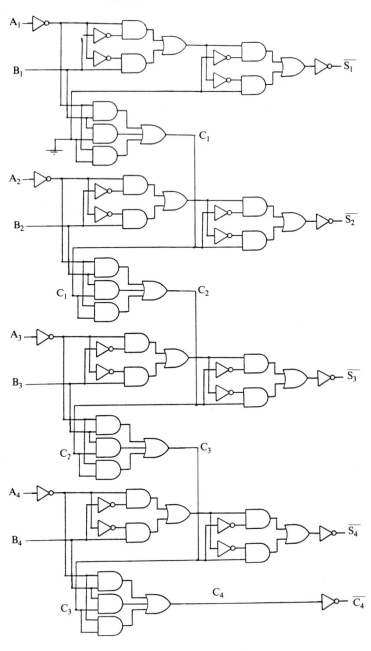

4 BIT BINARY SUBTRACTOR

achieve these and other data processing steps— for example, multiplication and division—and specialized variations for each of the binary codes that are used. The basic methods are the same, however, and the information presented in this section should provide a good starting point for more advanced study.

13.8 QUESTIONS AND PROBLEMS

1. Determine the decimal equivalents of the following binary numbers:
   (a) 10110101
   (b) 01001010
   (c) 11111111
   (d) 10101010

2. Determine the binary equivalents of the following decimal numbers:
   (a) 117
   (b) 254
   (c) 85
   (d) 203

3. Determine the BCD equivalents of the decimal numbers in Problem 2.

4. Determine the decimal equivalents of the following BCD numbers:
   (a) 1001 0100 0111
   (b) 0110 1000 0000
   (c) 0101 0010 1001 0110
   (d) 0011 0001 0111 0100

5. Determine the binary equivalents of the BCD numbers of Problem 4.

6. Express your name in the ASCII binary code.

7. Encode the following ASCII binary message:

1000100	1000111
1100001	1101111
1101101	1101111
1101110	1100100
	0100001

8. Write the hexadecimal equivalents of decimal 1 through 100.

9. Determine the hexadecimal equivalents of the binary patterns given in Problem 1.

10. Determine the hexadecimal equivalents of the decimal numbers in Problem 2.

11. Determine the decimal equivalents of the following hex codes:
    (a) $D63_{16}$
    (b) BAD_{16}
    (c) $C3B0_{16}$
    (d) $D4A8_{16}$

12. Write your name in the hex equivalent of the ASCII code.

13. Write the octal equivalents of decimal 1 through 50.

14. Determine the octal equivalents of the decimal numbers in Problem 2.

15. Determine the decimal equivalents of the following octal codes:
    (a) 732_8
    (b) 677_8
    (c) 423_8
    (d) 156_8

16. Add the following binary numbers and compare the results with their decimal equivalents:
    (a)   11001011
         $+$ 01101010

    (b)   10111010
         $+$ 01110111

    (c)   11010111
         $+$ 01011101

    (d)   11101011
         $+$ 11010111

17. Change the sign in Problems 19a through 19d and subtract the binary numbers given. Compare the results with their decimal equivalents.

18. Repeat Problem 17 using the method of subtraction by complement.

19. Using the logic flow diagram of Figure 13.22, trace the input and output logic values for each gate for the following 4 bit addition and verify your result:

    $$\begin{array}{rl} A = & 1011 \\ +B = & +1101 \\ \hline S = & \end{array}$$

20. Using the logic flow diagram of Figure 13.23, trace the input and output logic values for each gate for the following 4 bit subtraction and verify your result:

    $$\begin{array}{rl} A = & 1101 \\ -B = & -1011 \\ \hline D = & \end{array}$$

21. Verify that the Boolean expressions of Equations 3, 4, and 5 yield the truth table for the full adder.

Part
VI

POWER SUPPLIES

404

DIRECT CURRENT POWER SUPPLIES IN ELECTRONICS

Every conceivable electronic system, in applications as diverse as medicine, industry, law enforcement, government, and education, has a common requirement. Each needs a power source that furnishes direct current to their components (Figure 1). Essentially all electronic components, including transistors and integrated circuits, must be powered by a voltage source that provides a stable dc current. The importance of the power supply cannot be overstressed. It is central to the operation of every system and is critical to reliable operation. When a component fails in one section of a system, the remaining sections may continue to operate; but, when a component fails in the power source, the entire electronic system becomes inoperative.

One type of dc power source is the battery. Batteries are vital for portable applications, such as radios, walkie-talkies, hand calculators, and flashlights. However, eventually a battery runs down and must be replaced or recharged. The alternative source of power is the 115 V alternating voltage available from wall outlets,

DIRECT CURRENT VOLTAGE AND ELECTRONIC DEVICES

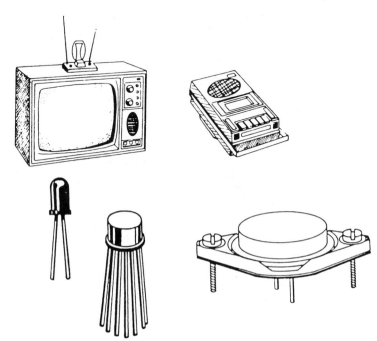

FIGURE 1. Nearly every electronic device and system require dc voltage for operation. Part VI describes the design and operation of a variety of low-voltage power supplies suitable for electronic systems.

FIGURE 2. The source of power for most dc power supplies is the wall outlet. The purpose of the power supply is to convert the outlet's ac voltage to a stable dc voltage while dissipating a minimum of waste heat.

PURPOSE OF DIRECT CURRENT POWER SUPPLIES

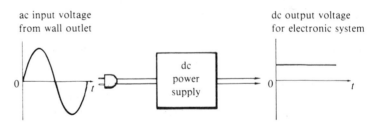

shown in Figure 2 as a sine wave. Because electronic devices require a source of low dc voltage, it is necessary to convert the 115 V alternating voltage to a lower dc voltage by means of a **dc power supply**, sometimes called a **rectifier** or **ac-to-dc converter.**

The voltage and current requirements of most modern electronic systems are generally small. In fact, the electronics industry has tried very hard to minimize the voltages and currents used by their systems. Low currents can be turned on and off faster than high ones, and lower-power electronic devices perform better at higher frequencies and speeds than do higher-power devices. In addition, low power consumption generates less waste heat and permits longer operation under battery power.

Thus, dc power supply voltages are generally less than 40 V. Typical standard values are 5 V, 12 V, 15 V, and 24 V. The dc current capability of most power supplies is also less than 1 A. A 500 mA maximum is quite common.

While requirements on the magnitude of the voltage and current from a dc power supply are quite modest, stability requirements are quite demanding. The ideal dc power supply has three essential characteristics:

1. It will supply a pure dc voltage that does not vary with time. That is, there is no alternating component of voltage, called *ripple*, riding on the dc voltage.
2. The magnitude of the pure dc voltage does not change, no matter how much current or power is drawn from it and no matter how the 115 V line voltage might change.
3. It does not generate any internal power that dissipates useless energy in itself.

In Part VI we look at a variety of practical dc power supplies and compare them with these ideal characteristics.

Chapter

14

UNREGULATED POWER SUPPLIES

14.1 OBJECTIVES

Following the completion of Chapter 14, you should be able to:

1. Identify each terminal of a wall outlet and explain its connection to the incoming power system at the fuse box.
2. Identify each terminal of an ac line cord and explain its connection to an electronic system.
3. Determine the peak voltage, peak-to-peak voltage, and period of an ac line voltage, given its rms voltage and frequency.
4. Describe the construction and principles of operation of a transformer.
5. Determine the output voltage and current of a transformer, given its turns ratio, input voltage, and current.
6. Determine the power rating of a transformer, given its input (or output) current and voltage ratings.
7. Describe the construction of multitap and multiwinding transformers and explain their different applications.

8. Explain the purpose and location of a fuse in an electronic system and explain its principal characteristics: current rating, voltage rating, and blow characteristic.
9. Determine the proper fuse for an electronic system, given the maximum allowable primary voltage and secondary current.
10. Explain the principle of operation of a rectifier diode using its V–I characteristic curve.
11. Draw a half-wave rectifier circuit and explain the principles of its operation.
12. Determine the load current and voltage (peak and average) of a half-wave rectifier, given the rms input voltage and load resistance.
13. Explain the current and voltage ratings of a rectifier diode.
14. Describe the grounding and output terminal connections for floating, single-polarity, and laboratory power supplies.
15. Explain the voltage ratings (nominal cell voltage, end-of-charge voltage and end-

FIGURE 14.1.   The power supply is an ac-to-dc converter. The source of ac is generally a 115 V ac wall outlet.

AC-TO-DC CONVERTER

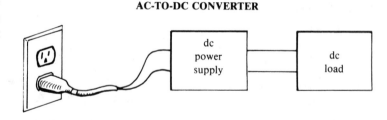

point voltage) and rated capacity (ampere-hours) of rechargeable batteries.

16. Design a half-wave charging circuit (recharge or trickle) for a battery, given the battery's voltage ratings and capacity.
17. Draw a full-wave rectifier circuit (FWB or FWCT) and explain the principles of its operation.
18. Determine the output load current and voltage (peak and average) of a full-wave rectifier (FWB or FWCT), given the rms input voltage and load resistance.
19. Design a full-wave charging circuit (recharge or trickle) for a battery, given the battery's voltage ratings and capacity.

14.2 ALTERNATING CURRENT VOLTAGE SOURCES

The source of electricity for most electronic systems is the 115 V ac wall outlet. The purpose of the power supply is to convert the 115 V ac source voltage to a lower dc voltage with sufficient current to power the electronic system (Figure 14.1).

Wall Outlets

The familiar 115 V ac wall outlet receptacle is pictured in Figure 14.2. It has three slots. The narrow slot is connected by a **black wire** to a fuse or circuit breaker in the central distribution panel. There, the black wire is connected to the power company's 115 V **ac "hot" wire.**

Thus, the narrow slot is at 115 V ac with respect to ground. The other two slots carry no voltage and are connected together to ground at the distribution panel.

The term **ground** must be clearly understood because it is this point that provides the safety connection in case of an accidental shorting of the 115 V ac voltage. Theoretically, ground means a connection where there is zero potential. It is therefore a reference connection with respect to which all voltages can be measured. Ideally, the earth is at zero potential because a charge stored in one location will quickly disperse—typically, within milliseconds—to zero. Thus, the earth is generally established as the true ground connection for all electrical systems.

The U-shaped terminal of the wall outlet is connected to the earth ground. This connection is made by a **green wire** and is called **safety ground, third wire ground,** or **chassis ground.** In Figures 14.2 and 14.3, a bare wire or metallic pipe is shown wired directly from the U-shaped terminal to a water pipe that is buried in the ground. This ground wire is connected to three points: the metal frame of the fuse box, the metal box that contains the wall outlet, and the metal frame of the wall socket. Thus, any metal of an electrical system can be safely touched because it is essentially at zero potential. The third safety wire conducts no current unless a fault occurs.

The wide slot of the 115 V wall outlet shown in Figure 14.2 is the **neutral,** or **common,** connection and is connected by a **white**

115 V WALL OUTLET

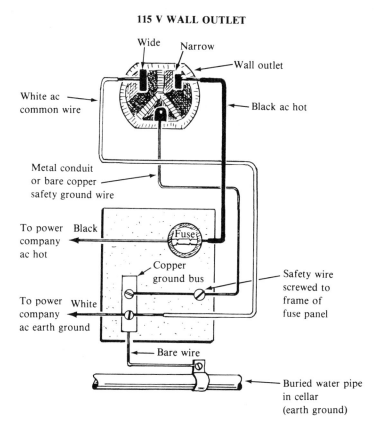

FIGURE 14.2. The wall outlet has three connections. The narrow slot is connected by a black wire to the hot 115 V ac power at the fuse box. The wide slot is called neutral, or common, and is connected by a white wire to ground. The U-shaped slot is the safety ground and is connected by a heavy copper wire directly to a water pipe buried in the ground.

WALL OUTLET CONDUCTORS

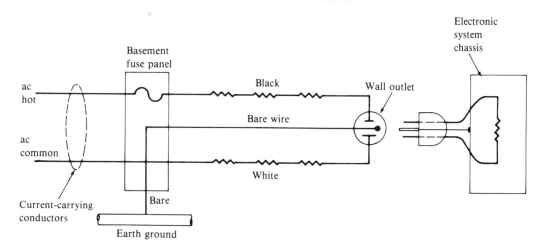

FIGURE 14.3. The current is carried by the black ac hot and white common wires. The safety ground wire carries no current except in case of accidental shorting.

EUROPEAN OUTLET AND PLUG

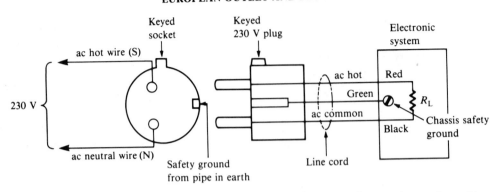

FIGURE 14.4.   European outlets differ in appearance from American outlets and provide 230 V ac at 50 Hz. However, they have the same three connections: ac hot, ac common, and safety ground.

wire. The neutral connection is sometimes mistakenly referred to as "ac ground." However, this wide terminal is not at ground potential because return currents can flow along this ac neutral wire to the true ac earth ground. As shown in Figure 14.3, current through the ac neutral wire will develop a voltage drop across the wire resistance. This voltage drop places the wide slot of the outlet, as well as the white wire, somewhat above earth ground potential.

Wall outlets in countries other than the United States are somewhat different from that shown in Figure 14.2. A typical European wall outlet is shown in Figure 14.4. It has three connections, as does the American system: one hot wire, one neutral wire, and a safety ground that is connected by a spring contact on the outside of the plug. The voltage on the hot wire, however, is 230 V ac, compared to the American standard of 115 V ac.

Line Cord

The **line cord,** or power cord, has a three-pronged plug that fits into the wall outlet. This plug is connected to three color-coded wires, as shown in Figure 14.5. The black and white wires connect to the system load, such as an amplifier or a television set. The load is mounted on a metal chassis to which the third green wire is bolted to complete a connection from chassis to earth. If the hot wire is accidentally connected to the metal chassis, it will cause a short circuit from the high-voltage wire to ground. The short circuit will cause a fuse to blow and consequently disconnect the high voltage. Anyone touching the chassis, therefore, will be protected from electric shock.

Line Voltage Magnitude and Frequency

The ac voltage between current-carrying conductors of a wall outlet is specified in terms of **root-mean-square** (rms) **volts.** The wall outlet voltage in the United States is normally $V_{rms} = 115$ V rms, although actual voltage measurement may vary from 90 V to 125 V. Wall outlet voltages in Europe, Asia, and Africa are nominally 230 V, but deviations from 210 V to 250 V are common.

Voltages lower than the nominal value may result if an outlet is located far from the fuse box. When large currents are drawn from the outlet, the resistance of the line wire will cause a voltage drop in the line and thereby lower the

THREE-WIRE 115 V LINE CORD

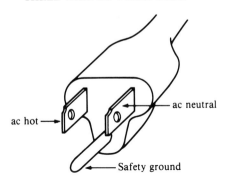

FIGURE 14.5. The three-wire 115 V line cord has two-current-carrying wires: black for ac hot and white for ac common. The third, a green safety wire, connects the metal chassis of an electrical system to earth.

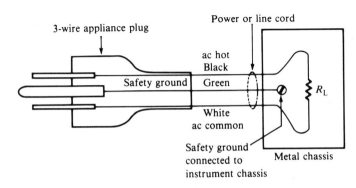

voltage available at the wall outlet. Also, the power company may occasionally have to supply an unusually large amount of current—for example, to power air conditioning on a very hot day—and the additional current flowing through the power grid conductors at such times similarly lowers the voltage available at the fuse box. This condition is known commonly as a *brownout*.

On an oscilloscope, the line voltage appears as a sine wave that can be completely described by its maximum or **peak voltage value** V_p and its frequency f. As shown in Figure 14.6, the sine wave peak voltage for 115 V rms is $V_p = 162$ V.

A portable ac voltmeter measures the rms value of a wall outlet voltage, as shown in Fig-

ure 14.6. The peak value V_p of a sine wave can be calculated from the rms voltmeter measurement V_{rms} by the relationship:

$$V_p = \sqrt{2} V_{rms}$$

or, equivalently:

$$V_p = 1.414 V_{rms} \tag{1}$$

Alternatively, if we know the peak value of voltage—for example, from an oscilloscope measurement—we can calculate V_{rms} as follows:

$$V_{rms} = \frac{1}{\sqrt{2}} V_p$$

115 V LINE VOLTAGE

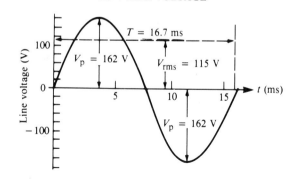

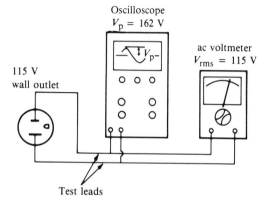

FIGURE 14.6. The wall outlet voltage is 115 V ac rms at a frequency of 60 Hz, which is the value measured with an ac voltmeter. The sinusoidal wave shape can be observed with an oscilloscope. The peak value of a 115 V ac voltage is seen to be $V_p = 162$ V.

or, equivalently:

$$V_{rms} = 0.707V_p \qquad (2)$$

The peak-to-peak excursion of line voltage V_{pp} is simply twice the value of V_p, or:

$$V_{pp} = 2V_p \qquad (3)$$

The sine wave frequency in American sys-

tems is 60 Hz—that is, 60 cycles of a sine wave per second. Thus, one cycle has a time duration equal to 1/60 s. The frequency of systems in Europe, Asia, and Africa is generally 50 Hz with a period of 1/50 s. The full time period T for one sine wave cycle and the frequency f are related by:

$$f = \frac{1}{T} \qquad \text{or} \qquad T = \frac{1}{f} \qquad (4)$$

where f is in hertz and T is in seconds.

EXAMPLE 1: A sine wave has a peak-to-peak value of 310 V and a frequency of 60 Hz. Find its (a) peak value, (b) rms value, and (c) time period for one cycle.

Solution: (a) From Equation 3:

$$V_p = \frac{1}{2}V_{pp} = \frac{1}{2}(310 \text{ V}) = \textbf{155 V}$$

(b) From Equation 2:

$$V_{rms} = 0.707V_p = 0.707(155 \text{ V}) = \textbf{110 V}$$

(c) From Equation 4:

$$T = \frac{1}{f} = \frac{1}{60 \text{ Hz}} = 0.0167 \text{ s} = \textbf{16.7 ms}$$

By convention, the first half-cycle from 0 ms to 8.3 ms is called the **positive half-cycle** because the hot wire is positive with respect to the ac common. The second half-cycle from 8.3 ms to 16.7 ms is the **negative half-cycle** because the hot wire is negative with respect to ac common (Figure 14.6).

EXAMPLE 2: A wall outlet is specified to supply 230 V at 50 Hz. Find the (a) period,

(b) peak voltage, and (c) peak-to-peak voltage.

Solution: (a) From Equation 4:

$$T = \frac{1}{f} = \frac{1}{50\ \text{Hz}} = 0.020\ \text{s} = \textbf{20 ms}$$

(b) From Equation 2:

$$V_\text{p} = 1.41 V_\text{rms} = 1.41 \times 230\ \text{V} = \textbf{324 V}$$

(c) From Equation 3:

$$V_\text{pp} = 2V_\text{p} = 2 \times 324\ \text{V} = \textbf{648 V}$$

14.3 TRANSFORMERS

Most dc power supplies must provide dc voltages that are smaller than the ac line voltage. The first stage of a power supply reduces the 115 V ac wall voltage to a smaller ac voltage. This reduction is accomplished by a **transformer.**

The construction of a transformer is shown in Figure 14.7a. A thinly insulated copper wire is wound around a laminated steel core. A second layer of wire that is electrically insulated from the first also surrounds the core. More heavily insulated lead wires are connected to the winding wire ends and brought out. The schematic symbol for an iron-core transformer is shown in Figure 14.7b.

Typical transformers for dc power supplies are shown in Figure 14.7c. They have a minimum of four leads. Two wires with black insulation identify the **primary winding** of the transformer. These wires are generally connected to the source of the ac voltage. Two wires with different color insulation — for example, green — identify the **secondary winding.** These wires are connected to the load.

When an alternating voltage is applied to one transformer winding, an alternating voltage

TRANSFORMERS

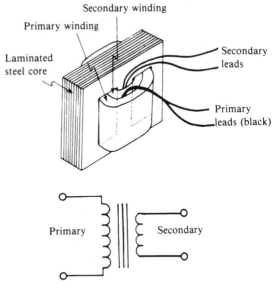

(a) Transformer Construction and Symbol

(b) Typical Power Supply Transformers

FIGURE 14.7. The first stage of a low-voltage power supply is a transformer that reduces 115 V ac to a lower ac voltage. Typical power supply transformers are illustrated along with their construction and schematic symbol. Note that the secondary winding is electrically isolated from the primary winding.

appears in the other transformer winding. The voltage in the secondary winding occurs because the ac primary voltage causes an ac current in the primary winding that in turn causes a changing magnetic field in the transformer core. The changing magnetic field induces an ac voltage in the secondary winding. The metal

TURNS RATIO

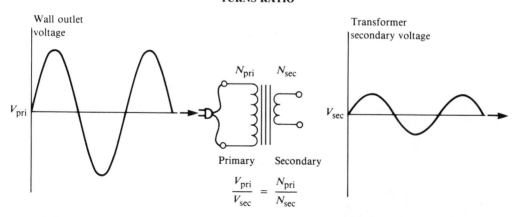

FIGURE 14.8.   In a step-down transformer, the ac voltage of the secondary winding is less than that at the primary. The amount of voltage reduction is directly proportional to the ratio of the number of turns in the primary and secondary windings.

core ensures that most of the magnetic flux caused by the primary winding links with the secondary winding.

Note that there is no direct electrical connection between the primary and secondary windings. The absence of a conduction path between primary and secondary is one of the most important properties of the transformer, namely, **electrical isolation.**

There are two basic types of transformers: (voltage) step-down transformers and (voltage) step-up transformers. For a **step-down transformer,** the secondary voltage is less than the primary voltage. For a **step-up transformer,** the secondary voltage is greater than the primary voltage. Step-down transformers are more widely used because most electronic systems require less than the available 115 V line voltage. Step-up transformers are used primarily for special high-voltage requirements—for example, for the plates of the cathode-ray tube in oscilloscopes and TV sets.

This part of the book deals primarily with low-voltage applications and step-down transformers. However, the principles are equally valid for high-voltage, step-up applications.

Turns Ratio

The voltage that is measured at the secondary winding depends on both the **primary voltage** and the turns ratio—that is, the ratio of the number of turns of wire of the primary and secondary windings. If the secondary winding has a smaller number of turns than the primary, a smaller voltage will appear across the secondary winding. The result is a step-down transformer (Figure 14.8).

It is not necessary to know the number of turns on each transformer winding, but it is necessary to know the turns ratio. From this ratio, we can calculate the secondary voltage that will result from any primary voltage. The ratio of primary voltage V_{pri} to secondary voltage V_{sec} can be found from the ratio of the number of turns on the primary winding N_{pri} and secondary winding N_{sec} with the formula:

$$\frac{V_{pri}}{V_{sec}} = \frac{N_{pri}}{N_{sec}} \qquad (5)$$

Note that this measurement gives the turns ratio

EXAMPLE 3

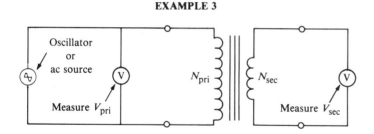

FIGURE 14.9. The turns ratio of a transformer can be determined by measuring V_{pri} and V_{sec} with an ac voltmeter and using Equation 5.

$N_{\text{pri}}/N_{\text{sec}}$, but it does not tell the value of either N_{pri} or N_{sec}.

> **EXAMPLE 3:** The ac primary voltage of a filament transformer is 115 V at 60 Hz. The secondary voltage is measured at 6.3 V rms. (a) Is this a step-up or a step-down transformer? (b) Find the turns ratio. (c) What secondary voltage would result if the primary were connected to 91 V?
>
> **Solution:** (a) It is a **step-down transformer** because V_{sec} is smaller than V_{pri}.
> (b) Using Equation 5:
>
> $$\frac{N_{\text{pri}}}{N_{\text{sec}}} = \frac{V_{\text{pri}}}{V_{\text{sec}}} = \frac{115\ \text{V}}{6.3\ \text{V rms}} = \mathbf{18.2}$$
>
> (c) Rewriting Equation 5, we get:
>
> $$V_{\text{sec}} = V_{\text{pri}} \times \frac{N_{\text{sec}}}{N_{\text{pri}}} = 91\ \text{V} \times \frac{1}{18.2}$$
> $$= \mathbf{5\ V}$$
>
> Figure 14.9 illustrates this result.

Primary and Secondary Currents

Power transformers are very efficient when supplying their rated current. Ideally, the input power to the primary winding P_{in} should equal the output load power P_{L} delivered from the secondary winding. In practice, the transformer delivers more than 90% of P_{in} to the load, so it

STEP-DOWN VOLTAGE TRANSFORMER

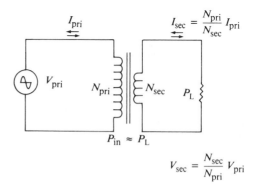

$$V_{\text{sec}} = \frac{N_{\text{sec}}}{N_{\text{pri}}} V_{\text{pri}}$$

FIGURE 14.10. A typical transformer delivers more than 90% of its input power to the load. Therefore, if the voltage is stepped down by a transformer, the secondary current will be stepped up. The stepped-up current is inversely proportional to the turns ratio. See Equation 10.

can be assumed to be nearly ideal. Thus, in Figure 14.10:

$$P_{\text{in}} \approx P_{\text{L}} \tag{6}$$

The input power is the product of the primary current I_{pri} and the primary voltage V_{pri}, or:

$$P_{\text{in}} = I_{\text{pri}} V_{\text{pri}} \tag{7}$$

The output power is the product of secondary current I_{sec} and secondary voltage V_{sec}:

$$P_{\text{L}} = I_{\text{sec}} V_{\text{sec}} \tag{8}$$

Substituting Equations 7 and 8 into Equation 6 gives:

$$I_{pri} V_{pri} = I_{sec} V_{sec} \qquad \text{(9a)}$$

or, expressed as ratios:

$$\frac{V_{pri}}{V_{sec}} = \frac{I_{sec}}{I_{pri}} \qquad \text{(9b)}$$

In order for the primary and secondary powers to remain equal, the products of their respective currents and voltages must be equal, as indicated by Equation 9a. Thus, *in a voltage step-down transformer, the secondary current will be stepped up.*

Another way of explaining this principle involves Equation 9b. Suppose that the ratio of V_{pri} to V_{sec} is 10 to 1. Then, the ratio of I_{pri} to I_{sec} must be 1 to 10. That is, if the secondary voltage is one-tenth the primary voltage, then the secondary current is ten times the primary current.

Since V_{pri} and V_{sec} are determined by the turns ratio, Equations 5 and 9b can be combined to show that I_{pri} and I_{sec} are also related by the turns ratio:

$$\frac{N_{pri}}{N_{sec}} = \frac{V_{pri}}{V_{sec}}$$

and

$$\frac{V_{pri}}{V_{sec}} = \frac{I_{sec}}{I_{pri}}$$

Therefore:

$$\frac{N_{pri}}{N_{sec}} = \frac{I_{sec}}{I_{pri}} \qquad \text{(10a)}$$

If we know the turns ratio, we can use Equation 10a to calculate either the primary or the secondary current if the other is known. By rewriting Equation 10a, we get:

$$I_{sec} = \frac{N_{pri}}{N_{sec}} I_{pri} \qquad \text{(10b)}$$

or

$$I_{pri} = \frac{N_{sec}}{N_{pri}} I_{sec} \qquad \text{(10c)}$$

Equations 9 and 10 are used to calculate the size of the fuses required to protect the transformer against short circuits.

Power, Voltage, and Current Ratings

Buying a tranformer for a power supply usually begins with the index of an electronics parts catalog or the catalog of a transformer manufacturer. Expect to see a variety of transformers categorized according to particular applications — for example, audio, microphone, pulse, and so forth. Transformers that are suitable for power supplies can be identified by any of the following descriptions:

— Power transformer,
— Filament transformer (control/rectifier),
— Rectifier transformer,
— Universal (rectifier) transformer,
— Rectifier/control transformer.

Once the power transformer listings have been located, the first transformer rating to look for is the **primary voltage rating.** The most common primary voltage rating is 115 V or 230 V, which corresponds to the wall outlet voltage. The second important transformer characteristic is the **frequency rating**, typically 25 Hz, 50 Hz to 60 Hz, or 400 Hz. For wall outlet use, select a transformer rated for 115 V and 50 Hz to 60 Hz. The third transformer ratings are the **secondary voltage and current ratings** that are listed under "Secondary Volts" and "Secondary Amps" (Figure 14.11).

An important rating that may not be listed is one that describes how much power the transformer can deliver to the secondary load. This rating is the **volt-amperes**, or VA, rating of the

TYPICAL POWER SUPPLY TRANSFORMER SPECIFICATIONS

RECTIFIER, CONTROL AND
FILAMENT TRANSFORMERS
WITH SINGLE SECONDARY: ALL PRIMARIES 50/60 HZ.§

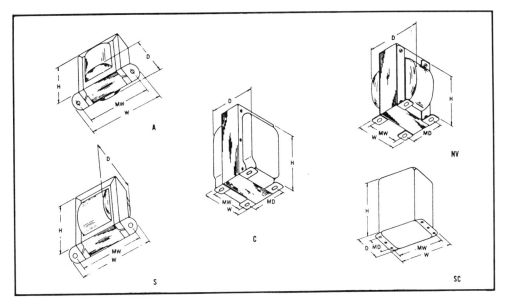

Sec-tion	PART NO.	Style	Secondary ▲		Primary Volts	Insulation Test RMS Volts*	Termination		Case Dimension			Mounting Dimension		Wt. Lbs
			Volts	Amps.			Pri.	Sec.	H	W	D	MW	MD	
1	P-8628	A	2.5C.T.	0.3	117	1500	Leads	Leads	1¼	2⅛	1⅝	1¾	—	0.3
	P-8629	A	2.5	1.0	117	1500	Leads	Leads	1⅜	2⅜	1⅜	2	—	0.4
	P-4026	A	2.5	1.5	117	2500	Leads	Leads	1⅝	2⅞	1⅝	2⅜	—	0.7
	P-4082	SC	2.5C.T.	2.5	117/107	2500	Leads	Leads	2¹¹⁄₁₆	2¹¹⁄₁₆	2¼	2⅜	1½	1.5
	P-6133†	S	2.5C.T.	5.0	117	7500	Leads	Leads	2¾	3⅛	2¼	2¹³⁄₁₆	—	1.5
	P-3024†	C	2.5C.T.	10.0	117/107	2500	Leads	Leads	3⅛	2½	2¾	2	1¾	2.5
	P-6454	S	2.5C.T.	10.0	117/107	7500	Leads	Leads	3⅛	3⅞	2½	3⅛	—	2 5
	P-3060	NV	2.5C.T.	10.0	117	10000	Lugs	Lugs	3⅞	2¹¹⁄₁₆	2⅜	2¼	1⅞	2 5
2	P-6467	A	5.0C.T.	3.0	117	2500	Leads	Leads	2	3¼	2⅛	2¹³⁄₁₆	—	1 4
	P-6455	S	5.0C.T.	6.0	117/107	2000	Leads	Leads	2¾	3⅛	2⅛	2¹³⁄₁₆	—	2
	P-3062†	NV	5.0C.T.	6.0	117	2500	Lugs	Lugs	3⅛	2½	2¼	2	2	2 0
	P-5000†	C	5.0C.T.	6.0	117/107	2500	Leads	Leads	3⅛	2½	3	2	2	3.1
	P-6135	NV	5.0C.T.	10.0	117	2500	Leads	Leads	3⅛	2½	2¾	2	2⅜	3 0
	P 6433†	N	5.0C.T.	15.0	117	2500	Leads	Leads Lugs	3⅛	2½	2¾	2	2¼	3 1
	F-5'5	SC	5.0C.T	20.0	115/230	2500	Terms	Terms	4¹¹⁄₁₆	4⅞	3¹¹⁄₁₆	3⅞	2¾	6 5
	P-6492	C	5.0C.T	30.0	117	2500	Leads	Leads	4¹¹⁄₁₆	3¾	4	3	2¹³⁄₁₆	7 1
	P-6468†	C	5.0C.T.	30.0	117/107	2500	Lugs	Lugs	4¼	3⅞	3⅞	2¾	2¹¹⁄₁₆	6 1
3	P 8385		6.3C.T	0.3	117	1500	Leads	Leads	1¼	2⅛	1⅜	1¾	—	0 3
	P-6465	A	6.3C.T.	0.6	117	1500	Leads	Leads	1⅜	2⅜	1½	2	—	0 4
	P-8705	A	6.3C.T.	0.6	230	1500	Leads	Leads	1⅜	2⅞	2	2	—	0 1
	P-8389	A	6.3	1.0	117	1500	Leads	Leads	1⅝	2⅞	1⅝	2⅜	—	0 6
	P 6134		6.3C.T.	1.2	117	3000	Leads	Leads	1⅝	2⅞	1⅞	2⅜	—	0 8
	P-8190	A	6.3	1.2	117	5000	Leads	Leads	2	3½	1⅞	2¹³⁄₁₆	—	1 0
	P-8191	A	6.3	1.2	6.3	5000	Leads	Leads	2	3½	2	2¹³⁄₁₆	—	1
	P-5011†	NV	6.3C.T.	3.0	117	2500	Lugs	Lugs	3⅛	2½	2½	2	1¼	2 5

*May be operated from a 400 Hz source with no change in output ratings
* Insulation Test Voltage is the allowable RMS working voltage plus 1000 volts.
‡ Has electrostatic shield
▲ R.M.S. values.

FIGURE 14.11.   Power supply transformers may be listed by several names: rectifier, control, filament, and so on. Important ratings are primary voltage and frequency (115 V rms at 50 Hz to 60 Hz) and secondary rms voltage and current. The power rating in volt-amperes can be calculated from the secondary ratings. See Equation 11.

MULTITAP TRANSFORMERS

FIGURE 14.12. Multitap transformers have wires connected to several taps on the primary and secondary windings and can provide a selection of secondary voltages depending on which taps are connected to the ac source and load.

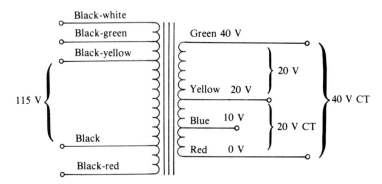

transformer. If the secondary voltage and current are in phase, the VA rating corresponds to the *number of watts that can be delivered to the secondary winding.*

The VA rating can be obtained indirectly in two steps. First, look for the secondary voltage rating in rms volts. Next, look for the maximum secondary current rating under "Secondary Amps." In this part of the book, the symbol V_{sec} (V) is used for rms secondary voltage and I_{sec} (A) is used for rms secondary current. The VA rating of the transformer, then, is:

$$\text{volt-amperes} = V_{sec}I_{sec} \qquad \textbf{(11)}$$

where V_{sec} = secondary rms volts and I_{sec} = secondary rms amperes.

There is an exception to the secondary current ratings for universal rectifier transformers. The maximum secondary current rating may be given in **dc amperes** and will depend on what type of power supply is constructed. This rating is discussed further in Chapter 15.

EXAMPLE 4: What is the VA rating of a transformer that is rated at 115 V/12.6 V with a maximum secondary current of 3 A?

Solution: From Equation 11, we have:

$$\text{volt-amperes} = V_{sec}I_{sec} = 12.6 \text{ V} \times 3 \text{ A}$$
$$= \textbf{37.8 VA}$$

Multivoltage Transformers

The cost of a transformer is generally lower if it is a stock unit and not custom-made to give a specific turns ratio. However, it is often desirable to have more than one turns ratio. The practical choices in terms of cost versus availability of secondary voltages are the multitap and multiwinding transformers.

Figure 14.12 shows a typical **multitap transformer**. Wires are brought out from several taps on the primary winding. All primary wires are either black or black with a colored stripe. The secondary winding may also be tapped to permit connecting a different number of turns in the secondary winding to the load.

To determine how the transformer taps must be connected to obtain a specific secondary voltage, the manufacturer supplies a specification sheet. Typical specification sheet data are shown in Figure 14.13. For example, the first entry says that if 115 V is connected across the black and black-yellow primary wires, the following will be measured:

PRIMARY AND SECONDARY CONNECTIONS FOR F91X TRANSFORMER

Connect 115 V across Primary Wires	rms Secondary Voltages at Terminals			
	Green-Yellow-Red	Green-Blue	Yellow-Blue-Red	Blue-Red
Black-yellow and black	40 V CT	30 V CT	20 V CT	10 V
Black-yellow and black-red	38 V CT	28.5 V CT	19 V CT	9.5 V
Black-green and black	34 V CT	25.5 V	17 V CT	8.5 V
Black-green and black-red	32 V CT	24 V	16 V CT	8 V
Black-white and black	30 V CT	22.5 V	15 V CT	7.5 V
Black-white and black-red	28 V CT	21 V	14 V CT	7 V

FIGURE 14.13. The proper primary and secondary connections for various secondary voltages are given in the transformer's specification sheet. The table gives connections for the F91X transformer represented in Figure 14.12.

— 10 V rms across the secondary blue and red wires,
— 20 V rms across the secondary yellow and red wires,
— 30 V rms across the secondary green and blue wires,
— 40 V rms across the secondary green and red wires.

These voltage levels are shown in Figure 14.12.

On the other hand, if 115 V is connected across the black-red and black-yellow wires, N_{pri} will increase and thereby increase the ratio of N_{pri}/N_{sec}. The result is a larger step-down voltage ratio V_{pri}/V_{sec} or a smaller secondary voltage for a given primary voltage. Thus, the blue-red secondary voltage will be reduced from 10 V to 9.5 V (Figure 14.13, row 2).

Note the terminology "20 V CT" on the secondary winding diagrammed in Figure 14.12. This winding will not only develop 20 V rms across both its terminals, but it also has a **center tap** (CT) that is a third wire brought out from its center. Thus, the 20 VCT winding is made of two 10 V windings connected in series with three terminals, one from each end and one from the center.

MULTIWINDING TRANSFORMERS

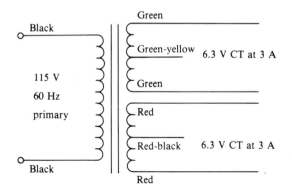

FIGURE 14.14. A multiwinding transformer has more than one primary and/or secondary winding. Various secondary voltages can be obtained by connecting the windings in series aiding or series opposition.

A typical **multiwinding transformer** is shown in Figure 14.14. It has two separate secondary windings wound on the same core as the primary winding. There is no conduction path between any winding, so they are isolated from each other. Each secondary winding can furnish

FIGURE 14.15. A special multiwinding transformer is made for instruments that may be powered from either a 115 V or a 230 V outlet. For 115 V operation, the two primary windings are connected in parallel (connect 2 to 1 and 3 to 4). For 230 V operation, they are connected in series (connect 2 to 3).

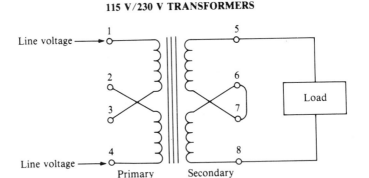

115 V/230 V TRANSFORMERS

$V_{sec} = 6.3$ V rms independently of the other winding.

The windings of a multiwinding secondary can be connected in either series aiding or series opposition to obtain a variety of voltages and current capabilities, in the same way that dc batteries can be connected in series aiding or series opposition. However, we must know the **polarity** of the transformer windings, just as we must know the battery polarities before connecting them together.

For power supply that may be plugged into either a 115 V or a 230 V outlet, a special multiwinding transformer is available. This type has two primary windings (and may also have multiple secondary windings). The schematic of a 115 V to 230 V, 50 Hz to 60 Hz dual-primary/dual-secondary transformer is shown in Figure 14.15. The 115 V primary windings are connected in parallel for 115 V operation or in series for 230 V operation. These transformers are installed in instruments that might be used with either the American 115 V, 60 Hz system or the European 230 V, 50 Hz system. A simple switch at the rear of such instruments changes the primary winding connection to either parallel or series depending on whether the power source is a 115 V or 230 V outlet.

14.4 FUSES

No electrical instrument or circuit is immune from an accidental short circuit. A complete or partial short circuit lowers the circuit resistance and thereby causes an increase in current. According to the I^2R law, unless the increase in current is prevented, excessive heat will be generated and fire may result. An inexpensive device that will limit the current and also disconnect the short-circuited device from its power source is the **fuse**. Two common types of fuse are shown in Figure 14.16, together with their circuit symbol.

The fuse is a conductor element that will carry only a specified maximum-rated current. When the fuse conducts more than its rated current, it heats up and finally melts, opening the circuit and preventing further current flow. The time it takes to open-circuit depends on the amount of excess current.

The fuse is always installed between the black hot wire of the line cord and the connection to the primary winding of the transformer. Thus, when the fuse blows, the transformer is immediately disconnected from the ac power source. Generally, fuses are easily removable from their holders so that they can be quickly

FUSES AND SYMBOL

AGC 1/500 A to 1/32 A   1/16 A to 2 A
MGB 1/16 A and 1/8 A

Normal Blow

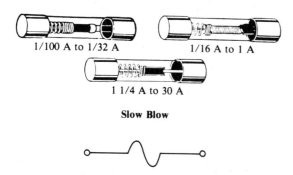

1/100 A to 1/32 A   1/16 A to 1 A

1 1/4 A to 30 A

Slow Blow

FIGURE 14.16. Power supplies are generally fused to protect the transformer and other components from burning out in case of accidental short or overload. The fuse is connected between the hot wire of the line cord and the transformer primary and may be either normal blow or slow blow.

replaced once the fault has been diagnosed and corrected.

Fuse Specifications

Fuses are specified according to three quantities: current rating, voltage rating, and blow characteristic. The **current rating** is expressed in amperes and is the continuous current that a fuse can conduct without melting in an ambient temperature of 25°C. Standard current ratings available in instrument fuses range from 1 mA to 30 A.

The **voltage rating** gives the circuit voltage that the fuse will interrupt for any value of short-circuit current up to 10,000 A. Standard voltage ratings are 32 V, 125 V, and 250 V.

The **blow characteristic** can be one of two types: normal blow or slow blow. A **normal**

blow fuse will blow about 10 ms after being subjected to a current five times its ampere rating. For the same five-times-current overload, it takes a **slow blow** fuse about 2 s, or about 200 times longer than a normal blow fuse, to open-circuit.

Fuse Selection

The fuse protects the primary winding of the transformer and every other circuit element. The maximum secondary transformer current required for an application is generally known, and the corresponding maximum primary rms current can be calculated (Equation 10c). It is good practice to allow for about a 10% to 50% overload current on the secondary to minimize the nuisance of blowing a fuse with a momentary overload. The procedure for determining the proper current rating for a fuse is as follows:

1. Obtain the value for the normal full load (FL) secondary rms current $I_{sec(FL)}$.
2. Calculate the maximum allowable overload secondary current $I_{sec(max)}$. For example, for a 50% overload allowance:

$$I_{sec(max)} = 1.5 I_{sec(FL)}$$

3. Calculate the maximum primary current from Equation 9b:

$$I_{pri(max)} = \frac{V_{sec}}{V_{pri}} I_{sec(max)}$$

4. Select a fuse with a current rating equal to or slightly greater than $I_{pri(max)}$.

The fuse's voltage rating should equal or exceed the circuit voltage.

Normal blow fuses are chosen for resistive loads to protect primarily against short circuits. If high current surges like those encountered in capacitor or motor circuits are expected, a slow blow fuse should be used. In general, a slow

LITTELFUSE TYPE 3AG SLOW BLOW FUSES

Electrical Characteristics	
Rating	**Blow Time**
110%	4 hours, minimum
135%	1 hour, maximum
200%	5 seconds, minimum

Catalog Number	Ampere Rating	Voltage Rating	Catalog Number	Ampere Rating	Voltage Rating
313.010	1/100	250	313 01.2	1 2/10	250
313.031	1/32	250	313 1.25	1 1/4	250
313.040	4/100	250	313 01.5	1 1/2	250
313.062	1/16	250	313 01.6	1 6/10	250
313.100	1/10	250	313 01.8	1 8/10	250
			313 002	2	250
313.125	1/8	250	313 2.25	2 1/4	250
313.150	15/100	250	313 02.5	2 1/2	250
313.175	.175	250	313 02.8	2 8/10	250
313.187	3/16	250	313 003	3	250
313.200	2/10	250	313 03.2	3 2/10	250
313.250	1/4	250	313 004	4	250
313.300	3/10	250	313 005	5	250
313.375	3/8	250	313 6.25	6 1/4	250
313.400	4/10	250	313 007	7	250
313.500	1/2	250	313 008	8	32
313.600	6/10	250	313 010	10	32
313.700	7/10	250	313 015	15	32
313.750	3/4	250	313 020	20	32
313.800	8/10	250	313 025	25	32
313.001	1	250	313 030	30	32

FIGURE 14.17. Fuses in the 3AG series are commonly used in electronic instruments. The table gives typical ampere and voltage ratings for slow blow 3AG fuses.

blow is the correct choice if protection is needed against *temporary* overload currents in excess of 50% of normal full load current.

One widely used fuse type is the 3AG series. Typical current and voltage ratings for 3AG slow blow fuses are given in Figure 14.17. Normal blow 3AG fuses have similar ratings.

EXAMPLE 5: A transformer is rated at 115 V/12.6 V with a secondary current rating of 3 A rms. What size fuse should be chosen to blow when the secondary current reaches 3 A?

Solution: To determine the desirable fuse design, we begin with $I_{sec(max)} = 3$ A. Then, we find the primary current from Equation 9b:

$$I_{pri(max)} = \frac{V_{sec}}{V_{pri}} I_{sec(max)} = \frac{12.6 \text{ V}}{115 \text{ V}} \times 3 \text{ A}$$
$$= 0.11 \times 3 \text{ A} = 0.33 \text{ A}$$

HALF-WAVE RECTIFIER

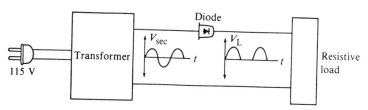

FIGURE 14.18. The simplest way to produce a dc voltage is to eliminate the negative half-cycle of the sine wave by adding a diode between the transformer secondary winding and the load. This circuit is called a half-wave rectifier.

Next, we determine the peak primary voltage from Equation 1:

$$V_{\text{pri(peak)}} = 1.41 V_{\text{rms}} = 1.41 \times 115$$
$$= 162 \text{ V}$$

From the list in Figure 14.17, we select the 3AG fuse with a voltage rating that just exceeds the peak primary voltage and peak primary current. The correct choice is number **313.375, 3/8 A, 250 V**.

In practical transformers, the primary winding draws a magnetizing current at all times, even when the secondary is drawing no current. The primary magnetizing current is neither sinusoidal nor in phase with the primary voltage and is typically on the order of $\frac{1}{4}$ A pp, or about 0.08 A rms. For this reason, the fusing current calculated from Equation 9b may be in error by a few percent. Thus, the calculated fuse selection should be tested at maximum overload and the final selection made by trial and error.

14.5 HALF-WAVE DIODE RECTIFIER

Once the 115 V ac wall outlet has been reduced by a transformer, the secondary ac voltage must be converted to a dc voltage. The simplest way to do this is to eliminate the negative half-cycle of the sine wave by means of a semiconductor

device called a **rectifier diode**, or simply a diode. As shown in Figure 14.18, the transformer secondary voltage is alternating. On the other side of the diode, however, only one half-cycle of the sine wave passes on to the load. The diode conducts current in one direction, but blocks current in the other. The circuit shown in Figure 14.18 is called a **half-wave diode rectifier**.

The load voltage from a half-wave rectifier has only one polarity; hence, it is dc. However, its magnitude changes considerably, causing a pulsating dc current in the load. Methods for reducing these pulsations are discussed in succeeding sections. For many applications, however, these pulsations are acceptable and the half-wave rectifier is an adequate dc power supply.

V–I Characteristics of a Diode

General-purpose rectifier diodes come in various shapes. Figure 14.19 shows a representative selection and their schematic symbol. The arrowhead on the symbol points in the direction in which the diode conducts conventional current. The diode terminals are called the **anode** and the **cathode**. The arrowhead in the illustration indicates that current flows toward the cathode by pointing to the cathode terminal. If the diode symbol is not stamped on the case, a colored band may identify the cathode terminal.

The V–I characteristic of a typical silicon rectifier diode is shown in Figure 14.20. When the anode voltage is greater than about 0.6 V

RECTIFIER DIODES AND SYMBOL

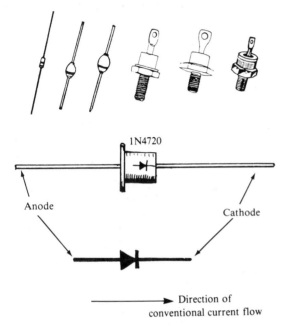

FIGURE 14.19. Typical rectifier diodes are shown along with their schematic symbol. The diode's anode and cathode can be identified from markings on the diode case.

with respect to the cathode, the diode is said to be **forward biased** and conducts current easily. As shown in Figure 14.21a, the positive terminal of the battery causes the current I_{fb} to flow through the diode in the forward-conducting direction, as indicated by the arrowhead.

The forward-bias voltage V_{fb} of the silicon diode remains relatively constant at about 0.6 V. Note in Figure 14.20 that V_{fb} changes by less than 0.1 V as I_{fb} increases from 0.1 A to 1.0 A. The circuit current can be expressed as:

$$I_{\text{fb}} = \frac{V - V_{\text{fb}}}{R_{\text{L}}} = \frac{V - 0.6}{R_{\text{L}}} \tag{12}$$

When a diode is reverse biased, the reverse current in the circuit I_{rb} is very nearly zero, as shown in Figure 14.21b. The diode is essentially an open circuit. Thus, all of the applied voltage

V appears across the diode as the reverse-bias voltage V_{rb}.

Figure 14.20 permits a comparison of both the diode's forward and reverse characteristics. When forward biased, the diode exhibits a small 0.6 V drop, but otherwise acts like a short circuit. When reverse biased, the diode blocks current flow and acts essentially like an open circuit. This $V–I$ characteristic is the basis for using the diode in the half-wave rectifier.

Principles of Operation

The operation of the diode is directly analogous to the operation of a half-wave rectifier. As shown in Figure 14.22a, when the positive half-cycle voltage of V_{sec} exceeds about 0.6 V, the diode becomes forward biased. Load voltage V_{L} then has the same polarity as V_{sec}, but is slightly lower because of the small diode drop of 0.6 V:

$$V_{\text{L}} = V_{\text{sec}} - 0.6\,\text{V} \tag{13}$$

If the peak value of V_{sec} exceeds about 10 V, the constant 0.6 V drop is negligible and V_{L} can be considered to equal V_{sec}.

During the negative half-cycle (Figure 14.22b), V_{sec} reverse-biases the diode. Because the reverse current of the diode is zero, no current flows through R_{L} and the load voltage is zero. Nearly all of the applied voltage V_{sec} is dropped across the diode, and none appears across the load.

Load Voltages and Currents

The operation of a half-wave rectifier can be observed most conveniently with an oscilloscope. Figure 14.23 shows typical results for a low-voltage transformer with a 6.3 V CT secondary. One-half of the 6.3 V CT secondary winding gives the secondary voltage of 3.15 V, which is not large with respect to the 0.6 V forward diode drop. Thus, the small drop is visible. For most applications, the secondary voltage is much larger than 0.6 V, so the notch in the waveform cannot be seen.

ELECTRICAL BEHAVIOR OF A SILICON DIODE RECTIFIER

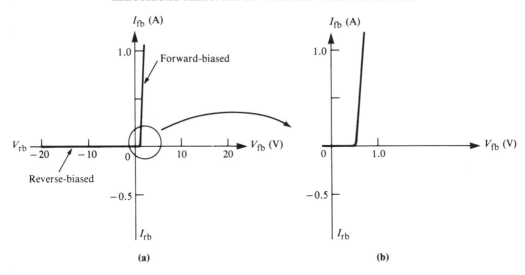

(a)   (b)

FIGURE 14.20. The electrical behavior of a silicon diode rectifier is described by its V–I characteristic. When forward biased (anode positive), it conducts current easily above 0.6 V. When reverse biased, it blocks current flow. The region between zero and 0.6 V is shown enlarged in (b).

The peak value of the secondary voltage is found from Equation 1:

$$V_{sec(peak)} = 1.41 V_{sec(rms)} = 1.41 \times 3.15 \text{ V}$$
$$= 4.4 \text{ V}$$

The peak value of voltage across the load $V_{L(peak)}$ is found from Equation 13:

$$V_{L(peak)} = V_{sec(peak)} - 0.6 \text{ V}$$
$$= 4.4 \text{ V} - 0.6 \text{ V} = 3.8 \text{ V}$$

Load voltage and diode voltage are shown in Figure 14.23. Note that it takes about 0.4 ms for the positive half-cycle of V_{sec} to rise from 0 V to 0.6 V. The diode does not become forward biased, and no output load voltage appears across R_L until $t = 0.4$ ms. At every instant, the sum of the diode voltage drop and the load voltage equals the secondary voltage, which is in agreement with Equation 13.

The load current I_L has the same shape as load voltage V_L (Figure 14.24). According to Ohm's law, V_L and R_L determine the load current:

BEHAVIOR OF A SILICON DIODE IN A CIRCUIT

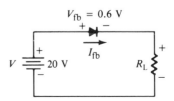

(a) Forward-Biased Diode: $V_L = 19.4$ V

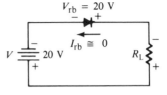

(b) Reverse-Biased Diode: $V_L = 0$ V

FIGURE 14.21. When a silicon diode is connected in series with a resistive load and forward biased, all but 0.6 V of the supply voltage appears across the load. When the diode is reverse biased, nearly all of the supply voltage is dropped across the diode and none appears across the load.

HALF-WAVE RECTIFIER OPERATION

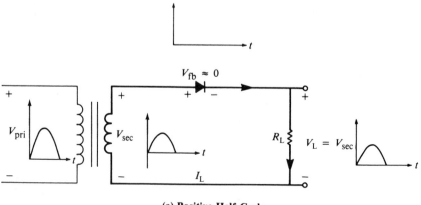

(a) Positive Half-Cycle

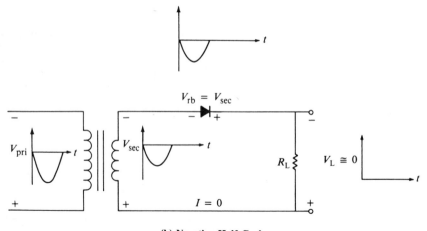

(b) Negative Half-Cycle

FIGURE 14.22.   When a diode is connected to form a half-wave rectifier, it is forward biased during the positive half-cycle of the sine wave and reverse biased during the negative half-cycle. Thus, the load voltage is essentially V_{sec} during the positive half-cycle and zero during the negative half-cycle.

TYPICAL HALF-WAVE RECTIFIER CIRCUIT

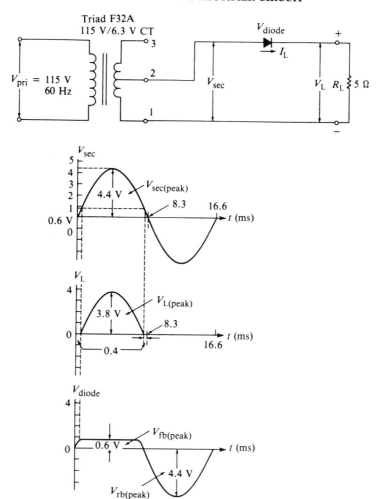

FIGURE 14.23.   The figure shows the voltage waveforms across the transformer second-ary, load resistor, and diode for a half-wave rectifier as seen by an oscilloscope. A low 6.3 V secondary makes the small 0.6 V diode drop observable.

FIGURE 14.24. The average values of load voltage and current are the values read by a dc voltmeter and ammeter. They are 0.318 times the peak values of the sine wave.

AVERAGE VALUES

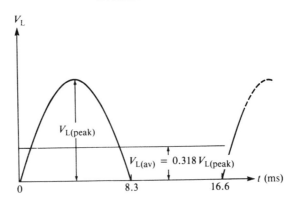

(a) Average Value of Load Voltage

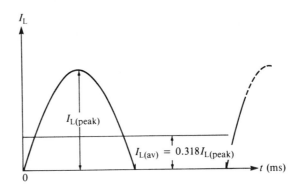

(b) Average Value of Load Current

$$I_L = \frac{V_L}{R_L} \quad (14)$$

It is sometimes necessary to calculate the *average* value of the load voltage or load current. This average value is the value that a dc voltmeter or ammeter measures. The average value of load voltage is given by:

$$V_{L(av)} = 0.318 \times V_{L(peak)} \quad (15)$$

The peak value of V_{sec} can be calculated from the transformer's rms secondary voltage rating using Equation 2 and then the peak load

voltage can be calculated from Equation 13. It then follows that the average dc load current is related to the peak load current by:

$$I_{L(av)} = 0.318 \times I_{L(peak)} \quad (16)$$

EXAMPLE 6: The circuit diagram in Figure 14.23 gives $V_{sec} = 3.15$ V and $R_L = 10 \, \Omega$. Find (a) the dc load voltage, (b) the peak load current, and (c) the dc load current

Solution: (a) From Figure 14.23, we know $V_{L(peak)} = 3.8$ V. From Equation 15:

$$V_{L(av)} = 0.318 \times V_{L(peak)}$$
$$= 0.318 \times 3.8 \text{ V} = \textbf{1.2 V}$$

(b) From Equation 14:

$$I_{L(peak)} = \frac{V_{L(peak)}}{R_L} = \frac{3.8 \text{ V}}{10 \text{ }\Omega} = \textbf{0.38 A}$$

(c) From Equation 16:

$$I_{L(av)} = 0.318 \times I_{L(peak)}$$
$$= 0.318 \times 0.38 \text{ A} = \textbf{0.12 A}$$

The same load current that is conducted through R_L is conducted through the transformer secondary and diode. This fact helps to determine the proper diode for the maximum required load current.

Rectifier Diode Ratings

The diodes for power supplies are designed specifically for operation as rectifiers. They are generally made of silicon, which has a higher current-carrying capacity than other materials—for example, germanium. Thus, rectifier diodes are often designated **silicon rectifiers** in electronic parts catalogues and manufacturers' data sheets. From the section on silicon rectifier diodes, information about only two electrical characteristics is normally needed: the average forward current rating and the peak inverse voltage rating (Figure 14.25).

The **average forward current rating** is symbolized by I_o or I_{av} and is the maximum average value of half-wave current that the diode can safely conduct without damage. Thus, a diode rated for $I_o = 1$ A can conduct a maximum average current of 1 A. Its maximum peak current can be calculated from Equation 16:

$$I_o = 0.318 \times I_{L(peak\,max)}$$

$$1 \text{ A} = 0.318 I_{L(peak\,max)}$$
$$I_{L(peak\,max)} = \frac{1 \text{ A}}{0.318} = 3.14 \text{ A}$$

The **peak inverse voltage rating** (PIV) is the maximum reverse-bias voltage that a diode can withstand without burning out. The maximum reverse-bias voltage rating is sometimes given another designation by the manufacturer. If a PIV listing cannot be found in a catalogue, check these common designations:

$$V_{rrm} = \text{peak repetitive reverse voltage}$$
$$\text{PIV} = \text{peak inverse voltage}$$
$$V_{rb} = \text{peak reverse-bias dc voltage}$$
$$\text{PRV} = \text{peak reverse voltage}$$

Note that each of these designations refers to *peak* values. The same maximum reverse voltage may be specified in terms of the rms value, however. For example:

$$V_{rb(rms)} = \text{rms maximum reverse-bias voltage}$$

14.6 POWER SUPPLY GROUNDING

Safety Ground

A half-wave rectifier has *two* dc output terminals labeled positive and negative. Remember that they are electrically isolated from the ac line voltage. Also, recall that the ac wall outlet has *three* terminals, one of which is at earth potential—the safety ground (see Figures 14.2 to 14.5).

The ac power is connected from the wall outlet to the rectifier by a three-wire line cord and plug, as shown in Figure 14.5. The black hot wire of the line cord is connected to the fuse either directly or through an on/off switch. The white ac neutral wire of the line cord is connected to one side of the transformer primary

FIGURE 14.25. Rectifier diodes are generally made of silicon. They are rated by their average forward current (I_o or I_{av}) and peak inverse voltage (PIV, V_{rrm}, PRV, and so on).

TYPICAL RECTIFIER DIODE RATINGS

Silicon Rectifiers

Type No.	Data File No.	Package	$V_{rrm(max)}$ (V)	I_o @ $T_{A(max)}$ (A)	(°C)
Standard—Hermetic and plastic lead-type packages					
1N440B	5	DO-1	100	0.25	150
1N441B	5	DO-1	200	0.25	150
1N442B	5	DO-1	300	0.25	150
1N443B	5	DO-1	400	0.25	150
1N536	3	DO-1	50	0.25	150
1N537	3	DO-1	100	0.25	150
1N538	3	DO-1	200	0.25	150
1N539	3	DO-1	300	0.25	150
1N540	3	DO-1	400	0.25	150
1N547	3	DO-1	600	0.25	150

Low-Voltage Rectifiers

I_o (A)	PRV			
	50	100	200	400
1	IN4001	IN4002	IN4003	IN4004
1	CB05	CB10	CB20	CB40
3	HAB005	HAB010	HAB020	HAB040

to provide a return path for ac current (Figure 14.26). The green safety wire of the line cord should be electrically connected to the metal chassis and any metal covers or front plates. This connection insures that any metal surface that can be contacted by a person is at earth potential. The point on the chassis where the green ground wire is connected is known as the **chassis ground.** It is represented on a schematic diagram by the symbol ⏚ .

Floating Power Supplies

When neither of the rectifier output terminals is connected to ground, the power supply is classified as a **floating power supply.** The dc terminals are electrically equivalent to a battery in that neither terminal of a battery is normally wired to ground. For example, the power supply diagrammed in Figure 14.26 is neither positive nor negative. However, if we chose to make all voltage measurements with respect to the negative terminal, the negative terminal would be the reference for, or common to, all measurements. The negative terminal is then called the **power common** or **dc return.** Unfortunately, common connections are often confused with the term *ground.* Thus, the power common sometimes is erroneously called ground, even though it is not connected to earth. It may, in fact, be at a high potential with respect to earth.

Similarly, if all voltage measurements were taken with respect to the positive dc terminal,

FLOATING POWER SUPPLY

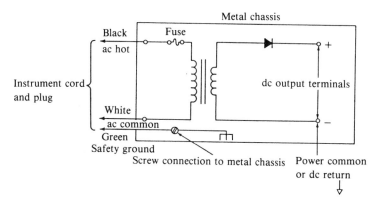

FIGURE 14.26. The figure shows line cord and output terminal connections for a floating power supply. Since neither output terminal is grounded, the power supply can be either positive or negative, depending on which terminal is taken to be the power common.

the positive terminal would be called power common. Power common connections are identified by the symbol ⏀ . This symbol indicates that all dc measurements are taken with respect to this point.

Single-Polarity Power Supplies

In many applications, only a single-polarity power supply is needed. Also, for safety, it is good practice to electrically connect the dc power common terminal to earth, as well as to the chassis. This precaution involves connecting the power common terminal to the green safety wire and to chassis ground.

A single-polarity, positive voltage supply is shown in Figure 14.27. A connection from the green safety ground to the negative dc terminal places this terminal at earth potential. All voltage measurements are now positive with respect to the grounded negative terminal. The grounded terminal is identified by the symbol ⏚ . If either the hot ac wire or the positive dc terminal is accidentally short-circuited to the chassis, the fuse will blow and no shock hazard exists.

The connection for a single-polarity negative voltage supply is also shown in Figure 14.27. Here, the positive terminal is used as a common terminal and is also grounded to earth

by the green safety wire. Note that the green safety wire does *not* conduct dc current from the rectifier: dc current is conducted only through the load.

Laboratory Power Supplies

In laboratory power supplies, the green safety wire is brought out to a terminal labeled *ground* or marked with the symbol ⏚ so that the supply can be used as a negative, positive, or floating source (Figure 14.28). If there is no connection between ground and the + and − terminals, the power supply will be floating. If the − terminal is connected to safety ground, it is a positive supply. If the + terminal is connected to a safety ground, it is a negative supply.

14.7 DESIGNING A HALF-WAVE BATTERY CHARGER

The pulsating output waveform of the half-wave rectifier is far from the ideal dc voltage expected from a power supply. Nonetheless, it is dc, having only one current direction, and the circuit has the advantage of minimal cost be-

FIGURE 14.27. A floating power supply can be converted to either a positive supply or a negative supply by connecting the negative or positive terminals, respectively, to the safety ground.

ALTERNATIVE FLOATING POWER SUPPLY CONNECTIONS

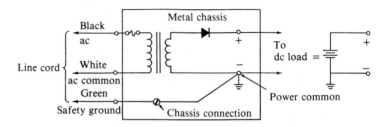

(a) Positive Voltage Supply

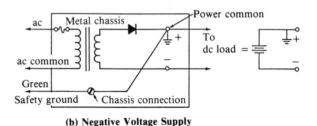

(b) Negative Voltage Supply

FIGURE 14.28. A laboratory power supply has all three terminals at the output so that the experimenter has the option of making it a floating, positive, or negative supply. Note that no voltage appears across the output terminals and ground without a ground connection.

LABORATORY POWER SUPPLY

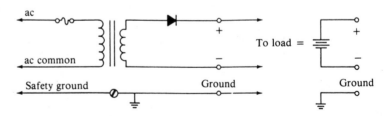

cause it requires only two basic components: a transformer and a diode.

One application in which the pulsations are not a problem and for which cost is a significant issue is battery chargers. Many modern portable electronic systems, such as calculators and radios, operate on rechargeable batteries. Because these items are very competitively priced, every effort is made to reduce their cost and the cost of the battery recharger accessory.

The half-wave rectifier is ideally suited for this application.

The basic circuit for a **half-wave battery charger** is shown in Figure 14.29. The primary requirement is that the secondary transformer voltage must be larger than the battery voltage in order to pump charging current into the battery. During the positive half-cycle (Figure 14.29a), the diode conducts charging current into the battery, but it prevents the battery from

HALF-WAVE RECTIFIER AS A BATTERY CHARGER

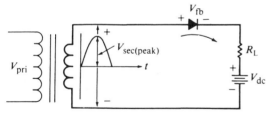

(a) Positive Half-Cycle

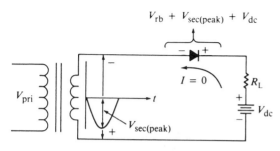

(b) Negative Half-Cycle

FIGURE 14.29. The half-wave rectifier can be used as a battery charger if $V_{sec(peak)}$ is greater than the battery voltage V_{dc}. When V_{sec} is positive and greater than V_{dc}, the diode conducts a charging current. The amount of charging current is determined by resistor R_L.

discharging back into the transformer during the negative half-cycle (Figure 14.29b).

Determining Charging Voltage

In order to design a battery charger, we must determine the voltages and currents that the charger must supply. There are three distinct voltages for rechargeable batteries: nominal cell voltage, end-point voltage, and end-of-charge voltage.

The terminal voltage of a purchased one-cell battery is a value specified by the manufacturer as the **nominal cell voltage** V_{dc}. For example, Figure 14.30 lists the nominal cell

voltage of a lead-acid cell as 2.1 V. A battery consists of one or more individual cells connected in series to yield higher voltages. For example, in an automobile battery, six 2.1 V lead-acid cells in series are used to generate a nominal battery voltage of 12.6 V.

The **end-point voltage** V_{dis} is the lowest acceptable cell voltage from a discharged battery. Once a cell voltage has reached the end-point voltage, recharging is necessary.

While a cell is being charged, its terminal voltage increases beyond the nominal cell voltage. When a cell becomes fully recharged, its voltage has a value called the **end-of-charge voltage** V_{ch}. After the cell has been removed from the charger, its voltage drops slowly toward the nominal cell voltage. Cell voltage values for alkaline-manganese, nickel-cadmium, and lead-acid rechargeable cells are given in Figure 14.30.

Determining Charging Current

The amount of charge that a battery can deliver depends on its size, material, and construction. Its supply capability is specified by the manufacturer as a rated **capacity** given in units of **ampere-hours** (A-h). The A-h rating tells how much current a battery can supply and for how long before the voltage drops to its end-point voltage. For example, a CF500 nickel-cadmium cell rated at 0.5 A-h means that a current of 0.1 A can be drawn from it for 5 h (0.1 A $\times$ 5 h = 0.5 A-h) before its 1.25 V nominal cell voltage drops to an end-point voltage of 1.1 V. Typical A-h ratings are shown in Figure 14.31 for the familiar 1.5 V AA, C, and D batteries.

The recommended **average charging current** I_{ch} is about one-tenth the A-h rating. Thus, for a known A-h rating, we can calculate I_{ch} from:

$$I_{ch} = \frac{\text{A-h rating}}{10 \text{ h}} \qquad (17)$$

FIGURE 14.30. To design a battery charger, three battery voltage values must be known: V_{dc}, V_{ch}, and V_{dis}. The table gives values for the most common rechargeable battery types. Note that these values are for single cells.

RECHARGEABLE BATTERY VOLTAGES

Battery Materials	Nominal Cell Voltage V_{dc} (V)	End-of-Charge Voltage V_{ch} (V)	End-Point Voltage V_{dis} (V)
Alkaline-manganese	1.5	1.7	1.1
Nickel-cadmium	1.25	1.45	1.1
Lead-acid	2.1	2.6	2.0

AMPERE-HOURS RATING FOR RECHARGEABLE BATTERIES

1.5 V	1.5 V	1.5 V
Size AA	Size C	Size D

Battery Type	V dc	A-h
AA	1.5	0.5
C	1.5	1.2
D	1.5	1.2

FIGURE 14.31. The average charging current is determined by the battery's ampere-hours (A-h) rating. The table gives values for three common types of rechargeable batteries.

The charge current in Equation 17 is for a completely discharged battery. To fully charge a completely discharged battery, this current should be applied to the battery for about 14 h. The extra 4 h are required because the battery will not convert 100% of the charge current to chemical energy.

Batteries will self-discharge very slowly when idle. If a charged battery must be stored for a long period of time, its charge can be maintained indefinitely with a **trickle charge current** $I_{trickle}$. The required average trickle charge current $I_{trickle}$ is about one-fiftieth of the A-h rating:

$$I_{trickle} = \frac{\text{A-h rating}}{50 \text{ h}} \qquad (18)$$

The trickle charge current can be left on indefinitely (for days).

Selecting the Transformer

The transformer shown in Figure 14.29 should provide a secondary voltage that is large with respect to the battery voltage. This precaution will ensure that the charging current is relatively constant while the battery voltage is changing from its end-point voltage to its end-of-charge voltage. Typically, the rms voltage of V_{sec} should be approximately twice the battery's end-of-charge voltage V_{ch}, and the transformer's secondary rms current rating should be about twice I_{ch}:

$$\begin{aligned} V_{sec} &\geq 2V_{ch} \\ I_{sec} &\geq 2I_{ch} \end{aligned} \qquad (19)$$

Selecting the Diode

The diode should have an average current rating I_o equal to or greater than the charging current I_{ch}. The diode's PIV rating should exceed the sum of the battery's end-of-charge voltage and the peak secondary voltage:

$$I_o \geq I_{ch}$$
$$\text{PIV} \geq V_{sec(peak)} + V_{ch} \tag{20}$$

Selecting the Fuse

The fuse is placed in series with the primary. Its current rating should equal or exceed the primary current value that in turn should correspond to a secondary current of about $2I_{ch}$. The primary fusing current is found from the inverse turns ratio V_{sec}/V_{pri}:

$$I_{fuse} = 2I_{ch} \frac{V_{sec}}{V_{pri}} \tag{21}$$

Selecting the Current-Limiting Resistor

A resistor R_L placed in series with the battery limits the current to the value required to charge the battery (see Figure 14.29). It must have the proper ohmic value and power capability. The resistance of R_L can be calculated from:

$$R_L = \frac{V_{sec} - V_{ch}}{2.5I_{ch}} \tag{22}$$

where R_L = current-limiting resistance
         in ohms
     I_{ch} = charging current in amperes,
         calculated from Equation 17
    V_{sec} = rms voltage of secondary
         winding
    V_{ch} = end-of-charge
         voltage of the battery

The power rating for R_L is determined from the maximum secondary current $2I_{ch}$:

$$P \geq (2I_{ch})^2 R_L \tag{23}$$

where P = power rating in watts
     I_{ch} = charging current in amperes
    R_L = current-limiting resistance
         in ohms

These design equations are not rigid specifications, but general guides to selecting components. A relatively wide range of secondary voltages and charging currents is adequate for charging a battery.

Design Example: 4.5 V Battery Charger

The design procedure just outlined is illustrated by the following example.

EXAMPLE 7: A rechargeable alkaline-manganese dioxide battery #563 is rated at 4.5 V and 2.5 A-h. Design a half-wave battery charger.

Solution: Figure 14.30 gives the alkaline-manganese nominal cell voltage as 1.5 V. Therefore, the number of cells in the battery can be calculated from:

$$\text{number of cells} = \frac{\text{battery voltage}}{\text{voltage per cell}}$$
$$= \frac{4.5 \text{ V}}{1.5 \text{ V}} = 3 \text{ cells}$$

The end-point, or discharged, voltage V_{dis} for each cell can be found in Figure 14.30. Then:

$$V_{dis} = \text{number of cells} \times V_{dis} \text{ per cell}$$
$$= 3 \times 1.1 \text{ V} = 3.3 \text{ V}$$

The end-of-charge voltage V_{ch} can be determined the same way:

$$V_{ch} = \text{number of cells} \times V_{ch} \text{ per cell}$$
$$= 3 \times 1.7 \text{ V} = 5.1 \text{ V}$$

The charging current is calculated from Equation 17:

$$I_{ch} = \frac{\text{A-h}}{10 \text{ h}} = \frac{2.5 \text{ A-h}}{10 \text{ h}} = 0.25 \text{ A}$$

The transformer secondary voltage should be equal to or greater than twice the end-of-charge voltage, according to Equation 19:

$$V_{sec} \geq 2 \times V_{ch} \geq 2 \times 5.1 \text{ V} \geq 10.2 \text{ V}$$

The secondary current rating should be equal to or exceed twice I_{ch}, again according to Equation 19:

$$I_{sec} \geq 2I_{ch} \geq 2 \times 0.25 \text{ A} \geq 0.5 \text{ A}$$

Therefore, we select a transformer with a standard secondary voltage of **12.6 V** and an rms secondary current rating greater than **0.5 A**. The **F32A transformer** shown in Figure 14.23 is satisfactory if its 6.3 V secondaries are wired in series aiding.

The average current rating I_o of the diode should exceed I_{ch}, according to Equation 20:

$$I_o \geq 0.25 \text{ A}$$

Its PIV rating should exceed the sum of $V_{sec(peak)}$ and V_{ch} (Equation 20):

$$\text{PIV} \geq V_{sec(peak)} + V_{ch}$$
$$\geq (1.4)(12.6 \text{ V}) + (3)(1.7 \text{ V})$$
$$\approx 23 \text{ V}$$

The **1N4720 diode rated at 3 A and 200 PIV** is more than adequate for our battery charger.

The primary fusing current, from Equation 21, is:

$$I_{fuse} = 2I_{ch}\frac{V_{sec}}{V_{pri}}$$

For a 115 V primary, the fuse current would be:

$$I_{fuse} = 2(0.25 \text{ A}) \times \frac{12.6 \text{ V}}{115 \text{ V}} = 0.055 \text{ A}$$

We should select the nearest standard fuse. Figure 14.17 tells us to choose **3AG 1/16 A at 250 V.**

The value of the current-limiting resistor R_L is found from Equation 22:

$$R_L = \frac{V_{sec} - V_{ch}}{2.5I_{ch}} = \frac{12.6 \text{ V} - 5.1 \text{ V}}{2.5(0.25 \text{ A})}$$
$$= 12 \text{ }\Omega$$

The power rating of R_L is found from Equation 23:

$$P \geq (2I_{ch})^2 R_L \geq (2 \times 0.25 \text{ A})^2 \times 12 \text{ }\Omega$$
$$\geq 3 \text{ W}$$

Select the nearest standard **5 W or 10 W resistor of 10 Ω or 15 Ω**. The final design is shown in Figure 14.32.

The procedure for converting this charger to a trickle charger is simple. Calculate $I_{trickle}$ from Equation 18 and then substitute this value for I_{ch} in Equations 22 and 23 to determine the trickle charger current-limiting resistance $R_{L(trickle)}$.

The value of the trickle charge current is found from Equation 18:

$$I_{trickle} = \frac{2.5 \text{ A-h}}{50 \text{ h}} = 50 \text{ mA}$$

From Equation 22, the value of the resistance is:

$$R_L = \frac{V_{sec} - V_{ch}}{2.5I_{trickle}} = \frac{12.6 \text{ V} - 5.1 \text{ V}}{2.5(0.050 \text{ A})}$$
$$= 60 \text{ }\Omega$$

EXAMPLE 7

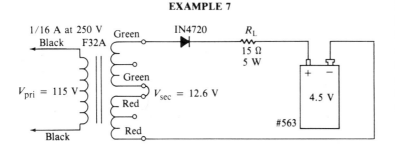

FIGURE 14.32. The half-wave rectifier can be used as a battery charger to charge a 4.5 V alkaline-manganese battery. The secondaries are wired for 12.6 V rms, and a 15 Ω, 5 W resistor is placed in series with the battery to set the charging current.

Note that $R_{L(trickle)}$ should equal $5R_L$.

The power rating of $R_{L(trickle)}$ is found from Equation 23:

$$P \geq (2I_{trickle})^2 R_{L(trickle)}$$
$$= (2 \times 0.050 \text{ A})^2 \times 60 \text{ Ω} \geq 0.6 \text{ W}$$

Therefore, we choose a stock **1 W carbon resistor** with the value nearest 60 Ω, namely, **68 Ω**.

A charger with both full- and trickle-charge capability can be constructed by installing both resistors and a two-position switch: one position supplying a full-charge current via the 15 Ω resistor and the other position supplying charge current via the 68 Ω resistor (Figure 14.33).

4.5 V TRICKLE CHARGER

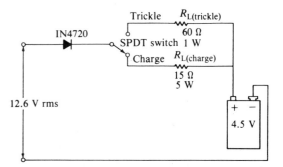

FIGURE 14.33. The charger designed in Example 7 can be modified to also provide a trickle charge current by adding a 60 Ω, 1 W resistor and a two-position switch.

14.8 FULL-WAVE DIODE RECTIFIER

An obvious disadvantage of the half-wave rectifier is that it supplies current to the load only during one half-cycle of the sine wave voltage. Thus, the current capabilities of both the transformer and the wall outlet are not fully utilized. To obtain more current, we must use both half-cycles of the ac supply voltage. A circuit that has this capability is called a **full-wave rectifier**.

There are two types of full-wave rectifiers:

the **full-wave bridge rectifier** (FWB) and the **full-wave center-tapped rectifier** (FWCT). Here, the use of the full-wave bridge rectifier for single-value power supplies is discussed. FWB rectifiers are generally used for this application because the transformer generates less heat for the same dc current and is usually less expensive than the transformer required for a FWCT rectifier.

Full-Wave Bridge Rectifier

A full-wave bridge rectifier is shown in Figure 14.34. The secondary of the transformer is connected to opposite corners of four diodes ar-

FULL-WAVE BRIDGE RECTIFIER

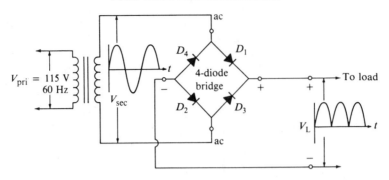

FIGURE 14.34. A full-wave bridge rectifier employs a four-diode bridge that essentially inverts the negative half-cycle of the secondary voltage. This circuit yields twice the dc voltage and current of a half-wave rectifier.

ranged in the shape of a diamond. This configuration is called a **diode bridge**. The load is connected to the other two corners of the bridge. The diode bridge can be made from four separate diodes, as shown in Figure 14.35a, or purchased as a single unit, as shown in Figure 14.35b.

To understand how the full-wave bridge rectifier works, assume that the transfomer secondary voltage V_{sec} is going through its positive half-cycle, as shown in Figure 14.36a. V_{sec} is a 24 V rms voltage with a maximum value of:

$$V_{sec(peak)} = 24 \times 1.4 = 34 \text{ V}$$

When V_{sec} rises above the threshold value of two diodes in series (about 1.2 V), diodes D_1 and D_2 become forward biased and begin to conduct current through the load resistor R_L.

The load voltage V_L will be slightly less than V_{sec} due to the two diode voltage drops. For example, when V_{sec} reaches its peak value, $V_{sec(peak)}$ in Figure 14.36a, the load voltage V_L will reach its peak value $V_{L(peak)}$:

$$V_{L(peak)} = V_{sec(peak)} - 1.2 \text{ V}$$
$$= 34 \text{ V} - 1.2 \text{ V} = 32.8 \text{ V} \quad \text{(24)}$$

Observe that diodes D_3 and D_4 are reverse biased in Figure 14.36a during the positive half-cycle of V_{sec}.

When V_{sec} goes through its negative half-cycle (Figure 14.36b), the polarity of the voltage applied to the bridge terminals is reversed. Diodes D_3 and D_4 then become forward biased, directing the load current through R_L in the same direction as in Figure 14.36a. Since current flows through the load resistor in the same direction for both half-cycles of V_{sec}, the load voltage must have the *same* polarity. V_L is still a pulsating dc, but it contains twice as many pulses per second as a half-wave rectifier.

In most applications, V_L is considered to be a fully rectified sine wave, even though a small notch exists in V_L when V_{sec} is less than 1.2 V. By neglecting the diode drops, we can make the approximation that the peak secondary voltage $V_{sec(peak)}$ is also the peak load voltage $V_{L(peak)}$. This approximation allows us to relate the peak load voltages directly to the secondary rms and peak voltages of the transformer:

$$V_{L(peak)} \approx V_{sec(peak)}$$
$$V_{L(peak)} \approx 1.41 V_{sec(rms)} \quad \text{(25)}$$

ALTERNATIVE DIODE BRIDGE ASSEMBLIES

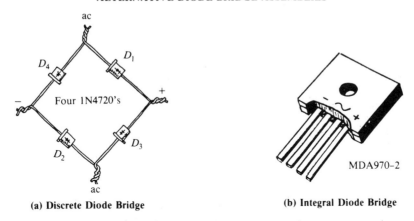

(a) Discrete Diode Bridge

(b) Integral Diode Bridge

FIGURE 14.35.   The four-diode bridge can be constructed from four discrete diodes wired together, but it is also available as a single integral assembly.

PRINCIPLES OF FULL-WAVE RECTIFIER OPERATION

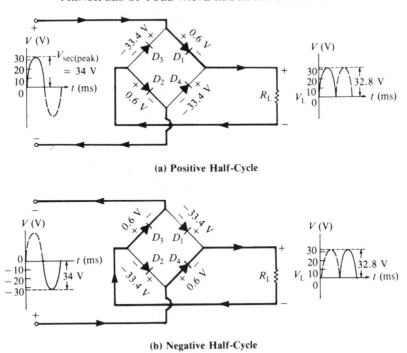

(a) Positive Half-Cycle

(b) Negative Half-Cycle

FIGURE 14.36.   During the positive half-cycle of the secondary, diodes D_1 and D_2 are forward biased and conduct current through the load. During the negative half-cycle, diodes D_3 and D_4 are forward biased and conduct current through the load in the same direction. Thus, both cycles of the sine wave produce a positive voltage across the load.

AVERAGE VALUES

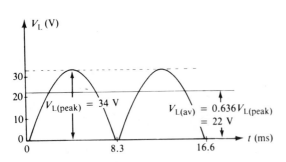

(a) Average Value of Load Voltage

(b) Average Value of Load Current

FIGURE 14.37. The average values of the load voltage and load current are the values read by a dc meter. They are 0.636 times the peak values of the sine wave. Compare with Figure 14.24 for a half-wave rectifier. Also see Example 8.

The voltage measured by a dc voltmeter is the average of V_L—that is, $V_{L(av)}$—and it is related to the peak value of the fully rectified sine wave by:

$$V_{L(av)} = 0.636 V_{L(peak)} \qquad (26)$$

Similarly, the average dc load current $I_{L(av)}$ is related to the peak load current by:

$$I_{L(av)} = 0.636 I_{L(peak)} \qquad (27)$$

The dc output values in Figure 14.37 are twice those obtained from a half-wave bridge rectifier (Equations 15 and 16).

EXAMPLE 8: A full-wave bridge rectifier has a transformer secondary voltage of 24 V rms. Neglecting diode drops, calculate (a) the peak voltage, (b) the average dc load voltage, and (c) the average dc load current if $R_L = 1 \text{ k}\Omega$.

Solution: (a) From Equation 25, we have:

$$V_{L(peak)} = 1.4 V_{sec(rms)} = 1.41 \times 24 \text{ V}$$
$$= \textbf{34 V}$$

See Figures 14.36 and 14.38.
  (b) From Equation 26:

$$V_{L(av)} = 0.636 V_{L(peak)} = 0.636 \times 34 \text{ V}$$
$$= \textbf{22 V}$$

  (c) From Ohm's law:

$$I_{L(av)} = \frac{V_{L(av)}}{R_L} = \frac{22 \text{ V}}{1 \text{ k}\Omega} = \textbf{22 mA}$$

Full-Wave Center-Tapped Rectifier

For the full-wave center-tapped rectifier, the transformer must have a center-tapped secondary, as shown in Figure 14.38a. An equivalent center-tapped secondary can be made by connecting two equal-voltage secondary windings in series aiding, as in Figure 14.38b. *Note:* The

FULL-WAVE CENTER-TAPPED RECITIFIER

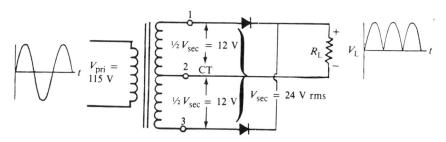

(a) FWCT from Center-Tapped Transformer

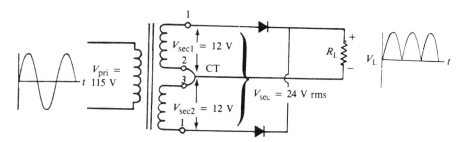

(b) FWCT from Transformer with Two Equal Secondaries

FIGURE 14.38. A full-wave center-tapped transformer requires a center-tapped second-ary winding (or two equal secondaries in series aiding) and only two diodes to produce a fully rectified dc load voltage.

secondary voltage of a center-tapped trans-former is specified in terms of the voltage be-tween the outer terminals, not the center tap. For example, in Figure 14.38a, the transformer is specified as 117 V/24 V CT, which means that the secondary voltage is 24 V rms between terminals 1 and 3. The voltage between center-tapped terminal 2 and either terminal 1 or 3 is $\frac{1}{2}V_{sec}$, or 12 V rms.

The FWCT rectifier operates like two half-wave rectifiers connected in parallel, as shown in Figure 14.39a. During the positive half-cycle, terminal 1 of the transformer is positive with re-spect to terminal 2, and terminal 3 is negative with respect to terminal 2. The half–secondary voltage $\frac{1}{2}V_{sec}$ must rise to a value greater than 0.6 V to forward-bias diode D_1. When $\frac{1}{2}V_{sec}$ ex-

ceeds 0.6 V, current flows through D_1 and load R_L to develop a load voltage V_L. The peak load voltage is expressed by:

$$V_{L(peak)} \approx \frac{1}{2}V_{sec(peak)} \qquad \textbf{(28)}$$

where $V_{sec(peak)}$ is the peak secondary voltage be-tween the transformer's outer terminals (*not* the center tap), and the diode voltage drop is ne-glected. Diode D_2 is reverse biased during the positive half-cycle shown in Figure 14.39a by the half–secondary voltage between terminals 3 and the center-tap terminal 2.

The negative half-cycle conditions for the FWCT rectifier are shown in Figure 14.39b. The secondary voltage polarities are reversed so that the half-wave rectifier formed by trans-

PRINCIPLES OF FWCT RECTIFIER OPERATION

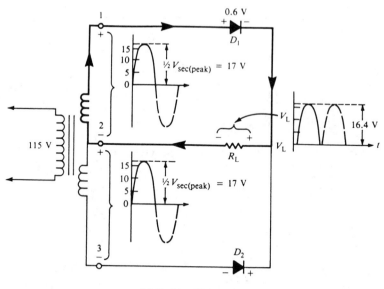

(a) Positive Half-Cycle

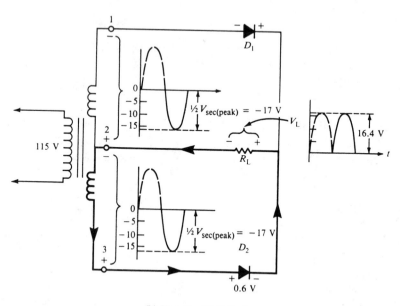

(b) Negative Half-Cycle

FIGURE 14.39.   FWCT rectifier operation can be seen by redrawing the circuit as shown. D_1 conducts current through R_L during the positive half-cycle, and D_2 conducts current through R_L in the same direction during the negative half-cycle. The resulting load voltage is half that of a FWB rectifier.

former terminals 2 and 3 and diode D_2 conduct current in the same direction through R_L. Thus, V_L is a pulsating dc voltage whose peak value is found from Equation 28 and whose average value is found from Equation 26.

EXAMPLE 9:   What are (a) the peak load voltage, (b) the average dc load voltage, and (c) the average dc load current obtained from a FWCT rectifier with a secondary winding rated at 24 V CT and $R_L = 1$ kΩ?

Solution: (a) By definition, the transformer has a 24 V rms secondary winding. Therefore, $V_{sec} = 24$ V and $\frac{1}{2}V_{sec} = 12$ V rms. The peak secondary voltage is 1.41×24 V ≈ 34 V. From Equation 28:

$$V_{L(peak)} = \frac{1}{2}V_{sec(peak)} = \frac{1}{2}(34 \text{ V}) = \mathbf{17 \text{ V}}$$

(b) From Equation 26:

$$V_{L(av)} = 0.636V_{L(peak)} = 0.636 \times 17 \text{ V}$$
$$= \mathbf{11 \text{ V}}$$

(c) From Ohm's law:

$$I_{L(av)} = \frac{V_{L(av)}}{R_L} = \frac{11 \text{ V}}{1 \text{ k}\Omega} = \mathbf{11 \text{ mA}}$$

By comparing Examples 8 and 9, we can see the differences between a FWB rectifier and a FWCT rectifier that uses a transformer with 24 V CT winding. The FWCT will use the secondary with the center tap and two diodes to deliver 11 V at 11 mA dc. The FWB uses the secondary without the center tap and four diodes to deliver 22 V at 22 mA dc.

14.9 DESIGNING A FULL-WAVE BATTERY CHARGER

In Example 7, a half-wave rectifier was used to design a battery charger that could supply up to

0.25 A. For larger charging currents, a full-wave rectifier, either FWB or FWCT, is generally used. In this section, a 1 A trickle charger will be designed for a standard lead-acid automobile battery.

The design equations for a **full-wave battery charger** are somewhat different from those of a half-wave charger due to the fact that both half-cycles of the sine wave contribute to the charging current. For example, the diode current rating needs to be only half as great as for a half-wave rectifier because each diode pair conducts only half the charging current per sine wave cycle. In the following design development, compare the design equations with their half-wave counterparts in Example 7 (Equations 17 to 23). Also, recall that the equations serve only as guides to component selection, with a wide range of secondary voltages and charging currents possible.

Design Example: Auto-Battery Trickle Charger

The design procedure is illustrated by the following example.

EXAMPLE 10:   Design a full-wave battery charger for a nominal 12.6 V lead-acid automobile battery with a rated capacity of 100 A-h.

Solution: Determining charger requirements: A 12 V automobile battery is made of six lead-acid cells in series. Each lead-acid cell has an end-of-charge voltage of about 2.6 V (Figure 14.30) so that the total end-of-charge voltage will be:

$$V_{ch} = 6 \text{ cells} \times \frac{2.6 \text{ V}}{\text{cell}} = 15.6 \text{ V}$$

Similarly, the end-point voltage of an automobile battery is about 12.0 V:

$$V_{dis} \approx 12.0 \text{ V}$$

The recommended trickle charge rate for a lead-acid battery is a current that will charge the battery in about 100 h to 125 h. Thus, the required trickle charge current is:

$$I_{ch(trickle)} = \frac{\text{A-h rating}}{100 \text{ h}} \tag{29}$$

For a 100 A-h rating, this equation gives:

$$I_{ch(trickle)} = \frac{100 \text{ A-h}}{100 \text{ h}} = 1.0 \text{ A}$$

Selecting the transformer: The transformer should have an rms secondary voltage equal to or greater than twice the end-of-charge voltage:

$$V_{sec} \geq 2V_{ch} \tag{30}$$

Its rms secondary current rating should be approximately $1.5I_{ch}$ (compare with Equation 19):

$$I_{sec} \geq 1.5I_{ch} \tag{31}$$

Substituting the values of V_{ch} and I_{ch} gives:

$$V_{sec} \geq 31.2 \text{ V}$$
$$I_{sec} \geq 1.5 \text{ A}$$

Thus, the transformer should have a secondary rated at about **31 V** and **1.5 A** or more.

Selecting the diodes: The diodes should have an average current rating of:

$$I_o \geq \tfrac{1}{2}I_{ch} \tag{32}$$

Note that this rating is half that required for a half-wave rectifier, according to Equation 20. Its PIV rating should exceed the value of $V_{sec(peak)}$:

$$PIV \geq V_{sec(peak)} + V_{ch} \tag{33}$$

This rating is also about half the value required for a half-wave rectifier (compare Equation 20) because V_{sec} always appears across two diodes in series.

Substituting in the values of I_{ch}, $V_{sec(peak)}$, and V_{ch} gives:

$$I_o \geq 0.5 \text{ A}$$
$$PIV \geq (1.4)(31.2 \text{ V}) + 15.6 \text{ V} \geq 59 \text{ V}$$

The **IN4720 diode rated at 3 A and 200 PIV** used in Example 6 would be more than adequate.

Selecting the fuse: The fuse to be placed in series with the primary hot lead should have a current rating of:

$$I_{fuse} = 1.5I_{ch}\frac{V_{sec}}{V_{pri}} \tag{34}$$

Compare Equation 34 with Equation 21.

Substituting in values gives:

$$I_{fuse} = 1.5(1.0 \text{ A}) \times \frac{31 \text{ V}}{115 \text{ V}} = 0.40 \text{ A}$$

According to Figure 14.17, a suitable choice would be the **3AG 0.4 A at 250 V**.

Selecting the current-limiting resistor: The approximate value of R_L is determined from:

$$R_L = \frac{V_{sec} - V_{ch}}{1.25I_{ch}} \tag{35}$$

and its power rating from:

$$P \geq (1.5I_{ch})^2 R_L \tag{36}$$

Compare these equations with Equations 22 and 23.

Substituting in values gives:

EXAMPLE 10

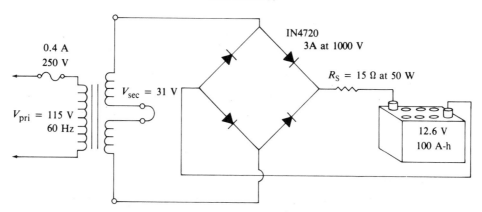

FIGURE 14.40.   The FWB rectifier can be used to make a trickle charger for a 12.6 V lead-acid automobile battery. The required secondary voltage is about 31 V, and a 15 Ω, 50 W resistor is placed in series with the battery to set the charging current.

$$R_{\mathrm{L}} = \frac{31.2\ \mathrm{V} - 15.6\ \mathrm{V}}{1.25(1.0\ \mathrm{A})} = 12.5\ \Omega$$

$$P \geq (1.5 \times 1.0\ \mathrm{A})^2 \times 12.5\ \Omega \geq 28\ \mathrm{W}$$

We should select the nearest available stock resistor, such as:

$$R_{\mathrm{L}} = \mathbf{15\ \Omega\ at\ 50\ W}$$

The complete circuit is shown in Figure 14.40.

14.10 QUESTIONS AND PROBLEMS

1. Draw a typical three-terminal ac wall outlet, label each terminal, and show its connection to the power system at the fuse box.
2. Draw a typical three-terminal ac line cord, label each terminal, and show its connection to an electronic system.
3. During a brownout, the line voltage from a power supply drops to 95 V. What is the peak voltage and peak-to-peak voltage?
4. If the line voltage is 115 V rms at 60 Hz, find the secondary voltage of transformers with the following turns ratio:

    (a) $\dfrac{N_{pri}}{N_{sec}} = 2.875$

    (b) $\dfrac{N_{pri}}{N_{sec}} = 4.80$

    (c) $\dfrac{N_{pri}}{N_{sec}} = 12.8$

5. If the maximum secondary current is 1.3 A, what will be the maximum primary current for the transformers in Problem 4?
6. What is the minimum power rating required for each of the transformers in Problems 4 and 5?
7. What is the proper fuse for each of the transformers in Problems 4 and 5? (Refer to Figure 14.17.)
8. Using the F91X multitap transformer shown in Figure 14.41, draw a half-wave rectifier circuit that will provide an rms secondary voltage of 40 V to a load resistance of 8.0 Ω. (See Figure 14.13 for proper primary and secondary transformer connections.) Show the current direction through the load resistor.
9. What is the load current and voltage (peak and average) of the rectifier in Problem 8? What fuse should be used?
10. Repeat Problems 8 and 9 for an rms secondary voltage of 22.5 V.
11. Design a half-wave battery charger that will charge four nickel-cadmium rechargeable D cells. (See Figures 14.30 and 14.31 for data on nickel-cadmium cells.) Draw the complete circuit showing component values and voltage polarities.
12. Using the F91X multitap transformer shown in Figure 14.41, draw a full-wave bridge rectifier circuit that will provide an rms secondary voltage of 32 V to a load resistance of 12 Ω. Show the current direction through the load resistor.
13. What is the load current and voltage (peak and average) of the rectifier in Problem 12? According to Figure 14.17, what fuse should be used?
14. Repeat Problems 12 and 13 for an rms secondary voltage of 19 V.
15. Design a full-wave battery charger that will charge six nickel-cadmium rechargeable D cells (see Figures 14.30 and 14.31 for data). Draw the finished circuit showing component values and voltage polarities.

F91X MULTITAP TRANSFORMER

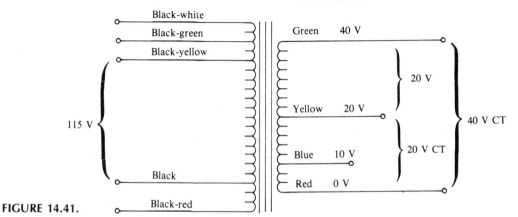

FIGURE 14.41.

Chapter 15

FILTERED POWER SUPPLIES

15.1 OBJECTIVES

Following the completion of Chapter 15, you should be able to:

1. Draw a capacitor-filtered rectifier circuit and explain the principles of its operation.
2. Describe the construction of an electrolytic filter capacitor and explain its voltage rating.
3. Explain the ripple voltage specification of dc power supplies including its dependence on capacitor value and load current.
4. Determine the correct filter capacitor for a dc power supply, given its ripple voltage and load current requirements.
5. Explain the voltage regulation specification of a dc power supply.
6. Determine the voltage regulation of a dc power supply, given its average no-load and full-load voltages.
7. Design a filtered power supply with specified output voltage, output current, ripple voltage, and load regulation.

15.2 CAPACITOR-FILTERED RECTIFIER

Most electronic applications require dc power supplies that deliver a steady, rather than pulsating, dc voltage. The voltage pulsations of both half- and full-wave rectifiers can be significantly reduced by adding a filter between the rectifier and the load. The filter smoothes the pulsations to give a dc load voltage that fluctuates considerably less (Figure 15.1).

Generally, the full-wave rectifier is preferable for a filtered power supply. A filtered half-wave rectifier is used only when load current requirements are below a few milliamperes.

The most inexpensive and most widely used power supply filter is the **filter capacitor.** The load voltage of a full-wave bridge rectifier is filtered simply by installing a large-value electrolytic capacitor across the rectifier output and load terminals (Figure 15.1). The load voltage V_L in this system resembles a dc voltage, although it does contain a small alternating rip-

FILTERED POWER SUPPLY

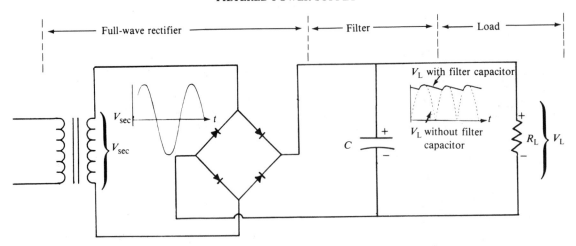

FIGURE 15.1. The voltage pulsations from a rectifier can be significantly reduced by placing a capacitor across the load. The result is called a capacitor-filtered power supply. The small voltage pulsations that remain are called ac ripple.

ple voltage component. The following section describes how the capacitor changes the rectified sine wave voltage into this more steady dc voltage.

Principles of Operation

When a capacitor is placed across the load resistor, the load voltage V_L has the wave shape shown in Figure 15.2. During each half-cycle of full-wave rectifier operation, two distinct actions occur in the capacitor. For the time intervals A–B and C–D in Figure 15.2, the transformer furnishes current directly to the load while simultaneously charging the capacitor C. During time intervals B–C and D–E, the charged capacitor furnishes current to the load while the transformer is essentially disconnected by the rectifier diodes. The charging action of C is analyzed in Figure 15.3.

During the positive half-cycle, diodes D_1 and D_2 are forward biased whenever V_{sec} exceeds V_L (Figure 15.3a). These diodes then conduct current both to charge C and to provide

current I_L to the load R_L. In Figure 15.2, this action occurs during time interval A–B.

When V_{sec} is less than the capacitor voltage V_L, all of the diodes are reversed biased. As shown in Figure 15.4b, C is essentially disconnected from the transformer. The capacitor then furnishes the entire load current by discharging through R_L. In Figure 15.2 this action happens during time interval B–C, when the load voltage is seen to drop gradually as the capacitor discharges.

During the negative half-cycle (Figure 15.4), V_{sec} reverses in polarity. When V_{sec} again exceeds the value of the dc voltage on the discharging capacitor (Figure 15.4a), the other pair of diodes, D_3 and D_4, become forward biased. As a result, V_{sec} is reconnected across both C and R_L, the capacitor is charged, and the load current is supplied. This activity takes place during time interval C–D in Figure 15.2. When V_{sec} falls below V_L during the negative half-cycle, as shown in Figure 15.4b, the diodes are again reverse biased, and the capacitor again supplies current to the load.

FILTERED POWER SUPPLY OPERATION

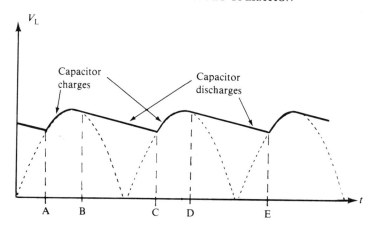

FIGURE 15.2. During the intervals A–B and C–D, the transformer provides the load current while simultaneously charging the capacitor. During intervals B–C and D–E, the load current is provided solely by the discharging capacitor.

POSITIVE HALF-CYCLE

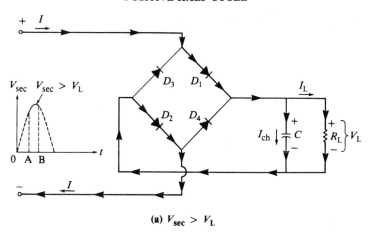

(a) $V_{sec} > V_L$

FIGURE 15.3. Shown are the load current loops during the positive half-cycle of the input sine wave for $V_{sec} > V_L$ and $V_{sec} < V_L$. In (a), diodes D_1 and D_2 conduct current to the load and capacitor. In (b), the capacitor provides load current.

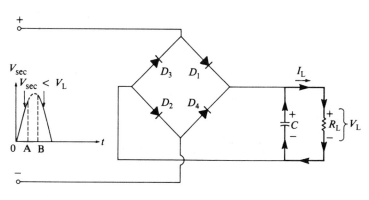

(b) $V_{sec} < V_L$

FIGURE 15.4. The figure shows the load current loops during the negative half-cycle for $-V_{sec} > V_L$ and $-V_{sec} < V_L$. In (a), diodes D_3 and D_4 conduct current to the load and capacitor. In (b), the capacitor provides load current.

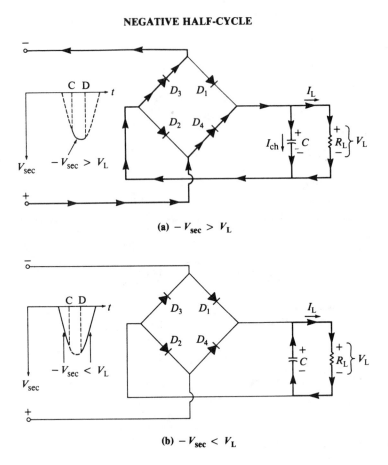

NEGATIVE HALF-CYCLE

(a) $-V_{sec} > V_L$

(b) $-V_{sec} < V_L$

The capacitor filter action can be summarized as follows:

1. During each half-cycle of the full-wave rectifier, the diodes connect the transformer secondary to R_L and C when V_{sec} is larger than V_L. They disconnect the secondary from R_L and C when V_{sec} is less than V_L.
2. When V_{sec} is less than V_L, the load current is furnished by the partial discharge of the capacitor.
3. When V_{sec} is greater than V_L, the transformer furnishes current directly to the load and also replaces the charge on the capacitor.

Types of Filter Capacitors

Large-value capacitors are needed for filter rectifiers. Typical values are 500 μF or more. Thus, **electrolytic capacitors** are used, because they give the greatest capacitance for the least cost.

Electrolytic capacitors are constructed as shown in Figure 15.5. Two metal foil conductors are separated by an insulating dielectric-oxide film and a liquid or solid electrolytic spacer. The two most common types of electrolytic capacitors are the aluminum electrolytic and the tantalum electrolytic, depending on the composition of the conductor. Tantalum capacitors are more vibration resistant and occupy

ELECTROLYTIC CAPACITOR AND SYMBOL

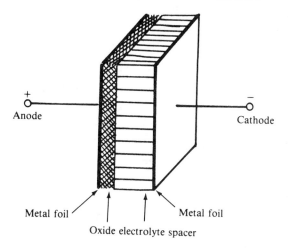

(a) Typical Construction

(b) Schematic Symbol

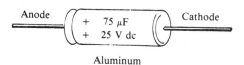

Aluminum

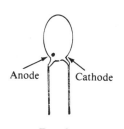

Tantalum

(c) Typical Capacitors

FIGURE 15.5. Filter capacitors are generally the electrolytic type that are polarized, having an anode and a cathode. The figure shows the construction, schematic symbol, and typical styles of aluminum and tantalum capacitors.

less space than aluminum capacitors for the same capacitance. However, aluminum electrolytic capacitors cost less and are available in a greater range of sizes.

Capacitor Ratings

An important characteristic of electrolytic capacitors is that they are polarized. They have a positive terminal called the **anode** and a negative terminal called the **cathode**. These terminals must be connection to the + and − sides, respectively, of the full-wave rectifier. As shown in Figure 15.5c, the anode of an aluminum electrolytic capacitor is marked with a + symbol, and the anode of a tantalum capacitor is marked with either a + or a dot of paint, usually silver or white. The circuit schematic of the capacitor, Figure 15.5b, represents the + anode with a straight line and the − cathode with a curved line.

Caution: Never connect an ac voltage or a reverse-biased dc voltage across an electrolytic capacitor. A reverse voltage of more than a few volts will heat up an electrolytic capacitor and permanently damage it. The capacitor may even explode.

Capacitors are rated by both the magnitude of their capacitance in microfarads and the normal **dc working voltage** (WVDC). For example, a 500 μF, 50 WVDC capacitor can be used in a circuit where the capacitor voltage will never exceed 50 V dc. Nominal values for the magnitude of capacitance generally have a large tolerance: + 100% to − 50%.

15.3 POWER SUPPLY CHARACTERISTICS

Alternating Current Ripple

Since the filter capacitor must supply load current during most of each half-cycle, it will dis-

OUTPUT RIPPLE VOLTAGE

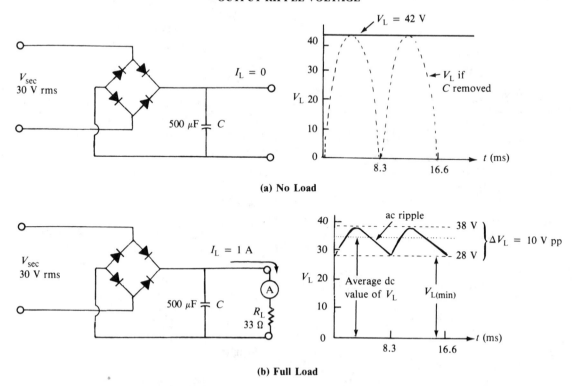

(a) No Load

(b) Full Load

FIGURE 15.6. Without a load, the capacitor has no discharge path and the output voltage is a steady dc. A resistive load, however, causes the capacitor to discharge through the load to provide the load current. This action produces a small ac ripple superimposed on the average dc output voltage.

charge. The amount of discharge depends on how much load current it must supply. For example, in Figure 15.6a, there is no load resistor, so there is no load current. The load voltage, therefore, is a constant dc voltage because the capacitor has no path for discharge.

In Figure 15.6b, however, the presence of a 33 Ω load resistor results in an average load current of 1 A. The capacitor must furnish most of this load current by discharging through R_L. Thus, the capacitor voltage and, consequently, the load voltage must drop somewhat during the capacitor's discharge. This small change in load voltage is called **ripple voltage.**

The ripple voltage is superimposed on the

dc value of V_L and is generally specified in terms of its **peak-to-peak value** ΔV_L. For example, in Figure 15.6b, $\Delta V_L = 10$ V pp. Alternately, the ripple voltage can be specified by its **rms value** V_r. Note that V_r is the rms value of the ac ripple voltage only. V_r does not include the dc value of V_L. More specifically, V_r is the value measured by an ac voltmeter that has a series capacitor to block the dc component of V_L. The rms and peak-to-peak ripple voltages are related by:

$$\Delta V_L = 3.5 V_r \qquad (1)$$

The ripple voltage increases in direct proportion to increases in load current. For example, as Figure 15.7a shows, the ripple voltage ΔV_L doubles from 5 V to 10 V when I_L is doubled from 0.5 A to 1.0 A.

The ripple voltage also depends on capacitor size. Larger-value capacitors store more charge than smaller ones. Therefore, the voltage across the larger capacitors drops less than smaller values while furnishing the same load current. In Figure 15.7b, the ripple voltage is halved from 10 V to 5 V when C is doubled from 500 μF to 1000 μF. Thus, the ripple voltage *increases as the load current increases, but decreases as the capacitor size increases.*

The ripple voltage specification of a commercial power supply may be given as any one of the following: the rms ripple voltage V_r, the peak-to-peak ripple voltage ΔV_L, or the percent ripple. **Percent ripple** is defined as the ratio of the rms ripple voltage V_r at full load to the full-load dc voltage. It is always expressed at full load current. In equation form:

$$\text{percent ripple} = \frac{V_{r(FL)}}{V_{L(av)FL}} \times 100\% \qquad \textbf{(2)}$$

EXAMPLE 1: Suppose a filtered full-wave rectifier has a full-load peak-to-peak ripple voltage of 5 V and a dc load voltage of 33 V. Find its percent ripple.

Solution: First, we find the rms ripple from Equation 1:

$$V_r = \frac{5 \text{ V}}{3.5} = 1.4 \text{ V}$$

Then, using Equation 2, we get:

$$\text{percent ripple} = \frac{1.4 \text{ V}}{33 \text{ V}} \times 100\% = \textbf{4.3\%}$$

RIPPLE VOLTAGE DEPENDENCE ON LOAD CURRENT AND CAPACITOR SIZE

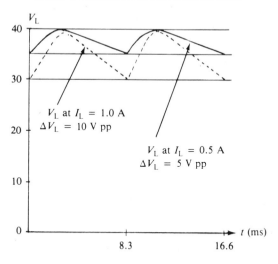

(a) Ripple Voltage vs. Load Current

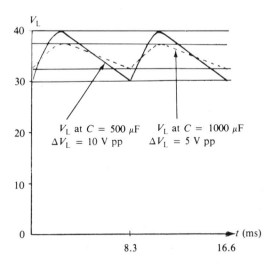

(b) Ripple Voltage vs. Capacitor Size

FIGURE 15.7. The magnitude of the ac ripple increases with increasing load current because the capacitor discharges to a lower voltage. The ac ripple decreases with increasing capacitor size because larger capacitors have more stored charge.

ESTIMATING RIPPLE VOLTAGE

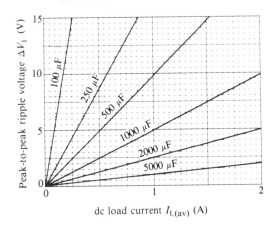

FIGURE 15.8. An estimate of the ripple voltage for various load currents and capacitor values can be obtained from this graph. It applies to both FWB and FWCT filtered rectifiers and to line frequencies of 50 Hz to 60 Hz.

An *estimate* of the magnitude of the ripple voltage in a filtered full-wave rectifier for various dc load currents and filter capacitors can be made with the aid of Figure 15.8. For example, if $I_L = 1$ A and $C = 1000$ μF, Figure 15.8 shows that $\Delta V_L = 5$ V. The graph applies to both FWB and FWCT filtered rectifiers and to line voltage frequencies of 50 Hz to 60 Hz. It is quite useful as a design aid.

EXAMPLE 2: Suppose we have a full load current of 1 A and need to hold the peak-to-peak ripple voltage to 5 V. What value capacitor should be used?

Solution: On Figure 15.8, we locate the intersection of the horizontal line $\Delta V_L = 5$ V and the vertical line $I_L = 1$ A. Reading the required capacitance from the closest (larger) capacitance curve, we find $C = \mathbf{1000}$ $\mu\mathbf{F}$.

It is difficult to predict the maximum ripple voltage allowable for a given application. Some inexpensive radios will sound the same if the rectifier has little or no filter capacitance. The reason is that the small loudspeaker usually cannot respond to the low oscillation frequency of the ripple voltage. On the other hand, power supplies for integrated circuits or instrumentation circuits may require ripple voltages of less than a few millivolts. The only true measure of allowable ripple voltage for a given application is found by trial and error.

Voltage Regulation

Voltage regulation is concerned with the dependence of the *average* value of the dc load voltage on the load current. Recall that the capacitor discharges to a lower voltage when the load current increases (Figure 15.9). Thus, the average, or dc, value of the *load voltage* across both the capacitor and the load resistor must also *decrease as the load current increases.*

There is a second mechanism that also acts to reduce the load voltage. During the capacitor charging time (A–B in Figure 15.2a and C–D in Figure 15.4a), the rectifier must replace the charge on the capacitor. The capacitor charge current is typically three times the load current. For example, in Figure 15.10, the discharging capacitor current I_{dis} provides an average load current of 1 A during time intervals B–C and D–E. Because the discharge time is about three times the charge time, the capacitor charge current I_{ch} supplied by the transformer must be about three times the discharge current. This relationship is shown in Figure 15.10 for time intervals A–B and C–D. Thus, when the diodes are conducting, the transformer must supply nearly four times the load current in order to both supply the load current I_L and provide current to recharge the capacitor, $3I_L$.

The large transformer current is conducted through the resistance of the secondary wind-

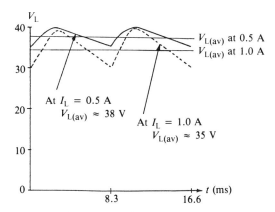

VOLTAGE REGULATION

FIGURE 15.9. Because the capacitor discharges to a lower value with increasing load current, the average load voltage also decreases. Voltage regulation is a measure of the decrease in average, or dc, load voltage with load current.

ing. The resulting I–R voltage drop in the secondary winding further reduces the available secondary voltage. Thus, the capacitor voltage will reduce to a lower peak value as the load current increases, and there will be a corresponding decrease in the average output dc voltage.

The dc load voltage of a filtered full-wave rectifier is a *maximum at no load current* and a *minimum at full load current*. This conclusion is verified by examining a **voltage regulation curve** — that is, a graph of the dc load voltage versus the load current for a power supply.

A typical dc voltage regulation curve is shown in Figure 15.11. Operation at no load current is shown by point N, where $I_L = 0$ A and the no-load voltage is 42 V. The no-load voltage is set by the peak voltage of the transformer secondary minus the two diode drops. In most cases, we can neglect the diode voltage, so the no-load voltage is:

$$V_{L(av)NL} = V_{sec(peak)} = 1.4V_{sec(rms)} \qquad (3)$$

In Figure 15.11, operation at full load current is marked by point F, where the full load voltage drops to an average value of 34 V when I_L increases to 1 A.

The voltage regulation of a power supply is specified by the percent load regulation. **Percent load regulation** gives the approximate increase in dc load voltage as load current decreases from full load. In equation form:

$$\text{percent load regulation}$$
$$= \frac{V_{L(av)NL} - V_{L(av)FL}}{V_{L(av)FL}} \times 100\% \qquad (4)$$

EXAMPLE 3: A 24 V power supply is specified to have 10% load regulation as its load current changes from 2.0 A to 0 A. Find (a) the change in load voltage and (b) $V_{L(av)NL}$.

Solution: (a) The 24 V power supply rating refers to the output voltage at full load. Therefore, $V_{L(av)FL} = 24$ V. Substituting into Equation 4, the change in voltage from no load to full load is:

$$10\% = \frac{V_{L(av)NL} - V_{L(av)FL}}{24 \text{ V}} \times 100\%$$

$$V_{L(av)NL} - V_{L(av)FL} = \textbf{2.4 V}$$

(b) Solving for $V_{L(av)NL}$ gives:

$$V_{L(av)NL} = 2.4 \text{ V} + 24 \text{ V} = \textbf{26.4 V}$$

It is difficult to predict accurately a value for the full load voltage before building and testing a filtered full-wave rectifier, because the change in $V_{L(av)}$ from no load to full load depends on capacitor size, load current, type of diode, transformer construction, line voltage, and even the lengths of the line cord and wiring at the wall outlet. However, we can estimate the

FIGURE 15.10. The discharging capacitor supplies load current for about three times as long as the transformer. Thus, when it is on, the transformer must supply about $4I_L$ in order to provide both load current I_L and capacitor charge current I_{ch}. These high currents in the secondary windings produce IR voltage drops that further reduce the load voltage with increasing I_L. Compare with Figures 15.3 and 15.4.

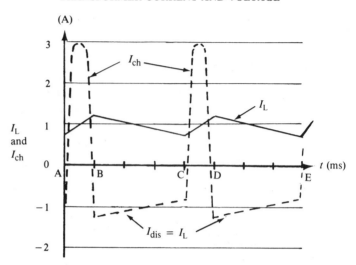

TRANSFORMER CURRENT AND VOLTAGE

full load voltage for a particular design by estimating a probable percent voltage regulation. Then, after building the circuit and testing it, we can correct the design based on the test data. A reasonable first estimate assumes a load regulation of 20%. Substituting this value into Equation 4 gives:

$$20\% = \frac{V_{L(av)NL} - V_{L(av)FL}}{V_{L(av)FL}} \times 100\%$$

Solving for $V_{L(av)FL}$ gives:

$$V_{L(av)FL} \approx 0.8 V_{L(av)NL} \tag{5}$$

Equation 5 is a reasonable first approximation to use in predicting the full-load dc voltage for full load currents in the range of 0.5 A to 2.0 A.

EXAMPLE 4: If the no-load voltage in Figure 15.11 is 42 V, estimate the full load voltage at 1 A.

Solution: Equation 5 gives an approximate value for the full load voltage:

$$V_{L(av)FL} \approx 0.8(42 \text{ V}) \approx 34 \text{ V}$$

Because the load voltage from a filtered rectifier depends strongly on the load current, filtered rectifiers are called **unregulated power supplies.** If we need a voltage that remains constant despite variations in load current, we must add a voltage regulator section between the load and the unregulated power supply. This procedure is discussed in Chapter 16.

Another disadvantage of the unregulated

FIGURE 15.11. The voltage regulation of a power supply is given as a graph of output voltage versus load current. The smaller the difference between $V_{L(av)NL}$ and $V_{L(av)FL}$, the better the voltage regulation.

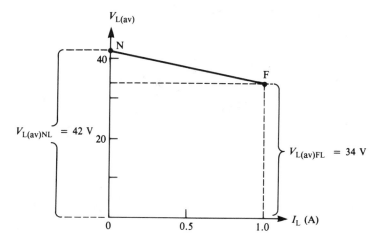

VOLTAGE REGULATION CURVE

supply is that we usually cannot get exactly the voltage we need. This problem is the topic of the next section.

15.4 DESIGNING A FILTERED POWER SUPPLY

Selecting the Transformer

In order to choose a transformer for a filtered FWB rectifier, the following information is needed:

— The **line voltage** at the wall outlet where the transformer will be used,
— The maximum ac rms **secondary current** at full load,
— The ac rms **secondary voltage** required by the transformer at full load.

The wall outlet voltage is normally 115 V or 230 V, and the transformer primary voltage rating should correspond to the wall outlet voltage.

For an FWB rectifier, the transformer's sec-

ondary rms current rating should equal or exceed 1.8 times the maximum full-load dc current:

$$I_{sec} \geq 1.8 I_{L(av)FL} \qquad (6)$$

For most applications, the full load dc voltage required at the full load current will be known. Equations 4 and 5 can then be used to predict the required secondary rms voltage rating for the transformer. The procedure is illustrated in Example 5.

EXAMPLE 5: An amplifier requires a 24 V dc supply voltage when drawing an average full load current of 1 A. What is the required secondary voltage rating?

Solution: From Equation 5, we estimate the no-load average voltage:

$$V_{L(av)NL} = \frac{V_{L(av)FL}}{0.8} = \frac{24 \text{ V}}{0.8} = 30 \text{ V}$$

Equation 3 gives the required rms secondary voltage:

$$V_{sec(rms)} = \frac{V_{L(av)NL}}{1.41} = \frac{30\ V}{1.41} = \textbf{21 V}$$

We would choose a standard transformer with a secondary voltage rating near **24 V rms.**

EXAMPLE 6: What is the required secondary current rating of the transformer in Example 5?

Solution: From Equation 6, we have:

$$I_{sec} \geq 1.8 I_{L(av)FL} \geq 1.8 \times 1\ A = \textbf{1.8 A}$$

We would choose a transformer with a 24 V secondary rated at **2.0 A rms.**

Selecting the Fuse

A fuse should be installed in series with the primary winding. Its fusing current rating should correspond to the rms primary current that occurs when the filtered power supply is furnishing full-load dc current. In equation form:

$$I_{fuse} \geq 1.8\ \frac{V_{sec}}{V_{pri}} \times I_{L(av)FL} \qquad (7)$$

EXAMPLE 7: What value of fuse should be installed in series with the primary winding in Example 6?

Solution: From Equation 7, we have:

$$I_{fuse} \geq 1.8\left(\frac{V_{sec}}{V_{pri}}\right) \times I_{L(av)FL}$$
$$\geq 1.8\left(\frac{24\ V}{115\ V}\right) \times 1\ A = \textbf{0.375 A}$$

We would select a fuse rated at 4/10 A (see Figure 14.17).

Selecting the Diodes

The PIV voltage rating of each diode should exceed the peak secondary voltage of the transformer. The average diode current rating I_o should be greater than one-half the dc full load current:

$$PIV \geq V_{sec(peak)}$$
$$I_o \geq {}^{1}\!/_{2} I_{L(av)FL} \qquad (8)$$

EXAMPLE 8: Select a diode for the FWB rectifier in Example 7.

Solution: We can find the peak secondary voltage from Equation 1 of Chapter 14:

$$V_{sec(peak)} = 1.4 V_{sec(rms)} = 1.4 \times 24\ V$$
$$= 34\ V$$

According to Equation 8, the average diode current rating I_o should be greater than:

$$I_o = {}^{1}\!/_{2} I_{L(av)FL} \geq {}^{1}\!/_{2} \times 1\ A \geq 0.5\ A$$

Therefore, we would select a diode rated for $I_o \geq \textbf{0.5 A}$ and **PIV $\geq$ 34 V.**

Selecting the Capacitor

Sometimes, the maximum ac ripple voltage that can be tolerated in the filtered power supply is not known. For example, suppose we want a 9 V power supply to listen to a portable radio when its battery runs down. The radio's manufacturer would not normally furnish data on the maximum allowable ripple because it was intended that the radio be powered solely by a

EXAMPLES 5 THROUGH 9

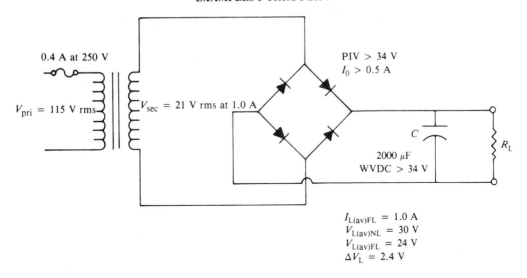

FIGURE 15.12.   Shown is a typical filtered power supply circuit.

battery. Therefore, we would have to make a trial estimate. If we assume that the peak-to-peak ripple voltage ΔV_L should not exceed one-tenth of the dc full load voltage $V_{L(av)}$, we would proceed as follows:

1. Calculate ΔV_L from $\Delta V_L = 0.1 V_{L(av)FL}$.
2. Knowing $I_{L(av)FL}$ and ΔV_L, use Figure 15.8 to obtain a value for the filter capacitor.
3. Build the filtered rectifier and listen to the radio.
4. If the radio operates satisfactorily, keep halving the capacitor size (to lower the cost) while maintaining satisfactory operation. Each time the capacitor is halved, the ripple will double.
5. If a low-frequency buzz is heard, double the size of the filter capacitor to halve the ripple. Repeat as necessary until satisfactory operation occurs.

If the maximum allowable ΔV_L at full load current is known, the capacitance can be found directly from Figure 15.8. The capacitor's voltage rating WVDC should exceed the peak sec-

ondary transformer voltage of the FWB rectifier:

$$\text{WVDC} > V_{\text{sec(peak)}} \qquad (9)$$

EXAMPLE 9:   Select a capacitor for the FWB rectifier in Example 8. Assume an acceptable percent ripple of 10%.

Solution: The full load voltage is 24 V. For a 10% ripple, the result is a peak-to-peak ripple voltage of:

$$\Delta V_L = 0.1 V_{L(av)FL} = 0.1 \times 24 \text{ V} = 2.4 \text{ V}$$

The full load current was given as:

$$I_{L(av)FL} = 1.0 \text{ A}$$

Figure 15.8 prescribes a filter capacitor of about:

$$C = 2000 \; \mu\text{F}$$

We find the capacitor's voltage rating from Equation 9:

$$\text{WVDC} > V_{sec(peak)} > 34 \text{ V}$$

The complete circuit design for Examples 5 through 9 is shown in Figure 15.12.

15.5 QUESTIONS AND PROBLEMS

1. Using the F91X transformer shown in Figure 15.13, draw a full-wave capacitor rectifier circuit that has an rms secondary voltage of 34 V. (See Figure 14.13 for primary and secondary connections.)

2. What value filter capacitor should be used for the following maximum values of dc load current and peak-to-peak ripple? (Refer to Figure 15.8.)

   (a) $I_{L(av)} = 1.0$ A and $\Delta V_L = 5$ V

   (b) $I_{L(av)} = 1.5$ A and $\Delta V_L = 9$ V

   (c) $I_{L(av)} = 2.0$ A and $\Delta V_L = 6$ V

3. What should be the full load voltage for dc power supplies with the following no-load voltage and voltage regulation specifications?

   (a) $V_{L(av)NL} = 15$ V; regulation = 5%

   (b) $V_{L(av)NL} = 40$ V; regulation = 10%

   (c) $V_{L(av)NL} = 9$ V; regulation = 2%

4. Design a capacitor-filtered power supply that would power a 9 V transistor radio with a maximum (full-load) current requirement of 500 mA. Assume a voltage regulation of 10% and an acceptable peak-to-peak ripple of $\Delta V_L = 0.5$ V.

MULTITAP TRANSFORMERS

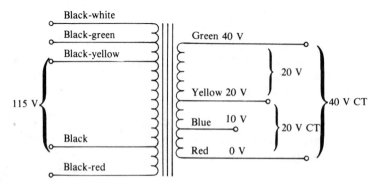

FIGURE 15.13.

Chapter 16

REGULATED POWER SUPPLIES

16.1 OBJECTIVES

Following the completion of Chapter 16, you should be able to:

1. Explain the purpose of IC voltage regulators and describe at least five of their desirable features.
2. Explain the four principal ratings of IC voltage regulators: output voltage, output current, minimum dropout voltage, and maximum junction temperature.
3. Explain the heat dissipation by a semiconductor device in terms of thermal resistance and give an equation that describes the process.
4. Determine the maximum power that a semiconductor device can dissipate, given its maximum allowable junction (or case) temperature, thermal resistance characteristics, and ambient temperature.
5. Explain the purpose of a heat sink and describe its principal characteristics.
6. Select a heat sink for a semiconductor device, given its case style, maximum allowable junction (or case) temperature, power dissipation, and ambient temperature.
7. Draw a fixed-voltage, IC-regulated, power supply circuit and explain the principles of its operation.
8. Design a fixed-voltage, IC-regulated power supply with a specific output voltage.
9. Draw an adjustable, IC-regulated, power supply circuit and explain the principles of its operation.
10. Design an adjustable, IC-regulated power supply with a specific range of output voltages.
11. Draw a dual-polarity, IC-regulated, power supply circuit using an FWCT rectifier and explain the principles of its operation.
12. Design a dual-polarity, IC-regulated power supply with specific positive and negative output voltages.
13. Determine the heat dissipation of an IC voltage regulator in a power supply circuit

and select a heat sink and heat-absorbing resistors that will keep the regulator at a safe operating temperature.

16.2 IC VOLTAGE REGULATORS

The filtered full-wave rectifier has three significant disadvantages:

1. It is difficult, if not impossible, to design a power supply with a specific dc voltage value, even for a restricted range of load currents.
2. The load voltage will always have an ac ripple component. The ripple can be minimized by increasing the filter capacitance, but if the ripple must be very small (millivolts), the required capacitance will be extremely large and expensive.
3. The dc load voltage will always decrease somewhat with increasing load current due to poor voltage regulation.

All three disadvantages can be nearly eliminated by adding an integrated-circuit (IC) voltage regulator, as shown in Figure 16.1. Modern IC regulators are low in cost, and their cost is generally offset by the saving realized through the use of a less-expensive filter capacitor. A dollar's worth of regulator will reduce ripple by more than an additional dollar's worth of capacitance.

IC regulators are also easy to use. Most have only three terminals—input, output, and common—and can be connected directly to the output of a filtered rectifier, as shown in Figure 16.2. The output of the rectifier becomes the input to the regulator V_{in}, and the output of the regulator V_o is connected directly across the load.

The availability of a wide variety of standard output voltages eliminates any voltage selection problem. If a nonstandard voltage is

IC VOLTAGE REGULATORS AND SYMBOL

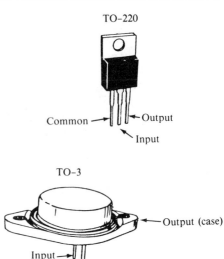

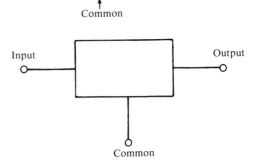

FIGURE 16.1.   The output voltage of a filtered full-wave rectifier is difficult to set precisely, has unwanted ripple, and has poor voltage regulation. These difficulties can be nearly eliminated by adding an inexpensive IC voltage regulator. The figure shows typical styles and the schematic symbol.

needed, there are adjustable regulators that require only one or two resistors for precise, simple adjustment to the voltage required.

Uses of IC Regulators

A voltage-regulated power supply is usually required both for linear ICs, such as op amps,

TYPICAL IC VOLTAGE REGULATOR CONNECTION

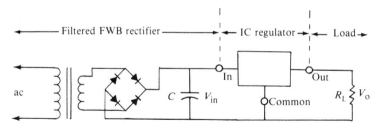

FIGURE 16.2. An IC regulator is generally a three-terminal device, with input, output, and common terminals connected directly between the output of a filtered rectifier and the load.

and for digital logic. Local regulators are also used in subassemblies of larger systems to tailor the voltages to the electrical characteristics of specific circuits and devices. Each printed circuit board may have its own regulator, thereby improving isolation and reducing signal coupling between circuit boards.

In addition, local regulators can be used to eliminate the voltage drops encountered on long power leads from a single, large common supply and to protect sensitive components (such as ICs) from the large transients that can occur on the supply leads. For example, in an automobile, turning on or off the starter produces large load dumps on the power supply leads. A regulator can be used to filter out these transients in the same manner as it reduces ripple. In short, the use of integrated circuits will probably require a voltage-regulated supply.

Features of IC Regulators

Regulation of the load voltage by an IC regulator is exceptionally good. Typically, its dc terminal voltage changes by only tens of millivolts, even when the load current changes by amperes (Figure 16.3). In a 9 V regulator, for example, the load voltage regulation will be only 50 mV

TYPICAL VOLTAGE REGULATION WITH IC VOLTAGE REGULATOR

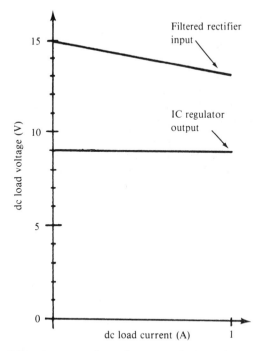

FIGURE 16.3. The voltage regulation of an IC regulator is exceptionally good. In the above example, the output voltage changes by less than 50 mV from no load to full load.

**TYPICAL RIPPLE REJECTION WITH
IC VOLTAGE REGULATORS**

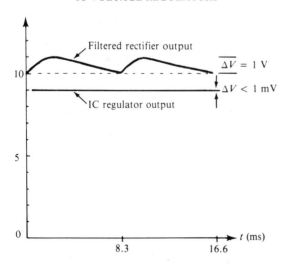

FIGURE 16.4. The ripple rejection of IC regulators is also good, generally better than 1000. For example, if the ripple at the regulator input is 1 V, the ripple at the output will be less than 1 mV.

for a load current change of 1 A. Thus, the dc load voltage will decrease by 50 mV—for example, from 9.050 V to 9.00 V—as the load current changes from 0 A to 1.0 A.

The ability of an IC regulator to reduce ripple is expressed as ripple rejection. The **ripple rejection ratio** in decibels (db) is given by:

$$\text{ripple rejection (db)} = 20 \log \frac{\Delta V_{\text{in}}}{\Delta V_{\text{o}}} \quad \text{(1)}$$

where ΔV_{in} is the input ripple voltage and ΔV_{o} is the output ripple voltage. We can express ΔV_{in} and ΔV_{o} as either rms values *or* peak-to-peak values provided that we use the same measure for both.

A typical IC regulator has a ripple rejection value greater than 60 db. Since every 20 db reduction of ripple is equivalent to a division by 10, a 40 db rejection means reduction by a factor of 100, and a 60 db rejection means reduc-

SUMMARY OF IC VOLTAGE REGULATOR TYPES

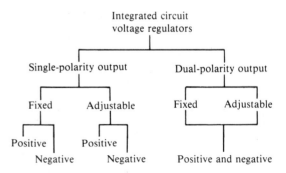

FIGURE 16.5. The general types of IC regulators available are single-polarity output (positive or negative) and dual-polarity output, and both are available as either fixed or adjustable.

tion by a factor of 1000. Thus, the 60 db of ripple rejection of a typical IC regulator signifies the reduction of any input ripple voltage by a factor of 1000 before it reaches the output (Figure 16.4).

IC voltage regulators have other desirable features. They generally contain **overload protection.** If their load terminals are short-circuited, internal circuitry limits the load current automatically to prevent it from burning out. In many types of IC regulators, the **maximum load current is adjustable** so that the IC regulators can be used as a constant-current source. Most IC regulators also have a **thermal shutdown** feature. An internal circuit senses the regulator's temperature; if it overheats, the output current is automatically reduced. When the cause of overheating has been corrected, the regulator will cool down and then resume operation. In the following sections, these features of IC regulator operation are examined in detail.

Types of IC Regulators

To select an IC voltage regulator for a given application, we must know the load current and voltage requirements, the types of IC regulators

SINGLE-POLARITY FIXED-VOLTAGE REGULATORS

Positive		Negative	
Number	Rating	Number	Rating
7805	+5 V	7905	−5 V
7806	+6 V	7906	−6 V
7808	+8 V	7908	−8 V
7812	+12 V	7912	−12 V
7815	+15 V	7915	−15 V
7818	+18 V	7918	−18 V
7824	+24 V	7924	−24 V

FIGURE 16.6. Typical examples of fixed single-polarity regulators are the 7800 series (positive output) and the 7900 series (negative output) manufactured by Fairchild. The table gives values of the standard voltages available.

DROPOUT VOLTAGE SPECIFICATION

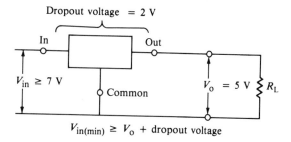

$$V_{in(min)} \geq V_O + \text{dropout voltage}$$

FIGURE 16.7. For the regulator to operate properly, the input voltage must always be greater than the sum of the regulator's output and dropout voltages. The dropout voltage for IC regulators is typically 2 V to 3 V.

available, and the electrical characteristics of the unregulated power supply. Figure 16.5 shows the classification of IC voltage regulators by type of output voltage. A **fixed-voltage regulator** has three terminals and can supply one fixed value of output voltage. One type of fixed-voltage regulator is designed to give only a positive output voltage; another, only a negative voltage. A **dual-polarity voltage regulator** can supply both positive and negative voltages. All of these devices are also available as **adjustable voltage regulators** for which the output voltage can be varied.

The 7800 series comprises typical fixed *positive* regulators. They can furnish over 1 A at the standard voltages listed in Figure 16.6. The 7900 series of fixed *negative* regulators furnishes 1 A at essentially the same standard negative voltages. In each case, the last digits of the identification number indicate the fixed-voltage output. For example, the 7805 has a fixed output of 5 V and the 7905 has a fixed output of −5 V. If an application matches a standard voltage, simply select the appropriate IC number from the list. If the required supply voltage is not a standard value, then an adjustable IC voltage regulator is needed.

Fixed-voltage regulators can be converted to provide an adjustable output voltage by the addition of external circuits. However, for higher performance, it is advisable to select a regulator that has been specifically designed to be adjustable. A popular positive three-terminal adjustable regulator is the LM317 that requires only two external resistors to set the output voltage to any value between 1.2 V and 37 V.

16.3 REGULATOR RATINGS

Minimum Dropout Voltage

The input voltage to an IC voltage regulator must be larger than its output voltage or the regulator cannot function properly. On a manufacturer's data sheet, the minimum difference between input voltage from the full-wave filtered rectifier and the regulator output is specified as the **dropout voltage** or **minimum input–output differential voltage**. A typical value at full load current is 2 V to 3 V, as shown in Figure 16.7.

The minimum input voltage from the fil-

DETERMINING RECTIFIER SUPPLY VOLTAGE REQUIREMENT

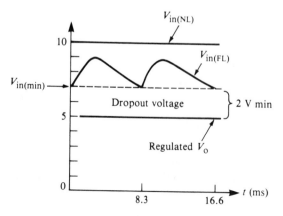

FIGURE 16.8. The lowest instantaneous value of the rectifier output voltage is the minimum value of the ripple at full load. This value must be at least 2 V greater than the regulator's output voltage.

tered rectifier should always exceed the regulator output voltage by at least the dropout voltage:

$$V_{in(min)} \geq V_o + \text{dropout voltage} \qquad (2)$$

The approximately 2 V limitation is determined by the lowest *instantaneous* value of rectifier output voltage due to ripple. Recall that the rectifier's output voltage will be lowest, and ripple voltage highest, at full load. As shown in Figure 16.8, the minimum filtered rectifier voltage is 7 V at full load current. This filtered rectifier would be satisfactory for a regulator output voltage of 5 V or less.

Current Ratings

The output current rating of an IC regulator is specified as a guaranteed output current up to a maximum value. For example, for a particular regulator from the 7800 series, the output terminal can furnish up to 1.5 A. If the load current demand exceeds 1.5 A, an **internal current-limiting circuit** will hold the maximum output current at 1.5 A. If it did not, excessive, brief, overload currents could burn out internal transistors and connecting leads in the IC.

In addition, an internal **thermal shutdown circuit** senses the IC's temperature. When the internal temperature exceeds a prespecified value (generally 175°C), the thermal shutdown circuit automatically reduces the load current to maintain a constant IC temperature below its maximum rating. When the overload current is removed, the IC cools down and the thermal shutdown circuit automatically places the regulator back in operation. Current ratings for IC voltage regulators, as well as operating data on thermal overload and current-limiting circuitry, must be obtained from manufacturer's specifications.

Temperature Ratings

A fundamental operating limit on any electronic device is its maximum temperature. The typical maximum temperature limit for an IC regulator is the same as for most transistor and other semiconductor devices—that is, between 125°C and 175°C. The place where heat damage first occurs is the semiconductor junction deep within the device. Generally, we cannot see or get at the junction without destroying or altering the device. The maximum operating temperature for this unseen junction is specified by the manufacturer as **maximum junction temperature** $T_{J(max)}$. Alternatively, the manufacturer may specify a range of allowable junction temperatures under a data heading such as "Storage Temperature Range" or "Junction Temperature Range." A typical range is $-55°C$ to 175°C, the highest temperature being $T_{J(max)}$.

Heat Dissipation

The heat developed within the IC regulator is equal to the electrical power used by the IC in operation (Figure 16.9). To determine this

VOLTAGE REGULATOR POWER DISSIPATION

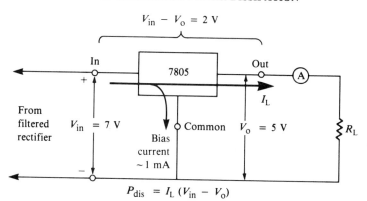

FIGURE 16.9. During operation, the regulator dissipates internal electrical power that raises its temperature. The maximum power that can be developed in the IC is determined by the maximum acceptable temperature of the internal semiconductor junctions, which is typically 175°C. The power developed is the product of the load current I_L and the input–output differential voltage ($V_{in} - V_o$).

value, we measure the dc input voltage V_{in} and output voltage V_o. Their difference, $V_{in} - V_o$, is the input–output differential voltage. Since the load current I_L is conducted between these terminals, the electrical power P_{dis} developed within the IC is:

$$P_{dis} = I_L(V_{in} - V_o) \qquad (3)$$

where P_{dis} is in watts if I is in amperes and V is in volts.

In Figure 16.9, the power converted into heat is:

$$P_{dis} = 1 \text{ A}(7 \text{ V} - 5 \text{ V}) = 2 \text{ W}$$

The power dissipated by the small bias current that flows between the input and common terminals is very small (1 mA × 7 V = 7 mW) and can be neglected.

Thermal Resistance

Once we know how much power is dissipated as heat, we must ensure that the junction temperature of the device will not exceed $T_{J(max)}$. Figure 16.10 shows the path of heat flow away from the hot junction at temperature T_J through the case to cooler surroundings, called the ambient, at temperature T_A. The **ambient** is usually air at

POWER DISSIPATION EQUATION

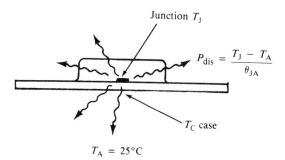

$$P_{dis} = \frac{T_J - T_A}{\theta_{JA}}$$

$$T_A = 25°C$$

FIGURE 16.10. Electrical power developed in the semiconductor junctions appears as heat that flows to the case, where it is dissipated into the air. The rate at which the heat flows P_{dis} is determined by the temperature difference between the junction and the ambient ($T_J - T_A$) and the thermal resistance of the path θ_{JA}.

room temperature, generally taken to be 25°C (77°F).

The flow of heat in a thermal circuit is analogous to the flow of current in an electrical circuit. Current flow I corresponds to the heat flow P_{dis}. Current flow through an electrical resistance R is caused by a voltage difference $V_x - V_y$. From Ohm's law:

$$V_x - V_y = RI$$

Similarly, a temperature difference $T_x - T_y$ between two points causes a flow of heat P_{dis} between the points. The rate at which the heat flows is determined by a corresponding **thermal resistance**. Thermal resistance is symbolized by the Greek letter theta (θ). Subscripts indicate the locations between which the thermal resistance exists. For example, θ_{JA} means the thermal resistance between junction and ambient. The heat flow equation between the two points is given by the **power dissipation equation:**

$$T_x - T_y = P_{dis}\theta_{xy} \qquad (4)$$

For heat flow between a regulator's junction and the ambient, Equation 4 becomes:

$$T_J - T_A = P_{dis}\theta_{JA} \qquad (5)$$

As in the case of electrical resistance, thermal resistances in series add to give a total effective resistance. Thus, the thermal resistance from the junction-to-ambient θ_{JA} is the sum of the thermal resistance from the junction-to-case θ_{JC} and the thermal resistance from case-to-ambient θ_{CA}, or:

$$\theta_{JA} = \theta_{JC} + \theta_{CA} \qquad (6)$$

See Figure 16.11. The heat flow Equation 4 can therefore be written:

$$T_J - T_A = (\theta_{JC} + \theta_{CA})P_{dis} \qquad (7)$$

Equations 4 and 5 indicate that the units of thermal resistance must be °C/W in order for the equation to balance. A thermal resistance of $\theta_{JA} = 1°C/W$ means that a temperature difference of $T_J - T_A = 1°C$ will cause a power of $P_{dis} = 1$ W to be dissipated from the junction into the ambient air.

Actual thermal resistance values of θ_{JC} and θ_{CA} for different semiconductor cases are specified by the semiconductor manufacturer. The values for θ_{JC} and θ_{CA} vary somewhat from device to device, but generally they lie in the range

THERMAL RESISTANCE

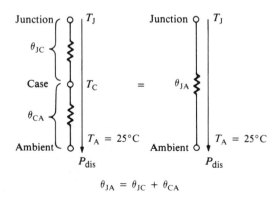

$$\theta_{JA} = \theta_{JC} + \theta_{CA}$$

FIGURE 16.11. Thermal resistances in series add to give an effective total resistance. Thus, the thermal resistance of the heat flow path from the junction-to-ambient θ_{JA} is the sum of the resistances from the junction-to-case θ_{JC} and from the case-to-ambient θ_{CA}.

of 1°C/W to 5°C/W for θ_{JC} and 30°C/W to 60°C/W for θ_{CA}. For example, the TO–3 (series 7800) IC regulator manufactured by Fairchild has a thermal resistance between junction and case of $\theta_{JC} = 4°C/W$ and a thermal resistance between case and ambient of $\theta_{CA} = 35°C/W$ (Figure 16.12a). The Fairchild thermal plastic TO–220 case has $\theta_{JC} = 2°C/W$ and $\theta_{CA} = 50°C/W$ (Figure 16.12b). Thermal resistances for both packages of the Fairchild 7800 series IC regulator can be summarized as follows:

TO–3 package: $\theta_{JC} = 4°C/W$
$\qquad\qquad\qquad\quad \theta_{CA} = 35°C/W$
Therefore: $\theta_{JA} = \theta_{JC} + \theta_{CA}$
$\qquad\qquad\qquad\quad = 4° + 35°C/W$
$\qquad\qquad\qquad\quad = 39°C/W$

TO–220 package: $\theta_{JC} = 2°C/W$
$\qquad\qquad\qquad\qquad \theta_{CA} = 50°C/W$
Therefore: $\theta_{JA} = 52°C/W$

The maximum recommended operating junc-

IC REGULATOR CASE STYLES

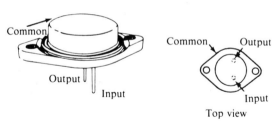

(a) TO–3: $\theta_{JA} = 39°C/W$

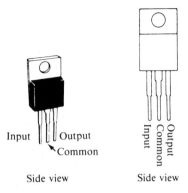

(b) TO–220: $\theta_{JA} = 52°C/W$

FIGURE 16.12. IC regulators come in several case styles. For the TO–3 and the TO–220 units, the junction-to-ambient thermal resistances are $\theta_{JA} = 39°C/W$ and $\theta_{JA} = 52°C/W$, respectively. The lower the thermal resistance, the more power that can be dissipated for the same junction temperature.

tion temperature for both packages is specified at 125°C. They can be stored and operated at 150°C, but this temperature increases the chance of failure.

Maximum Power Dissipation

The **maximum power dissipation** $P_{dis(max)}$ that any semiconductor can handle is determined by its thermal resistance, ambient temperature, and maximum allowable junction temperature as shown by rewriting Equation 5:

$$P_{dis(max)} = \frac{T_{J(max)} - T_A}{\theta_{JA}} \tag{8}$$

Since $T_{J(max)}$ and θ_{JA} are fixed quantities for the device, $P_{dis(max)}$ depends only on the ambient temperature. The higher the ambient temperature, the less power that can be dissipated. Semiconductor manufacturers generally specify $P_{dis(max)}$ for a particular ambient temperature, usually 25°C. If $T_{J(max)}$ and θ_{JA} are also specified, then you can calculate $P_{dis(max)}$ at other ambients.

For semiconductors that will dissipate a large amount of heat, the maximum power rating is also specified for a specific **case temperature** T_C, usually 50°C. The equation that relates case temperature to maximum power and junction temperature is:

$$P_{dis(max)} = \frac{T_{J(max)} - T_C}{\theta_{JC}} \tag{9}$$

As in Equation 8, the higher the case temperature, the less power the device will dissipate.

To eliminate the need for calculations, some manufacturers also give a **maximum power dissipation curve**. This curve shows the maximum allowable power dissipation at any case (or ambient) temperature. For example, in Figure 16.13, the power dissipation curve shows that an IC in a TO–3 package can dissipate the maximum power rating of 15 W at *case* temperatures below 65°C. At case temperatures above 65°C, the amount of power that can safely be dissipated decreases, becoming zero at $T_{J(max)} = 125°C$. The same regulator in a TO–220 case can dissipate 15 W up to a case temperature of 85°C before it becomes derated. This power dissipation curve is simply a graph of Equation 9. The slope of the curve is $1/\theta_{JC}$.

In Figure 16.14, the maximum allowable power at any *ambient* temperature is shown for

FIGURE 16.13. The maximum power that a semiconductor device can dissipate at a given case temperature is given by a maximum power dissipation curve. The curves shown are for Fairchild 7800 series IC regulators in both TO-3 and TO-220 cases. Each curve is simply a graph of Equation 9 for $T_{J(max)} = 125°C/W$ and the appropriate value of θ_{JC}.

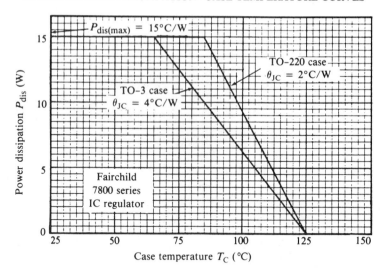

MAXIMUM POWER DISSIPATION- CASE TEMPERATURE CURVES

FIGURE 16.14. Shown are the maximum allowable power dissipation versus ambient air temperature curves for Fairchild 7800 series IC regulators in TO-3 and TO-220 cases. The lower thermal resistance from junction to air of the TO-3 case means that it can dissipate more power at a given air temperature.

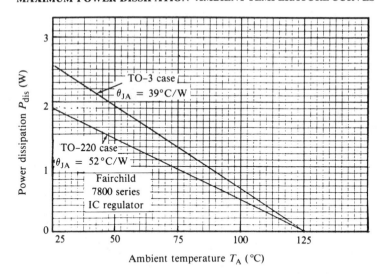

MAXIMUM POWER DISSIPATION-AMBIENT TEMPERATURE CURVES

the TO-3 and TO-220 cases. Observe that the TO-3 case can dissipate more power at a given ambient temperature because it has a lower thermal resistance between junction and ambient. The following examples illustrate the use of both the maximum power dissipation curves and the thermal resistance equations.

EXAMPLE 1: How much power can a 7805 voltage regulator in a TO-3 case dissipate in *ambients* of 25°C, 85°C, and 125°C if the maximum junction temperature rating is 125°C?

Solution: Using Equation 8 and $\theta_{JC} = 39°C/W$ for a TO–3 case, we have:

$$P_{dis(max)} = \frac{T_{J(max)} - T_A}{\theta_{JA}}$$

At $T_A = 25°C$:

$$P_{dis(max)} = \frac{(125 - 25)°C}{39°C/W} = \textbf{2.6 W}$$

At $T_A = 85°C$:

$$P_{dis(max)} = \frac{(125 - 85)°C}{39°C/W} = \textbf{1 W}$$

At $T_A = 125°C$:

$$P_{dis(max)} = \frac{(125 - 125)°C}{39°C/W} = \textbf{0 W}$$

The same results can be obtained without calculation by referring to the maximum power dissipation curve shown in Figure 16.14.

EXAMPLE 2: How much power can a 7805 voltage regulator in a TO–220 case dissipate for *case* temperatures of 95°C and 125°C if the junction temperature is at the maximum of 125°C?

Solution: Using Equation 9 and $\theta_{JC} = 2°C/W$ for a TO–220 case, we find:
At $T_C = 95°C$:

$$P_{dis(max)} = \frac{T_{J(max)} - T_C}{\theta_{JC}} = \frac{(125 - 95)°C}{2°C/W}$$
$$= \textbf{15 W}$$

At $T_C = 125°C$:

$$P_{dis(max)} = \frac{(125 - 125)°C}{2°C/W} = \textbf{0 W}$$

The same solution can be found by checking Figure 16.13.

EXAMPLE 3: A 7805 voltage regulator in a TO–3 case dissipates 2 W at room temperature (25°C). (a) Will its junction temperature exceed the maximum rating of 125°C? (b) Could the case be touched?

Solution: (a) Equation 5 and $\theta_{JA} = 39°C/W$ for a TO–3 case give:

$$T_J = T_A + \theta_{JA}P_{dis}$$
$$= 25°C + (39°C/W)2\ W = \textbf{103°C}$$

Therefore, T_J **does not exceed** $T_{J(max)} = $ **125°C.**
   (b) Using Equation 4 and $\theta_{JC} = 4°C/W$ for a TO–3 case, we have:

$$T_C = T_J - \theta_{JC}P_{dis}$$
$$= 103°C - (4°C/W)2\ W = \textbf{95°C}$$

This temperature is very nearly that of boiling water, so it would be **inadvisable to touch the case** even when the semiconductor junction is operating satisfactorily.

If the semiconductor manufacturer does not specify θ_{JC} and instead gives the maximum power dissipation curve, θ_{JC} can be calculated from the curve using the technique set forth in Example 4.

EXAMPLE 4: Calculate θ_{JC} for a TO–3 case from the maximum power dissipation curve shown in Figure 16.13.

Solution: The curve of Figure 16.13 gives $T_{J(max)} = 125°C$ because the allowable power dissipation P_{dis} is zero ($T_C = T_{J(max)}$). We then obtain the maximum case temperature for some particular power dissipation—for example, $T_C = 65°C$ at $P_{dis(max)} = 15$ W. Finally, using Equation 9, we find:

$$\theta_{JC} = \frac{T_{J(max)} - T_C}{P_{dis(max)}} = \frac{(125 - 65)°C}{15\ \text{W}}$$
$$= 4°C/W$$

The next example shows how to predict whether a regulator will function under a proposed overload.

EXAMPLE 5: Assume that a regulator must dissipate 5 W in an ambient of 25°C. Could a TO–3 package be used?

Solution: Equation 5 gives the junction temperature of the TO–3 case:

$$T_J = T_A + \theta_{JA}P_{dis}$$
$$= 25°C + (39°C/W)5\ \text{W} = 215°C$$

Since T_J exceeds $T_{J(max)} = 125°C$, a **TO–3 package should not be used.** If it were used, the regulator would sustain damage without internal thermal overload protection. The thermal overload circuit reduces the load current and, consequently, the load voltage to a value that will maintain the junction below 175°C. The circuit will not function properly, but at least it will not burn up.

16.4 HEAT SINKS

In order for the IC regulator in Example 4 to operate at 5 W, its junction temperature must be kept below $T_{J(max)}$. The most practical cooling method is to attach a heat radiator that dissipates more heat into the air. Heat radiators for electronic devices are called **heat sinks**, or **heat exchangers**. From Figure 16.14, we can

INCREASING POWER DISSIPATION WITH HEAT SINKS

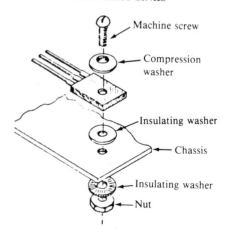

(a) TO–220 Case

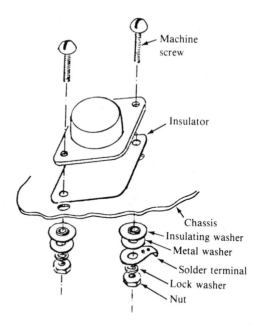

(b) TO–3 Case

FIGURE 16.15. The amount of power that a semiconductor device can safely dissipate can be increased by adding a heat sink. The simplest method of heat sinking is to bolt the device to the chassis. If the case of the device is one of its electrical connections, an electrically insulating washer must be used as shown.

WAKEFIELD HEAT SINKS

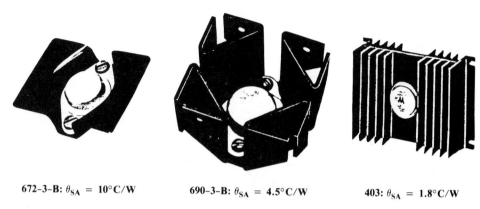

672-3-B: $\theta_{SA} = 10°C/W$ 690-3-B: $\theta_{SA} = 4.5°C/W$ 403: $\theta_{SA} = 1.8°C/W$

FIGURE 16.16.   Separate heat sinks that bolt directly to the IC case are also available. The figure shows several styles and gives thermal resistance values for the heat-sink-to-air θ_{SA} of each.

conclude that heat sinks are required whenever the power dissipation must exceed about 2 W for the TO–220 and 3 W for the TO–3 packages. The process of adding a radiator to a semiconductor for more power dissipation is called **heat sinking**.

A simple way to heat-sink a TO–3 or TO–220 semiconductor package is to bolt it to the metal chassis of the electronic system that then acts as a heat radiator (Figure 16.15). Heat is conducted from the case to the chassis, where it is dissipated to the ambient. The chassis essentially increases the IC's effective volume and surface area, permitting greater heat dissipation and cooler IC operation.

The case of an IC regulator in a TO–3 package is also its input terminal, so it must be connected to one side of the filtered full-wave rectifier. Thus, an electrical connection to the case is made with a solder terminal lug, metal screw, washers, and nuts. The case must be electrically insulated from the chassis, however, by means of an insulating spacer and shoulder washer.

A variety of heat sinks are also available commercially (Figure 16.16). Typical values for the thermal resistance of a heat sink to air are $\theta_{SA} = 1°C/W$ to $5°C/W$, which is on the same order as θ_{JC} of the semiconductor and about ten times less than θ_{CA}.

A model for heat flow from a heat-sink semiconductor case is shown in Figure 16.17. The value of θ_{CA} is given by:

$$\theta_{CA} = \theta_{CS} + \theta_{SA} \qquad (10)$$

where θ_{CS} represents the thermal resistance of the gap between the case of the semiconductor and the surface of the heat sink and θ_{SA} is the thermal resistance of the heat sink to ambient. θ_{CS} can be minimized by filling the air gaps between the two surfaces with a thin coat of electrically insulating liquid or paste called **thermal joint compound**. If an insulating washer must be used, it should be coated on both sides with the compound before installation. A typical value for θ_{CS} with thermal joint compound is $0.5°C/W$ or less. Since θ_{SA} is only about $3°C/W$, the heat sink acts essentially as a ther-

FIGURE 16.17. A heat sink acts as a low thermal resistance shunt from the semiconductor case to the air, thus permitting greater power dissipation. The required value of heat sink thermal resistance θ_{SA} for a given power dissipation can be calculated from Equation 11. See Example 6.

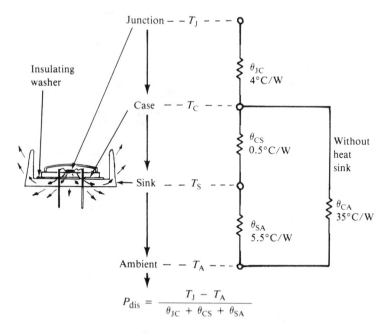

POWER DISSIPATION EQUATION FOR TO-3 IC CASE
WITH HEAT SINK

$$P_{dis} = \frac{T_J - T_A}{\theta_{JC} + \theta_{CS} + \theta_{SA}}$$

mal shunt across the relatively high case to ambient thermal resistance of the case alone.

Selecting a Heat Sink

When choosing a heat sink for an application, the ambient temperature, the required power to be dissipated, and the thermal resistance of the device will usually be known. We must then calculate the maximum allowable thermal resistance between sink and ambient θ_{SA} and choose a heat sink with a rated θ_{SA} equal to or less than the calculated value. To calculate θ_{SA}, we use an equation developed from the model in Figure 16.17:

$$T_J - T_A = P_{dis}(\theta_{JC} + \theta_{CS} + \theta_{SA})$$
$$\theta_{SA} = \frac{T_J - T_A}{P_{dis}} - \theta_{JC} - \theta_{CS} \qquad (11)$$

EXAMPLE 6: An IC regulator in a TO-3 case must dissipate 10 W in an ambient of 25°C. The IC manufacturer specifies $\theta_{JC} = 4°C/W$ and $T_{J(max)} = 125°C$. What size heat sink is required?

Solution: We assume $\theta_{CS} = 0.5°C/W$. From Equation 11, we have:

$$\begin{aligned}
\theta_{SA} &= \frac{T_J - T_A}{P_{dis}} - \theta_{JC} - \theta_{CS} \\
&= \frac{(125 - 25°C/W)}{10 \text{ W}} \\
&\quad -4°C/W - 0.5°C/W \\
&= \mathbf{5.5°C/W}
\end{aligned}$$

Thus, we should choose a heat sink that fits the TO-3 case whose $\theta_{SA} \leq 5.5°C/W$.

HEAT SINK TEMPERATURE-POWER DISSIPATION CURVES

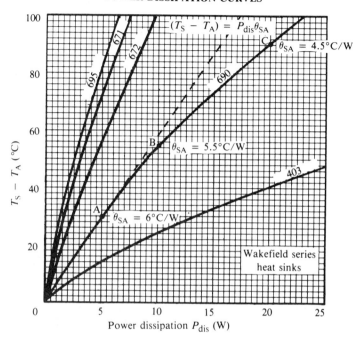

FIGURE 16.18. A heat sink manufacturer may give actual data describing the heat sink temperature rise above ambient for various power dissipations. These data give more accurate results than using the power dissipation equation because θ_{SA} is not a fixed value. The data here show that θ_{SA} actually decreases with increasing temperature. See Example 7.

The heat sink manufacturer may not give θ_{SA} directly, but instead provide a graph of its thermal behavior showing the temperature rise of the heat sink as a function of its power dissipation. A typical **heat sink characteristic** for the Wakefield series is shown in Figure 16.18. Note that this graph describes thermal behavior for the *natural convection* condition — that is, when the heat sink is simply sitting out in the open air. This condition is in contrast to *forced convection* in which air is forced past the surface of the sink by a fan. Forced convection further reduces the value of θ_{SA}.

The heat sink characteristic curve is a plot of actual data describing the increase in heat sink temperature above the ambient ($T_S - T_A$) for various power dissipations in the semiconductor attached to it. Note that it is not exactly a straight line. The power dissipation equation between the sink and the ambient gives:

$$T_S - T_A = P_{dis}\theta_{SA} \qquad (12)$$

which is a straight line passing through $T_S - T_A = 0$ and having a slope of θ_{SA}. Thus, the power dissipation equation is only an *approximate* mathematical expression of the heat sink's behavior for various powers. The fact that the actual data curve downward from the straight line means that θ_{SA} is not a fixed value for the heat sink, but decreases somewhat as the sink temperature goes up. Thus, the heat sink performs somewhat better at higher temperatures.

To determine the actual thermal resistance of the heat sink for a given power dissipation, enter the horizontal axis of the characteristic curve at the power dissipation required — for example, 10 W. Proceed vertically to the intersection of the curve and read the case temperature rise above ambient, $T_S - T_A = 55°C$. Then, calculate θ_{SA} from Equation 12:

$$\theta_{SA} = \frac{T_S - T_A}{P_{dis}} = \frac{55°C}{10\ W} = 5.5°C/W$$

This result indicates that the 690 heat sink would just be satisfactory for the IC regulator in Example 6.

EXAMPLE 7: In Example 5, the junction temperature of a TO-3 package went to 215°C when dissipating 5 W. What will its junction temperature be if the 690 sink is added?

Solution: From Figure 16.18, $\theta_{SA} = 6°C/W$ at $P_{dis} = 5$ W. Using Equation 11, we get:

$$T_J = T_A + P_{dis}(\theta_{JC} + \theta_{CS} + \theta_{SA})$$
$$= 25°C + 5\ W(4 + 0.05 + 6)°C/W$$
$$= \mathbf{77.5°C}$$

Note that adding the heat sink has reduced the junction temperature from 215°C to 77.5°C, thus allowing the IC to operate.

Usually, the semiconductor manufacturer gives the thermal resistance θ_{JC} for its device and the heat sink manufacturer gives θ_{SA} for its heat sink because neither knows what combination the circuit designer will use. As shown in Figure 16.19, however, some semiconductor manufacturers provide a single curve that combines *both* the thermal characteristics of their device and a heat sink. For example, in Figure 16.19, the heat dissipation curve for the TO-3 case in ambient air with no heat sink is a replot of Figure 16.14 with the P_{dis} axis marked off in logarithmic increments rather than linear increments.

Like Figure 16.14, Figure 16.19 indicates that a TO-3 case with no heat sink can dissipate up to about 2.6 W in an ambient of 25°C and that there is less dissipation at higher ambient temperatures. The second curve indicates that a

MAXIMUM POWER DISSIPATION–AMBIENT TEMPERATURE CURVES

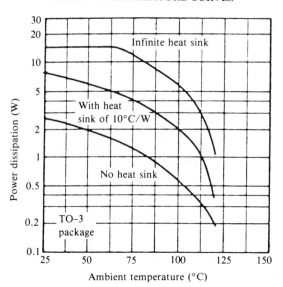

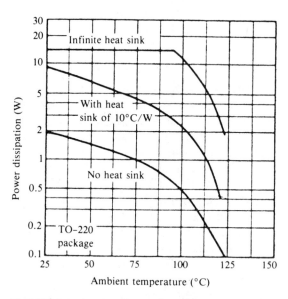

FIGURE 16.19. Some semiconductor manufacturers give data on the maximum power dissipation of their device when mounted on heat sinks with various values of θ_{SA}. These data make the selection of a heat sink a simple task.

TO–3 case on a 10°C/W heat sink can handle up to about 6 W of power in an ambient of 25°C. The third curve represents data for an infinite heat sink. This one for $\theta_{SA} = 0$ means that the case is always maintained at 25°C. Data for heat sinks with values of θ_{SA} less than 10°C/W lie between the latter two curves.

Making a Heat Sink

The power ratings for TO–3 and TO–220 case ICs on commercial heat sinks are generally reliable because the manufacturers of both ICs and heat sinks give actual data for their products. You can make your own heat sink, however, from aluminum sheet, or you can use the chassis, though the thermal resistance in these cases is not precisely predictable. An approximate rule of thumb is that a heat sink made from a 4″ × 4″ square of 3/32″ aluminum sheet allows a TO–3 or TO–220 to dissipate about 10 W in an ambient of 25°C.

16.5 DESIGNING A REGULATED POWER SUPPLY

A design example will illustrate the use of IC regulators and heat sink selection with a regulated power supply. The problem is to design a power supply that can deliver 5 V to a microprocessor system that requires up to 1 A of full load current. The ripple voltage should not exceed 10 mV rms. The design procedure is:

1. Select an IC regulator that meets the specification.
2. Design a filtered rectifier to supply the regulator.
3. Select a heat sink for the IC regulator if one is required.

Selecting the IC Regulator

The choice for this example is the 7805 IC regulator because it is a standard, inexpensive IC that will furnish 5 V at 1 A. Its specification sheet is given in Figure 16.20. Alternate possibilities are the LM340–5 and 109 that will also deliver 1 A at 5 V with the same excellent performance.

Calculating Rectifier Output Voltage

The determining factor in designing the full-wave rectifier is the input voltage required by the rectifier. The minimum instantaneous output voltage of the rectifier must exceed the highest regulated output voltage of the 5 V regulator plus the dropout voltage (Figure 16.21). The specification sheet gives 2 V as the dropout voltage for the 7805 regulator. Thus, $V_{L(min)}$ of the filtered rectifier is found from Equation 2:

$$V_{L(min)} = V_o + \text{dropout voltage}$$
$$= 5\,V + 2\,V = 7\,V$$

The minimum filtered rectifier output voltage $V_{L(min)}$ occurs at full load current and is the difference between the average rectifier output voltage at full load $V_{L(av)FL}$ and half the peak-to-peak ripple voltage ΔV_L (Figure 16.21):

$$V_{L(min)} = V_{L(av)FL} - \frac{\Delta V_L}{2} \tag{13}$$

This equation can be used to determine $V_{L(av)FL}$ of the rectifier, if ΔV_L is known. The ripple voltage can be estimated from Figure 15.8 if a filter capacitor value is assumed. Because the maximum load current is 1 A, we make a trial selection for the filter capacitor of 1000 μF. The maximum peak-to-peak ripple voltage of the filtered full-wave rectifier for a 1000 μF capacitor, according to Figure 15.8, is $\Delta V_L = 5$ V. The

SPECIFICATION SHEET FOR FAIRCHILD μA7800 SERIES VOLTAGE REGULATORS

μA7800 SERIES
3-TERMINAL POSITIVE VOLTAGE REGULATORS
FAIRCHILD LINEAR INTEGRATED CIRCUITS

GENERAL DESCRIPTION — The μA7800 series of monolithic 3-Terminal Positive Voltage Regulators is constructed using the Fairchild Planar* epitaxial process. These regulators employ internal current limiting, thermal shutdown and safe area compensation, making them essentially indestructible. If adequate heat sinking is provided, they can deliver over 1 A output current. They are intended as fixed voltage regulators in a wide range of applications including local (on card) regulation for elimination of distribution problems associated with single point regulation. In addition to use as fixed voltage regulators, these devices can be used with external components to obtain adjustable output voltages and currents.

- OUTPUT CURRENT IN EXCESS OF 1 A
- NO EXTERNAL COMPONENTS
- INTERNAL THERMAL OVERLOAD PROTECTION
- INTERNAL SHORT CIRCUIT CURRENT LIMITING
- OUTPUT TRANSISTOR SAFE AREA COMPENSATION
- AVAILABLE IN THE TO-220 AND THE TO-3 PACKAGE
- OUTPUT VOLTAGES OF 5, 6, 8, 8.5, 12, 15, 18, AND 24 V

CONNECTION DIAGRAMS
TO-220 PACKAGE
(TOP VIEW)
PACKAGE OUTLINE GH
PACKAGE CODE   U

TO-3 PACKAGE
(TOP VIEW)
PACKAGE OUTLINE GJ
PACKAGE CODE K

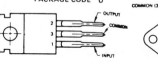
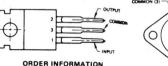

ORDER INFORMATION

OUTPUT VOLTAGE	TYPE	PART NO.
5 V	μA7805C	μA7805UC
6 V	μA7806C	μA7806UC
8 V	μA7808C	μA7808UC
8.5 V	μA7885C	μA7885UC
12 V	μA7812C	μA7812UC
15 V	μA7815C	μA7815UC
18 V	μA7818C	μA7818UC
24 V	μA7824C	μA7824UC

ORDER INFORMATION

OUTPUT VOLTAGE	TYPE	PART NO.
5 V	μA7805	μA7805KM
6 V	μA7806	μA7806KM
8 V	μA7808	μA7808KM
8.5 V	μA7885	μA7885KM
12 V	μA7812	μA7812KM
15 V	μA7815	μA7815KM
18 V	μA7818	μA7818KM
24 V	μA7824	μA7824KM
5 V	μA7805C	μA7805KC
6 V	μA7806C	μA7806KC
8 V	μA7808C	μA7808KC
8.5 V	μA7885	μA7885KC
12 V	μA7812C	μA7812KC
15 V	μA7815C	μA7815KC
18 V	μA7818C	μA7818KC
24 V	μA7824C	μA7824KC

ABSOLUTE MAXIMUM RATINGS

Input Voltage (5 V through 18 V)		35 V
(24 V)		40 V
Internal Power Dissipation		Internally Limited
Storage Temperature Range		−65°C to +150°C
Operating Junction Temperature Range	μA7800	−55°C to +150°C
	μA7800C	0°C to +150°C
Lead Temperature (Soldering, 60 s time limit) TO-3 Package		300°C
(Soldering, 10 s time limit) TO-220 Package		230°C

ELECTRICAL CHARACTERISTICS
(V_{IN} = 10 V, I_{OUT} = 500 mA, 0°C < T_J < 125°C, C_{IN} = 0.33 μF, C_{OUT} = 0.1 μF unless otherwise specified)

PARAMETER	CONDITIONS		MIN	TYP	MAX	UNITS
Output Voltage	T_J = 25°C		4.8	5.0	5.2	V
Line Regulation	T_J = 25°C	7 V < V_{IN} < 25 V		3	100	mV
		8 V < V_{IN} < 25 V		1	50	mV
Load Regulation	T_J = 25°C	5 mA < I_{OUT} < 1.5 A		15	100	mV
		250 mA < I_{OUT} < 750 mA		5	50	mV
Output Voltage	7 V < V_{IN} < 20 V 5 mA < I_{OUT} < 1.0 A P < 15 W		4.75		5.25	V
Quiescent Current	T_J = 25°C			4.2	8.0	mA
Quiescent Current Change	with line	7 V < V_{IN} < 25 V			1.3	mA
	with load	5 mA < I_{OUT} < 1.0 A			0.5	mA
Output Noise Voltage	T_A = 25°C, 10 Hz < f < 100 kHz			40		μV
Ripple Rejection	f = 120 Hz, 8 V < V_{IN} < 18 V		62	78		dB
Dropout Voltage	I_{OUT} = 1.0 A, T_J = 25°C			2.0		V
Output Resistance	f = 1 kHz			17		mΩ
Short Circuit Current	T_J = 25°C			750		mA
Peak Output Current	T_J = 25°C			2.2		A
Average Temperature Coefficient of Output Voltage	I_{OUT} = 5 mA, 0°C < T_J < 125°C			−1.1		mV/°C

FIGURE 16.20. Shown are portions of the specification sheet for the 7805 IC regulator used in the design example.

SELECTING TRANSFORMER FOR IC VOLTAGE REGULATOR

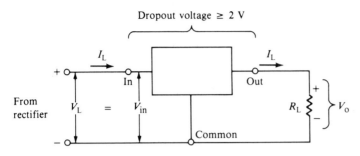

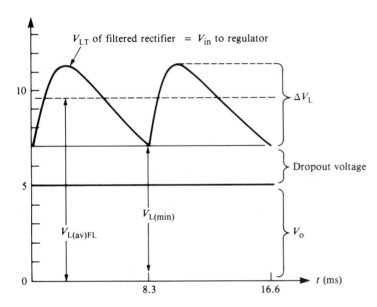

FIGURE 16.21. The key factor in selecting the transformer is that the rectifier's output voltage must never be less than the regulator output voltage plus its dropout voltage. The minimum value of V_L occurs at full load when $V_{L(av)}$ is at its lowest value and ripple is greatest.

full-load dc voltage of the filtered rectifier can now be found from Equation 13:

$$V_{L(av)FL} = V_{L(min)} + \frac{\Delta V_L}{2} = 7\text{ V} + \frac{5\text{ V}}{2}$$
$$= 9.5\text{ V}$$

Finally, assuming a voltage regulation of 20%, we can estimate the corresponding no-load dc voltage of the filtered rectifier from Equation 5 in Chapter 15:

$$V_{L(av)FL} \approx 0.8 V_{L(av)NL}$$

or

$$V_{L(av)NL} \approx 1.2 V_{L(av)FL} \approx 1.2(9.5\text{ V}) \approx 12\text{ V}$$

This value can be used in selecting the rectifier components.

Selecting the Transformer

A full-wave bridge rectifier is selected for simplicity and cost. The transformer rms secondary rating can be calculated from Equation 3 in Chapter 15:

$$V_{sec(rms)} = \frac{V_{L(av)NL}}{1.4} = \frac{12\text{ V}}{1.4} = 8.6\text{ V}$$

The rms secondary current rating is found from Equation 6 in Chapter 15:

$$I_{sec} \geq 1.8 I_{L(av)FL} = 1.8(1 \text{ A}) = 1.8 \text{ A}$$

We select a standard transformer with a secondary voltage rating of at least 9 V and a current rating of 2 A or more. The primary voltage rating should be 115 V or 230 V depending on the line voltage. The F32A transformer shown in Figure 14.23 with a 12.6 V secondary is an acceptable choice for 115 V ac. Using this transformer, we have $V_{sec(rms)} = 12.6$ V.

Selecting the Fuse

The primary fuse current is calculated from Equation 7 in Chapter 15:

$$I_{fuse} \geq 1.8 \left(\frac{V_{sec}}{V_{pri}} \right) \times I_{L(av)FL}$$

For $V_{pri} = 115$ V and $V_{sec} = 12.6$ V:

$$I_{fuse} = 1.8 \left(\frac{12.6 \text{ V}}{115 \text{ V}} \right) \times 1 \text{ A} = 0.2 \text{ A}$$

Allowing for a 50% overload, we select from Figure 14.17 a 3/10 A standard slow blow fuse for 115 V operation.

Selecting the Diodes

From Equation 8 in Chapter 15, we know that the minimum average diode current rating I_o for a filtered full-wave rectifier is one-half the full load current and that the PIV rating should exceed $V_{sec(peak)}$. From Equation 3 in Chapter 15:

$$V_{sec(peak)} = 1.4 V_{sec(rms)} = 1.4 \times 12.6 \text{ V}$$
$$= 18 \text{ V}$$

Therefore, according to Equation 8 in Chapter 15:

$$I_o \geq \tfrac{1}{2} I_{L(av)FL} \geq \tfrac{1}{2}(1 \text{ A}) \geq \tfrac{1}{2} \text{ A}$$
$$PIV \geq V_{sec(peak)} \geq 18 \text{ V}$$

The 1N4720, rated at $I_o = 3$ A and PIV = 100 V, is more than adequate.

Selecting the Filter Capacitor

The value of the filter capacitor has already been selected as 1000 μF. Its voltage rating should exceed $V_{sec(peak)}$, or 18 V. Therefore, we select an electrolytic capacitor with $C = 1000$ μF and a working voltage rating of WVDC = 50 V.

Selecting the Heat Sink

The value of $V_{L(av)NL}$ for the F32A transformer with $V_{sec(rms)} = 12.6$ V is the same as $V_{sec(peak)}$:

$$V_{L(av)NL} = V_{sec(peak)} = 18 \text{ V}$$

For an assumed 20% voltage regulation, Equation 5 in Chapter 15 gives the actual value of $V_{L(av)FL}$ as:

$$V_{L(av)FL} \approx 0.8 V_{L(av)NL} \approx 0.8(18 \text{ V}) \approx 14 \text{ V}$$

The input voltage to the regulator, then, would be $V_{in} = 14$ V when it conducts 1 A, and its output would be 5 V. The maximum power dissipation is calculated from Equation 3:

$$P_{dis} = I_L(V_{in} - V_o) = 1 \text{ A}(14 \text{ V} - 5 \text{ V})$$
$$= 9 \text{ W}$$

Since P_{dis} is more than the rated 3 W for a TO–3 case and 2 W for a TO–220 case, a heat sink is required.

We select a 7805 in a TO–3 case. Figure 16.13 gives $\theta_{JC} = 4°$C/W, and assume $\theta_{CS} = 0.5°$C/W. Let $T_{J(max)} = 125°$C for a longer life and assume the ambient to be 25°C. The thermal resistance of the heat sink can then be calculated from Equation 11:

$$\theta_{SA} = \frac{T_J - T_A}{P_{dis}} - \theta_{JC} - \theta_{CS}$$
$$= \frac{125°C - 25°C}{9 \text{ W}} - 4.5°C/W$$
$$\approx 6.6°C/W$$

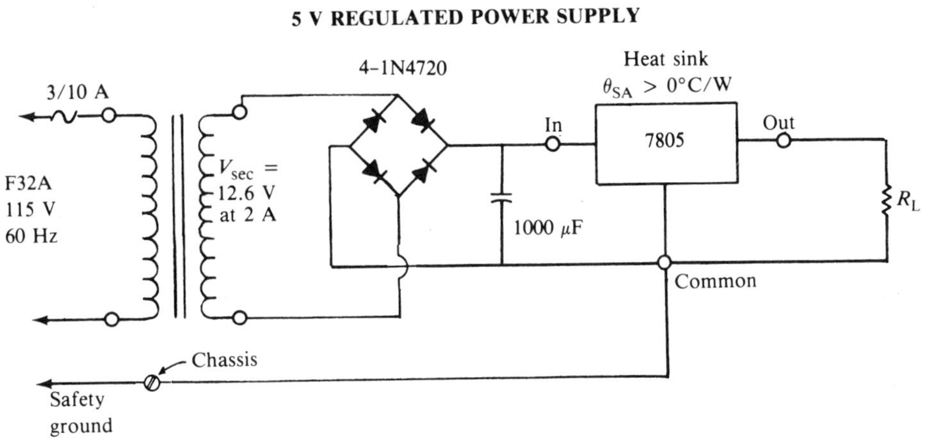

5 V REGULATED POWER SUPPLY

FIGURE 16.22.   Shown is the final design for the 5 V regulated power supply.

The Wakefield 690–3–B, whose thermal characteristics are given in Figure 16.18, is sufficient. Its θ_{SA} at $P_{dis} = 9$ W is less than 6°C/W, which is smaller than the requirement of 6.6°C/W. A schematic of the complete design is shown in Figure 16.22.

16.6 ADJUSTABLE VOLTAGE REGULATORS

There are many situations in which an **adjustable voltage regulator** is preferable. For example, the supply voltage required may not correspond to one of the available standard values. Or, perhaps, a precise reference voltage, such as 5.00 V, is needed. A 7805 would not be suitable in this case because its output voltage produces only a "typical" value of 5.0 V (see the specification sheet in Figure 16.20). According to the specification sheet, the output voltage of a given regulator may be anywhere between 4.8 V and 5.2 V. Also, a regulated voltage that is continuously variable might be desirable — for example, in a laboratory power supply. A similar application is the programmable power supply,

in which the IC regulator is made to vary over a range of predetermined voltages.

All fixed regulators of the 7800 and 7900 series (and their equivalents) can be made adjustable by adding a few extra components. However, a regulator designed to be adjustable will exhibit superior performance and give a wider choice of output voltages. Two examples are the LM117 and the LM317 three-terminal, adjustable *positive* regulators. These regulators are easy to use because they have only three terminals: input, output, and common (Figure 16.23). The LM117 is a premium regulator whose performance is guaranteed over a wide temperature range (−55°C to +150°C). The LM317 is somewhat less expensive, but it operates satisfactorily between 0°C and 125°C.

Two other examples are the LM137 and LM337 three-terminal, adjustable *negative* regulators that have almost the same characteristics as the positive LM317. Although a positive regulator can be made to furnish a negative voltage (and a negative one made to furnish a positive voltage), this practice is not recommended. The analysis and design discussions that follow focus primarily on the positive voltage LM317, but they also apply to the negative voltage LM337.

**LM317 POSITIVE ADJUSTABLE
VOLTAGE REGULATOR AND SYMBOL**

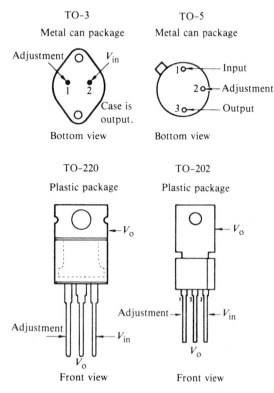

(a) **Positive Adjustable Voltage Regulators**

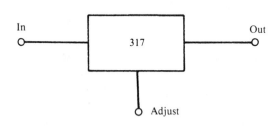

(b) **Schematic Symbol**

FIGURE 16.23. Three-terminal, adjustable regulators are used in applications requiring a precise or nonstandard voltage or when the voltage must be variable. The ones shown are LM317 positive regulators, but a negative-type LM337 is also available.

16.7 LM317 ADJUSTABLE POSITIVE REGULATOR

Performance Specifications

The specification sheet for the National Semiconductor LM317 voltage regulator is given in Figure 16.24. It shows that the LM317 can supply any output voltage between 1.2 V and 37 V at currents that exceed 1.5 A. It features thermal overload protection, as well as current-limiting protection, and is available in both TO–3 and TO–220 cases.

Like fixed-output regulators, the LM317 adjustable regulator needs a minimum input–output differential, or dropout, voltage of 3.0 V. Unlike the fixed–output voltage, the LM317 requires a minimum load current of 5 MA for satisfactory operation. Its load regulation is typically 0.1%, and it has a ripple rejection ratio of 80 db, or 10,000.

As for the fixed-voltage regulator, the thermal resistance between junction and case θ_{JC} is typically 1°C/W to 5°C/W, and the maximum junction temperature is about 125°C. Therefore, the maximum power dissipation for a given ambient temperature can be estimated from curves like those shown in Figure 16.14. For example, at room temperature, the TO–3 case can handle about 2.5 W and the TO–220, about 2 W. Both TO–3 and TO–220 cases can dissipate about 10 W on a Wakefield 690 series heat sink ($\theta_{SA} = 5.5°C/W$) and about 5 W on a Wakefield 672 series heat sink ($\theta_{SA} = 15°C/W$).

Principles of Operation

The principle behind setting the output voltage for an adjustable regulator is quite simple (Figure 16.25). The regulator always maintains a constant reference voltage V_{ref} between its output and adjustment terminals. For example, in the adjustable LM317 shown in Figure 16.24, $V_{ref} = 1.2$ V. The values R_1 and V_{ref} set the

SPECIFICATION SHEET FOR NATIONAL SEMICONDUCTOR LM317 VOLTAGE REGULATOR

absolute maximum ratings

Power Dissipation	Internally limited
Input–Output Voltage Differential	40V
Operating Junction Temperature Range	
LM117	-55°C to $+150^\circ$C
LM217	-25°C to $+150^\circ$C
LM317	0°C to $+125^\circ$C
Storage Temperature	-65°C to $+150^\circ$C
Lead Temperature (Soldering, 10 seconds)	300°C

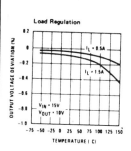

Load Regulation

electrical characteristics (Note 1)

PARAMETER	CONDITIONS	LM117/217 MIN	LM117/217 TYP	LM117/217 MAX	LM317 MIN	LM317 TYP	LM317 MAX	UNITS
Line Regulation	$T_A = 25^\circ$C, $3V \leq V_{IN} - V_{OUT} \leq 40V$ (Note 2)		0.01	0.02		0.01	0.04	%/V
Load Regulation	$T_A = 25^\circ$C, $10\,mA \leq I_{OUT} \leq I_{MAX}$							
	$V_{OUT} \leq 5V$, (Note 2)		5	15		5	25	mV
	$V_{OUT} \geq 5V$, (Note 2)		0.1	0.3		0.1	0.5	%
Adjustment Pin Current			50	100		50	100	μA
Adjustment Pin Current Change	$10\,mA \leq I_L \leq I_{MAX}$ $2.5V \leq (V_{IN}-V_{OUT}) \leq 40V$		0.2	5		0.2	5	μA
Reference Voltage	$3 \leq (V_{IN}-V_{OUT}) \leq 40V$, (Note 3) $10\,mA \leq I_{OUT} \leq I_{MAX}$. $P \leq P_{MAX}$	1.20	1.25	1.30	1.20	1.25	1.30	V
Line Regulation	$3V \leq V_{IN} - V_{OUT} \leq 40V$, (Note 2)		0.02	0.05		0.02	0.07	%/V
Load Regulation	$10\,mA \leq I_{OUT} \leq I_{MAX}$. (Note 2)							
	$V_{OUT} \leq 5V$		20	50		20	70	mV
	$V_{OUT} \geq 5V$		0.3	1		0.3	1.5	%
Temperature Stability	$T_{MIN} \leq T_j \leq T_{MAX}$		1			1		%
Minimum Load Current	$V_{IN}-V_{OUT} = 40V$		3.5	5		3.5	10	mA
Current Limit	$V_{IN}-V_{OUT} \leq 15V$							
	K and T Package	1.5	2.2		1.5	2.2		A
	H and P Package	0.5	0.8		0.5	0.8		A
	$V_{IN}-V_{OUT} = 40V$							
	K and T Package		0.4			0.4		A
	H and P Package		0.07			0.07		A
RMS Output Noise, % of V_{OUT}	$T_A = 25^\circ$C, $10\,Hz \leq f \leq 10\,kHz$		0.003			0.003		%
Ripple Rejection Ratio	$V_{OUT} = 10V$, $f = 120\,Hz$		65			65		dB
	$C_{ADJ} = 10\mu F$	66	80		66	80		dB
Long-Term Stability	$T_A = 125^\circ$C		0.3	1		0.3	1	%
Thermal Resistance, Junction to Case	H Package		12	15		12	15	$^\circ$C/W
	K Package		2.3	3		2.3	3	$^\circ$C/W
	T Package		5			5		$^\circ$C/W
	P Package		12			12		$^\circ$C/W

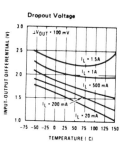

Dropout Voltage

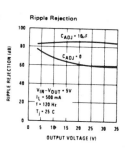

Ripple Rejection

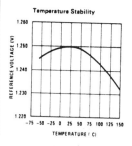

Temperature Stability

Note 1: Unless otherwise specified, these specifications apply $-55^\circ C \leq T_j \leq +150^\circ C$ for the LM117, $-25^\circ C \leq T_j \leq +150^\circ C$ for the LM217 and $0^\circ C \leq T_j \leq +125^\circ C$ for the LM317; $V_{IN}-V_{OUT} = 5V$ and $I_{OUT} = 0.1A$ for the TO-5 package and $I_{OUT} = 0.5A$ for the TO-3 package and TO-220 package. Although power dissipation is internally limited, these specifications are applicable for power dissipations of 2W for the TO-5 and 20W for the TO-3 and TO-220. I_{MAX} is 1.5A for the TO-3 and TO-220 package and 0.5A for the TO-5 package.

Note 2: Regulation is measured at constant junction temperature. Changes in output voltage due to heating effects must be taken into account separately. Pulse testing with low duty cycle is used.

Note 3: Selected devices with tightend tolerance reference voltage available.

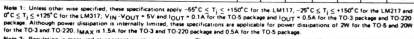

FIGURE 16.24.   Shown is the specification sheet of the National Semiconductor LM317 adjustable positive voltage regulator. Its performance characteristics are similar to those of the fixed 7800 series except that the output voltage can be varied between 1.2 V and 37 V.

FIGURE 16.25. An adjustable regulator maintains a constant reference voltage V_{ref} between its output and adjustment terminals. The output voltage is set by the voltage divider R_1 and R_2 according to Equation 14. The minimum output voltage is $V_{ref} = 1.2$ V. Capacitor C_1 prevents voltage oscillations in the regulator.

OPERATION OF AN ADJUSTABLE VOLTAGE REGULATOR

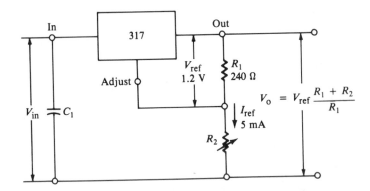

value of the reference current I_{ref} at:

$$I_{ref} = \frac{V_{ref}}{R_1}$$

Typically, $R_1 = 240 \, \Omega$, so for $V_{ref} = 1.2$ V, $I_{ref} = 5$ mA.

The reference current I_{ref} flows through both R_1 and a variable resistor R_2 that sets the value of V_o according to Ohm's law:

$$V_o = I_{ref}(R_1 + R_2)$$

or

$$V_o = V_{ref}\left(\frac{R_1 + R_2}{R_1}\right) \tag{14}$$

$$R_1 + R_2 = R_1\left(\frac{V_o}{V_{ref}}\right)$$

$$R_2 = R_1\left(\frac{V_o}{V_{ref}} - 1\right)$$

$$= 0.24\left(\frac{6.2}{1.2} - 1\right) = 1.0 \text{ k}\Omega$$

In practice, R_1 would be a fixed 240 Ω resistor, but R_2 would be a 5 kΩ potentiometer, as shown in Figure 16.25. By appropriately adjusting the 5 kΩ potentiometer, we could get a specific output voltage of anywhere from 1.2 V at $R_2 = 0 \, \Omega$ to 26.2 V at $R_2 = 5$ kΩ. Note that V_o *cannot be adjusted to 0 V or to any other voltage below V_{ref} without additional circuitry.*

EXAMPLE 8: Suppose we want a fixed output voltage of $V_o = 6.2$ V. If the voltage regulator has a value of $V_{ref} = 1.2$ V, what values should we use for R_1 and R_2?

Solution: Normally, we would select the value $R_1 = 240 \, \Omega$. Then we would calculate the value of R_2 from Equation 14:

$$V_o = V_{ref}\left(\frac{R_1 + R_2}{R_1}\right)$$

EXAMPLE 9: In Figure 16.25, $V_{ref} = 1.2$ V. Resistor $R_1 = 240 \, \Omega$ and R_2 is a 5 kΩ potentiometer. What output voltages result when R_2 is adjusted to (a) 0 Ω and (b) 5 kΩ? (c) What value of R_2 will give an output voltage of 5.0 V?

Solution: (a) Using Equation 14, we get:

DUAL-POLARITY POWER SUPPLY USING FIXED-VALUE REGULATORS

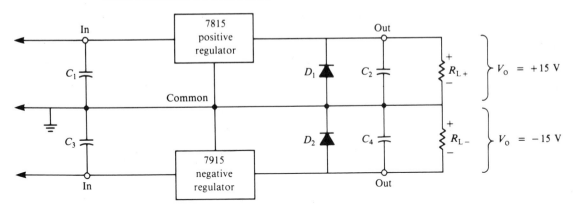

FIGURE 16.26. A common electronic power supply has a dual-polarity output. We can construct one from two fixed regulators, as shown, or use a dual-polarity tracking regulator designed specifically for this purpose. In the figure, D_1 and D_2 are 1N4720s, and C_1, C_2, C_3, and C_4 are 1 μF solid tantalum.

$$V_o = \frac{1.2 \text{ V}}{R_1}(R_1 + 0) = \textbf{1.2 V}$$

(b) Equation 14 gives:

$$V_o = \frac{1.2 \text{ V}}{240}(240 + 5000) = \textbf{26.2 V}$$

(c) Rewriting Equation 14, we have:

$$R_2 = R_1\left(\frac{V_o}{V_{\text{ref}}} - 1\right) = 240\left(\frac{5.0 \text{ V}}{1.2 \text{ V}} - 1\right)$$
$$= \textbf{760 } \Omega$$

An input capacitor C_1 may also be required to prevent oscillation if the LM317 is located more than a few inches from the rectifier's filter capacitor. If so, it should be inserted directly between the input terminal (the case) and the power supply common. The manufacturer recommends that C_1 be a 0.1 μF disk ceramic capacitor or a 1 μF solid tantalum capacitor.

16.8 DUAL-POLARITY TRACKING REGULATORS

Many electronic devices, including the operational amplifier, require dual-polarity supplies that deliver voltages of equal mangitude but opposite sign. One solution for high-current applications is to use two fixed-voltage regulators, as shown in Figure 16.26. The two diodes will insure regulator startup if a common load is connected directly across the + and − outputs. A disadvantage of the two-regulator approach for equal but opposite output voltages is that the magnitudes of the output voltages may not be exactly equal. For example, V_o of the 7815 could be + 15.5 V, whereas − V_o of the 7915 could be − 14.5 V.

The problem of achieving dual-polarity voltages that are equal in magnitude is solved by the *tracking regulator*. In a tracking regulator, the negative regulator voltage is set by its internal circuitry, and an additional external

SPECIFICATION SHEET FOR RAYTHEON ± 15 V DUAL-TRACKING REGULATOR

FIXED ±15V DUAL-TRACKING VOLTAGE REGULATORS

4195

DESIGN FEATURES

- ±15V Op-Amp power at reduced cost and component density
- Thermal shutdown at T_i = +175°C in addition to short-circuit protection

- Output currents to 100mA
- May be used as single output regulator with up to +50V output
- Available in TO-66, TO-99, and 8-pin plastic mini-DIP

The RM4195 and RC4195 are dual polarity tracking regulators designed to provide balanced positive and negative 15 volt output voltages at currents to 100mA. These devices are designed for local "on-card" regulation eliminating distribution problems associated with single point regulation. The regulator is intended for

ease of application. Only two external components are required for operation (two 10μF bypass capacitors).

The device is available in three package types to accommodate various applications requiring economy, high power dissipation, and reduced component density.

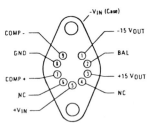

TK (TO-66) Power Package
(Bottom View)

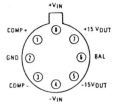

T Metal Can Package
(Top View)

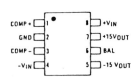

DN Dual In-line
(Top View)

PARAMETER	CONDITIONS	PACKAGE			UNITS
		DN	T (TO-99)	TK (TO-66)	
Power Dissipation	T_A = 25°C	0.6	0.8	2.4	W
	T_C = 25°C		2.1	9	
Thermal Resistance	ϕj-C		70	17	°C/W
	ϕj-A	210	185	62	

Balanced Output (V$_O$ = ±15V)

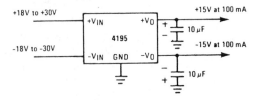

FIGURE 16.27.   Shown is the specification sheet for the 4195, ± 15 V dual-tracking regulator. This regulator is simple to use, requiring only two external capacitors.

FWCT RECTIFIER FOR USE WITH DUAL-POLARITY VOLTAGE REGULATORS

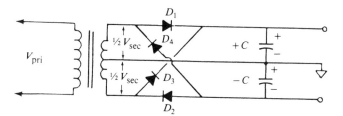

FIGURE 16.28. The filtered full-wave center-tapped rectifier is generally used with a dual-polarity voltage regulator. The basic circuit design for an FWCT rectifier is shown in the figure.

circuit serves to force the positive regulator to track the negative's output voltage, or vice versa.

Also available are a variety of IC **dual-polarity tracking regulators**. Some provide tracking voltages that can also be made adjustable by adding a few external resistors. The dual-tracking regulators that are easiest to use, however, are the fixed dual-tracking regulators that are discussed here. Standard fixed tracking regulators can be purchased that furnish ± 12 V and ± 15 V.

16.9 4195, ±15 V DUAL-TRACKING VOLTAGE REGULATOR

Raytheon's 4195 dual-polarity tracking regulator has been chosen for analysis because it is easy to use and relatively inexpensive. As shown on its specification sheet (Figure 16.27), its dual output voltages are $+15$ V and -15 V with respect to ground. It can supply output currents up to 100 mA with a load regulation of 5 mV and a ripple rejection of 75 db. The negative voltage tracks the positive, typically within ± 50 mV, provided that the power dissipation rating is not exceeded. The 4195 has a 3 V dropout voltage, so it requires minimum instantaneous input voltages from the filtered rectifier of $15 + 3 = 18$ V and $-15 + (-3) = -18$ V.

The maximum input voltage ratings are 30 V for the positive input and -30 V for the negative input. Note that it comes in three types of packages that differ from the TO–3 and TO–220 studied earlier. The thermal characteristics of each of the packages are given in Figure 16.27.

The input voltages to the regulator may be obtained from two separate filtered FWB rectifiers via two transformers or from a single transformer with two secondaries. However, a more economical and standard rectifier for this purpose uses only one center-tapped transformer secondary. This type of filtered rectifier is examined next.

16.10 FWCT FILTERED RECTIFIER

A center-tapped transformer plus a four-diode bridge and two filter capacitors make an ideal filtered full-wave rectifier for dual-polarity power supplies that require equal voltages (Figure 16.28). Current directions during the peak of the positive half-cycle of V_{pri} are shown in Figure 16.29a. Diode D_1 conducts load current I_{L+} through $+C$ and R_{L+} to develop V_{L+}. Diode D_2 conducts current I_{L-} through $-C$ and R_{L-} to develop V_{L-}.

During the negative half-cycle (Figure 16.29b), diodes D_3 and D_4 are forward bias steer current in the same direction through both

OPERATION OF FWCT RECTIFIER

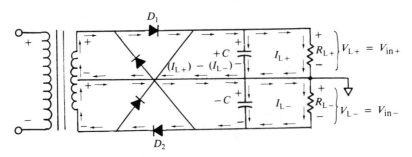

(a) Positive Half-Cycle

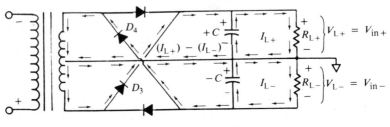

(b) Negative Half-Cycle

FIGURE 16.29. During the positive half-cycle of the secondary, diodes D_1 and D_2 are forward biased, conducting current I_{L+} and I_{L-} through R_{L+} and R_{L-} to develop V_{in+} and V_{in-}. The return path for both $I_{L+} = I_{L-}$ is the center tap–common connection. If $I_{L+} = I_{L-}$, the net current in this wire is 0. During the negative half-cycle, D_3 and D_4 conduct to produce the same result.

±15 V DUAL-TRACKING REGULATED POWER SUPPLY

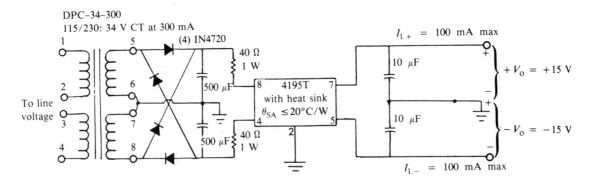

FIGURE 16.30. Shown is the circuit diagram for a ± 15 V dual-tracking regulated power supply. The 40 Ω heat-absorbing resistors and 209AB heat sink are just sufficient for a maximum load current of 100 mA from both outputs simultaneously.

R_{L+} and R_{L-}. Note that the peak reverse-bias voltage across the diodes equals the sum of V_{L+} and V_{L-}. Notice also that load currents I_{L+} and I_{L-} flow in opposite directions through the common ground wire. If the load voltages are equal and the load resistors are equal, the load currents will be equal. Thus, I_{L+} and I_{L-} will balance and give zero current in the returning ground. Also, if the two load currents are equal, the current in both halves of the secondary will be equal, and the circuit will act like an FWB rectifier rather than like an FWCT rectifier. This fact is important in selecting the proper transformer for an FWCT rectifier. In designing a power supply, values for the rms voltage and current ratings for the transformer secondary winding must be determined. These ratings are based on power supply operation at full load voltage and current. At full load current from both outputs, $I_{L+} = I_{L-}$ and the FWCT rectifier acts like an FWB rectifier.

Thus, the FWB rectifier design equations can be used.

16.11 ±15 V OPERATIONAL AMPLIFIER POWER SUPPLY

Op amps generally require ±15 V for operation with a current requirement of about ±7 mA—that is, 2 mA for operation plus a 5 mA output current. The 4195 can deliver ±15 V at full load currents of 100 mA per side. Thus, the regulator can power as many as fourteen op amps. A typical circuit diagram using an FWCT rectifier is shown in Figure 16.30. Note the two 40 Ω resistors between the rectifier and the regulator. These resistors are used to drop excess rectifier voltage and, hence, dissipate excess regulator heat.

16.12 QUESTIONS AND PROBLEMS

1. How much power can a 7812 voltage regulator in a TO–3 case dissipate at the following ambient temperatures assuming that the maximum junction temperature rating is 125°C: (a) T_A = 20°C, (b) T_A = 40°C, and (c) T_A = 90°C?

2. Repeat Question 1 for a 7908 voltage regulator in a TO–220 case.

3. What heat sink thermal resistance would the following semiconductors require? Assume $T_{J(max)}$ = 125°C, θ_{CS} = 0.5°C/W, and an ambient of 25°C in each case.
    (a) θ_{JC} = 6°C/W and $P_{dis(max)}$ = 3 W
    (b) θ_{JC} = 4°C/W and $P_{dis(max)}$ = 7.5 W
    (c) θ_{JC} = 5.3°C/W and $P_{dis(max)}$ = 12 W
    (d) θ_{JC} = 2.6°C/W and $P_{dis(max)}$ = 17 W

4. Design a 7 V fixed-voltage, regulated power supply that can deliver a maximum of 750 mA at full load with a ripple voltage less than 20 mV. Use the F91X transformer shown in Figure 14.12 and select the proper primary and secondary connections from Figure 14.13. Draw the complete circuit showing component values.

5. Design an adjustable-voltage, regulated power supply with a voltage range of 1.2 V to 20 V that can deliver a load current of 1.2 A. Use the F91X transformer diagrammed in Figure 14.12 and select the proper primary and secondary connections from Figure 14.13. Draw the complete circuit showing component values.

6. Design a ± 15 V op amp power supply using the 4195 dual-tracking regulator. Use the F91X transformer shown in Figure 14.12. Draw the complete circuit showing component values.

7. Determine the heat dissipation of each of the IC voltage regulators in Problems 4, 5, and 6 and state whether a heat sink or a heat-absorbing resistor is required. If so, determine the specifications (θ_{SA} for the heat sink and the resistance and power rating for the resistors).

Part VII

ALTERNATING CURRENT, VOLTAGE, AND IMPEDANCE

ALTERNATING CURRENT SIGNALS IN ELECTRONICS

The primary purpose of most electronic circuits is to process information. The most obvious applications are calculators and computer systems whose sole activity is data processing; but other areas, such as communication, measurement, and control, also have a need for these systems.

In a communication system like a telephone or TV, for example, the electronic system accepts information in one form at one location, transmits it, and then transforms it into another form at another location. In a measurement and control application, electronic circuits receive information about the behavior of a physical system from a transducer and convert it into a form that can be displayed on an output device, such as a meter or recorder, or used to drive a control system, such as a switch or motor.

The basic structure of a measurement and control system is shown schematically in Figure 1, along with typical examples of each block. Note that the electronic circuit components of the system perform two tasks: processing the signal and supplying power. Power supply circuits were discussed in Part VI, and various signal-processing circuits have been described throughout the book to illustrate applications of basic electronics concepts.

The flow and processing of information in a system like that shown in Figure 1 can take several different forms, each of which

FIGURE 1. A principal application of electronic circuits is information processing. In a measurement and control system, the circuits receive information about a physical source from a transducer and convert it to a form that can be displayed on an output device or used to control the source behavior. Alternating current signal processing has many advantages in such applications because it has three independent parameters for carrying information and because it can reject noise.

MEASUREMENT AND CONTROL SYSTEM BLOCK DIAGRAM

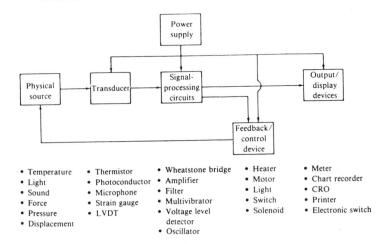

• Temperature	• Thermistor	• Wheatstone bridge	• Heater	• Meter
• Light	• Photoconductor	• Amplifier	• Motor	• Chart recorder
• Sound	• Microphone	• Filter	• Light	• CRO
• Force	• Strain gauge	• Multivibrator	• Switch	• Printer
• Pressure	• LVDT	• Voltage level detector	• Solenoid	• Electronic switch
• Displacement		• Oscillator		

utilizes a different set of electrical quantities. The simplest form, called **digital signal processing**, is the basis of modern computers and calculators. Digital electronic systems convert physical information into a series of voltage pulses, typically 5 V, and the electronic circuits then process the information in this form. This concept was introduced in Part V.

The two other types of signal-processing circuits are termed **analog** and are based on either dc or ac principles. **Direct current signal processing**, the simpler of the two, has two parameters to convey the information: voltage and polarity. That is, a dc voltage is completely described by two quantities—voltage magnitude and polarity—and the value of both can be used to convey information about a system.

For example, in Part IV, we saw how to use a zero-crossing detector, which detected a change in polarity from + to −, to activate a switch. Other circuits convert resistance changes—for example, from a thermistor—into voltage changes whose amplitude is proportional to a physical quantity, such as temperature.

The most complex signal-processing method is **ac signal processing**. It provides three parameters for conveying information. Alternating current signals are based on the properties of sine waves, and sine waves are completely described by three quantities: peak amplitude, frequency, and phase. The last quantity is the relative point in time when a sine wave peaks.

In measurement and control applications, ac signal processing has many advantages over dc. First, it is a richer source of information because it provides three parameters for conveying information. Equally important, ac signals are much less susceptible to interference and noise. Every electronic circuit adds small ac and dc voltages to the signal being processed. These "noise" voltages are much more easily removed from, or filtered out of, an ac signal. Consequently, most electronic measurement and control systems are based on the principles and behavior of ac circuits.

Another major application is sound. Sound is essentially a combination of sinusoidal pressure waves. These pressure waves are converted into sinusoidal voltages by a microphone. Then, ac electronic circuits amplify, transmit, and/or process these voltages in a wide variety of ways, some of which are explored in this part of the book.

The other major application of ac circuits is radio wave communication that includes radio and TV transmissions and radar (Figure 2). Such basic topics as amplitude modulation (AM) and frequency modulation (FM) all belong to this wide-ranging field.

**SYSTEMS WITH ALTERNATING
CURRENT SIGNALS**

Clearly, the whole range of circuits and applications cannot be covered here, but the basic principles of ac circuits and the basic methods of analyzing and describing ac circuit behavior are introduced. Chapter 19 examines how these principles are applied to a primary class of ac circuits: filters. Filter circuits play a major role in most ac signal-processing systems whether their function is electronic measurement and control, sound, or radio communication.

FIGURE 2.   Applications of ac circuits include a wide variety of sound systems, such as record players, tape recorders, and telephones, and all radio wave communication systems, such as radio and TV transmission and radar. This part of the book introduces the basic principles of ac circuits as a preparation for more advanced study of these applications.

Chapter 17

IMPEDANCE AND PHASE

17.1 OBJECTIVES

Following the completion of Chapter 17, you should be able to:

1. Explain the basis of ac circuit analysis in terms of Fourier's theorem.
2. Explain the difference between analyses of time-varying signals made in the time and in the frequency domains.
3. Mathematically express and graph a sinuosidal signal (current or voltage), given its amplitude, frequency, and phase.
4. Determine the mathematical sum of a sine and cosine wave function of the same frequency, given their amplitudes and frequency.
5. Describe the general frequency response to a series RC network and explain it in terms of the network's response to square waves of various periods.
6. Calculate the impedance of a capacitor at any frequency f, given its capacitance C and the frequency.

7. Explain and mathematically express the relationship between current and voltage in a capacitor.
8. Determine the current (voltage) in a capacitor, given the applied voltage (current) and the value of the capacitance, and graph the two quantities.
9. Determine the equivalent impedance of a series RC network at any frequency and use it to explain the frequency response of each element.
10. Determine the peak amplitudes and phase of the current and voltage for each component of a series RC network, given the amplitude and frequency of the applied voltage.
11. Determine the equivalent impedance of a parallel RC network at any frequency and use it to explain the frequency response of each component.
12. Determine the peak amplitude and phase of the current and voltage for each compo-

nent of a parallel *RC* network, given the amplitude and frequency of the applied voltage.

13. Calculate the impedance of an inductor at any frequency *f*, given its inductance *h* and the frequency.

14. Explain and mathematically express the relationship between current and voltage in an inductor.

17.2 ALTERNATING CURRENT CIRCUIT ANALYSIS

The term **alternating current,** or ac, can be applied to almost every time-varying signal that changes its polarity periodically. Thus, square waves, triangle waves, sawtooth waves, sine waves, modulated waves, and the like, all qualify as ac signals. In ac circuit analysis, however, attention is confined to sine waves. Thus, when we speak of "ac circuit analysis," we mean that we want to know how a circuit responds to sine waves. That is, if the supply voltage to a circuit is a sine wave with a particular amplitude and frequency, we want to know what the resulting voltage will look like at a certain pair of output terminals.

Fourier's Theorem

The reason that sine waves are so important derives from a law of nature that is described by **Fourier's theorem.** This theorem states that any periodic signal, no matter how complicated, can be expressed as the sum of sine waves. Thus, if we had a summing amplifier and a lot of sine wave generators, we could create any desired periodic signal. By properly selecting the amplitudes and frequencies of each of the sine wave generators and adding their outputs together, we can generate any periodic signal—square waves, triangle waves, sawtooth waves, and so on. This principle is the basis for mod-

ern music synthesizers that create a wide variety of musical tones from a combination of sine waves.

The more complicated the waveform, of course, the greater the number of sine waves required to create it. A square wave, for example, requires an infinite number of sine waves (Figure 17.1). Clearly, the electronic creation of a square wave by summing individual sine waves is not a practical method. However, the power of Fourier's theorem is that there are mathematical (and, hence, computer) methods for determining the sine wave amplitudes and frequencies necessary to reproduce any given signal.

Frequency Domain

For our purposes, the value of Fourier's theorem is that it lets us concentrate our attention on sine waves. If we know how an ac circuit behaves to sine wave input voltages at all frequencies, then, in principle, we know how it will respond to any periodic signal.

Fortunately, when a sine wave voltage is applied to a circuit, the frequency of the voltage that appears at any other point in the circuit, such as the output terminals, will be the same. Thus, we needn't worry about frequency changes. What does change, of course, is the amplitude. Thus, much like dc circuits, one of the central problems of ac circuit analysis is to determine the amplitude (typically, the peak amplitude) of the output voltage V_o for a given amplitude input voltage V_{in}.

Unfortunately, the ratio of V_o/V_{in} differs for input voltages of different frequencies. Hence, we must determine V_o/V_{in} for all possible input frequencies. That is, we need some expression that relates V_o to both the amplitude V_{in} and the frequency *f* of the input signal.

Analyses of this type are said to be analyses in the **frequency domain.** It simply means that we are interested in how things change with fre-

FOURIER'S THEOREM

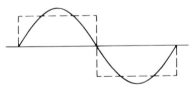

(a) $V(t) = V \sin \omega t$

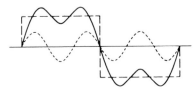

(b) $V(t) = V(\sin \omega t + 1/3 \sin 3\omega t)$

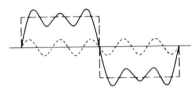

(c) $V(t) = V(\sin \omega t + 1/3 \sin 3\omega t$
$+ 1/5 \sin 5\omega t)$

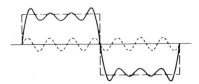

(d) $V(t) = V(\sin \omega t + 1/3 \sin 3\omega t$
$+ 1/5 \sin 5\omega t$
$+ 1/7 \sin 7\omega t)$

FIGURE 17.1. Analysis of ac circuits is primarily concerned with how these circuits respond to sine waves. Its foundation is Fourier's theorem that states that any periodic signal can be built up from sine waves. Shown here is an approximate construction of a square wave from sine waves. To produce an exact square wave would require an infinite number of sine waves.

quency, in contrast to our concern in Part III with behavior in the **time domain** – that is, with the way signals change with time. Results of analysis in the time domain are related to those in the frequency domain, but the relationship is beyond the scope of this book.

Frequency Response

Another way of describing the ac behavior of circuit networks is in terms of its **frequency response.** How does the circuit respond to sine waves of different frequencies? In this case, "respond" means, What is the ratio of output to input amplitude V_o/V_{in} for sine waves at all possible frequencies?

The frequency response of a circuit (or system) can be presented in a number of different ways. For fairly simple circuits, it can be given as a **mathematical equation** in which V_o/V_{in} is expressed mathematically as a function of sine wave frequency f. Much of Part VII is devoted to determining the equations for basic circuit connections, as was done earlier for simple series and parallel connections of resistors (see Part II).

In these chapters, the basic circuit connections include a capacitor, primarily in a simple series or simple parallel connection with a resistor. The behavior of simple series and parallel connections of a capacitor with a resistor form the heart of ac circuit analysis.

For more complex circuits or systems, such as an amplifier, the mathematical expressions are unwieldy, if they can be derived at all. In these cases, the frequency response is generally given as a graph of V_o/V_{in} versus f. For the circuit shown in Figure 17.2, for example, the graph is that shown in Figure 17.3. **Frequency response graphs** are important measures of circuit performance, particularly for audio and communication systems where we want to be certain that all of the available information (music or data) is faithfully reproduced.

FIGURE 17.2. When a sine wave is applied to an ac circuit, the voltages at the output terminals have the same frequency, but they may differ in amplitude. The value of V_o/V_{in} may also be different for different frequencies. Primarily, ac circuit analyses focus on determining the values of V_o/V_{in} for all possible sine wave frequencies in a circuit. The ratio V_o/V_{in} versus frequency is termed the circuit's frequency response.

ALTERNATING CURRENT CIRCUIT RESPONSE TO SINE WAVES

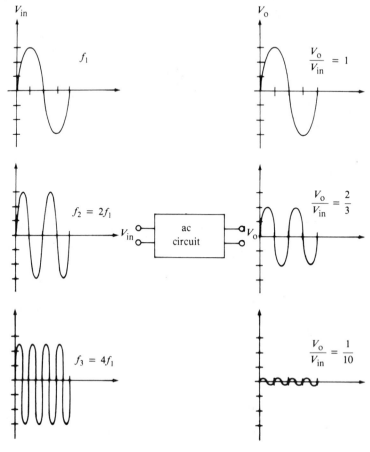

17.3 REVIEW OF SINE WAVES

Because sine waves are central to ac circuit analysis, the chapter begins with a brief review of sine wave characteristics. Here are gathered in one place the properties of sine waves that will be needed later on.

Amplitude and Frequency

The mathematical equation that describes a sine wave is:

$$V(t) = V \sin 2\pi ft \tag{1}$$

As shown in Figure 17.4, $V(t)$ is the actual amplitude of the sine wave at any time t, and V is the peak amplitude, both positive and negative, that the sine wave attains. The value of $V(t)$ oscillates between $+V$ and $-V$ with the frequency f according to the sine function. The sine wave repeats itself (is periodic) with a period T where:

FREQUENCY RESPONSE GRAPH

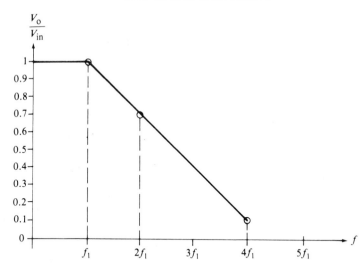

FIGURE 17.3. The frequency response of an ac circuit can be given as a mathematical equation or as a graph. Much of the work in this part of the book involves deriving the mathematical equations that describe basic ac circuits, such as series and parallel connections of resistors and capacitors. For more complex ac circuits, the behavior is expressed more simply as a graph, as shown for the circuit in Figure 17.2.

$$T = \frac{1}{f} \qquad (2)$$

GENERALIZED SINE WAVE VOLTAGE

Thus, every T seconds, the sine wave undergoes a complete sine wave cycle. Note that when t changes by one period T, the argument of the sine wave, $2\pi ft$, changes by 2π.

For convenience in writing Equation 1, the quantity $2\pi f$ is frequently combined and represented by the Greek letter omega (ω). Equation 1 (and all sine wave expressions) will henceforth be written as:

$$V(t) = V \sin \omega t \qquad (3)$$
$$\text{where } \omega = 2\pi f \qquad (4)$$

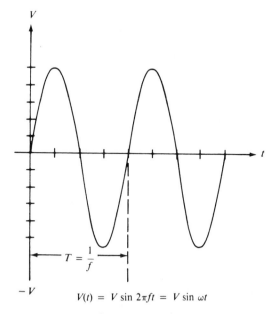

$$V(t) = V \sin 2\pi ft = V \sin \omega t$$

FIGURE 17.4. The figure shows a generalized sine wave voltage and its mathematical expression. The voltage magnitude $V(t)$ oscillates between peak amplitudes $\pm V$ according to the sine function with a period T. The oscillation frequency f in hertz (cycles per second) equals $1/T$. For simplicity, the frequency is expressed as $\omega = 2\pi f$.

Phase

Part III stated that a sine wave is fully described by its amplitude V and frequency f (see Equation 1 in Chapter 7). When comparing two sine waves, we must include another factor: phase.

PHASE

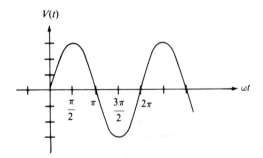

(a) $V(t) = V \sin \omega t$

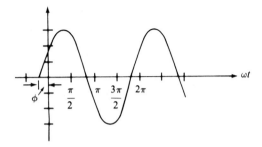

(b) $V(t) = V \sin(\omega t + \phi)$

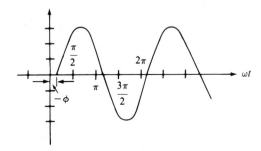

(c) $V(t) = V \sin(\omega t - \phi)$

FIGURE 17.5. A sine wave (a) is completely described by three quantities: peak amplitude (V) frequency (ω), and phase (ϕ). The phase expresses the relative displacement in time between two sine waves. Positive values of phase shift a sine wave to earlier times (b), and negative values shift it to later times (c).

Although two sine waves may be identical in appearance, their oscillations may be displaced from each other in time (Figure 17.5). In ac circuit analysis, this consideration is particularly important. When a sine wave is the input signal to a circuit, the output sine wave always has the same frequency, but the phase, as well as the amplitude, may be different. In comparing V_o with V_{in}, therefore, we must be aware of shifts in phase.

The **phase** of a sine wave is represented by the Greek letter phi (ϕ) and is added to the sine wave expression:

$$V(t) = V \sin(\omega t + \phi) \qquad (5)$$

The units of ϕ, like those of ωt, are generally the angular units called **radians**, though they may be expressed in **degrees**. Recall that:

$$360° = 2\pi \text{ radians}$$
$$180° = \pi \text{ radians}$$
$$90° = \pi/2 \text{ radians}$$
$$57.3° = 1 \text{ radian} \qquad (6)$$

As various phase angles ϕ are added to ωt in Equation 5, the sine function shifts along the time axis, as shown in Figure 17.5. *Positive values of ϕ result in shifts to the left (earlier times), and negative values of ϕ produce shifts to the right (later times).* Note that when $\phi = \pm 2\pi$, the value of the sine wave is identical to that at $\phi = 0$. Also, certain positive and negative values of ϕ yield identical results. For example, $\phi = +\pi$ and $-\pi$ are identical, $\phi = \pi/2$ and $-3\pi/2$ are identical, and so on. For this reason, all possible shifts between two sine waves can be fully represented by values of ϕ between $-\pi$ and $+\pi$.

In addition, the sine functions for certain values of ϕ are mathematically equivalent to other sine functions, as well as to the cosine function. These relations are summarized below and shown in Figure 17.6. Take a moment to verify the equivalence of these trigonometric identities.

$$A \sin\left(\omega t + \frac{\pi}{2}\right) = A \cos \omega t \qquad (7)$$

$$A \sin(\omega t - \pi) = -A \sin \omega t \qquad (8)$$

$$A \sin\left(\omega t - \frac{\pi}{2}\right) = -A \cos \omega t \qquad (9)$$

Trigonometric Relations

To compare sine waves with different phase angles, we need some additional mathematical tools from trigonometry. The most important tool for our purposes is the trigonometric identity:

$$A \sin \omega t \pm B \cos \omega t$$
$$= C \sin(\omega t + \phi) \qquad (10)$$
$$\text{where } C^2 = A^2 + B^2 \qquad (11)$$
$$\tan \phi = \frac{\pm B}{A} \qquad (12)$$

These relationships can be used in two ways. First, a sine wave with a phase angle ϕ can be expressed as the sum of a sine wave and a cosine wave at the same frequency. Alternatively, a sine wave and a cosine wave at the same frequency ω add to give a simple sine wave at the same frequency ω, but shifted in phase by an angle ϕ. Equations 11 and 12 express the amplitude and phase angle of the resulting sine wave in terms of the peak amplitudes of the sine and cosine waves.

Figure 17.7 gives a geometric representation of Equations 10, 11, and 12. The ampli-

TRIGONOMETRIC IDENTITIES

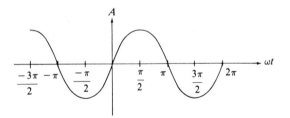

(a) $A \sin \omega t$

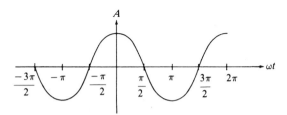

(b) $A \sin\left(\omega t + \frac{\pi}{2}\right) = A \cos \omega t$

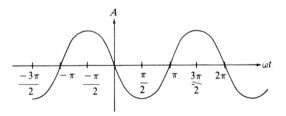

(c) $A \sin(\omega t - \pi) = -A \sin \omega t$

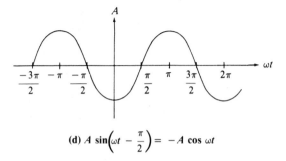

(d) $A \sin\left(\omega t - \frac{\pi}{2}\right) = -A \cos \omega t$

FIGURE 17.6. Sine functions for certain values of ϕ are mathematically equivalent to other sine and cosine functions. These four examples will be encountered again later.

ADDING SINE AND COSINE FUNCTIONS

$$A \sin \omega t \pm B \cos \omega t = C \sin(\omega t + \phi)$$

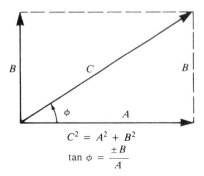

$$C^2 = A^2 + B^2$$
$$\tan \phi = \frac{\pm B}{A}$$

FIGURE 17.7.   Another key trigonometric identity expresses the sum of a sine and a cosine function of the same frequency. The result is a sine function at the same frequency, but shifted in phase. The graph is a geometrical representation of the amplitude and phase of the sine function in terms of the amplitudes of the original waves.

tude of the sine function A is a vector of length A at an angle of $0°$, and the amplitude of the cosine function B is a vector of length B at an angle of $90°$. This graph is a representation of Equation 7, which states that a sine function shifted in phase by $\pi/2$ is a cosine function. Note also that a sine function shifted in phase by $-\pi/2$ is a negative cosine function (Equation 9).

The amplitude of the resulting sine function of Equation 10 is the hypotenuse of the right triangle, formed by the vector sum of A and B. Equation 11, therefore, is simply an expression of the Pythagorean theorem for right triangles. Similarly, the phase angle ϕ is the angle between the sine function at $0°$ and the vector sum, where the tangent of ϕ is given by Equation 12. Figure 17.7 can be very helpful in remembering Equations 10, 11, and 12.

EXAMPLE 1:   A sine wave voltage with a peak amplitude of 6.0 V and at a frequency of 1 kHz and a cosine voltage at the same frequency, but with a peak amplitude of 9.0 V, are inputs to a summing amplifier. What is the output voltage?

Solution: The value of ω for a frequency of 1 kHz is found from Equation 4 to be:

$$\omega = 2\pi f = 2(3.14)(1000 \text{ Hz}) = 6280 \text{ Hz}$$

Thus, the two input voltages are given by:

$$V_1 = 6.0 \text{ V} \sin(6280t)$$
$$V_2 = 9.0 \text{ V} \cos(6280t)$$

The resulting output voltage can be found from Equations 10, 11, and 12 where:

$$A = 6.0 \text{ V}$$
$$B = 9.0 \text{ V}$$

Using Equation 11 to find the amplitude of the resulting sine wave, we have:

$$C^2 = A^2 + B^2 = 6^2 + 9^2 = 117 \text{ V}^2$$
$$C = \sqrt{117 \text{ V}^2} = 10.8 \text{ V}$$

Equation 12 gives the phase angle ϕ:

$$\tan \phi = \frac{B}{A} = \frac{9.0 \text{ V}}{6.0 \text{ V}} = 1.5$$
$$\phi = 56°$$

Using Equation 10, we get the result:

$$6.0 \text{ V} \sin(6280t) + 9.0 \text{ V} \cos(6280t)$$
$$= \mathbf{10.8 \text{ V} \sin(6280t + 56°)}$$

The result is shown in Figure 17.8 and can be verified by adding the amplitudes of V_1 and

EXAMPLE OF SINE-COSINE FUNCTION ADDITION

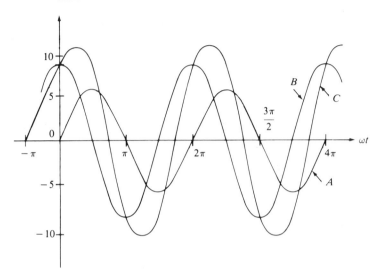

FIGURE 17.8. The figure shows a typical addition of a sine function *A* and a cosine function *B* to form a new sine wave *C* shifted in phase. See Example 1.

V_2 at any point in time and seeing that their sum at that time is equal to that of the new wave. Take a moment and check several points to verify that this procedure works.

17.4 BEHAVIOR OF SERIES *RC* NETWORKS

The study of ac circuit begins with an analysis of a simple circuit discussed earlier: the series *RC* combination. In Part III, we looked at the time response of this circuit to a step voltage input (Figure 17.9). When a step voltage is applied to the *RC* network, the voltage across the capacitor that was initially zero gradually increases toward the amplitude of the step voltage with a time constant given by the product of *R* and *C*:

$$\tau = RC \qquad (13)$$

The voltage across the resistor, on the other hand, begins at the amplitude of the step func-

tion and gradually decreases toward zero as the capacitor charges. By Kirchhoff's voltage law (KVL), the sum of the voltage across the resistor and capacitor always equals the applied voltage.

The current in the circuit is directly proportional to the voltage across the resistor according to Ohm's law. Hence, it, too, begins at its maximum value and gradually decreases to zero as the capacitor becomes fully charged.

Square Wave Behavior

If a square wave is applied to a series *RC* circuit, as shown in Figure 17.10, the capacitor will alternately charge and discharge at a rate determined by the time constant τ. Whether or not the capacitor voltage reaches the amplitude of the square wave before it begins to discharge is determined by the relative values of τ for the *RC* circuit and the period *T* of the square wave.

If *T* is long compared to τ, as shown in Figure 17.10a, the capacitor voltage will oscillate between the peak values of the square wave. However, if *T* is short compared to τ, as shown in Figure 17.10b, the peak amplitude of the ca-

STEP VOLTAGE RESPONSE OF *RC* CIRCUIT

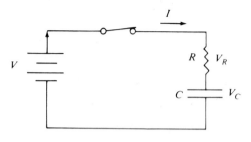

(a) Series *RC* Connection

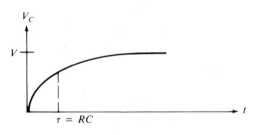

(b) Capacitor Voltage Rise

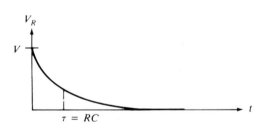

(c) Resistor Voltage Decrease

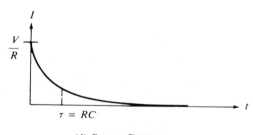

(d) Current Decrease

FIGURE 17.9. The study of ac circuits begins with the series *RC* connection (a). Its time response to a step voltage is shown. The capacitor voltage increases to the applied voltage with a time constant $\tau = RC$ (b). Meanwhile, the resistor voltage (c) and current (d) decrease to zero with the same time constant.

pacitor voltage will be less than that of the square wave. Note also in Figure 17.10 the amplitude of the voltage across the resistor for the two cases of $T \gg \tau$ and $T \ll \tau$.

Sine Wave Behavior

The results shown in Figure 17.10 form the basis of understanding the behavior of series *RC* circuits for sine waves of different frequencies. Essentially, the series *RC* circuit acts as a frequency-dependent voltage divider. For sine waves with frequencies small compared to τ— that is, $T \gg \tau$—most of the voltage will be dropped across the capacitor, and very little will be dropped across the resistor. For sine waves with frequencies large compared to τ—that is, $T \ll \tau$—most of the voltage will be dropped across the resistor, and very little will be dropped across the capacitor.

Frequency Response

Earlier, it was noted that what we want to know is how the amplitudes of a sine wave change for all frequencies in an ac circuit. This relationship is shown in a frequency response graph, like the one for the series *RC* circuit shown in Figure 17.11. Note that the frequency axis is logarithmic in order to display the behavior over a wide range of frequencies.

Figure 17.11 shows the ratio of the peak amplitude of the capacitor voltage and the applied voltage V_C/V as a function of frequency and the ratio of the peak amplitude of the resistor voltage and the applied voltage V_R/V as a function of frequency. Recall that the voltages across the capacitor and the resistor are sine waves at the same frequency as the applied voltage. The figure shows that above about 10 kHz, all of the voltage is dropped across the resistor, and below about 300 Hz, all of the voltage is dropped across the capacitor. Between 300 Hz and 10 kHz, some voltage appears across each component.

Imagine the result of applying a voltage

SQUARE WAVE RESPONSE OF *RC* CIRCUIT

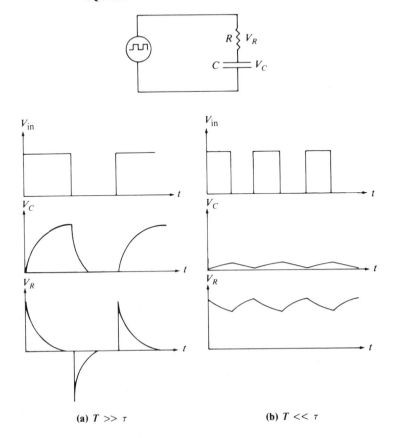

FIGURE 17.10. The response of a series *RC* circuit to square waves of different periods forms the basis for understanding its response to sine waves. For $T \gg \tau$ (low-frequency sine waves), most of the voltage is dropped across the capacitor, but for $T \ll \tau$ (high-frequency sine waves), most of the voltage is dropped across the resistor.

(a) $T \gg \tau$ **(b)** $T \ll \tau$

with many frequencies, such as an audio signal, to the series *RC* network. For the right values of *R* and *C*, high-frequency tones would appear across the resistor and low-frequency tones would appear across the capacitor. This behavior is the basis for audio filters and tone controls.

The next task is to get a mathematical expression that describes the behavior shown in Figure 17.11. With that expression, we will be able to calculate what values of *R* and *C* will produce this type of behavior.

17.5 IMPEDANCE OF A CAPACITOR

For dc currents and voltages, the behavior of a resistor is described by Ohm's law:

$$V_R = IR \qquad (14)$$

Ohm's law is equally valid when V and I vary with time and can be expressed by:

$$V_R(t) = I_R(t)R \qquad (15)$$

FIGURE 17.11. The sine wave response of an ac circuit is generally given as a frequency response graph that gives the ratio of the peak amplitude of the resistor and capacitor voltage to that of the applied voltage for all frequencies. At low frequencies, all of the voltage is dropped across the capacitor, and, at high frequencies, it is dropped across the resistor.

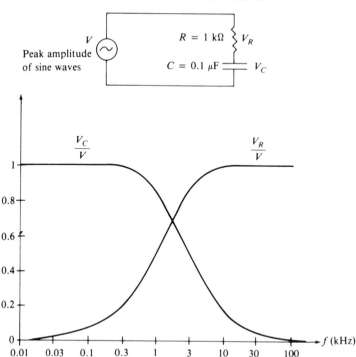

If the voltage $V_R(t)$ is sinusoidal, then Equation 15 becomes:

$$V_R \sin \omega t = R I_R \sin \omega t \qquad (16)$$

which says that if the voltage across a resistor varies sinusoidally, then the current through the resistor will also vary sinusoidally, at the same frequency as the voltage *and with the same phase*.

What is needed now is an equation like Equation 15 or 16 that gives the relation between current and voltage for a capacitor. To obtain such a relationship requires the techniques of calculus. The derivation is presented here, but remember that it is more important to understand what the results mean than how they are derived.

To begin, recall from Part III that the fundamental characteristic of a capacitor is that it

can store charge. The relation between the charge Q stored by a capacitor whose capacitance is C and the resulting voltage V_C that appears across the capacitor is:

$$Q = C V_C \qquad (17)$$

When a capacitor is charging, current flows in the capacitor circuit to cause the charge on the capacitor to change in either a positive or a negative way. The current I is defined as the rate at which the charge flows with time. By calculus, this current is defined as:

$$I = \frac{dQ}{dt} \qquad (18)$$

where dQ represents the change in charge on the capacitor that occurs during any very small period of time dt.

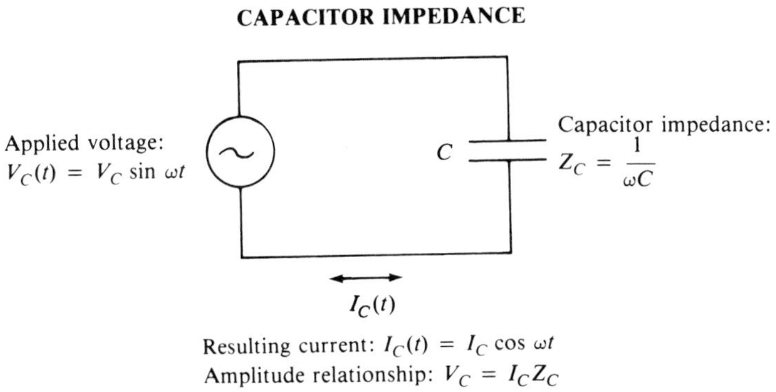

CAPACITOR IMPEDANCE

Applied voltage:
$V_C(t) = V_C \sin \omega t$

C

Capacitor impedance:
$Z_C = \dfrac{1}{\omega C}$

$I_C(t)$

Resulting current: $I_C(t) = I_C \cos \omega t$
Amplitude relationship: $V_C = I_C Z_C$

FIGURE 17.12. When a sinusoidal voltage is applied to a capacitor, the capacitor presents an impedance to current flow given by $Z_C = 1/\omega C$. The result is a current that varies with the cosine and whose amplitude is given by $I_C = V_C/Z_C$. Hence, capacitive impedance plays much the same role in ac circuits as does resistance in dc circuits.

Substituting for the relation between a capacitor's charge and its voltage in Equation 17 gives:

$$I_C = \frac{d(CV_C)}{dt} \tag{19}$$

In an ac circuit, the capacitance C does not change with time, but the voltage V_C does. Thus, we can rewrite Equation 19 as:

$$I_C(t) = C\left(\frac{dV_C(t)}{dt}\right) \tag{20}$$

This is the basic relation between voltage and current for a capacitor, much as Ohm's law is for a resistor (Equation 15).

Equation 20 is not very useful for solving circuit problems, however. We need algebraic expressions for the current and voltage. The following paragraphs discuss how these expressions can be obtained from Equation 20.

Let us begin by assuming that a sinusoidal voltage of amplitude V_C and frequency f is applied to the capacitor (Figure 7.12). This voltage is given by:

$$V_C(t) = V_C \sin \omega t \tag{21}$$

Substitution of Equation 21 into Equation 20 gives the resulting current:

$$I_C(t) = C\left(\frac{dV_C \sin \omega t}{dt}\right) \tag{22}$$

Because V_C is a constant, we can bring it outside the differential to get:

$$I_C(t) = CV_C\left(\frac{d \sin \omega t}{dt}\right) \tag{23}$$

From calculus, we know that:

$$\frac{d \sin \omega t}{dt} = \omega \cos \omega t \tag{24}$$

Hence, Equation 23 becomes:

$$I_C(t) = \omega CV_C \cos \omega t \tag{25}$$

This expression can be rewritten as:

$$I_C(t) = I_C \cos \omega t \tag{26}$$
$$\text{where } I_C = \omega CV_C \tag{27}$$

Equation 27 for a capacitor can be made to look like Ohm's law for a resistor if we write it as:

$$V_C = I_C Z_C \qquad (28)$$

$$\text{where } Z_C = \frac{1}{\omega C} \qquad (29)$$

or

$$Z_C = \frac{1}{2\pi f C} \qquad (30)$$

CONSISTENT UNITS FOR IMPEDANCE

Quantity	Basic Set of Units	Equivalent Set of Units
C	Farads (F)	Microfarads (μF)
ω	Hertz (Hz)	Kilohertz (kHz)
Z	Ohms (Ω)	Kilohms (kΩ)

$$Z_C = \frac{1}{\omega C} \qquad \Omega = \frac{1}{\text{Hz} \times \text{F}} \qquad \text{k}\Omega = \frac{1}{\text{kHz} \times \mu\text{F}}$$

FIGURE 17.13. In calculating Z_C, the basic units for capacitance C, frequency ω, and impedance Z_C are farads, hertz, and ohms, respectively. An equivalent set that makes calculation simpler in most situations is microfarads (10^{-6} F), kilohertz (kHz), and kilohms (kΩ). When using Equation 28, however, recall that if Z_C is in kilohms, then current will be in milliamperes.

Thus, for an applied capacitor voltage whose amplitude is V_C and whose frequency is f, Equation 28 gives the amplitude of the resulting current I_C.

The quantity Z_C is a fundamental ac quantity for a capacitor. It represents the extent to which the capacitor impedes, or slows down, the flow of current and, therefore, is called the **impedance.** As in the case of resistance, the greater the impedance Z_C, the smaller the amplitude of the current I_C for a given applied voltage V_C, and vice versa. Unlike resistance, however, the impedance is not constant, but rather changes with the frequency of the applied voltage.

Units of Z_c

If C is measured in farads and f in hertz, then Z_C will be in ohms. Note that the units of Z_C must be the same as those used for resistance in order to satisfy Equation 28. However, be aware that Z_C is *not a resistance* but an impedance and that dc circuit analysis methods cannot be used directly with Z_C. This concept will become clearer in subsequent discussion.

There is another set of consistent units that can be used to make calculations somewhat easier for most practical situations. As noted in the discussion of Ohm's law, circuit resistances are normally expressed in kilohms and circuit currents are generally in milliamperes. This fact was used to simplify Ohm's law calculations by noting that these two units were consistent. If I

is in milliamperes and R is in kilohms, V will be in volts. In a similar fashion, note that circuit capacitors are generally in units of microfarads (1 μF = 10^{-6} F) and frequencies are generally in kilohertz (10^3 Hz). Kilohms (kΩ), microfarads (μF), and kilohertz (kHz) also form a consistent set of units for Equation 29:

$$\text{k}\Omega = \frac{1}{\text{kHz} \times \mu\text{F}}$$

$$10^3 \, \Omega = \frac{1}{10^3 \, \text{Hz} \times 10^{-6} \, \text{F}}$$

$$10^6 \, \Omega = \frac{10^6}{\text{Hz-F}} \qquad (31)$$

or

$$\Omega = \frac{1}{\text{Hz-F}} \qquad (32)$$

Figure 17.13 summarizes the units of impedance. Using these units greatly simplifies calculations involving Equation 30, as the following examples illustrate.

FREQUENCY RESPONSE OF A CAPACITOR

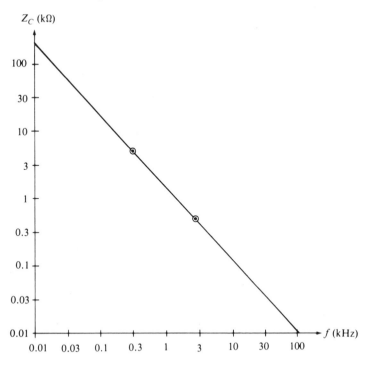

FIGURE 17.14. The variation of impedance with frequency for a typical capacitor is shown (see Example 3). Both axes are logarithmic in order to cover a wide range of values. Note that at high frequencies, the impedance is essentially zero, but at low frequencies, it is large.

Calculating Z_C

EXAMPLE 2: What is the impedance of a 0.10 μF capacitor at frequencies of (a) 300 Hz and (b) 3000 Hz?

Solution: The impedance can be calculated with Equation 30. We will use the consistent set of units, kilohertz, microfarads, and kilohms. (a) For $f = 0.30$ kHz:

$$Z_C = \frac{1}{2\pi fC}$$

$$= \frac{1}{2 \times 3.14 \times 0.30 \text{ kHz} \times 0.10 \ \mu\text{F}}$$

$$= 5.3 \text{ k}\Omega$$

(b) For $f = 3.0$ kHz:

$$Z_C = \frac{1}{6.28 \times 3.0 \text{ kHz} \times 0.10 \ \mu\text{F}}$$

$$= 0.53 \text{ k}\Omega$$

Note that increasing the frequency by a factor of 10 decreases the impedance by the same factor. This general behavior is shown in Figure 17.14 over a wide range of frequencies. Both scales are logarithmic in order to cover a wide range of frequencies.

EXAMPLE 3: What size capacitor has the same impedance at 1000 Hz that a 0.10 μF capacitor has at 5000 Hz?

Solution: In Example 2, when $C = 0.10\ \mu\text{F}$ and $f = 5$ kHz, $Z_C = 0.32$ kΩ. For this value of impedance and a frequency of 1 kHz, Equation 30 can be used to calculate the capacitance. Rewriting Equation 30 gives:

$$C = \frac{1}{2\pi f Z_C}$$
$$= \frac{1}{6.28 \times 1.0\ \text{kHz} \times 0.32\ \text{k}\Omega}$$
$$= \mathbf{0.50\ \mu\text{F}}$$

Note that decreasing the frequency by a factor of 5 increases the capacitance by a factor of 5.

EXAMPLE 4: At what frequency does the impedance of a 0.15 μF capacitor have the same magnitude as a 2.7 kΩ resistor?

Solution: Using Equation 29, we have:

$$Z_C = 2.7\ \text{k}\Omega = \frac{1}{\omega C}$$

$$\omega = \frac{1}{2.7\ \text{k}\Omega \times 0.15\ \mu\text{F}}$$

$$2\pi f = 2.5\ \text{kHz}$$

$$f = \frac{2.5}{6.28} = \mathbf{0.40\ \text{kHz}}$$

17.6 PHASE OF CAPACITOR VOLTAGE

Thus far, the ac analysis of capacitor behavior appears to be exactly like Ohm's law for a resistor except that the value of its effective resistance — its impedance — changes with frequency. In fact, the situation is more compli-

cated. In the analysis of the voltage and current in a capacitor, we found that if the voltage is given by Equation 21:

$$V_C(t) = V_C \sin \omega t$$

then the resulting current is given by Equation 26:

$$I_C(t) = I_C \cos \omega t$$

This expression can be rewritten using the trigonometric identity, Equation 7:

$$I_C(t) = I_C \sin\left(\omega t + \frac{\pi}{2}\right) \tag{33}$$

This expression shows that the current in a capacitor leads its voltage by $\pi/2$ radians, or 1/4 period. The waveforms are shown in Figure 17.15. Note that the sinusoidal current peaks $\pi/2$ radians earlier in time than the voltage. Or, put another way, *the voltage in a capacitor lags behind its current by $\pi/2$ radians.*

The fact that the capacitor voltage lags the circuit current can be understood by recalling an earlier discussion of the response of a series *RC* network to a step voltage. When a series *RC* network is charged by a positive step voltage, as shown in Figure 17.9, the initial capacitor charging current is at its peak value when the capacitor voltage is zero. As the capacitor voltage increases toward its peak value, the capacitor current decreases toward zero. Hence, the voltage and current are out of phase, which is exactly the sinusoidal behavior shown in Figure 17.15. The voltage and current are out of phase by $\pi/2$ radians.

The analysis therefore demonstrates two characteristics of a capacitor. First, the impedance of a capacitor varies with the frequency and, second, the voltage across a capacitor lags behind its current by $\pi/2$ radians. These two properties are central to the performance of ac circuits and, consequently, to the remainder of the discussion.

PHASE OF CAPACITOR VOLTAGE

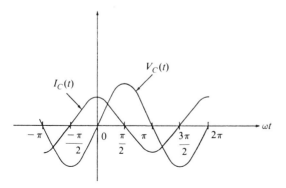

FIGURE 17.15. The capacitor has two important characteristics. One is that its impedance to sinusoidal voltages varies with frequency. The second is that the voltage in a capacitor always lags behind its current by 90° ($\pi/2$ radians), as shown in this phase diagram. These properties are central to ac circuit behavior.

To summarize, the sinusoidal (ac) current and voltage of a capacitor are governed by two rules:

1. The voltage across a capacitor lags behind the current by 90°. If the current is expressed by (Equation 26):

$$I_C(t) = I_C \cos \omega t$$

then the voltage is (Equation 21):

$$V_C(t) = V_C \sin \omega t$$

2. The peak amplitude of the capacitor voltage and the peak amplitude of the current are related by (Equation 28):

$$V_C = I_C Z_C$$

where the magnitude of the impedance Z_C depends on the frequency as follows (Equation 29):

$$Z_C = \frac{1}{\omega C}$$

EXAMPLE 5: If the voltage across a 0.010 μF capacitor is 2.0 V pp at 1200 Hz, how large is the current amplitude?

Solution: A peak-to-peak voltage of 2.0 V is equivent to a voltage amplitude of half that value, or:

$$V_C = 1.0 \text{ V}$$

The capacitor's impedance can be found from Equation 30:

$$
\begin{aligned}
Z_C &= \frac{1}{2\pi f C} \\
&= \frac{1}{2 \times 3.14 \times 1.2 \text{ kHz} \times 0.010 \ \mu\text{F}} \\
&= 13.3 \text{ k}\Omega
\end{aligned}
$$

With this value, we can find the current amplitude from Equation 28:

$$I_C = \frac{V_C}{Z_C} = \frac{1.0 \text{ V}}{13.3 \text{ k}\Omega} = \mathbf{0.075 \text{ mA}}$$

EXAMPLE 6: If the current in a circuit containing a 0.15 μF capacitor is given by:

$$I(t) = 6.3 \text{ mA} \cos(9.4 \text{ kHz} \times t)$$

what is the voltage?

Solution: From the argument of the sine wave, we see that the circuit frequency is:

$$\omega = 9.4 \text{ kHz}$$

At this frequency, the capacitor impedance is given by Equation 29:

$$
\begin{aligned}
Z_C &= \frac{1}{\omega C} \\
&= \frac{1}{9.4 \text{ kHz} \times 0.15 \ \mu\text{F}} = 1.4 \text{ k}\Omega
\end{aligned}
$$

Now, using Equation 28, we can find the amplitude of the voltage:

$$V_C = I_C Z_C = 6.3 \text{ mA} \times 1.4 \text{ k}\Omega = 8.8 \text{ V}$$

According to Equation 21, then, the capacitor voltage is:

$$V_C(t) = 8.8 \text{ V} \sin(9.4 \text{ kHz} \times t)$$

17.7 ANALYZING SERIES RC NETWORKS

Now that the relationship between current and voltage for both resistors and capacitors is known, the behavior of *RC* networks can be analyzed. The analysis begins with the series *RC* connection that was discussed earlier. The problem is: If a sinusoidal (ac) voltage is applied across a series *RC* network, what will be the circuit current and the voltage across each element? We know that the waveforms will be sinusoidal, but what will be their amplitudes and phase relations?

Figure 17.16b shows the waveform of a typical *RC* circuit (Figure 17.16a). Shown are the applied voltage $V(t)$ and the voltages across the capacitor $V_C(t)$ and the resistor $V_R(t)$. Recall that the circuit current $I(t)$ and the resistor voltage are in phase and related by Equation 15:

$$I_R(t) = \frac{V_R(t)}{R}$$

Adding Voltage Amplitudes

Note in Figure 17.16b that the sum of the *peak* amplitudes of the resistor and capacitor voltages exceeds that of the applied voltage:

$$V > V_C + V_R \tag{34}$$

At any instant of time, however, the sum of the *instantaneous* values of resistor voltage and capacitor voltage does equal the value of applied voltage at that instant. That is, by KVL, we can write:

$$V(t) = V_C(t) + V_R(t) \tag{35}$$

This equation is not particularly useful, however, because sinusoidal voltages are described by their peak values—not by their instantaneous values. Hence, we need an expression that relates the peak values of V_R and V_C to the peak value of V, as well as an expression for the phase shifts.

Because the phase angle is a relative quantity between waveforms, we can assume the phase angle of one wave function to be 0° and then determine relative phase angles of the other waveforms. Therefore, we will assume the current to be:

$$I(t) = I \sin \omega t \tag{36}$$

Note that this expression is a reversal of the normal practice. Generally, we start with the applied voltage V and try to determine the resulting current. In this case, we will assume a current $I(t)$ and work backward to determine an expression for the applied voltage. The processes are equivalent, as will be evident by the results.

Because the net circuit current flows through both the resistor and the capacitor, we must have:

$$I_R(t) = I_R \sin \omega t \tag{37}$$
$$I_C(t) = I_C \sin \omega t \tag{38}$$
$$\text{where} \quad I_R = I_C = I \tag{39}$$

For a resistor current given by Equation 36, the resistor voltage must be in phase and expressed by:

SINUSOIDAL VOLTAGE RESPONSE OF RC CIRCUIT

(a) Series RC Circuit

FIGURE 17.16. When a sinusoidal voltage is applied to a series RC circuit, the resulting voltages across the resistor and capacitor are as shown in the graph. Note that the instantaneous values of $V_R(t)$ and $V_C(t)$ add to give $V(t)$, but the peak values do not. Our task is to relate the peak values of voltage and current and to find the phase that describes this relationship.

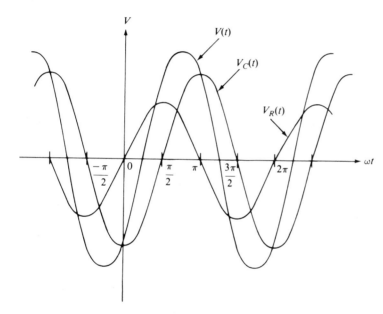

(b) Resulting Voltages

$$V_R(t) = V_R \sin \omega t \tag{40}$$
$$\text{where } V_R = IR \tag{41}$$

From an earlier discussion, we know that the capacitor voltage will lag behind its current by $\pi/2$. Hence, we can write:

$$V_C(t) = V_C \sin\left(\omega t - \frac{\pi}{2}\right) \tag{42}$$

which, by Equation 9, is equivalent to:

$$V_C(t) = -V_C \cos \omega t \tag{43}$$
$$\text{where } V_C = IZ_C = \frac{I}{\omega C} \tag{44}$$

Thus, we can write the applied voltage, Equation 35, as:

$$V(t) = V_R \sin \omega t - V_C \cos \omega t \tag{45}$$

Now all we need is a single sinusoidal expression for $V(t)$ to describe its phase relative to $I(t)$. We can find it by using the trigonometric identity of Equations 10, 11, and 12. Applying Equation 10 to Equation 45 gives:

$$V(t) = V \sin(\omega t + \phi) \tag{46}$$

where V and ϕ are given by Equations 11 and 12:

$$V^2 = V_R^2 + V_C^2 \tag{47}$$

$$\tan \phi = \frac{-V_C}{V_R} \tag{48}$$

The sinusoidal curves for these results are shown in Figure 17.16. Note that the phase angle ϕ is negative (Equation 48).

Equation 47 is the key result. It says that the voltage amplitudes add as squares. Though the instantaneous amplitudes follow KVL (Equation 35), the values we are interested in — the peak values of the sine waves — are related by Equation 47. Equations 47 and 48 can be expressed also in terms of the circuit values by using Equations 41 and 44 for V_R and V_C. Substitution gives:

$$V^2 = I^2 R^2 + I^2 Z_C^2$$

Solving for V gives:

$$V = I\sqrt{R^2 + Z_C^2} \tag{49}$$

Substituting in $Z_C = 1/\omega C$ gives:

$$V = I\sqrt{R^2 + \frac{1}{\omega^2 C^2}}$$

$$\tan \phi = \frac{-I Z_C}{IR}$$

Simplifying gives:

$$V = IR\sqrt{1 + \frac{1}{\omega^2 C^2 R^2}} \tag{50}$$

$$\tan \phi = \frac{-Z_C}{R} \tag{51}$$

Again substituting in $Z_C = 1/\omega C$, we have:

$$\tan \phi = \frac{-1}{\omega C R} \tag{52}$$

These results are the ones we want. The applied voltage is given by Equation 46:

$$V(t) = V \sin(\omega t + \phi)$$

and the resulting current is given by Equation 36:

$$I(t) = I \sin \omega t$$

where peak values of V and I are related by Equation 50 in terms of the circuit values R and C and the frequency ω:

$$V = IR\sqrt{1 + \frac{1}{\omega^2 C^2 R^2}}$$

$V(t)$ lags behind $I(t)$ by a phase angle given by Equation 52:

$$\tan \phi = \frac{-1}{\omega C R}$$

The voltage across the capacitor lags behind the circuit current and resistor voltage by 90°, as described by Equations 40 and 43, where Equations 41 and 44 give the peak values:

$$V_R(t) = V_R \sin \omega t$$

where $V_R = IR$

and

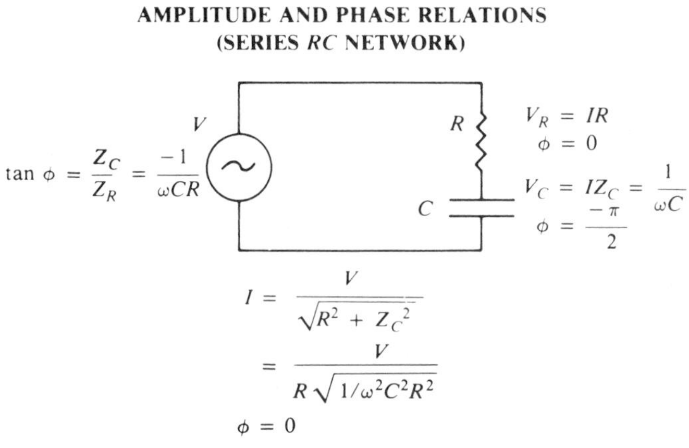

**AMPLITUDE AND PHASE RELATIONS
(SERIES *RC* NETWORK)**

$$\tan \phi = \frac{Z_C}{Z_R} = \frac{-1}{\omega CR}$$

$$V_R = IR$$
$$\phi = 0$$

$$V_C = IZ_C = \frac{1}{\omega C}$$
$$\phi = \frac{-\pi}{2}$$

$$I = \frac{V}{\sqrt{R^2 + Z_C^2}}$$

$$= \frac{V}{R\sqrt{1/\omega^2 C^2 R^2}}$$

$$\phi = 0$$

FIGURE 17.17. For analysis, let us assume that the resistor voltage, and hence the current, have no phase angle. Therefore, the capacitor voltage has a phase angle of $\pi/2$. The results of the analysis then yield the equations shown for the amplitudes of the resistor voltage, capacitor voltage, and current and for the phase angle of the applied voltage. Each equation is expressed in terms of the known amplitude and frequency of the applied voltage V and the circuit values R and C.

$$V_C(t) = -V_C \cos \omega t$$

where $V_C = IZ_C = \dfrac{I}{\omega C}$

These results are summarized in Figure 17.17.

Equivalent Impedance

In the analysis of resistor networks, we were able to determine an effective network resistance that would let us calculate the net circuit current for a given applied voltage. For two resistors R_1 and R_2 in series, the equivalent resistance was simply their sum:

$$R_{eq} = R_1 + R_2 \qquad (53)$$

The net circuit current could then be calculated from Ohm's law:

$$V = IR_{eq} \qquad (54)$$

Similarly, for a series *RC* network, we can write an expression like Equation 54 for the *peak* values of the applied voltage and resulting current in terms of an **equivalent circuit impedance** Z_{eq} (Figure 17.18). We simply rewrite Equation 49 as:

$$V = IZ_{eq} \qquad (55)$$

where Z_{eq}^2 is defined as:

$$Z_{eq}^2 = R^2 + Z_C^2 \qquad (56)$$

or

$$Z_{eq} = \sqrt{R^2 + Z_C^2} = \sqrt{R^2 + \frac{1}{\omega^2 C^2}}$$

Final simplification gives the result:

$$Z_{eq} = R\sqrt{1 + \frac{1}{\omega^2 C^2 R^2}} \qquad (57)$$

Note that for a series *RC* network, the impedances add as squares (Equation 56), in contrast to Equation 53 for a series resistor combination.

Equation 57 is exactly the expression sought for the purpose of explaining the frequency dependence of a series *RC* network (see Figure 17.11). When $\omega C \ll 1$ (low frequencies), then $(1/\omega C) \gg 1$. Also if $(1/\omega C) \gg R$, then we can neglect the value of R^2 to get:

$$Z_{eq} \approx \sqrt{\frac{1}{\omega^2 C^2}}$$

EQUIVALENT IMPEDANCE (SERIES RC)

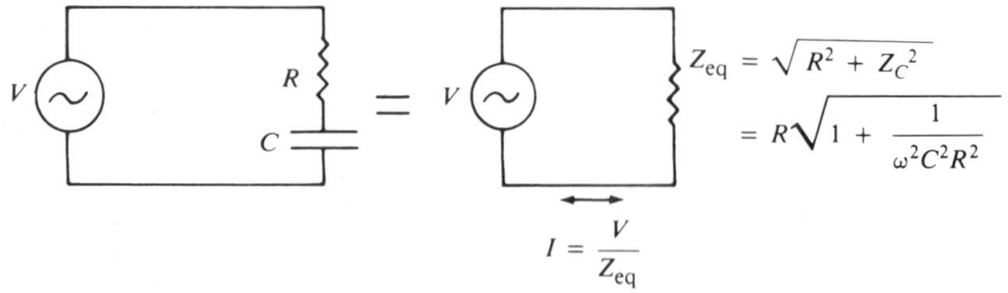

$$Z_{eq} = \sqrt{R^2 + Z_C^2}$$

$$= R\sqrt{1 + \frac{1}{\omega^2 C^2 R^2}}$$

$$I = \frac{V}{Z_{eq}}$$

FIGURE 17.18. As in the case of resistor networks, we can determine an equivalent impedance for the series RC connection. Knowing the equivalent impedance, we can calculate the peak value of the current from the peak value of applied voltage.

or

$$Z_{eq} \approx \frac{1}{\omega C} \qquad \text{for } \frac{1}{\omega C} \gg R \qquad \textbf{(58)}$$

That is, at low frequencies, and for sufficiently small values of capacitance C (or small values of resistance), we can ignore the resistor, and the equivalent impedance is simply that of the capacitor.

Similarly, for $\omega C \gg 1$, $(1/\omega C) \ll 1$, and we can neglect the value of $1/\omega^2 C^2$ in Equation 57 to get:

$$Z_{eq} \approx \sqrt{R^2}$$

or

$$Z_{eq} \approx R \qquad \text{for } \frac{1}{\omega C} \ll R \qquad \textbf{(59)}$$

Thus, at high frequencies, and for sufficiently large values of capacitance (or large values of resistance), we can ignore the capacitor. In these cases, the equivalent impedance is simply that of the resistor.

EXAMPLE 7: What is the equivalent impedance of a series RC network where $R = 1$ kΩ and $C = 0.10$ μF at the following frequencies?

f
0.10 kHz
0.30 kHz
1.0 kHz
3.0 kHz
10. kHz
30. kHz
100. kHz

Solution: Z_{eq} for a series RC connection is given by Equation 57:

$$Z_{eq} = \sqrt{\frac{1}{\omega^2 C^2} + R^2}$$

Substituting in $f = 0.1$ kHz, $C = 0.1$ μF, and $R = 1$ kΩ gives:

$$Z_{eq} = \sqrt{\frac{1}{(2\pi \times 0.1 \times 0.1)^2} + (1)^2}$$

$$= \sqrt{\frac{1}{39 \times 10^{-4}} + 1} = \sqrt{250 + 1}$$

$$= \sqrt{251} = 16 \text{ k}\Omega$$

Note that the value of $(1/\omega^2 C^2) \gg R^2$, so we can ignore the value of R^2 at 0.10 kHz.

Let us check the result for $f = 100$ kHz. Substitution gives:

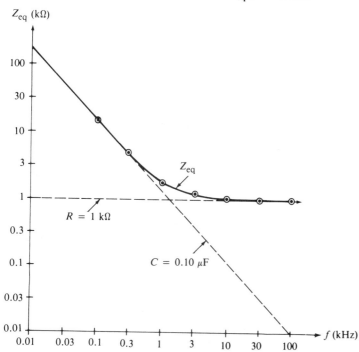

FREQUENCY DEPENDENCE OF Z_{eq} (SERIES RC)

FIGURE 17.19. This graph shows the frequency dependence of Z_{eq} for a series RC circuit. The equivalent impedance of a 1 kΩ resistor in series with a 0.10 μF capacitor is shown over a range of frequencies, along with the impedance of the resistor and capacitor alone. Note that at low frequencies, the equivalent impedance is essentially that of the capacitor and, at high frequencies, it is essentially that of the resistor. See Example 7.

$$Z_{eq} = \sqrt{\frac{1}{(2\pi \times 100 \times 0.1)^2} + (1)^2}$$

$$= \sqrt{\frac{1}{6.28 \times 10^2} + 1} = \sqrt{0.0003 + 1}$$

$$= \sqrt{1.0003} = 1 \text{ k}\Omega$$

Note that the value of $(1/\omega^2 C^2) \ll R^2$, so we can ignore the value of $(1/\omega^2 C^2)$ at 100 kHz.

Similar calculations for the intermediate values give the complete result:

f		Z_{eq}	
0.10	kHz	16.	kΩ
0.30	kHz	5.4	kΩ
1.0	kHz	1.9	kΩ
3.0	kHz	1.2	kΩ

10.	kHz	1.01	kΩ
30.	kHz	1.00	kΩ
100.	kHz	1.00	kΩ

These data are plotted in Figure 17.19 along with the data from Figure 17.14 for the impedance of a 0.10 μF capacitor alone. Take a moment to look at the results. Note that the equivalent impedance at low frequencies approaches that of the capacitor and at high frequencies, it approaches that of the resistor. The equivalent impedance makes a smooth transition from that of a pure 0.10 μF capacitor to that of a pure 1 kΩ resistor as the frequency increases. At about 1 kHz, the impedance of the capacitor is about the same as that of the resistor: 1 kΩ.

Voltage Divider Relations

Two resistors in series form a simple voltage divider. Similarly, a resistor and capacitor in series form an ac voltage divider, with the important feature that the voltage division is frequency dependent. This frequency dependence can be seen by writing expressions for the peak values of V_R and V_C in terms of R, C, and ω.

The peak value of resistor voltage is given by Equation 41:

$$V_R = IR$$

Substituting in Equation 55 gives:

$$V_R = V\left(\frac{R}{Z_{eq}}\right) \tag{60}$$

or, using Equation 57:

$$\frac{V_R}{V} = \frac{R}{R\sqrt{1 + (1/\omega^2 C^2 R^2)}}$$

Simplifying gives:

$$\frac{V_R}{V} = \frac{1}{\sqrt{1 + (1/\omega^2 C^2 R^2)}} \tag{61}$$

The peak value of capacitor voltage is given by Equation 44:

$$V_C = IZ_C$$

Substituting in Equations 55 and 56 gives:

$$V_C = V\left(\frac{Z_C}{Z_{eq}}\right) \tag{62}$$

or

$$\frac{V_C}{V} = \frac{1}{\omega C R\sqrt{1 + (1/\omega^2 C^2 R^2)}} \tag{63}$$

Simplifying gives:

$$\frac{V_C}{V} = \frac{1}{\sqrt{1 + \omega^2 C^2 R^2}} \tag{64}$$

At low frequencies $(1/\omega C) \gg R$ (or, equivalently, $\omega C R \ll 1$), and Equations 61 and 64 become, respectively:

$$V_R = V\left(\frac{1}{\sqrt{1/\omega^2 C^2 R^2}}\right) \approx 0$$

$$V_C = V\left(\frac{1}{\sqrt{1 + 0}}\right) = V$$

At high frequencies $(1/\omega C) \ll R$ (or, equivalently, $\omega C R \gg 1$), and Equations 61 and 64 become, respectively:

$$V_R = V\left(\frac{1}{\sqrt{1 + 0}}\right) = V$$

$$V_C = V\left(\frac{1}{\sqrt{\omega^2 C^2 R^2}}\right) \approx 0$$

Thus, at low frequencies, all of the voltage is dropped across the capacitor and, at high frequencies, all of the voltage is dropped across the resistor. This behavior over a range of frequencies for $V = 10.0$ V, $R = 1$ kΩ, and $C = 0.1$ μF is shown in Figure 17.20. Compare these results with those presented in Figure 17.11. The voltage divider equations, Equations 61 and 64, fully describe Figure 17.11.

Phase Relations

Finally, the phase relations between the various voltages must be considered. At the outset, the current was established as (Equation 36):

$$I(t) = I \sin \omega t$$

The resistor voltage was in phase with the current (Equation 40):

$$V_R(t) = V_R \sin \omega t$$

VOLTAGE DIVIDER BEHAVIOR

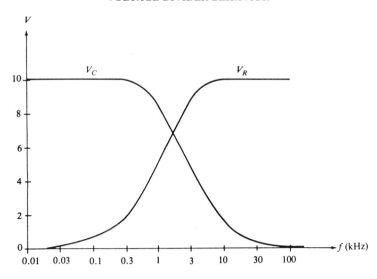

FIGURE 17.20.   A series *RC* network forms an ac voltage divider in which the voltage dropped across each component varies with frequency. As the figure shows, at low frequencies, all of the voltage is dropped across the capacitor and, at high frequencies, it is dropped across the resistor.

However, the capacitor voltage lagged behind the current by 90° (Equation 43):

$$V_C(t) = -V_C \cos \omega t$$

The applied voltage was then found to be (Equation 46):

$$V(t) = V \sin(\omega t - \phi)$$

where the phase angle ϕ was given by (Equation 52):

$$\tan \phi = \frac{-V_C}{V_R} = \frac{-1}{\omega CR}$$

At high frequencies $(1/\omega C) \ll R$ (or $\omega CR \gg 1$) and:

$$\tan \phi = 0$$
$$\phi = 0°$$

Equation 46 becomes:

$$V(t) = V \sin \omega t$$

Thus, for high frequencies, the applied voltage is in phase with the resistor voltage, which

agrees with the earlier conclusion that at high frequencies, the capacitor can be neglected.

Similarly, at low frequencies $(1/\omega C) \gg R$ (or $\omega CR \ll 1$) and:

$$\tan \phi = -\infty$$
$$\phi = \frac{-\pi}{2}$$

and Equation 46 becomes:

$$V(t) = V \sin\left(\omega t - \frac{\pi}{2}\right)$$

Using the trigonometric identity of Equation 9, we rewrite this equation as:

$$V(t) = -V \cos \omega t$$

Thus, for low frequencies, the applied voltage is in phase with the capacitor voltage, which agrees with the earlier finding that at low frequencies, the resistor can be neglected. As the frequency goes from low to high, the applied voltage shifts from being in phase with the capacitor to being in phase with the resistor.

We can represent the phase angle ϕ geome-

GEOMETRICAL REPRESENTATION
OF CIRCUIT VOLTAGES

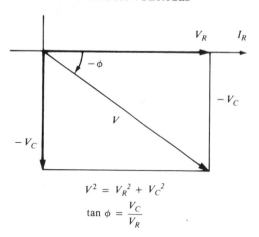

$$V^2 = V_R^2 + V_C^2$$

$$\tan \phi = \frac{V_C}{V_R}$$

FIGURE 17.21. Phase relations among voltage amplitudes can be represented geometrically as shown. The resistor voltage V_R is a vector with a phase angle of $0°$, and the capacitor voltage V_C is a vector with a phase angle of $-90°$. The applied voltage V is the vector sum of V_R and V_C at a phase angle $-\phi$. The algebraic expressions for V and ϕ can be derived from the figure.

trically, as shown in Figure 17.21. The resistor voltage is a vector of length V_R with a phase angle of $0°$, and the capacitor voltage is a vector of length V_C at $-90°$. The amplitude of the applied voltage is the magnitude of the vector sum of V_R and V_C with a phase angle $-\phi$. Note that if $\phi = -90°$ (low frequencies), $V = -V_C$, and if $\phi = 0°$ (high frequencies), $V = V_R$. For intermediate values, V is given by the Pythagorean theorem, Equation 47:

$$V^2 = V_C^2 + V_R^2$$

and, from Equation 48, the phase angle is seen to be:

$$\tan \phi = \frac{-V_C}{V_R}$$

Figure 17.21 provides a good way of remember-

ing the relations for phase angle and voltage amplitudes for a series RC circuit.

EXAMPLE 8: At what frequency will the voltage drop across an 8 kΩ resistor and a 0.02 μF capacitor be the same?

Solution: According to the question, we have:

$$V_R = V_C$$

We can find the frequency from the expressions for V_R and V_C (Equations 41 and 44):

$$V_R = IR$$

$$V_C = \frac{I}{\omega C}$$

Equating these expressions gives:

$$IR = \frac{I}{\omega C}$$

or

$$\omega = \frac{1}{RC}$$

or

$$f = \frac{1}{2\pi RC}$$

Substituting the values given, we find:

$$f = \frac{1}{6.28 \times 8 \text{ k}\Omega \times 0.02 \ \mu\text{F}} = 1 \text{ kHz}$$

The phase angle is given by Equation 48:

$$\tan \phi = \frac{-V_C}{V_R} = -1$$

$$\phi = -45° = \frac{-\pi}{4}$$

When $V_R = V_C$, we have the condition:

$$\omega RC = 1 \qquad (65)$$

The frequency for which Equation 65 is true has a special significance in ac circuits and is defined with the symbol f_o. From Equation 65, f_o is given by:

$$2\pi f_o RC = 1$$

Solving for f_o gives:

$$f_o = \frac{1}{2\pi RC} \tag{66}$$

At the frequency f_o:

$$V_C = V_R$$

and

$$\phi = \frac{\pi}{4}$$

For frequencies less than f_o, the circuit is primarily capacitive. Thus, for $f < f_o$:

$$\omega CR < 1$$
$$V_C > V_R$$
$$\phi > \frac{\pi}{4} \tag{67}$$

For frequencies greater than f_o, the circuit is primarily resistive. Thus, for $f > f_o$:

$$\omega CR > 1$$
$$V_C < V_R$$
$$\phi < \frac{\pi}{4} \tag{68}$$

The significance of the quantities f_o and ωCR can be easily understood by looking at the expressions for the effective impedance of the series RC circuit (Equation 56) and the voltage divider relations (Equations 61 and 64). The quantities f_o and ωCR will appear again and again as the analysis of ac circuits continues.

EXAMPLE 9: A 5.0 kHz signal voltage with an amplitude of 10.0 V is applied to a series RC circuit where $R = 2.2$ kΩ and $C = 0.020$ μF. What are the amplitudes of the circuit current, capacitor voltage, and resistor voltage, and what is the phase angle between the circuit current and voltage?

Solution: We define the circuit current by:

$$I(t) = I \sin \omega t$$

The circuit voltage, then, must be:

$$V(t) = V \sin(\omega t + \phi)$$

where ϕ is given by Equation 52:

$$\tan \phi = \frac{-1}{\omega CR}$$

The procedure for solving this problem is identical to that described for dc circuits in Part II. The first step is to find the equivalent impedance of the circuit, which is given by Equation 57:

$$Z_{eq} = \sqrt{\frac{1}{\omega^2 C^2} + R^2}$$

$$= \sqrt{\begin{array}{c} \dfrac{1}{(6.28 \times 5.0 \text{ kHz} \times 0.02 \text{ } \mu\text{F})^2} \\ + (2.2 \text{ k}\Omega)^2 \end{array}}$$

$$= \sqrt{2.53 + 4.83} = 2.7 \text{ k}\Omega$$

Using this value, we can find the amplitude of the circuit current from Equation 55:

$$V = IZ_{eq}$$

$$I = \frac{V}{Z_{eq}} = \frac{10.0 \text{ V}}{2.7 \text{ k}\Omega} = 3.7 \text{ mA}$$

The amplitude of the resistor and capacitor voltages can be found from Equations 41 and 44 or from the voltage divider equations, Equations 61 and 64:

EXAMPLE 9

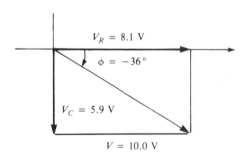

$V_R = 8.1$ V

$\phi = -36°$

$V_C = 5.9$ V

$V = 10.0$ V

FIGURE 17.22. Shown is the geometrical representation of voltage amplitudes and phase for a typical series *RC* circuit.

$V_R = IR = 3.7 \text{ mA} \times 2.2 \text{ k}\Omega$
$= \mathbf{8.1 \text{ V}}$

$V_C = \dfrac{I}{\omega C} = \dfrac{3.7 \text{ mA}}{6.28 \times 5.0 \text{ kHz} \times 0.02 \ \mu\text{F}}$
$= \mathbf{5.9 \text{ V}}$

The phase angle is determined from Equation 48:

$\tan \phi = \dfrac{-V_C}{V_R} = \dfrac{-5.9 \text{ V}}{8.1 \text{ V}} = -0.72$

$\phi = \mathbf{-36°}$

The geometrical representation of these results is shown in Figure 17.22.

EXAMPLE 10: A 10 kHz signal voltage has an amplitude of 5.0 V. What values of series resistor and capacitor will yield a voltage amplitude of 3.0 V across the resistor? What will be the phase shift between the circuit current and voltage?

Solution: The values of resistor and capacitor can be found from the voltage divider relation for the resistor voltage, Equation 61:

$\dfrac{V_R}{V} = \dfrac{1}{\sqrt{1 + (1/\omega^2 C^2 R^2)}} = \dfrac{3.0 \text{ V}}{5.0 \text{ V}} = 0.6$

Squaring both sides gives:

$\dfrac{1}{1 + (1/\omega^2 C^2 R^2)} = (0.6)^2 = 0.36$

$\dfrac{1}{0.36} = 1 + \dfrac{1}{\omega^2 C^2 R^2}$

$\dfrac{1}{\omega^2 C^2 R^2} = \dfrac{1}{0.36} - 1 = 1.8$

$R^2 = \dfrac{1}{1.8 \omega^2 C^2}$

We are free to select a value for either R or C and determine the other from the above equation. Let us select:

$C = \mathbf{0.010 \ \mu\text{F}}$

Substitution of this value gives:

$R^2 = \dfrac{1}{1.8(6.28 \times 10 \text{ kHz})^2(0.010 \ \mu\text{F})^2}$

$= 1.4$

$R = \mathbf{1.2 \text{ k}\Omega}$

The phase angle is given by Equation 52:

$\tan \phi = \dfrac{-1}{\omega CR}$

Substituting values gives:

$\tan \phi = \dfrac{-1}{6.28 \times 0.01 \ \mu\text{F} \times 1.2 \text{ k}\Omega}$

$= -13.3$

$\phi = \mathbf{-86°}$

17.8 ANALYZING PARALLEL *RC* NETWORKS

As in dc circuit analysis, the majority of practical situations involving ac circuits comprise two

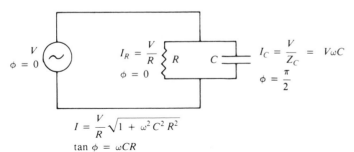

AMPLITUDE AND PHASE RELATIONS
(PARALLEL *RC* NETWORK)

$$I = \frac{V}{R}\sqrt{1 + \omega^2 C^2 R^2}$$

$$\tan \phi = \omega CR$$

FIGURE 17.23. A parallel *RC* circuit acts as an ac current divider. The figure gives expressions for the amplitudes of the resistor current, capacitor current, and net circuit current and for the phase angle between the net circuit current and the applied voltage. Note that the applied voltage is assumed to have a phase angle of 0° and that the net circuit current leads the applied voltage by an angle $+\phi$.

basic circuit connections. One is the series *RC* connection just examined, and the other is the parallel *RC* connection to be considered next. Unlike the series *RC* connection that acts as a frequency-dependent *voltage divider*, the parallel *RC* connection acts as a frequency-dependent *current divider*.

Adding Current Amplitudes

The parallel *RC* connection is shown in Figure 17.23. The problem to be considered is: If a sinusoidal voltage is applied across the parallel *RC* connection, what will be the net circuit current and the current through each element?

For this circuit, it is assumed that the applied voltage has no phase angle and is given by:

$$V(t) = V \sin \omega t \qquad (69)$$

Clearly, this voltage is the voltage that appears across both the resistor and the capacitor. Thus, we must have:

$$V_R(t) = V_R \sin \omega t \qquad (70)$$
$$V_C(t) = V_C \sin \omega t \qquad (71)$$

where $\quad V_R = V_C = V \qquad (72)$

From our knowledge of the ac behavior of resistors and capacitors, we know that the current through the resistor $I_R(t)$ will be in phase

with its voltage, while the current in the capacitor $I_C(t)$ will lead the voltage by 90°, or $\pi/2$ radians. Hence, we can write the following expressions for the two currents:

$$I_R(t) = I_R \sin \omega t \qquad (73)$$

$$\text{where } I_R = \frac{V}{R} \qquad (74)$$

and

$$I_C(t) = I_C \sin\left(\omega t + \frac{\pi}{2}\right) \qquad (75)$$

$$\text{where } I_C = \frac{V}{Z_C} = V\omega C \qquad (76)$$

By the trigonometric identity of Equation 7, we can rewrite the capacitor current as:

$$I_C(t) = I_C \cos \omega t \qquad (77)$$

By KCL, the net circuit current $I(t)$ must be the sum of the time-varying currents through the two elements:

$$I(t) = I_R \sin \omega t + I_C \cos \omega t \qquad (78)$$

As before, we can use the trigonometric identity of Equation 10 to get a simplified expression for the net circuit current:

$$I(t) = I \sin(\omega t + \phi) \qquad (79)$$

FIGURE 17.24. An equivalent impedance Z_{eq} can also be determined for the parallel RC network. Calculating Z_{eq} is generally an early step in calculating the amplitude of the net circuit current.

EQUIVALENT IMPEDANCE (PARALLEL RC)

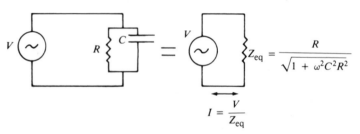

where:

$$I^2 = I_R^2 + I_C^2 \tag{80}$$

and

$$\tan \phi = \frac{I_C}{I_R} \tag{81}$$

Substituting Equations 74 and 77 for I_R and I_C reduces these expressions to:

$$I^2 = \left(\frac{V}{R}\right)^2 + (V\omega C)^2$$
$$= V^2 \left(\omega^2 C^2 + \frac{1}{R^2}\right) \tag{82}$$

Solving for V gives:

$$V = I\left(\frac{R}{\sqrt{1 + \omega^2 C^2 R^2}}\right) \tag{83}$$

and

$$\tan \phi = \frac{V/Z_C}{V/R} = \frac{R}{Z_C}$$

Substituting for Z_C gives:

$$\tan \phi = \omega CR \tag{84}$$

Note in Equation 80 that the current amplitudes for a parallel RC network also add as squares. This situation is different from the situation in which two resistors are in parallel and the current amplitudes simply add, but it is comparable to the series RC network where the *voltage* amplitudes add as squares (Equation 47).

Equivalent Impedance

Equation 83 for the relationship between the applied voltage and the resulting current can be simplified by defining an equivalent impedance for the parallel RC connection. That is, we can write Equation 83 as:

$$V = IZ_{eq} \tag{85}$$

where:

$$Z_{eq} = \frac{R}{\sqrt{1 + \omega^2 C^2 R^2}} \tag{86}$$

See Figure 17.24.

As in the case of the series connection, we can analyze the low- and high-frequency limits. At low frequencies, $\omega CR \ll 1$ and Equation 86 becomes:

$$Z_{eq} \approx R \quad \text{for } \omega CR \ll 1 \tag{87}$$

Note that this result is the opposite of the result for the series connection. At low frequencies, the capacitor has a large impedance that for the parallel connection, acts like an open circuit.

Hence, all of the current passes through the resistor.

At high frequencies, $\omega CR \gg 1$ and Equation 86 becomes:

$$Z_{eq} \approx \frac{R}{\omega CR}$$

$$Z_{eq} = \frac{1}{\omega C} = Z_C \qquad \text{for } \omega CR \gg 1 \qquad (88)$$

In this case, the capacitor has a very low impedance and acts like a short circuit across the resistor, so all of the current passes through the capacitor.

EXAMPLE 11: What is the equivalent impedance of a parallel RC network in which $R = 1 \text{ k}\Omega$ and $C = 0.1 \ \mu\text{F}$ at the following frequencies?

f
0.10 kHz
0.30 kHz
1.0 kHz
3.0 kHz
10. kHz
30. kHz
100. kHz

Solution: Z_{eq} for a parallel RC connection is given by Equation 86. Substituting in $f = 0.1$ kHz, $C = 0.1 \ \mu\text{F}$, and $R = 1 \text{ k}\Omega$ gives:

$$Z_{eq} =$$
$$\frac{1 \text{ k}\Omega}{\sqrt{1 + (6.28 \times 0.1 \text{ kHz} \times 0.1 \ \mu\text{F} \times 1 \text{ k}\Omega)^2}}$$
$$= \frac{1}{\sqrt{1 + (0.0628)^2}} = \frac{1}{\sqrt{1.0039}}$$
$$= 1 \text{ k}\Omega$$

which is the value of the resistor.
Substitution for $f = 100$ kHz gives:

$$Z_{eq} =$$
$$\frac{1 \text{ k}\Omega}{\sqrt{1 + (6.28 \times 100 \text{ kHz} \times 0.1 \ \mu\text{F} \times 1 \text{ k}\Omega)^2}}$$
$$= \frac{1}{\sqrt{1 + (6.28)^2}} \approx \frac{1}{62.8}$$
$$= 0.016 \text{ k}\Omega$$

which is the value of the capacitor at 100 kHz.

Similar calculations for the intermediate values give the complete result:

f	Z_{eq}
0.10 kHz	1.0 kΩ
0.30 kHz	0.99 kΩ
1.0 kHz	0.85 kΩ
3.0 kHz	0.47 kΩ
10. kHz	0.16 kΩ
30. kHz	0.053 kΩ
100. kHz	0.016 kΩ

These data are graphed in Figure 17.25 along with the data for the impedance of a 0.1 μF capacitor alone (see Figure 17.14). Note that at low frequencies, the equivalent impedance is determined by the resistor and, at high frequencies, by the capacitor.

Current Divider Relations

Just as in the case of two resistors in parallel, we can obtain current divider relations for the parallel RC connection. The ratio of the peak value of resistor current to the peak value of the total current can be found by combining Equations 74 and 85:

$$\frac{I_R}{I} = \frac{V/R}{V/Z_{eq}} = \frac{Z_{eq}}{R}$$

Substituting in for Z_{eq} gives:

FIGURE 17.25. The equivalent impedance of a 1 kΩ resistor in parallel with a 0.10 μF capacitor over a range of frequencies is shown along with the impedance of the resistor and capacitor alone. Note that it has the opposite behavior of a series *RC* network (Figure 17.19). See Example 11.

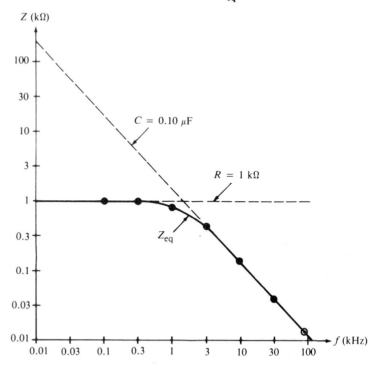

FREQUENCY DEPENDENCE OF Z_{eq} (PARALLEL *RC*)

$$\frac{I_R}{I} = \frac{1}{\sqrt{1 + \omega^2 C^2 R^2}} \tag{89}$$

Similarly, the ratio of the peak value of capacitor current to the peak value of the total current can be found from Equations 76 and 85:

$$\frac{I_C}{I} = \frac{V/Z_C}{V/Z_{eq}} = \frac{Z_{eq}}{Z_C} = \frac{\omega CR}{\sqrt{1 + \omega^2 C^2 R^2}}$$

Simplifying gives:

$$\frac{I_C}{I} = \frac{1}{\sqrt{1 + (1/\omega^2 C^2 R^2)}} \tag{90}$$

Looking at the frequency dependence, at

low frequencies, $\omega CR \ll 1$, and we have:

$$I_R \approx I$$
$$I_C \approx 0$$

All of the current passes through the resistor.
At high frequencies, $\omega CR \gg 1$ and:

$$I_R \approx 0$$
$$I_C \approx I$$

All of the current passes through the capacitor. The behavior over a full range of frequencies for $R = 1$ kΩ and $C = 0.1$ μF is illustrated in Figure 17.26.

Phase Relations

The final topic to consider for the parallel *RC* connection is phase relations. The relations for

CURRENT DIVIDER BEHAVIOR

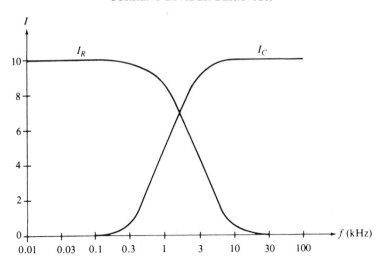

FIGURE 17.26. The figure shows the variation with frequency of the current in each branch of a parallel *RC* circuit. At low frequencies, most of the current passes through the resistor and, at high frequencies, it passes through the capacitor.

the voltages and currents are given by Equations 69, 73, 77, and 79. The applied voltage is:

$$V(t) = V \sin \omega t$$

which results in a net circuit current:

$$I(t) = I \sin(\omega t + \phi)$$

where (Equation 84):

$$\tan \phi = \omega CR$$

This equation says that the circuit current *leads* the applied voltage by an angle ϕ that is determined by the frequency ω and the values of R and C.

At high frequencies, when the capacitive impedance ($Z_C = 1/\omega C$) is small ($\omega CR \gg 1$), essentially all of the current passes through the capacitor, and:

$$\tan \phi \approx \infty$$

$$\phi \approx \frac{\pi}{2}$$

Substitution into Equation 76 gives:

$$I(t) = I \sin\left(\omega t + \frac{\pi}{2}\right)$$

which, by Equation 7, is equivalent to:

$$I(t) = I \cos \omega t \qquad \text{for } \omega CR \gg 1 \qquad \textbf{(91)}$$

Thus, the voltage lags the current by $\pi/2$, which is what is expected from a purely capacitive circuit.

On the other hand, at low frequencies, when the capacitive impedance is large ($\omega CR \ll 1$), essentially all of the current passes through the resistor, and:

$$\tan \phi \approx 0$$

$$\phi = 0$$

Substitution into Equation 79 gives:

$$I(t) = I \sin \omega t \qquad \text{for } \omega CR \ll 1 \qquad \textbf{(92)}$$

The voltage and current are in phase, which is what is expected from a purely resistive circuit.

The current through the resistor is:

$$I_R(t) = I_R \sin \omega t$$

which is always in phase with the applied voltage. The current through the capacitor is:

$$I_C(t) = I_C \cos \omega t$$

which always leads the applied voltage by $\pi/2$ radians.

The phase angle of the circuit current can be represented geometrically, as shown in Figure 17.27. The resistor current is a vector of length I_R with a phase angle of $0°$. The capacitor current is a vector of length I_C with a phase angle of $90°$. The amplitude of the net circuit current is the magnitude of the vector sum of I_R and I_C at a phase angle of ϕ. Note that the net circuit current leads the resistor current, but lags behind the capacitor current. When $\phi = 0°$ (high frequencies), $I = I_R$; when $\phi = 90°$ (low frequencies), $I = I_C$. For intermediate frequencies, I is given by the Pythagorean theorem (Equation 80):

$$I^2 = I_R^2 + I_C^2$$

and the phase angle is seen to be:

$$\tan \phi = \frac{I_C}{I_R}$$

Figure 17.27 provides a good way of remembering the relations for phase angle and current amplitudes for a parallel RC circuit.

GEOMETRICAL REPRESENTATION OF CIRCUIT CURRENTS

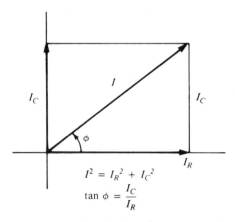

$$I^2 = I_R^2 + I_C^2$$

$$\tan \phi = \frac{I_C}{I_R}$$

FIGURE 17.27. The phase relations among current amplitudes can be represented geometrically as shown. Note that the net circuit current leads the resistor current, but lags behind the capacitor current. The algebraic relations for I and ϕ can be derived from the figure.

EXAMPLE 12:   At what frequency will the current amplitudes through the parallel connection of an 8 kΩ resistor and a 0.02 μF capacitor be the same? What is the phase angle at that frequency?

Solution: According to the question, we have:

$$I_R = I_C$$

Substituting expressions for I_R and I_C (Equations 74 and 76) gives:

$$\frac{V}{R} = V\omega C$$

$$\omega = \frac{1}{RC}$$

Substituting values gives:

$$f = \frac{1}{6.28 \times 8 \text{ k}\Omega \times 0.02 \ \mu\text{F}} = 1.0 \text{ kHz}$$

which is the frequency f_o.

The phase angle is given by Equation 81:

$$\tan \phi = \frac{I_C}{I_R} = 1$$

$$\phi = 45° = \frac{\pi}{4} \text{ radians}$$

Of course, the phase angle is immediately obvious from Figure 17.27.

EXAMPLE 13: A 5 kHz signal voltage with an amplitude of 10.0 V is applied to a parallel RC circuit in which $R = 2.2$ kΩ and $C = 0.020$ μF. What are the amplitudes of the circuit current, capacitor current, and resistor current? What is the phase angle between the circuit current and the voltage?

Solution: The applied voltage is defined as:

$$V(t) = V \sin \omega t$$

where $V = 10.0$ V and $f = 5$ kHz. Then, the circuit current is:

$$I(t) = I \sin(\omega t + \phi)$$

where (Equations 83 and 86):

$$I = \frac{V}{Z_{eq}}$$

$$Z_{eq} = \frac{R}{\sqrt{1 + \omega^2 C^2 R^2}}$$

and (Equation 84):

$$\tan \phi = \omega C R$$

Substituting values for Z_{eq} gives:

$$Z_{eq} = \frac{2.2 \text{ k}\Omega}{\sqrt{1 + (6.28 \times 5 \text{ kHz} \times 0.02 \, \mu\text{F} \times 2.2 \text{ k}\Omega)^2}}$$

$$= \frac{2.2 \text{ k}\Omega}{\sqrt{1 + (1.38)^2}} = 1.3 \text{ k}\Omega$$

Thus the amplitude of the circuit current is:

$$I = \frac{V}{Z_{eq}} = \frac{10.0 \text{ V}}{1.3 \text{ k}\Omega} = \textbf{7.7 mA}$$

Also, the phase angle is:

$$\tan \phi = 6.28 \times 5 \text{ kHz} \times 0.02 \, \mu\text{F}$$
$$\times 2.2 \text{ k}\Omega$$
$$= 1.38$$
$$\phi = \textbf{54}°$$

The amplitudes of the resistor and capacitor currents are given by Equations 74 and 76:

$$I_R = \frac{V}{R} = \frac{10.0 \text{ V}}{2.2 \text{ k}\Omega} = \textbf{4.5 mA}$$

$$I_C = \frac{V}{Z_C} = V\omega C$$
$$= 10.0 \text{ V} \times 6.28 \times 5 \text{ kHz} \times 0.02 \, \mu\text{F}$$
$$= \textbf{6.3 mA}$$

These results are shown in Figure 17.28.

EXAMPLE 14: A 10 kHz signal voltage has a current amplitude of 5.0 mA. What values of parallel resistor and capacitor will yield a current amplitude of 3.0 mA through the resistor? What will be the phase shift between the circuit current and voltage?

Solution: The values of resistor and capacitor can be found from the current divider relations for the resistor current:

$$\frac{I_R}{I} = \frac{1}{\sqrt{1 + \omega^2 C^2 R^2}} = \frac{3.0 \text{ mA}}{5.0 \text{ mA}} = 0.60$$

Squaring and inverting both sides gives:

$$1 + \omega^2 C^2 R^2 = \frac{1}{(0.60)^2}$$

$$\omega^2 C^2 R^2 = 2.8$$

$$R^2 = \frac{2.8}{\omega^2 C^2}$$

We are free to select a value for either R or

EXAMPLE 13

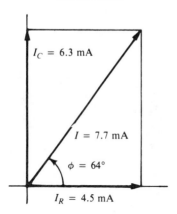

FIGURE 17.28. Shown is the geometrical representation of the voltage amplitude and phase for a typical *RC* circuit.

INDUCTORS AND SYMBOL

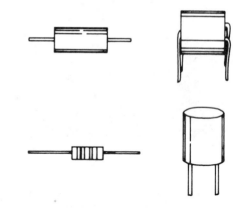

(a) Typical Inductors

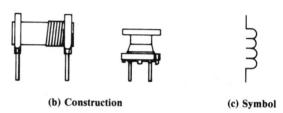

(b) Construction   **(c) Symbol**

FIGURE 17.29. Another frequency-dependent circuit component is the inductor. An inductor is a simple coil of wire that has frequency-dependent properties much like those of a capacitor. The ac properties of an inductor are characterized by its inductance *L*, which has units of henrys (H).

C and determine the other from the above equation. Let us select:

$$C = 0.010 \ \mu F$$

Substitution gives:

$$R^2 = \frac{2.8}{(6.28 \times 10 \text{ kHz} \times 0.01 \ \mu F)^2}$$

$$= \frac{2.8}{0.39} = 7.2$$

$$R = 2.7 \text{ k}\Omega$$

The phase angle is given by Equation 84:

$$\tan \phi = \omega CR$$
$$= 6.28 \times 10 \text{ kHz} \times 0.01 \ \mu F$$
$$\times 2.7 \text{ k}\Omega$$
$$= 1.70$$
$$\phi = 59°$$

17.9 INDUCTORS

One other frequency-dependent component is used in ac circuit design: the inductor. An **in-**

ductor is simply a coil of wire, but as a coil, it has electromagnetic properties that cause its ac behavior to vary with frequency. The electromagnetic properties of a coil are characterized by its inductance *L* and measured in **henrys** (H). Figure 17.29 shows typical inductors, along with the electronic symbol for an inductor.

From a practical standpoint, inductors are rarely used in modern ac circuit design (except in radio-frequency circuits that are beyond the scope of this book). They are relatively large and expensive, and they cannot be accurately manufactured over a wide range of inductance values. Furthermore, by using special circuits call *gyrators*, capacitors can be made to behave like inductors, so real inductors are rarely encountered. Nevertheless, inductors are basic circuit elements, and their ac characteristics are described briefly here.

Impedance of an Inductor

Like that of a capacitor, the effective impedance of an inductor Z_L varies with the frequency of the sine wave voltage applied to it. The magnitude of Z_L is expressed by:

$$Z_L = \omega L \tag{93}$$

If ω is expressed in kilohertz (kHz) and L in henrys (H), then Z_L will have units of kilohms (kΩ), as is the case for capacitive impedance.

EXAMPLE 15: What is the impedance of a 100 mH inductor at frequencies of 500 Hz and 5000 Hz?

Solution: The magnitude of the impedance is given by Equation 93:

$$Z_L = \omega L$$

We will use the consistent set of units of kilohms, kilohertz, and henrys, so for $f = 0.50$ kHz:

$$Z_L = 6.28 \times 50 \text{ kHz} \times 0.10 \text{ H}$$
$$= 0.31 \text{ k}\Omega = \textbf{310 } \Omega$$

For $f = 5.0$ kHz:

$$Z_L = 6.28 \times 50 \text{ kHz} \times 0.10 \text{ H} = \textbf{3.1 k}\Omega$$

The variation with frequency of a 0.1 H inductor over a range of frequencies is shown in Figure 17.30. Compare this graph with Figure 17.14 for a capacitor.

EXAMPLE 16: At what frequency will the impedance of a 50 mH inductor have the same magnitude as a 0.10 μF capacitor?

Solution: Equating the expressions for the impedance of inductors and capacitors (Equations 93 and 29) gives:

$$Z_L = Z_C$$
$$\omega L = \frac{1}{\omega C}$$
$$\omega^2 = \frac{1}{LC}$$
$$f^2 = \frac{1}{(6.28)^2 LC}$$

Substituting values yields:

$$f^2 = \frac{1}{(6.28)^2 \times 0.05 \text{ H} \times 0.1 \text{ } \mu\text{F}} = 5.07$$
$$f = \textbf{2.3 kHz}$$

Phase of Inductor Voltage

Just as a capacitor has a phase shift between its voltage and current, so does an inductor, but in the opposite direction. That is, if the current in an inductor is given by:

$$I_L(t) = I_L \sin \omega t \tag{94}$$

then, the inductor voltage must *lead* the current by a phase angle of $\pi/2$. Thus, the inductor voltage is:

$$V_L(t) = V_L \sin\left(\omega t + \frac{\pi}{2}\right) \tag{95}$$

which, by Equation 7, reduces to:

$$V_L(t) = V_L \cos \omega t \tag{96}$$
$$\text{where } V_L = I_L Z_L \tag{97}$$

FIGURE 17.30. The effective impedance of an inductor increases with frequency according to $Z_L = \omega L$. If frequency is measured in kilohertz and inductance in henrys, then the inductive impedance has units of kilohms. The figure shows the impedance of a 0.1 H inductor over a range of frequencies.

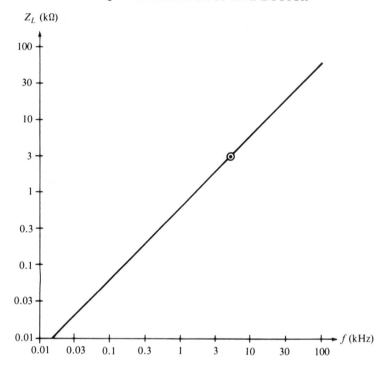

FREQUENCY RESPONSE OF AN INDUCTOR

These results can also be derived by calculus in the following way. The voltage across an inductor is proportional to the time rate of change of the current, where the constant of proportionality is the inductance L. Expressed mathematically:

$$V_L(t) = L\left(\frac{dI(t)}{dt}\right) \tag{98}$$

This expression that is the basic relation between current and voltage for an inductor is like Ohm's law for a resistor and Equation 20 for a capacitor. If the current is given by Equation 94, then we have:

$$V_L(t) = L\left(\frac{d}{dt}\right)(I_L \sin \omega t)$$

Performing the differentiation gives:

$$V_L(t) = \omega L I_L \cos \omega t \tag{99}$$

or

$$V_L(t) = V_L \cos \omega t \tag{100}$$

where the voltage amplitude V_L is defined as:

$$V_L = \omega L I_L \tag{101}$$

or

$$V_L = Z_L I_L \qquad (102)$$

and where the inductive impedance Z_L is defined as:

$$Z_L = \omega L \qquad (103)$$

These results are identical to the results given earlier.

Figure 17.31 shows the relationship between inductor voltage and current. Note that the inductor voltage *leads* the inductor current by $\pi/2$ radians. Compare with Figures 17.15 and 17.21 for a capacitor.

Figure 17.32 summarizes capacitor and inductor behavior. Take a moment to compare the two columns. Series and parallel combinations of an inductor with a resistor or with a capacitor yield a variety of frequency-dependent voltage and current dividers. The phase shifts between circuit current and voltage can vary over the full range from $-\pi/2$ to $+\pi/2$. The analysis of these circuits is identical to that given for series and parallel combinations of a resistor and a capacitor. Some of the possibilities will be encountered in the problems at the end of this section. But, since the inductor is encountered only in special circumstances, such as radio-frequency applications, the topic is not pursued further here.

PHASE OF INDUCTOR VOLTAGE

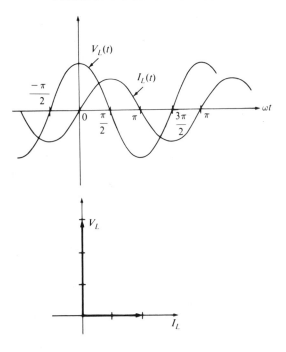

FIGURE 17.31. In contrast to the voltage across a capacitor, the voltage across an inductor leads its current by 90° or $\pi/2$ radians. The figure shows typical waveforms and their geometrical representation. Compare with Figures 17.15 and 17.21 for a capacitor.

FIGURE 17.32. The table compares the characteristics of capacitors and inductors. Their voltage–current behavior is opposite. A capacitor's voltage lags behind its current, whereas an inductor's voltage leads its current. Various combinations of R, L, and C can produce phase shifts from $-\pi/2$ to $+\pi/2$.

CAPACITOR-INDUCTOR COMPARISON

Quantity	Capacitor	Inductor
Impedance	$Z_C = \dfrac{1}{\omega C}$	$Z_L = \omega L$
Phase angle	$\phi_C = \dfrac{-\pi}{2}$	$\phi_L = \dfrac{\pi}{2}$
V-I relation	$I_C(t) = C\left(\dfrac{dV_C(t)}{dt}\right)$	$I_L = L\left(\dfrac{dV_L(t)}{dt}\right)$
Current	$I_C(t) = I_C \sin \omega t$	$I_L(t) = I_L \sin \omega t$
Voltage	$V_C(t) = V_C \sin\left(\omega t - \dfrac{\pi}{2}\right)$ $= -V_C \cos \omega t$	$V_L(t) = V_L \sin\left(\omega t + \dfrac{\pi}{2}\right)$ $= V_L \cos \omega t$

17.10 QUESTIONS AND PROBLEMS

1. Write the mathematical expression for and graph the following sinusoidal voltages and currents. For simplicity, use radians for the time axis.
   (a) $V = 10$ V, $f = 1$ kHz, $\phi = 0$
   (b) $V = 8.6$ mV, $f = 1.5$ kHz, $\phi = +90°$
   (c) $I = 25$ mA, $f = 25$ kHz, $\phi = -\pi/2$ radians

2. What is the sum of a sine wave current with a peak amplitude of 9.2 V and a frequency of 15 kHz and a cosine current of 12.5 V at the same frequency? Draw a geometrical diagram of the addition.

3. Repeat Problem 2 for a sine wave voltage with a peak amplitude of 8.5 mA and a cosine voltage with a peak amplitude of 4.0 mA, both at a frequency of 50 kHz.

4. Find the impedance of a 0.50 μF capacitor at the following frequencies: (a) 20 kHz, (b) 5000 Hz, and (c) 0.050 MHz.

5. At what frequency is the impedance of a 0.50 μF capacitor equal to that of the following resistors: (a) 22 kΩ, (b) 500 Ω, and (c) 0.22 MΩ?

6. What is the current in a 0.050 μF capacitor for an applied sinusoidal voltage with a peak amplitude of 4.5 V and a frequency of 5.0 kHz?

7. What is the voltage across a 0.15 μF capacitor if the current is sinusoidal with a peak amplitude of 0.10 A and a frequency of 900 Hz?

8. Find the capacitance that will yield a current amplitude of 25 mA for the following applied voltages:
   (a) $V(t) = 10$ V sin 1.5 kHz $\times t$
   (b) $V(t) = 200$ mV sin 800 Hz $\times t$
   (c) $V(t) = 6.5$ V cos 25 kHz $\times t$

9. Find the equivalent impedance of a series RC network at a frequency of 25 kHz for the following values of R and C:
   (a) $R = 25$ kΩ and $C = 0.050$ μF
   (b) $R = 25$ kΩ and $C = 5000$ pF
   (c) $R = 2500$ Ω and $C = 0.050$ μF

10. What is the equivalent impedance of a series RC network where $R = 4.7$ kΩ and $C = 0.035$ μF at the following frequencies? Graph the result and show the impedance of the resistor and capacitor alone.

f
0.10 kHz
0.30 kHz
1.0 kHz
3.0 kHz
10. kHz
30. kHz
100. kHz

11. What are the amplitudes of the resistor voltage, capacitor voltage, and the circuit current, and what is the phase angle between the circuit current and voltage for the circuit shown in Figure 17.33? Draw a geometrical representation of the result.

FIGURE 17.33.

12. A signal voltage has a frequency of 500 Hz and an amplitude of 20 mV. Design a circuit that will drop the voltage to an amplitude of 7 mV. What is the phase angle of the voltage?

13. For Problem 12, design a circuit that has a phase angle of $-54°$.

14. Find the equivalent impedance of a parallel RC network at a frequency of 25 kHz for the following values of R and C:
   (a) $R = 25$ kΩ and $C = 0.050$ μF
   (b) $R = 25$ kΩ and $C = 5000$ pF
   (c) $R = 2500$ Ω and $C = 0.050$ μF

15. What is the equivalent impedance of a parallel RC network where $R = 4.7$ kΩ and $C = 0.035$ μF at the following frequencies? Graph the result and show the impedance of the resistor and capacitor alone.

f
0.10 kHz
0.30 kHz
1.0 kHz
3.0 kHz
10. kHz
30. kHz
100. kHz

16. What are the amplitudes of the resistor current, capacitor current, and the net circuit current, and what is the phase angle between the circuit current and voltage for the circuit shown in Figure 17.34? Draw a geometrical representation of the result.

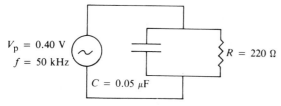

$V_p = 0.40$ V
$f = 50$ kHz

$C = 0.05$ μF

$R = 220$ Ω

FIGURE 17.34.

17. A 2.0 kHz signal voltage has a voltage amplitude of 300 mV. What values of parallel resistor and capacitor will yield a current amplitude of 0.20 mA through the capacitor? What will be the phase shift between the circuit current and voltage?

18. For Problem 17, design a circuit that will yield a capacitor current of 0.05 mA through the resistor and a phase shift of 30° between the circuit current and voltage.

19. Find the impedance of a 5 mH inductor at the following frequencies: (a) 0.070 MHz, (b) 30 kHz, and (c) 700 Hz.

20. Determine the frequency at which the impedance of a 300 mH inductor is equal to that of the following components: (a) $C = 0.020$ μF, (b) $R = 10$ kΩ, and (c) $Z = 300$ Ω.

Chapter

18

ANALYZING ALTERNATING CURRENT NETWORKS

18.1 OBJECTIVES

Following the completion of Chapter 18, you should be able to:

1. Describe the general strategy for analyzing ac circuits and identify the quantities to be determined.
2. Define the quantity s and properly express capacitive and inductive impedance in s-notation.
3. Evaluate expressions involving s and s^2 to get their magnitude and phase angle.
4. Draw a vector diagram of a generalized impedance showing the magnitude and phase angle of the net impedance and its component parts.
5. Analyze any simple series or parallel combinations of resistor, capacitor, and/or inductor (RC, RL, or LC) to find: (a) the equivalent impedance Z_{eq}, (b) the net circuit current I, (c) the voltage across and current through each component, and (d) the phase angle for each quantity.

6. Analyze any three impedance combinations of resistor, capacitor, and/or inductor to find: (a) the equivalent impedance Z_{eq}, (b) the net circuit current I, (c) the voltage across and current through each component, and (d) the phase angle for each quantity.
7. Draw a full Wien bridge circuit and express the special characteristics of its center voltage.
8. Draw an impedance bridge circuit and show how it can be used to measure an unknown capacitance, including the balance equations.
9. Identify Maxwell and Hay bridges by their circuit diagrams and explain their purpose and basic principles of operation.
10. Describe how the basic network theorems—Thevenin's theorem, superposition, and Norton's theorem—can be used to solve ac circuit problems.
11. Describe how to measure the frequency and

phase of an unknown ac signal using a dual-trace oscilloscope and a calibrated sine wave generator.

12. Describe how to generate a Lissajous figure on an oscilloscope and graph the figures that result from sine waves having the same frequency but different amplitudes and phase.

13. Measure the frequency and phase of an unknown signal, given the resulting Lissajous pattern (and rate of rotation, if any) and the frequency of the known sine wave input.

18.2 GENERAL STRATEGY

Chapter 17 looked in some detail at a very simple ac circuit: the combination of a single resistor and a single capacitor. The detailed analysis provided the basis for understanding the general behavior of ac circuits, particularly the importance of phase in determining the voltage and current for a particular circuit element. To analyze more complex circuits, simpler techniques are needed than those discussed in Chapter 17. These techniques are described here.

Let us begin with an outline of the information that we want from the analysis of an ac circuit. Recall from Part II that the goal is usually to determine the voltage across (or current through) a particular circuit component for a given input signal V_{in}. The component of interest is termed the *load* and is assumed to be connected to the output terminals. Hence, the normally desired quantities are the output circuit voltage V_o and current I_o at a single pair of terminals.

In ac circuit analysis, the load may be resistive, capacitive, or inductive, or it may be a combination of *R, C,* and *L*. Therefore, we must assume that the load is some generalized impedance *Z*. For this load impedance, we must

also determine the phase of the output voltage relative both to the input voltage and to the output current.

The general strategy for analyzing ac circuits is identical to that described for dc circuits except that we want to determine the phase characteristics (Figure 18.1). Thus, the general approach is to:

1. Determine the net equivalent impedance of the circuit Z_{eq}.

2. Use the equivalent impedance to determine the net circuit current $I(t)$ for a given voltage $V(t)$. Note that this determination also includes their relative phase angle ϕ.

3. Use the net circuit current to determine the voltage across and current through any circuit component, as well as the phase angle.

Recall that all of the circuit voltages and currents are sine waves and that sine waves are completely described by three quantities:

— Peak amplitude: V_p and I_p
— Frequency: f or ω
— Phase angle: ϕ

Therefore, we would like the analysis to deal only with these three quantities and not have to write repeatedly "sin($\omega t + \phi$)" for every algebraic manipulation. We would also like the mathematical techniques to be similar to those used for dc circuits so that we don't have to learn a whole new circuit analysis methodology.

18.3 s-NOTATION

The first simplification introduced is a mathematical technique to account for the phase information. Chapter 15 stated that currents, voltages, and impedances do not simply add, as they do in dc circuits. For example, in determining the equivalent impedance of a series resistor

FIGURE 18.1. When analyzing complex ac circuits, the general task is to determine the voltage and current for an output impedance Z, where Z can be a combination of R, C, and L. To simplify the analysis, we want to deal only with the independent parameters that describe these quantities — peak amplitude, frequency, and phase angle — and use methods similar to those learned for dc circuit analysis.

ALTERNATING CURRENT CIRCUIT ANALYSIS

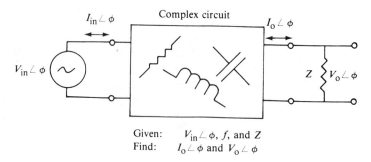

Given: $V_{in} \angle \phi$, f, and Z
Find: $I_0 \angle \phi$ and $V_0 \angle \phi$

and a series capacitor, it was found that their amplitudes added as squares:

$$|Z_{eq}|^2 = R^2 + |Z_C|^2 \tag{1}$$

This expression includes the absolute value symbol | | as a reminder that impedances have phase characteristics associated with them. A resistor has no phase angle and does not require an absolute value symbol, but a capacitor introduces a phase angle of $-90°$, and an inductor, a phase angle of $+90°$, so for these quantities, the symbol is necessary.

Defining s

To account for the phase characteristics, the quantity s is introduced and defined as follows:

$$s = j\omega = j2\pi f \tag{2}$$

Of course, ω ($= 2\pi f$) is the frequency. The quantity j is the mathematical quantity:

$$j = \sqrt{-1} \tag{3}$$

The valuable property of this quantity is that:

$$j^2 = -1 \tag{4}$$

Thus, the quantity s^2 becomes:

$$s^2 = s \times s = j\omega \times j\omega = j^2\omega^2$$

Substituting for j^2 gives:

$$s^2 = -\omega^2 \tag{5}$$

which is the negative of the normal squared quantity. The importance of this property will become evident as we go along.

Using the quantity s, we can define capacitive impedance Z_C and inductive impedance Z_L as follows:

$$Z_C = \frac{1}{sC} = \frac{1}{j\omega L} \tag{6}$$

$$Z_L = sL = j\omega L \tag{7}$$

Evaluating Expressions with s

The use of s allows us to retain the phase information of circuit components during the algebraic manipulations of the circuit analysis. We can regain the information in the final result by using Equation 5 to evaluate quantities involving s^2 and by using the following mathematical rule for evaluating quantities involving s:

$$|A \pm sB| = \sqrt{A^2 + (\omega B)^2} \qquad (8)$$

where the phase angle between A and B is given by:

$$\tan \phi = \pm \frac{\omega B}{A} \qquad (9)$$

Symbolically, the phase information is combined with the magnitude by writing:

$$A \pm sB = \sqrt{A^2 + (\omega B)^2} \angle \phi \qquad (10)$$

where $\angle \phi$ indicates a phase angle between A and B determined by Equation 9.

Therefore, the strategy is to treat terms with s and s^2 as "special" quantities distinct from those without s. In the final result, we can collect terms with s and s^2 and evaluate them using the rules of Equations 5, 8, and 9 to obtain the final magnitudes and phase angles.

Two other rules also help to simplify the final result. When multiplying and dividing expressions of the form of Equation 10, we can account for the phase information with the following rules:

$$C \angle \phi_1 \times D \angle \phi_2$$
$$= CD \angle (\phi_1 + \phi_2) \qquad (11)$$
$$\frac{C \angle \phi_1}{D \angle \phi_2} = \frac{C}{D} \angle (\phi_1 - \phi_2) \qquad (12)$$

Equation 11 states that when quantities with phase characteristics are multiplied, their product is the product of the magnitudes with a phase angle that is the *sum* of the individual phase angles.

Equation 12 states that when quantities with phase characteristics are divided, the result

s-NOTATION

Definitions
$s = j\omega = \sqrt{-1}\ \omega$
$Z_C = \dfrac{1}{sC}$
$Z_L = sL$

Rules for Evaluating s
$s^2 = -\omega^2$
$A \pm sB = \sqrt{A^2 + (\omega B)^2} \angle \phi$
where $\tan \phi = \pm \dfrac{\omega B}{A}$
$C \angle \phi_1 \times D \angle \phi_2 = CD \angle (\phi_1 + \phi_2)$
$\dfrac{C \angle \phi_1}{D \angle \phi_2} = \dfrac{C}{D} \angle (\phi_1 - \phi_2)$

FIGURE 18.2.   To simplify ac circuit analysis, the symbol s is used to keep track of the phase characteristics of capacitors and inductors. When the analysis is complete, the terms involving s and s^2 can be evaluated according to the rules shown to get the final peak amplitudes and phase angles.

is the division of the magnitudes with a phase angle that is the *difference* between the phase of the numerator and the phase of the denominator. A summary of s-notation is provided in Figure 18.2.

Capacitive and Inductive Impedance

To understand the meaning of s, let us evaluate the expressions for capacitive impedance and inductive impedance, Equations 6 and 7. Using Equation 6, capacitive impedance is:

$$Z_C = \frac{1}{sC} = \frac{1}{0 + sC}$$

Applying the rule of Equations 8 and 9 to the denominator gives:

$$Z_C = \frac{1}{\sqrt{0^2 + (\omega C)^2} \angle \phi}$$

$$= \frac{1}{\omega C \angle \phi} \qquad (13)$$

where $\quad \tan \phi = \dfrac{\omega C}{0} = \infty \qquad (14)$

Therefore, $\phi = 90°$ and Equation 13 can be written:

$$Z_C = \frac{1}{\omega C \angle 90°} \qquad (15)$$

Applying the rule of Equation 12 to bring the phase angle into the numerator gives the final result:

$$Z_C = \frac{1}{\omega C} \angle -90° \qquad (16)$$

This expression says that a purely capacitive impedance has a phase angle of $-90°$, which we know from Chapter 17 to be true.

Similarly, using Equation 7, inductive impedance is:

$$Z_L = sL = 0 + sL$$

Applying Equation 8 gives:

$$\begin{aligned} Z_L &= \sqrt{0^2 + (\omega L)^2} \angle \phi \\ &= \omega L \angle \phi \end{aligned} \qquad (17)$$

where $\quad \tan \phi = \dfrac{\omega L}{0} = \infty \qquad (18)$

Therefore, $\phi = 90°$ and Equation 17 can be written:

$$Z_L = \omega L \angle 90° \qquad (19)$$

This expression says that a purely inductive impedance has a phase angle of $90°$, which we also know from Chapter 17 to be true.

IMPEDANCE REPRESENTATION

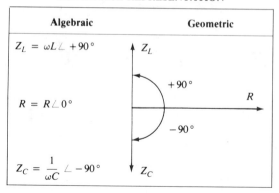

FIGURE 18.3. Using s-notation, inductive impedance is represented with a phase angle of $90°$ and capacitive impedance, with a phase angle of $-90°$. These quantities can be represented geometrically as vectors whose length is equal to the magnitude of the impedance at an angle equal to the phase angle.

These results can be represented geometrically, as shown in Figure 18.3. A pure capacitance appears as a vector pointing downward $(-90°)$ with a length equal to its magnitude $1/\omega C$. A pure resistance is a vector pointing to the right $(0°)$ with a length equal to its magnitude R. A pure inductance is a vector pointing upward $(90°)$ with a length equal to its magnitude ωL.

A generalized impedance consisting of some combination of R, C, and L has the form:

$$Z = (A \pm sB) \angle \phi \qquad (20)$$

The term without s—that is, A—is the resistance component, and the term with s—that is, B—is the impedance component, including capacitors and/or inductors. If the sign preceding s is negative, Equation 9 indicates that ϕ is negative, so the net equivalent impedance component must be capacitive. On the other hand, if the sign preceding s is positive, ϕ must be positive and the net equivalent impedance component must be inductive.

GENERALIZED IMPEDANCE

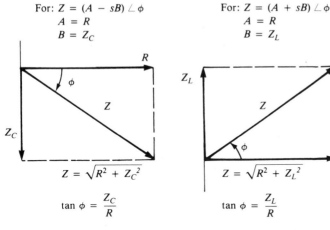

For: $Z = (A - sB) \angle \phi$
$A = R$
$B = Z_C$

$$Z = \sqrt{R^2 + Z_C{}^2}$$

$$\tan \phi = \frac{Z_C}{R}$$

(a) Net Capacitive Impedance

For: $Z = (A + sB) \angle \phi$
$A = R$
$B = Z_L$

$$Z = \sqrt{R^2 + Z_L{}^2}$$

$$\tan \phi = \frac{Z_L}{R}$$

(b) Net Inductive Impedance

FIGURE 18.4. A generalized impedance Z consists of a combination of R, C, and L and will have the form $Z = (A \pm sB) \angle \phi$. The A term is the net resistance component and B is the net impedance component. If the sign preceding B is $-$, then B is capacitive, Z_C. If it is $+$, then B is inductive, Z_L. The geometrical representations will help in remembering the equations for evaluating expressions with s.

A generalized impedance Z, therefore, is a vector at the angle ϕ whose magnitude is given by Equation 8. Equation 8 is simply the vector sum of a resistive vector of magnitude A at $0°$ and an impedance vector of magnitude ωB at either $-90°$ or $90°$ depending on the sign of s. Vector representation is shown in Figure 18.4 for a capacitive impedance (Figure 18.4a) and an inductive impedance (Figure 18.4b).

The geometrical representation shown in Figure 18.4 provides a good way of remembering Equations 8 and 9. The magnitude equation, Equation 8, is simply the application of the Pythagorean theorem:

$$Z = \sqrt{R^2 + Z_C^2} \tag{21}$$

or

$$Z = \sqrt{R^2 + Z_L^2}$$

and the phase angle ϕ is simply:

$$\tan \phi = \frac{-Z_C}{R}$$

or

$$\tan \phi = \frac{Z_L}{R} \tag{22}$$

where R is the resistive component of the impedance diagram, Z_C is the capacitive component at $-90°$, and Z_L is the inductive component at $90°$. Using s-notation and following the mathematical rules given above, ac circuits can be analyzed by the same methods used in Part II.

18.4 SERIES AND PARALLEL NETWORKS

Parallel RC Networks

The first example of the use of s-notation solves the parallel RC network problem analyzed in Chapter 17 using sine functions. We will then compare the results and, clearly, they should be the same.

The parallel RC network to be solved is shown in Figure 18.5. Using s-notation (Equation 6) we write the impedances of the two components:

$$R = R$$

$$Z_C = \frac{1}{sC}$$

FIGURE 18.5. In analyzing a parallel *RC* network, first find the equivalent impedance of the original current using *s*-notation (a). Then, Z_{eq} is used to determine the net circuit current *I* (b). The *s*-expression for *I* then gives the current divider relation directly (c).

PARALLEL *RC* NETWORK

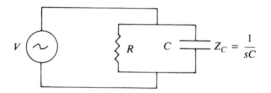

(a) Original Circuit

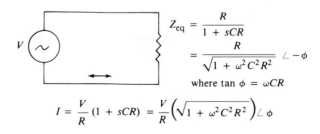

$$Z_{eq} = \frac{R}{1 + sCR}$$

$$= \frac{R}{\sqrt{1 + \omega^2 C^2 R^2}} \angle -\phi$$

where $\tan \phi = \omega CR$

$$I = \frac{V}{R}(1 + sCR) = \frac{V}{R}\left(\sqrt{1 + \omega^2 C^2 R^2}\right)\angle \phi$$

(b) Equivalent Circuit

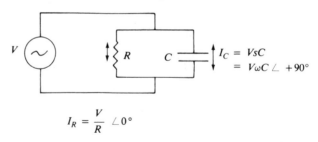

$$I_C = VsC$$
$$= V\omega C \angle +90°$$

$$I_R = \frac{V}{R} \angle 0°$$

(c) Current Divider Relations

The first step is to find the equivalent impedance of the circuit Z_{eq}. As discussed in Part II, for a parallel combination of components, the impedance was found to be:

$$\frac{1}{Z_{eq}} = \frac{1}{R} + \frac{1}{Z_C} = \frac{R + Z_C}{RZ_C} \qquad (23)$$

$$Z_{eq} = \frac{RZ_C}{R + Z_C} \qquad (24)$$

We can substitute in the values for R and Z_C given above:

$$Z_{eq} = \frac{R \times (1/sC)}{R + (1/sC)} = \frac{R}{1 + sCR} \qquad (25)$$

To evaluate this expression, the addition rule of Equation 8 is applied to the denominator:

$$Z_{eq} = \frac{R}{\sqrt{1 + \omega^2 C^2 R^2} \angle \phi} \qquad (26)$$

The phase angle in the denominator can be moved to the numerator according to the rule of Equation 12 so that:

$$Z_{eq} = \frac{R}{\sqrt{1 + \omega^2 C^2 R^2}} \angle -\phi \qquad (27)$$

where $\tan \phi = \omega CR$ $\qquad (28)$

This result is identical to the result in Equation 86 of Chapter 17. The phase angle is negative, indicating that the equivalent impedance is capacitive, as we would expect it to be.

The next step is to find the circuit current. Using Ohm's law, we have:

$$I = \frac{V}{Z_{eq}} \tag{29}$$

where V is the peak amplitude of the applied voltage and I is the peak amplitude of the net current. Substitution of Equation 25 for Z_{eq} gives:

$$I = \frac{V(1 + sCR)}{R} \tag{30}$$

which can be evaluated by applying Equation 8 to the term in the numerator:

$$I = \frac{V}{R} \sqrt{1 + \omega^2 C^2 R^2} \angle \phi \tag{31}$$

where $\tan \phi = \omega CR \tag{32}$

Compare these results with Equations 83 and 84 of Chapter 17.

Note in Equation 32 that ϕ is a positive angle, which means that the circuit current I *leads* the applied voltage V by the angle ϕ. Equivalently, if the current is assumed to have a phase angle of $0°$, the voltage V will *lag* I by the angle ϕ.

The results of Equation 30 can be looked at another way. Separating the s-notation and non-s factors in Equation 30 gives:

$$I = \frac{V}{R} + VsC \tag{33}$$

In this form, the first factor contains no s, so it must be resistive, and it is the current in the resistor:

$$I_R = \frac{V}{R} \angle 0° \tag{34}$$

The second factor does contain an s and is clearly the current in the capacitor:

$$I_C = VsC \tag{35}$$

Applying Equations 8 and 9 to Equation 35 gives:

$$I_C = V\omega C \angle 90° \tag{36}$$

The capacitor current leads the resistor current by $90°$. Compare these current divider relations with Equations 74 and 76 of Chapter 17.

Series *RC* Networks

For a series *RC* circuit (Figure 18.6), the equivalent impedance is simply the sum of the two impedances:

$$Z_{eq} = R + Z_C = R + \frac{1}{sC} \tag{37}$$

$$Z_{eq} = \frac{1 + sCR}{sC} \tag{38}$$

Applying the addition rule, Equation 8, to the numerator and denominator gives:

$$Z_{eq} = \frac{\sqrt{1 + \omega^2 C^2 R^2} \angle \phi_1}{\omega C \angle \phi_2} \tag{39}$$

where $\tan \phi_1 = \omega CR \tag{40}$

$\tan \phi_2 = \infty \tag{41}$

$\phi_2 = \frac{\pi}{2} \tag{42}$

Applying the division rule, Equation 12, gives:

$$Z_{eq} = \frac{\sqrt{1 + \omega^2 C^2 R^2}}{\omega C} \angle (\phi_1 - \phi_2)$$

$$= R\sqrt{1 + \frac{1}{\omega^2 C^2 R^2}} \angle \left(\phi_1 - \frac{\pi}{2}\right) \tag{43}$$

The magnitude of Z_{eq} corresponds to the results in Equation 57 of Chapter 17. The phase

FIGURE 18.6. The figure shows the stages in analyzing a series *RC* network. The procedure is the same as for a parallel *RC* network.

SERIES *RC* NETWORK

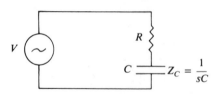

(a) Original Circuit

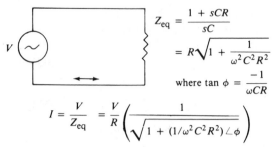

$$Z_{eq} = \frac{1 + sCR}{sC}$$

$$= R\sqrt{1 + \frac{1}{\omega^2 C^2 R^2}}$$

where $\tan \phi = \dfrac{-1}{\omega CR}$

$$I = \frac{V}{Z_{eq}} = \frac{V}{R}\left(\frac{1}{\sqrt{1 + (1/\omega^2 C^2 R^2)} \angle \phi}\right)$$

(b) Equivalent Circuit

$$V_R = V\left(\frac{R}{Z_{eq}}\right) = V\left(\frac{sCR}{1 + sCR}\right)$$

$$V_C = V\left(\frac{Z_C}{Z_{eq}}\right) = V\left(\frac{1}{1 + sCR}\right)$$

(c) Voltage Divider Relations

angle, however, requires some interpretation. Trigonometry tells us that if:

$$\tan \phi = \frac{a}{b}$$

then

$$\tan\left(\phi - \frac{\pi}{2}\right) = \frac{b}{a} \qquad (44)$$

Thus, we can rewrite Equation 43 as:

$$Z_{eq} = R\sqrt{1 + \frac{1}{\omega^2 C^2 R^2}} \angle \phi \qquad (45)$$

where $\tan \phi = \dfrac{-1}{\omega CR}$ $\qquad (46)$

In this form, the phase angle agrees with Equation 52 of Chapter 17.

The circuit current, according to Ohm's law, is:

$$I = \frac{V}{Z_{eq}} = \frac{V}{R}\left(\frac{1}{\sqrt{1 + (1/\omega^2 C^2 R^2)} \angle \phi}\right)$$

$$V = IR\sqrt{1 + \frac{1}{\omega^2 C^2 R^2}} \angle \phi \qquad (47)$$

which agrees with Equation 50 of Chapter 17.

The voltage divider equation can be found from:

$$\frac{V_C}{V} = \frac{Z_C}{Z_{eq}} \qquad (48)$$

$$\frac{V_R}{V} = \frac{R}{Z_{eq}} \qquad (49)$$

Substitution of expressions for Z_C and Z_{eq} gives:

$$\frac{V_C}{V} = \frac{1/sC}{(1 + sCR/sC)}$$

and simplifying gives:

$$\frac{V_C}{V} = \frac{1}{1 + sCR} \qquad (50)$$

and

$$\frac{V_R}{V} = \frac{R}{(1 + sCR)/sC}$$

and simplifying gives:

$$\frac{V_R}{V} = \frac{sCR}{1 + sCR} \qquad (51)$$

Evaluation of these expressions will yield the same results as Equation 61 and 64 of Chapter 17. This exercise is left as a problem.

Series *LC* Networks

While inductors are not common in modern electronic systems, the fact that their characteristics are opposite those of capacitors leads to some interesting behavior in combinations of inductors and capacitors. The following examples involve the simplest of the *LC* combinations: series and parallel connections.

For the series *LC* connection shown in Figure 18.7, we must determine the equivalent impedance. The equivalent impedance is given by:

$$Z_{eq} = Z_C + Z_L \qquad (52)$$

Substitution of Equations 6 and 7 for Z_C and Z_L gives:

$$Z_{eq} = \frac{1}{sC} + sL \qquad (53)$$

or

$$Z_{eq} = \frac{1 + s^2 LC}{sC} \qquad (54)$$

Evaluating this expression using Equations 8 and 9 gives:

$$Z_{eq} = \frac{1 - \omega^2 LC}{\omega C} \angle -90° \qquad (55)$$

This result is shown in Figure 18.7. When Z_C is greater than Z_L (Figure 18.7c), $\omega^2 LC < 1$ and the equivalent impedance is purely capacitive with a phase angle of $-90°$. However, when Z_C is less than Z_L (Figure 18.7d), $\omega^2 LC > 1$ and the equivalent impedance is purely inductive with a phase angle of $90°$. This situation is apparent in Equation 55 upon noting that when $Z_L > Z_C$, $\omega^2 LC > 1$ and the magnitude of Z_{eq} is negative. The sign of Z_{eq} can be made positive by adding $180°$ to the phase angle. This addition changes the angle from $-90°$ to $+90°$, which is the characteristic of a pure inductor.

Therefore, a capacitor and an inductor connected in series tend to cancel each other out. The equivalent impedance will be either a pure capacitor or a pure inductor depending on which impedance is larger. However, note what happens when:

$$\omega^2 LC = 1$$

or

$$\omega = \sqrt{\frac{1}{LC}} \qquad (56)$$

The condition $\omega = \sqrt{1/LC}$ occurs when $Z_C = Z_L$. We then have:

$$Z_{eq} = 0 \qquad (57)$$

FIGURE 18.7. In a series *LC* circuit, the capacitive and inductive impedances tend to cancel each other. When $Z_C > Z_L$, as in (c), Z_{eq} is purely capacitive at a phase angle of $-90°$. When $Z_C < Z_L$, as in (d), Z_{eq} is purely inductive at a phase angle of $90°$. When $Z_C = Z_L$, $Z_{eq} = 0$. The frequency for which $Z_C = Z_L$ is called the natural frequency of the circuit.

SERIES *LC* NETWORK

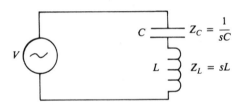

(a) Original Circuit

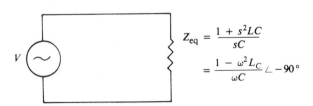

(b) Equivalent Circuit

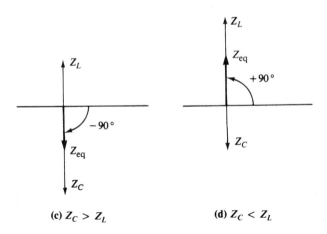

(c) $Z_C > Z_L$ **(d)** $Z_C < Z_L$

Thus, for any series *LC* connector, there is some frequency ω given by Equation 56 for which the equivalent impedance is zero. This frequency is called the **natural frequency** of the circuit and has important applications that are discussed in Chapter 19.

Parallel *LC* Networks

For the parallel *LC* connection shown in Figure 18.8, the equivalent impedance is given by:

$$\frac{1}{Z_{eq}} = \frac{1}{Z_C} + \frac{1}{Z_L} \tag{58}$$

Substitution of Equations 6 and 7 for Z_C and Z_L gives:

$$\frac{1}{Z_{eq}} = sC + \frac{1}{sL}$$

$$\frac{1}{Z_{eq}} = \frac{1 + s^2LC}{sL}$$

$$Z_{eq} = \frac{sL}{1 + s^2LC} \tag{59}$$

PARALLEL *LC* CIRCUIT

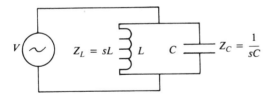

(a) Original Circuit

FIGURE 18.8. The equivalent impedance of a parallel *LC* circuit is opposite to that of a series *LC* circuit. When $Z_C > Z_L$, as in (c), Z_{eq} is purely inductive, and when $Z_C < Z_L$, as in (d), Z_{eq} is purely capacitive. At the natural frequency, when $Z_C = Z_L$, Z_{eq} becomes infinite.

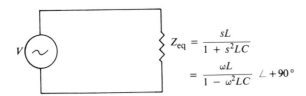

(b) Equivalent Circuit

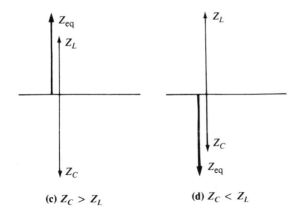

(c) $Z_C > Z_L$ **(d) $Z_C < Z_L$**

Using Equations 8 and 9 to evaluate this expression gives:

$$Z_{eq} = \frac{\omega L}{1 - \omega^2 LC} \angle +90° \tag{60}$$

In this case, when $Z_C > Z_L$, $\omega^2 LC < 1$ and the equivalent impedance is purely inductive—that is, the inductive impedance shunts the capacitive impedance. When $Z_C < Z_L$, $\omega^2 LC > 1$ and the equivalent impedance is purely capacitive—that is, the capacitive impedance shunts the inductor.

Note, however, that at the natural frequency, when:

$$\omega^2 LC = 1$$

or

$$\omega = \sqrt{\frac{1}{LC}} \tag{61}$$

the equivalent impedance becomes infinite:

$$Z_{eq} = \infty$$

This frequency-sensitive characteristic also has important applications that are described in Chapter 19.

It must be pointed out that there is no such thing as a pure inductor because inductors are made of coils of wire and the wire has some resistance. Hence, an inductor is always a series *LR* circuit. Whether or not the resistance of the wire will affect the behavior of the circuit can be determined only by adding it into the analysis. It can be done with the analysis methods developed here, but it is beyond the purpose of this book.

18.5 COMPLEX ALTERNATING CURRENT NETWORKS

As indicated, the techniques for solving ac circuits are the same as for dc circuits, except that terms involving s must be kept separate from those without s. Then, when the algebraic analysis is complete and the final expression for the desired quantity known, terms involving s^2 are evaluated using Equation 5 and terms involving s are evaluated according to Equations 8 and 9 to get the phase angle.

The analysis of complex ac networks is rarely required in practice, so excessive space will not be devoted to illustrating the many possible impedance combinations. Rather, a sampling of typical circuit network solutions is given here, in particular, those with specific properties that are useful for instrumentation applications. Chapter 19 illustrates one of the most common uses of ac networks: filters.

Three-Impedance Combinations

This section begins with analyses of combinations of three-impedance elements. Figure 18.9 shows four ways of connecting three-impedance elements Z_1, Z_2, and Z_3, where any of the three can be a resistor, a capacitor, or an inductor. In the following analyses, the possibilities are limited to resistors and capacitors, which are far more common than inductors.

Also given in Figure 18.9 are expressions for the net equivalent impedance Z_{eq} of the combination. These expressions were derived in Chapter 5 for similar combinations of resistors. The expressions are applicable to ac circuits provided that s-notation is properly used. The following examples illustrate the process.

EXAMPLE 1: If the applied ac voltage given in Figure 18.10 has a frequency of 1.5 kHz and a peak amplitude of $V = 15$ V, what is the voltage across each component?

Solution: The net equivalent impedance of the circuit is given by:

$$Z_{eq} = R_1 + Z_C + R_2 \tag{62}$$

where $R_1 = 10$ kΩ
$\qquad C = 0.01$ μF
$\qquad R_2 = 47$ kΩ

Thus:

$$Z_{eq} = R_1 + R_2 + \frac{1}{sC} \tag{63}$$

$$Z_{eq} = \frac{1 + s(R_1 + R_2)C}{sC} \tag{64}$$

The net circuit current is given by:

$$I = \frac{V}{Z_{eq}} \tag{65}$$

$$I = \frac{sVC}{1 + s(R_1 + R_2)C} \tag{66}$$

Thus, the voltage across R_1 is:

$$V_{R_1} = IR_1 \tag{67}$$

$$V_{R_1} = \frac{sVR_1C}{1 + s(R_1 + R_2)C} \tag{68}$$

This expression can be evaluated using Equations 8 and 9:

POSSIBLE COMBINATIONS OF THREE IMPEDANCES

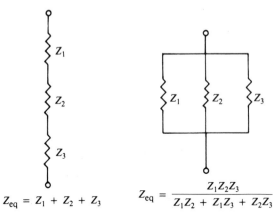

$$Z_{eq} = Z_1 + Z_2 + Z_3$$

(a) Series Connection

$$Z_{eq} = \frac{Z_1 Z_2 Z_3}{Z_1 Z_2 + Z_1 Z_3 + Z_2 Z_3}$$

(b) Parallel Connection

FIGURE 18.9. The techniques for analyzing complex ac circuits are the same as for dc circuits provided that terms involving s are kept separate and evaluated properly. For example, the figure shows four ways of connecting three-impedance elements Z_1, Z_2, and Z_3 and gives the expressions for the equivalent impedance derived in Part II for resistors. Z_{eq} can be directly evaluated using s-notation.

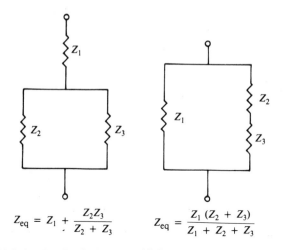

$$Z_{eq} = Z_1 + \frac{Z_2 Z_3}{Z_2 + Z_3}$$

(c) Series–Parallel Connection

$$Z_{eq} = \frac{Z_1 (Z_2 + Z_3)}{Z_1 + Z_2 + Z_3}$$

(d) Parallel–Series Connection

$$V_{R_1} = \frac{\omega V R_1 C \angle \phi_1}{\sqrt{1 + \omega^2 (R_1 + R_2)^2 C^2} \angle \phi_2} \qquad (69)$$

where $\tan \phi_1 = \infty$

$\phi_1 = 90°$

$\tan \phi_2 = \omega (R_1 + R_2) C \qquad (70)$

Substitution of values gives:

$V_{R_1} = \dfrac{6.28 \times 1.5 \times 15 \times 10 \times 0.01 \angle (90° - \phi_2)}{\sqrt{1 + [6.28 \times 1.5 (10 + 47) 0.01]^2}}$

$= \dfrac{13.3 \angle (90° - \phi_2)}{\sqrt{1 + 28.8}}$

$= 2.4 \text{ V} \angle (90° - \phi_2)$

$\tan \phi_2 = 6.28 \times 1.5 \text{ kHz}(10 \text{ k}\Omega + 47 \text{ k}\Omega) 0.01 \text{ μF}$

$= 5.37$

$\phi_2 = 79°$

FIGURE 18.10. The procedure for analyzing three impedances in series is illustrated. Note that to get the final voltage values, the intermediate expressions for Z_{eq} and I need not be evaluated.

EXAMPLE 1

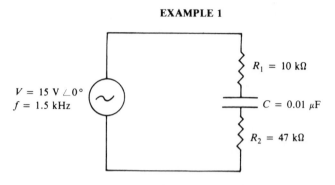

(a) Original Circuit

$V = 15 \text{ V} \angle 0°$
$f = 1.5 \text{ kHz}$

$R_1 = 10 \text{ k}\Omega$

$C = 0.01 \text{ }\mu\text{F}$

$R_2 = 47 \text{ k}\Omega$

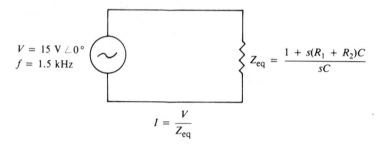

(b) Equivalent Circuit

$V = 15 \text{ V} \angle 0°$
$f = 1.5 \text{ kHz}$

$Z_{eq} = \dfrac{1 + s(R_1 + R_2)C}{sC}$

$I = \dfrac{V}{Z_{eq}}$

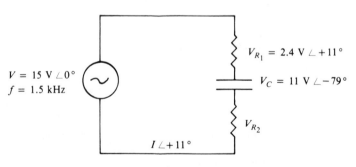

(c) Final Result

$V = 15 \text{ V} \angle 0°$
$f = 1.5 \text{ kHz}$

$V_{R_1} = 2.4 \text{ V} \angle +11°$

$V_C = 11 \text{ V} \angle -79°$

V_{R_2}

$I \angle +11°$

Therefore:

$$V_{R_1} = 2.4 \text{ V} \angle (90° - 79°)$$
$$= 2.4 \text{ V} \angle 11° \qquad (71)$$

We know that the voltage across the resistor is in phase with its current I. Therefore, the current I and the resistor voltage V_{R_1} must lead the applied voltage V by 11°. The sinusoidal expression for V_{R_1} is:

$$V_{R_1}(t) = 2.4 \text{ V} \sin(\omega t + 11°) \qquad (72)$$

The voltage across the capacitor is given by:

$$V_C = IZ_C$$
$$= \frac{sVC}{1 + s(R_1 + R_2)C} \times \frac{1}{sC} \qquad (73)$$
$$V_C = \frac{V}{1 + s(R_1 + R_2)C} \qquad (74)$$

Evaluation in terms of Equations 8 and 9 gives:

$$V_C = \frac{V}{\sqrt{1 + \omega^2(R_1 + R_2)^2 C^2} \angle \phi_2} \qquad (75)$$

where $\tan \phi_2 = \omega(R_1 + R_2)C \qquad (76)$

Substitution of values gives:

$$V_C = \frac{15 \text{ V}}{\sqrt{1 + [6.28 \times 1.5(10 + 47)0.01]^2} \angle \phi_2}$$
$$= \frac{15 \text{ V}}{5.46 \angle 79°} = 11 \text{ V} \angle -79° \qquad (77)$$

Again, we know that the capacitor voltage lags behind its current I by 90° and, from our results, we see that it also lags behind the applied voltage by 79°. Its sinusoidal expression is:

$$V_C = 11 \text{ V} \sin(\omega t - 79°) \qquad (78)$$

Determination of V_{R_2} is similar to that of V_{R_1}.

EXAMPLE 2: If the applied ac voltage shown in Figure 18.11 has a frequency of 1.5 kHz and a peak value of $V = 15$ V, what is the voltage across and current through each component?

Solution: The net equivalent impedance of the circuit is given by:

$$Z_{eq} = R_1 + \frac{R_2 Z_C}{R_2 + Z_C} \qquad (79)$$

where $R_1 = 10 \text{ k}\Omega$
$\qquad R_2 = 47 \text{ k}\Omega$
$\qquad C = 0.01 \text{ } \mu\text{F}$

Thus:

$$Z_{eq} = R_1 + \frac{R_2(1/sC)}{R_2 + (1/sC)}$$
$$= R_1 + \frac{R_2}{1 + sCR_2} \qquad (80)$$
$$Z_{eq} = \frac{R_1 + R_2 + sCR_1R_2}{1 + sCR_2} \qquad (81)$$

The net circuit current is given by:

$$I = \frac{V}{Z_{eq}} \qquad (82)$$
$$I = \frac{V(1 + sCR_2)}{R_1 + R_2 + sCR_1R_2} \qquad (83)$$

Using Equations 8 and 9 to evaluate this expression gives:

$$I = \frac{V\sqrt{1 + \omega^2 C^2 R_2^2} \angle \phi_1}{\sqrt{(R_1 + R_2)^2 + \omega^2 C^2 R_1^2 R_2^2} \angle \phi_2} \qquad (84)$$

where $\tan \phi_1 = \omega C R_2 \qquad (85)$

$$\tan \phi_2 = \frac{\omega C R_1 R_2}{R_1 + R_2} \qquad (86)$$

Substitution of values and reduction gives:

FIGURE 18.11. The procedure for analyzing a series–parallel connection of three impedances is shown. The same techniques can be applied to other complex ac networks.

EXAMPLE 2

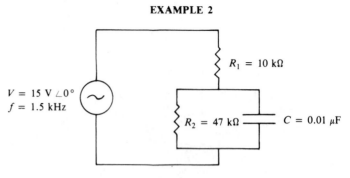

(a) Original Circuit

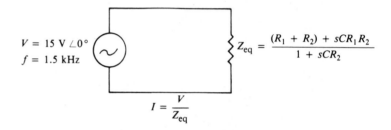

(b) Equivalent Circuit

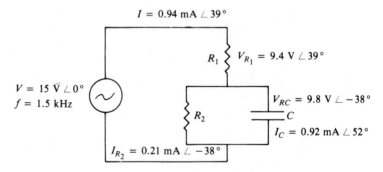

(c) Final Result

$$I = \frac{15\sqrt{1 + 19.6} \angle (\phi_1 - \phi_1)}{\sqrt{3249 + 1960}}$$

$$= \frac{15 \times 4.5}{72} \angle (\phi_1 - \phi_2)$$

$$= 0.94 \text{ mA} \angle (\phi_1 - \phi_2)$$

$$\tan \phi_1 = 6.28 \times 1.5 \times 0.01 \times 47 = 4.43$$

$$\phi_1 = 77°$$

$$\tan \phi_2 = \frac{6.28 \times 1.5 \times 0.01 \times 10 \times 47}{10 + 47}$$

$$= \frac{44.3}{57} = 0.78$$

$$\phi_2 = 38°$$

Therefore:

$$I = 0.94 \text{ mA} \angle (77° - 38°)$$

$$= \mathbf{0.94 \text{ mA} \angle 39°} \tag{87}$$

which is also the current through R_1.
The voltage across R_1 is:

$$V_{R_1} = IR_1$$

$$= 0.94 \text{ mA} \angle 39° \times 10 \text{ k}\Omega \tag{88}$$

$$V_{R_1} = \mathbf{9.4 \text{ V} \angle 39°} \tag{89}$$

The voltage across the parallel RC network can be found from:

$$V_{RC} = IZ_{RC} \tag{90}$$

where Z_{RC} is the equivalent impedance of the parallel RC network, as given by Equations 27 and 28:

$$Z_{RC} = \frac{R_2}{\sqrt{1 + \omega^2 C^2 R_2^2}} \angle -\phi \tag{91}$$

where $\quad \tan \phi = \omega C R_2 \tag{92}$

Substitution of values gives:

$$Z_{RC} = \frac{47 \text{ k}\Omega}{\sqrt{1 + (6.28 \times 1.5 \times 0.01 \times 47)^2}} \angle -\phi$$

$$= \frac{47 \text{ k}\Omega}{\sqrt{1 + 19.6}} \angle -\phi$$

$$= 10.4 \text{ k}\Omega \angle -\phi$$

$$\tan \phi = 6.28 \times 1.5 \text{ kHz} \times 0.01 \text{ }\mu\text{F} \times 47 \text{ k}\Omega$$

$$= 4.43$$

$$\phi = 77°$$

Therefore:

$$Z_{RC} = 10.4 \text{ k}\Omega \angle -77° \tag{93}$$

Substitution of the values for I and Z_{RC} into Equation 90 for V_{RC} gives:

$$V_{RC} = 0.94 \text{ mA} \angle 39°$$

$$\times 10.4 \text{ k}\Omega \angle -77°$$

$$= 9.8 \text{ V} \angle (39° - 77°)$$

$$= \mathbf{9.8 \text{ V} \angle -38°}$$

which is the voltage across both R_2 and C.
The current through R_2 is given by:

$$I_{R_2} = \frac{V_{RC}}{R_2} = \frac{9.8 \text{ V} \angle -38°}{47 \text{ k}\Omega} \tag{94}$$

$$I_{R_2} = \mathbf{0.21 \text{ mA} \angle -38°} \tag{95}$$

The current through C is given by:

$$I_C = \frac{V_{RC}}{Z_C} = \frac{9.8 \text{ V} \angle -38°}{1/\omega C \angle -90°}$$

$$= 9.8 \text{ V} \angle -38° \text{ } \omega C \angle +90° \tag{96}$$

Substitution of values gives:

$$I_C = 9.8 \text{ V} \times 6.28 \times 1.5 \text{ kHz}$$

$$\times 0.01 \text{ }\mu\text{F} \angle (-38° + 90°)$$

$$= \mathbf{0.92 \text{ mA} \angle 52°} \tag{97}$$

Wien Bridge

Of the many possible combinations of four impedances, one has some particularly interesting and valuable properties. This circuit, called a **Wien bridge**, is shown in Figure 18.12. The Wien bridge is a series RC network connected to a parallel RC network, and the values of the two Rs and Cs are the same. This simple circuit

WIEN BRIDGE CIRCUIT

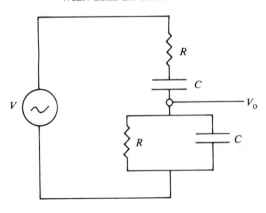

For $\omega CR = 1$: $V_0 = \tfrac{1}{3} V \angle 0°$

FIGURE 18.12. An interesting four-impedance combination is the Wien bridge circuit. For the special condition when $\omega CR = 1$, the voltage at the center is 1/3 the applied voltage ($V_0 = 1/3V$) and has the same phase as the applied voltage. See Examples 3 and 4.

forms the basis of several important circuits that are examined in the next chapter.

Earlier, we found the impedance of the series RC connection to be defined by Equation 38:

$$Z_1 = R + \frac{1}{sC} = \frac{1 + sCR}{sC}$$

and the parallel RC connection was given by Equation 25:

$$Z_2 = \frac{R}{1 + sCR}$$

Hence, the equivalent impedance of the Wien bridge is:

$$\begin{aligned} Z_{eq} &= Z_1 + Z_2 \\ &= \frac{1 + sCR}{sC} + \frac{R}{1 + sCR} \end{aligned} \tag{98}$$

Adding the fractions gives:

$$\begin{aligned} Z_{eq} &= \frac{(1 + sCR)^2 + sCR}{sC(1 + sCR)} \\ &= \frac{1 + 3sCR + (sCR)^2}{sC(1 + sCR)} \end{aligned} \tag{99}$$

Of particular interest is the voltage at the center of the series-parallel connection. This voltage is shown in Figure 18.12 as the Wien-bridge output voltage V_0. The value of V_0 can be found from the voltage divider relation for Z_1 and Z_2:

$$V_0 = V\left(\frac{Z_2}{Z_{eq}}\right) \tag{100}$$

Substituting Equations 25 and 99 for Z_2 and Z_{eq} gives:

$$\frac{V_0}{V} = \frac{R/(1 + sCR)}{[1 + 3sCR + (sCR)^2]/[sC(1 + sCR)]} \tag{101}$$

Simplifying this expression gives:

$$\frac{V_0}{V} = \frac{sCR}{1 + s^2C^2R^2 + 3sCR} \tag{102}$$

Substitution of $s^2 = -\omega^2$ gives:

$$\frac{V_0}{V} = \frac{sCR}{1 - \omega^2C^2R^2 + 3sCR} \tag{103}$$

Although Equation 103 is somewhat complicated, the presence of a minus sign in the denominator permits a significant simplification for a specific set of values. If we look at the case in which the product of frequency, resistance, and capacitance yields:

$$\omega CR = 1 \tag{104}$$

Then Equation 103 becomes:

$$\frac{V_0}{V} = \frac{sCR}{3sCR}$$

or

$$V_o = \tfrac{1}{3} V \qquad \text{for } \omega CR = 1 \qquad \textbf{(105)}$$

This equation says that when $\omega CR = 1$, the amplitude of the output voltage V_o is *one-third the amplitude of the applied voltage V.* Furthermore, because s does not appear in the voltage transfer equation, the output voltage has *the same phase as the applied voltage.* For any values of R and C for the Wien bridge, there must be some frequency for which $\omega CR = 1$ and the voltage transfer equation is Equation 105.

If a resistive voltage divider is added to the Wien bridge, as shown in Figure 18.13, for which the resistance ratio is 1 to 2, then the voltage at the center of the resistance voltage divider must also be:

$$V_o = V\left(\frac{R}{R + R}\right) = \tfrac{1}{3} V \qquad \textbf{(106)}$$

If the output voltage of the Wien bridge is now defined as the voltage between the center of the two arms, when $\omega CR = 1$, we have:

$$V_o = 0 \text{ V} \qquad \text{for } \omega CR = 1 \qquad \textbf{(107)}$$

Thus, the Wien bridge has a very interesting frequency-sensitive balance condition. For a given set of values of R and C, the output voltage will be zero and have no phase shift with the input voltage for one and only one frequency. This frequency-selective property is used in Chapter 19 to build a filter.

EXAMPLE 3: What is the current in the Wien bridge network when $\omega CR = 1$?

Solution: The current is given by:

$$I = \frac{V}{Z_{eq}} \qquad \textbf{(108)}$$

FULL WIEN BRIDGE

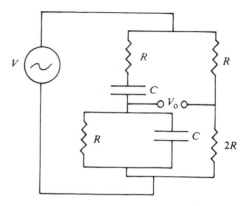

For $\omega CR = 1$: $V_0 = 0 \text{ V} \angle 0°$

FIGURE 18.13. If an **R–2R** voltage divider is added to the Wien bridge circuit, then the center voltage must be 0 V and has no phase shift. This circuit has important applications in the design of filters and oscillators.

Substituting Equation 99 for Z_{eq} gives:

$$I = \frac{VsC(1 + sCR)}{1 + (sCR)^2 + 3sCR}$$
$$= V\left(\frac{sC + s^2C^2R}{1 + s^2C^2R^2 + 3sCR}\right)$$

Lettings $s^2 = -\omega^2$ gives:

$$I = V\left[\frac{sC - \omega^2C^2R}{(1 - \omega^2C^2R^2) + 3sCR}\right] \qquad \textbf{(109)}$$

When $\omega CR = 1$, Equation 109 becomes:

$$I = V\left[\frac{sC - (1/R)}{3sCR}\right] = V\left(\frac{sCR - 1}{3sCR^2}\right)$$
$$= -V\left(\frac{1 - sCR}{3sCR^2}\right) \qquad \textbf{(110)}$$

Evaluating this expression in terms of Equations 8 and 9 gives:

$$I = -V\left(\frac{\sqrt{1 + \omega^2 C^2 R^2} \angle \phi_1}{3\omega C R^2 \angle \phi_2}\right) \tag{111}$$

where $\tan \phi_1 = -\omega CR$

$$\tan \phi_2 = 0 \tag{112}$$

Substituting $\omega CR = 1$ gives:

$$I = -V\left[\frac{\sqrt{1 + 1} \angle (\phi_1 - \phi_2)}{3R}\right]$$

$$\tan \phi_1 = -1$$

$$\phi_1 = -\frac{\pi}{4}$$

$$\tan \phi_2 = 0$$

$$\phi_2 = \frac{\pi}{2}$$

Therefore:

$$I = -\frac{\sqrt{2}}{3}\left(\frac{V}{R}\right) \angle \left(-\frac{\pi}{4} - \frac{\pi}{2}\right)$$

$$= -\frac{\sqrt{2}}{3}\left(\frac{V}{R}\right) \angle -\frac{3\pi}{4} \tag{113}$$

The $-$ sign can be incorporated into the phase angle because:

$$-1 = 1 \angle \pi \tag{114}$$

Therefore, Equation 113 becomes:

$$I = \frac{\sqrt{2}}{3}\left(\frac{V}{R}\right) \angle \pi - \frac{3\pi}{4}$$

$$= \frac{\sqrt{2}}{3}\left(\frac{V}{R}\right) \angle \frac{\pi}{4} \tag{115}$$

EXAMPLE 4: Using the expression for current found in Example 3, find the voltage at the center of the Wien bridge.

Solution: The voltage at the center of the Wien bridge is given by:

$$V_o = IZ_2 \tag{116}$$

Because we have an expression for I, we need only an expression for Z_2, which is Equation 25:

$$Z_2 = \frac{R}{1 + sCR}$$

Evaluation of this expression in terms of Equations 8 and 9 gave:

$$Z_2 = \frac{R}{\sqrt{1 + \omega^2 C^2 R^2} \angle \phi} \tag{117}$$

where $\tan \phi = \omega CR$

When $\omega CR = 1$, then:

$$\tan \phi = 1$$

$$\phi = \frac{\pi}{4}$$

and

$$Z_2 = \frac{R}{\sqrt{2}} \angle -\frac{\pi}{4} \tag{118}$$

Substituting Equations 115 and 118 into Equation 116 gives:

$$V_o = \frac{\sqrt{2}}{3}\left(\frac{V}{R}\right) \times \frac{R}{\sqrt{2}} \angle \left(\frac{\pi}{4} - \frac{\pi}{4}\right)$$

$$= \frac{1}{3} V \angle 0° \tag{119}$$

which is exactly what we found in Equation 105.

Note that even though the current is $\pi/4$ radians out of phase with the input voltage, the difference is exactly compensated for by a $\pi/4$ radian phase difference in Z_2 when $\omega CR = 1$. The result is that the voltage across Z_2 has no phase difference with the applied voltage.

Impedance Bridges

The final network to be examined is the impedance bridge circuit. The **impedance bridge** is like the Wheatstone bridge for resistors, except that the bridge components may include capacitors and inductors. Like the Wheatstone bridge, the impedance bridge has two important applications: the measurement of unknown values of impedance, primarily capacitors and inductors, and the transformation of transducer output signals into voltages that can be processed by electronic circuitry.

The generalized form of the impedance bridge is shown in Figure 18.14, where Z_1, Z_2, Z_3, and Z_x are impedances that can consist of a resistor, capacitor, or inductor alone or in combination. When used for measurement, one of the impedances is the unknown quantity to be measured Z_x. Two of the remaining arms, Z_1 and Z_2, are known, fixed values of impedance, while the third, Z_3, is a variable impedance. Z_3 is adjusted until the bridge is balanced—that is, until:

$$V_o = 0 \angle 0° \qquad (120)$$

To achieve balance, analysis of the circuit shows that the following condition is required:

$$\frac{Z_1}{Z_3} = \frac{Z_2}{Z_x} \qquad (121)$$

Solving for Z_x gives:

$$Z_x = \frac{Z_2 Z_3}{Z_1} \qquad (122)$$

This result and the analysis are identical to those for the Wheatstone bridge (see Part II). The difference, of course, is that the circuit components may include capacitors and inductors, so we must account for phase differences.

A common application of the impedance bridge is the **capacitance comparison bridge**, as shown in Figure 18.15. C_x is an unknown capacitor whose value is to be measured. Real capacitors have some resistance associated with them called the *leakage resistance*, and R_x rep-

IMPEDANCE BRIDGE

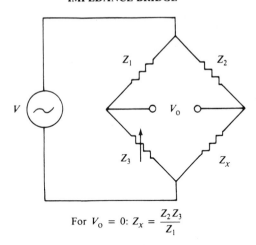

For $V_o = 0$: $Z_x = \dfrac{Z_2 Z_3}{Z_1}$

FIGURE 18.14. Another important four-impedance combination is the impedance bridge. Like the Wheatstone bridge, the impedance bridge is widely used for measuring unknown impedances and transforming impedance transducer changes into voltage changes.

CAPACITANCE COMPARISON BRIDGE

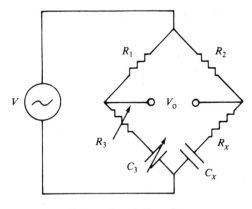

If $R_1 = R_2$ and $V_o = 0$,
then $R_x = R_3$ and $C_x = C_3$

FIGURE 18.15. A simple version of the impedance bridge is the capacitance comparison bridge. Because all capacitors have some leakage resistance R_x, it must be included in the analysis and in the balancing arm R_3. R_3 and C_3 are adjusted until $V_o = 0$. Then, if $R_1 = R_2$, $C_x = C_3$ and $R_x = R_3$.

resents this leakage resistance. Typically, R_x is quite large and varies from 1 MΩ to more than 1000 MΩ.

For balance, the variable arm of the bridge must include both a variable capacitor C_3 to balance C_x and a variable resistor R_3 to balance R_x. Generally, R_1 and R_2 are fixed, equal values of resistance, so that the analysis is as simple as possible.

The values of Z_1, Z_2, Z_3, and Z_x, therefore, are:

$$Z_1 = R_1$$
$$Z_2 = R_2$$
$$Z_3 = R_3 + \frac{1}{sC_3}$$
$$Z_x = R_x + \frac{1}{sC_x} \qquad (123)$$

Substitution of these values into the balance equation, Equation 122, gives:

$$R_x + \frac{1}{sC_x} = \frac{R_2}{R_1}\left(R_3 + \frac{1}{sC_3}\right)$$
$$R_x + \frac{1}{sC_x} = \frac{R_2 R_3}{R_1} + \frac{R_2}{R_1}\left(\frac{1}{sC_3}\right)$$
$$R_x - \frac{R_2 R_3}{R_1} = \frac{1}{s}\left(\frac{R_2}{R_1 C_3} - \frac{1}{C_x}\right) \qquad (124)$$

An inspection of Equation 124 shows that it will be satisfied if we set the term on the left-hand side of the equals sign and the term in parentheses equal to zero:

$$R_x - \frac{R_2 R_3}{R_1} = 0$$

and

$$\frac{R_2}{R_1 C_3} - \frac{1}{C_x} = 0 \qquad (125)$$

Solving for R_x and C_x gives the conditions on R_x and C_x necessary for balance and, hence, their measured values:

$$R_x = \frac{R_2 R_3}{R_1}$$
$$C_x = \frac{R_1 C_3}{R_2} \qquad (126)$$

Thus, C_3 and R_3 are adjusted until $V_o = 0$, and the unknown values of R_x and C_x are calculated according to Equation 126. Note that if $R_1 = R_2$, Equation 126 reduces to:

$$R_x = R_3$$
$$C_x = C_3 \qquad (127)$$

This result is what is expected and is identical to the results for the Wheatstone bridge.

Two other common forms of the impedance bridge are the **Maxwell bridge** and the **Hay bridge** shown in Figure 18.16. These bridges are used to measure values of unknown inductance L_x. Like capacitors, inductors have a resistance R_x associated with them that must be included in the analysis.

Note that a capacitor is the variable component used to achieve balance in these two bridges and that it is in the diagonally opposite arms of the bridge. Variable capacitors are used rather than variable inductors because they can be fabricated more precisely. To balance the phase shifts of an inductor, the capacitor must be in the diagonally opposite arm.

The balance conditions for the Maxwell bridge can be shown to be:

$$L_x = C_1 R_2 R_3$$
$$R_x = \frac{R_2 R_3}{R_1} \qquad (128)$$

The balance conditions for the Hay bridge are:

INDUCTANCE BRIDGES

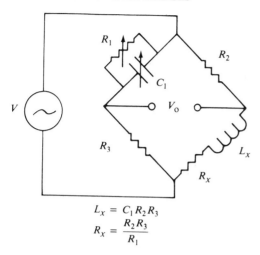

$$L_x = C_1 R_2 R_3$$

$$R_x = \frac{R_2 R_3}{R_1}$$

(a) Maxwell Bridge

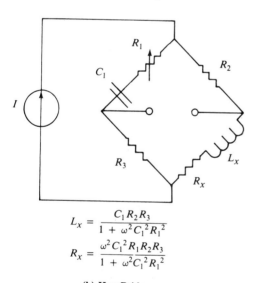

$$L_x = \frac{C_1 R_2 R_3}{1 + \omega^2 C_1^2 R_1^2}$$

$$R_x = \frac{\omega^2 C_1^2 R_1 R_2 R_3}{1 + \omega^2 C_1^2 R_1^2}$$

(b) Hay Bridge

FIGURE 18.16. Two types of inductance bridges are the Maxwell bridge and the Hay bridge. Note that a variable capacitor is used to balance an unknown inductor and that it must be placed in the diagonally opposite arm to balance the phase. Also, the balance expressions for the Hay bridge are frequency dependent.

$$L_x = \frac{C_1 R_2 R_3}{1 + \omega^2 C_1^2 R_1^2}$$

$$R_x = \frac{\omega^2 C_1^2 R_1 R_2 R_3}{1 + \omega^2 C_1^2 R_1^2} \qquad (129)$$

Note that the balance conditions of the Hay bridge are frequency dependent. Thus, the frequency of the applied voltage must be known. Also, different frequencies of applied voltage require different values of both C_1 and R_1 for balance.

18.6 NETWORK THEOREMS

Part II of the book introduced several network theorems that greatly simplify the analysis of complex resistive networks. These theorems include Thevenin's theorem, the principle of superposition, and Norton's theorem (Figure 18.17). We can also use these tools for ac circuits in which the applied voltages are sinusoids and the circuit components include capacitors and inductors. As with Ohm's law, we simply analyze the circuit according to the rules of the theorem, except that we use the impedance Z instead of the resistance R. After substituting the correct s-notation expressions for capacitive and inductive impedance into the final result, we evaluate the terms involving s and s^2, using Equations 5, 8, and 9, to get the final magnitudes and phase angles.

The use of the theorems is not illustrated here because this type of circuit analysis is required only in advanced courses. The procedure is quite straightforward, however, and proceeds according to the steps outlined above.

FIGURE 18.17. The network theorems used in dc circuit analysis are also applicable to the analysis of ac circuit networks provided that *s*-notation is used.

NETWORK THEOREMS

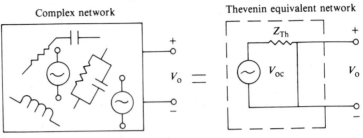

(a) Thevenin's Theorem

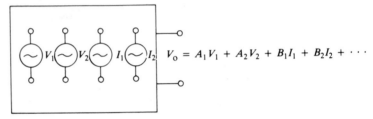

(b) Superposition

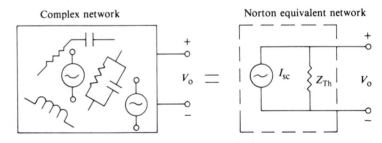

(c) Norton's Theorem

18.7 MEASURING FREQUENCY AND PHASE

Frequency Counters and Phase Meters

In the course of analyzing ac circuits, we have seen the importance of frequency and phase angle in determining circuit behavior. Because these quantities are so important, we must have ways of measuring them. Modern electronic instrumentation provides many ways to measure the frequency of a signal and its phase relative to other signals. Frequency counters and phase meters are commercially available, and they are extremely accurate, generally achieving accuracy to one part in a million (10^6) or better in frequency measurement. These instruments are relatively expensive, however, and are not commonly available in the laboratory. Also, the precision achieved by these instruments is rarely required. For most purposes, the common laboratory oscilloscope can be used to measure fre-

MEASURING FREQUENCY

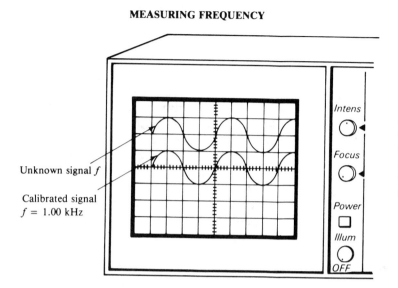

Unknown signal f

Calibrated signal
$f = 1.00$ kHz

Intens

Focus

Power

Illum

OFF

FIGURE 18.18. Frequency and phase are so important that we need ways of measuring them. Though frequency counters and phase meters are available, their high accuracy and cost mean they are not common laboratory tools. An oscilloscope is usually adequate, however. Here, the frequency of an unknown signal is measured by comparing it to that from a calibrated sine wave generator.

quency and phase to sufficient accuracy. The next section describes how the oscilloscope serves this purpose.

Oscilloscope Measurement

The simplest way of measuring frequency and phase is to use a dual-trace oscilloscope. To measure frequency, simply compare the waveform of the signal whose frequency is unknown to one whose frequency is known—for example, using a calibrated sine wave generator.

The signal of unknown frequency is fed into one input of the oscilloscope and the output of a sine wave generator into the other. The two inputs are given the same time/div deflection factors and are triggered to start at the same point in time, as shown in Figure 18.18. The frequency of the calibrated sine wave generator is then adjusted until the zero crossings of the two signals occur at exactly the same time. The frequency of the known signal that appears on the calibrated dial is equal to that of the unknown signal.

To measure the relative phase of two signals

at the same frequency, we similarly feed them into the two oscilloscope inputs and adjust the oscilloscope's triggering controls so that the two signals begin on the screen at the same point in time. Triggering the signal on one input from the signal at the other generally produces this condition. The relative phase between the two signals is determined by measuring the time Δt between comparable zero-crossing points. The phase difference ϕ, then, is:

$$\phi = \frac{\Delta t}{T} \times 360° \qquad (130)$$

where T is the period of the sine wave. In Figure 18.19, for example, the period T is:

$$T = 4.0 \text{ ms}$$

and the difference between zero crossings *in the same direction* is:

$$\Delta t = 1.0 \text{ ms}$$

Hence, the relative phase angle is:

$$\phi = \frac{1.0 \text{ ms}}{4.0 \text{ ms}} \times 360° = 90°$$

MEASURING PHASE

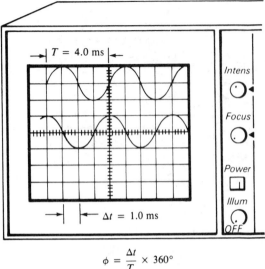

$$\phi = \frac{\Delta t}{T} \times 360°$$

FIGURE 18.19. To measure the phase difference between two signals, we trigger both to start at the same time on a dual-trace oscilloscope. The measured time Δt between comparable zero crossings can be used to calculate the phase difference. The accuracy of the measurement is only about 5%, however.

To determine the phase angle of a circuit current, we use the fact that the current and voltage of a resistor are in phase. Thus, we locate a resistor in the branch of the circuit whose current is to be measured and use the resistor voltage as the input current signal.

Of course, the measurement of frequency and phase using the methods outlined above is only as accurate as the reading taken from the oscilloscope screen. Generally, reading accuracy is about $\pm 5\%$. For somewhat greater accuracy, we can use another method based on the geometrical principle of Lissajous figures. This method has the advantage of requiring only a single-trace oscilloscope.

Method of Lissajous Figures

As described above, the method of Lissajous figures involves the comparison of two sine waves. In this case, however, one sine wave is fed into the vertical input as usual, but the other is used to drive the horizontal deflection plates in place of the time-base generator. The resulting trace is that of a vertically varying sine wave that is modulated horizontally by another sine wave whose frequency and/or phase may be different. These traces are called **Lissajous figures**. Let us examine what the traces look like and see what they tell us about the relative frequencies and phases of the two signals.

To begin, let us assume that the two signals have the same amplitude, frequency, and phase. Figure 18.20a shows the resulting oscilloscope trace. At $t = 0$, both signals are at 0 V and the trace is at $x = 0$, $y = 0$. At $t = 1$, both signals are at about 1 V and the beam is deflected up and to the right to about $x = 1$ V, $y = 1$ V. As time passes, the trace increases to its peak value at $x = 2$ V, $y = 2$ V and then gradually decreases through 0 V to $x = -2$ V, $y = -2$ V and back to 0 V. The resulting trace is a straight line at an angle of 45°.

If the signals are of the same frequency and phase but *different amplitudes*, the resulting trace also is a straight line but inclined at a different angle with respect to the x axis. In Figure 18.20b, for example, the peak amplitude of the signal applied to the x-axis input is twice that of the signal at the y-axis input and the angle is 60°.

If the signals are of the same frequency but *different phases*, the resulting figure is not a straight line. It opens into an ellipse. In Figure 18.20c, for example, the signals differ in phase by an angle of $\phi = 45° = \pi/4$ radians. The result is an ellipse whose major axis is tilted at an angle to the y axis. To obtain a point-by-point verification of the ellipse, we take x and y coordinates from the two sine waves for each point in time. When $\phi = 90°$, as in Figure 18.20d, the ellipse becomes a circle.

LISSAJOUS FIGURES

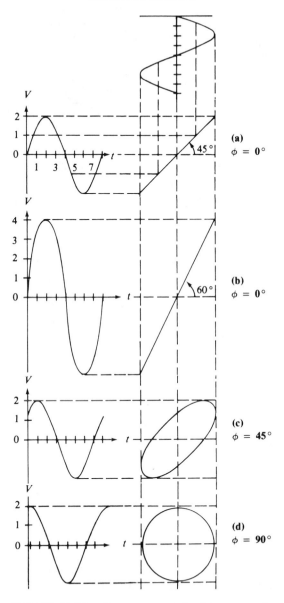

FIGURE 18.20. Frequency and phase can be measured quite accurately using the method of Lissajous figures. The unknown signal is put into the normal y-axis input, and a known sine wave is put into the sweep, or x-axis input. The oscilloscope figures that result for sine waves of the same frequency but different amplitudes and phase are illustrated here.

MEASURING PHASE ANGLES

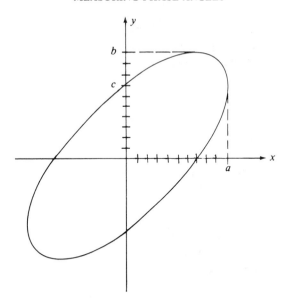

FIGURE 18.21. The Lissajous figure for two sine waves of the same frequency but different phase can be used to measure the phase difference. We simply measure the distances b and c for the figure and calculate ϕ from $\sin \phi = c/b$.

The value of the phase difference ϕ between the two signals can be measured from the oscilloscope screen as follows. The peak deflections of the oscilloscope figure in the x axis and y axis are measured to be a and b, respectively (Figure 18.21). These deflections, of course, are the peak amplitudes of the corresponding sine waves:

$$V_x(t) = a \sin \omega t \tag{131}$$
$$V_y(t) = b \sin(\omega t + \phi) \tag{132}$$

where ϕ is the assumed phase difference between the two signals.

When $t = 0$, Equations 131 and 132 become:

$$V_x(0) = 0 \tag{133}$$
$$V_y(0) = b \sin \phi \tag{134}$$

Figure 18.21 indicates that when $t = 0$, we can also measure the value of $V_y(0)$, which is called c in the figure. Note that the proper measurement of c requires that the ellipse be centered relative to the origin. With c known, we have:

$$V_y(0) = c = b \sin \phi \qquad (135)$$

and

$$\sin \phi = \frac{c}{b} \qquad (136)$$

Using Equation 136, we find that by measuring the values of c and b from the Lissajous figure in Figure 18.21, we can calculate the value of the phase angle. For example, in Figure 18.21:

$$b = 10 \text{ units}$$
$$c = 7.0 \text{ units}$$

Thus:

$$\sin \phi = \frac{7.0}{10} = 0.7$$
$$\phi = 44.5°$$

which is within the experimental error of the true value of 45°.

The shape of the Lissajous figure when the two signals have the same frequency is a stable ellipse (Figure 18.21). For *different phase angles*, the shape of the ellipse changes, as shown in Figure 18.20 for various values of ϕ. As Figure 18.20 shows, the Lissajous figures repeat themselves when the phase angle exceeds 180°. What this means is that for a measured phase angle ϕ, there is no way to tell which signal leads the other. For example, if V_x lags V_y by 15°, it will have the same Lissajous figure as when it leads V_y by 360° − 15° = 345°. Leading by 345° is the same as lagging by −15°.

Thus, although the Lissajous technique measures the magnitude of phase differences, it

EXAMPLES OF LISSAJOUS FIGURES

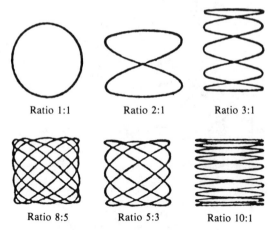

Ratio 1:1   Ratio 2:1   Ratio 3:1

Ratio 8:5   Ratio 5:3   Ratio 10:1

FIGURE 18.22.   Lissajous figures can also be used to measure frequency. The known signal frequency is adjusted until a stable Lissajous figure is obtained. When the figure is stable (does not rotate), the frequency of the unknown signal equals that of the known. For various exact integer differences between the two signals, a variety of interesting signals result, as shown.

does not yield the sign of the phase angle. Very often, however, examination of the circuit suggests whether the signal can be expected to lead or lag.

The Lissajous procedure is typically used to measure the phase shift produced by a circuit network. The input signal to a network is applied to one axis of an oscilloscope and the output signal to the other. The phase shift produced by the network between V_{in} and V_o can be measured as described above.

If the two signals to be compared have *different frequencies*, then a variety of results are possible. If the two signal frequencies are only slightly different, the ellipse will rotate in a direction determined by which frequency is greater. When they differ by exact integer values, quite complex, stable figures result, as shown in Figure 18.22. For intermediate frequency values, the figures are not stable and change shape continuously. The Lissajous figures shown in Figure 18.22 are quite interesting and easy to obtain with two signal generators and an oscilloscope.

The stable patterns of Figure 18.22 can also be used to measure the frequency of an unknown signal by comparison with a known signal, as described earlier. That is, the unknown signal is fed into one input (either x or y) of an oscilloscope, and the output of a calibrated sine wave generator is fed into the other. The frequency of the signal generator is then adjusted until a stable (not rotating) ellipse appears on the oscilloscope screen. When the ellipse is stable, the frequency of the unknown signal is exactly equal to that of the known signal. This method of measuring frequency can be quite accurate, particularly if the frequency of the known, calibrated source is accurate.

In fact, the measurement method is generally more accurate than adjusting the dial of the known signal source. No matter how carefully the frequency dial is set, the ellipse will gradually rotate. The rate at which the ellipse rotates, however, is proportional to the difference in frequency between the two signals. This fact will become obvious in the discussion that follows.

Suppose the unknown signal source frequency ω differs from the known signal generator frequency ω_o by a small amount $\Delta\omega$. That is:

$$\omega = \omega_o + \Delta\omega \tag{137}$$

If the signal generator sine wave is given by:

$$V_x(t) = a \sin \omega_o t \tag{138}$$

then the unknown signal source is:

$$V_y(t) = b \sin(\omega_o + \Delta\omega)t \tag{139}$$

or

$$V_y(t) = b \sin(\omega_o t + \phi) \tag{140}$$

where $\phi = \Delta\omega t \tag{141}$

Therefore, if we measure the time t re-

quired for the ellipse to undergo two full revolutions (the phase angle ϕ will change by 2π), then we can calculate the value of $\Delta\omega$ from:

$$\Delta\omega = \frac{2\pi}{t} \tag{142}$$

or the frequency difference Δf by:

$$\Delta f = \frac{\Delta\omega}{2\pi} \tag{143}$$

or

$$\Delta f = \frac{1}{t} \tag{144}$$

EXAMPLE 5: A signal whose frequency is unknown is measured against a calibrated signal generator by comparing their Lissajous figures. When the signal generator is set to exactly 2.00 kHz, the ellipse undergoes two complete rotations in 5.0 s. What is the frequency of the unknown signal?

Solution: We can calculate the difference between the two frequencies for $t = 5.0$ s using Equation 144:

$$\Delta f = \frac{1}{t} = \frac{1}{5.0 \text{ s}} = 0.20 \text{ Hz}$$

Whether the unknown signal is greater or less than the known signal by this amount can be determined from the direction of ellipse rotation. We purposely adjust the signal generator so that it is below the frequency of the unknown signal and note the direction of ellipse rotation (Figure 18.23). This direction determines the direction of rotation for the condition $f > f_o$.

Assuming that this measurement shows that the unknown frequency is slightly

HIGH-PRECISION MEASUREMENT

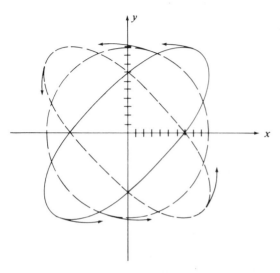

FIGURE 18.23. It is generally difficult to adjust the known signal accurately enough to get a stable figure. However, the time it takes to go through two complete cycles of rotation Δt is a measure of the difference between the known signal frequency and the unknown signal frequency. This measurement method is extremely accurate: $\pm 0.01\%$ or better.

greater than 2.00 kHz, we deduce that its frequency must be:

$$f = f_o + 0.20 \text{ Hz}$$
$$= 2000 \text{ Hz} + 0.20 \text{ Hz}$$
$$= \textbf{2000.20 Hz}$$

Note that the measurement accuracy is:

$$\frac{\Delta f}{f_o} = \frac{0.20 \text{ Hz}}{2000 \text{ Hz}} \times 100\% = 0.01\%$$

This result is quite accurate and requires that the frequency of the calibrated signal source be known to five significant figures in order to be a useful absolute frequency measurement. Only very expensive signal generators achieve such calibration accuracy. However, as a technique for measuring frequency *differences*, it is extremely valuable.

18.8 QUESTIONS AND PROBLEMS

1. Determine the equivalent impedance, including phase characteristics, of a series RC network at a frequency of 2.2 kHz for the following values of R and C. Draw a vector diagram of the result showing the components.
   (a) $R = 47$ kΩ and $C = 0.01$ μF
   (b) $R = 0.22$ MΩ and $C = 0.0025$ μF
   (c) $R = 8500$ Ω and $C = 0.050$ μF
2. Determine the equivalent impedance, including phase characteristics, of a series RC network where $R = 2.6$ kΩ and $C = 0.025$ μF at the following frequencies. Draw a vector diagram of the results showing the components.

f
0.1 kHz
0.3 kHz
1.0 kHz
3.0 kHz
10.0 kHz
30.0 kHz
100.0 kHz

3. Repeat Problem 1 for a parallel RC network.
4. Repeat Problem 2 for a parallel RC network.
5. Derive expressions for the voltage across each component of a series RC network (the voltage divider relations).
6. For the circuits of Figures 18.24 to 18.27, determine:
   (a) Equivalent impedance
   (b) Net circuit current
   (c) Voltage across and current through each impedance
   (d) Phase angle for each of the above quantities

   Draw a vector diagram of the circuit voltages.

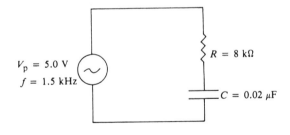

FIGURE 18.24.

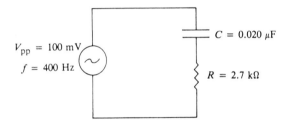

FIGURE 18.25.

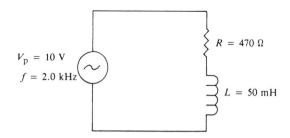

FIGURE 18.26.

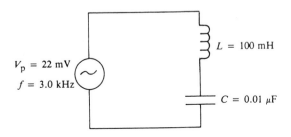

FIGURE 18.27.

7. At what frequency will the circuit shown in Figure 18.27 have an impedance of 0 Ω?
8. For the circuits shown in Figures 18.28 to 18.31, determine:
   (a) Equivalent impedance
   (b) Net circuit current
   (c) Voltage across and current through each impedance
   (d) Phase angle for each of the above quantities

   Draw a vector diagram of the circuit currents.

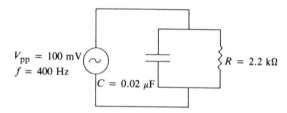

FIGURE 18.28.

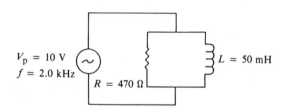

FIGURE 18.29.

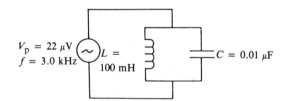

FIGURE 18.30.

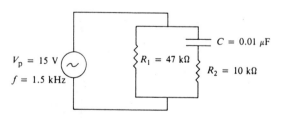

FIGURE 18.31.

9. At what frequency will the circuit of Figure 18.31 have an infinite impedance?

10. What is the voltage across and current through each component of the circuit shown in Figure 18.32?

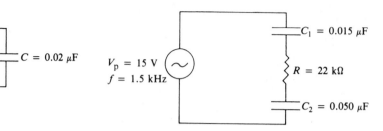

FIGURE 18.32.

11. What is the current through each component of the circuit shown in Figure 18.33?

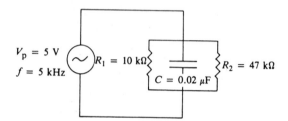

FIGURE 18.33.

12. What is the voltage across and current through each component in Figure 18.34?

FIGURE 18.34.

13. What is the voltage across and current through each component in Figure 18.35?

FIGURE 18.35.

14. What is the phase difference between the two signals shown in Figure 18.36?

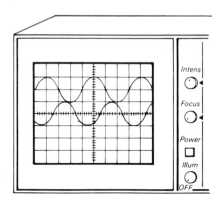

Time/Div = 0.5 ms

FIGURE 18.36.

15. What is the phase difference between the two signals that give the Lissajous figure shown in Figure 18.37?

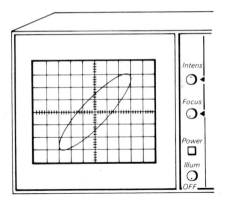

Time/Div = 2 ms

FIGURE 18.37.

16. The method of Lissajous figures is an accurate way to measure the frequency of an unknown signal. If the known signal has a frequency of 400.0 Hz and the Lissajous figure rotates at a rate of 3.0 s per revolution, what is the frequency of the unknown signal?

Chapter

19

FILTERS

19.1 OBJECTIVES

Following the completion of Chapter 19, you should be able to:

1. Describe the general purpose of a filter circuit and explain the four primary types of filters.
2. State the two ways of expressing the frequency characteristics of a filter circuit and give examples of each for a simple filter.
3. Determine the gain of a circuit in decibels (db), given the value of V_o/V_{in}, and vice versa.
4. Interpret a frequency response graph whose frequency axis is logarithmic and whose gain axis is given in decibels.
5. Explain the purpose of a Bode diagram and its relation to the frequency response graph it represents.
6. Determine the cutoff frequency and roll-off of a filter circuit, given its frequency response graph or Bode diagram.
7. Draw the circuit diagrams for simple low- and high-pass filters; give expressions for

their voltage transfer function, corner frequency, and roll-off; and draw their frequency response graph, generalized Bode diagram, and a generalized graph of a phase angle versus frequency.

8. State the requirement for producing the basic characteristics of a filter circuit and describe a method of doing it.
9. Describe the effects on the gain and corner frequency of loading simple low- and high-pass filters.
10. Give a generalized expression for the output of a two-stage filter in terms of the voltage transfer functions of each stage and state the requirement that establishes its validity.
11. Draw circuit diagrams for practical two-stage low-pass, high-pass, band-pass, and band-reject filters; give expressions for their voltage transfer function and corner frequency; and draw their Bode diagrams showing the proper roll-off.
12. Design low-pass, high-pass, band-pass, and

band-reject filters with a given roll-off and corner frequency.

13. Determine the Q factor of a narrow band filter from its Bode diagram or frequency response graph and explain its purpose in evaluating filter characteristics.

14. State the purpose of a tuned amplifier and draw a typical circuit diagram.

15. Identify twin T and bridge T filter circuits by their circuit diagrams and describe their general characteristics.

16. Explain the basic characteristics of an active filter and draw a generalized circuit diagram using an op amp.

17. Explain the general approach to designing an active filter with a given order using an op amp.

19.2 OVERVIEW

Chapter 17 described the basic characteristics of ac circuits. We saw how to determine the voltage across and current through various connections of a resistor and capacitor (and/or inductor) for applied sine waves of different frequencies, and two important properties of these circuits were discussed. The first is that the voltages and currents in each element depend on the frequency of the sine wave. The second is that phase shifts develop between the voltages and currents in the circuit and these shifts are also frequency dependent.

This chapter examines how these properties figure into the design of circuits for ac systems. The applications fall into two primary categories: filters and oscillators. Filter circuits are discussed here.

19.3 PURPOSE OF FILTER CIRCUITS

The primary purpose of a **filter circuit** is just what the name implies: to filter out sine waves whose frequencies are not the right value. There are four basic classes of filter circuits: low-pass, high-pass, band-pass, and band-reject. Each of these classes performs the function suggested by its name. A **low-pass filter** passes only sine waves with low frequencies—that is, all frequencies below some specified design value—and blocks sine waves of higher frequencies. A **high-pass filter** does the opposite: It passes sine waves whose frequencies are above some design value and rejects the ones below that value.

Band-pass and band-reject filters are also similar but opposite. A **band-pass filter** passes sine waves whose frequencies are within a certain band of values and rejects all sine waves whose frequencies are above or below the frequency band. A **band-reject filter** rejects sine waves whose frequencies lie within a narrow band, but passes all others.

By *pass* is meant that the amplitude of the sine wave at a given frequency will be the same leaving the filter circuit as it was when it entered—that is, $V_o/V_{in} = 1$. Similarly, *reject* means that the amplitude of the sine waves at a given frequency will be essentially zero upon leaving the filter circuit—that is, $V_o/V_{in} = 0$.

The behavior of filter circuits for sine waves of different frequencies is represented by frequency response curves, shown schematically in Figure 19.1 for the four basic filter types. These graphs are plots of V_o/V_{in} as a function of sine wave frequency. As seen in the figure, the transition from pass to reject is a gradual one. One of the key tasks of a filter designer is to make that transition as sharp as possible. Some of the techniques used to sharpen the transition are discussed throughout the chapter.

Filter circuits are most commonly seen in audio systems, though they are central to the design of most other electronic systems, particularly radio-frequency (rf) systems and measurement instrumentation. In audio systems, the tone controls are basically high- and low-pass filters. The treble control is a high-pass

TYPES OF FILTERS

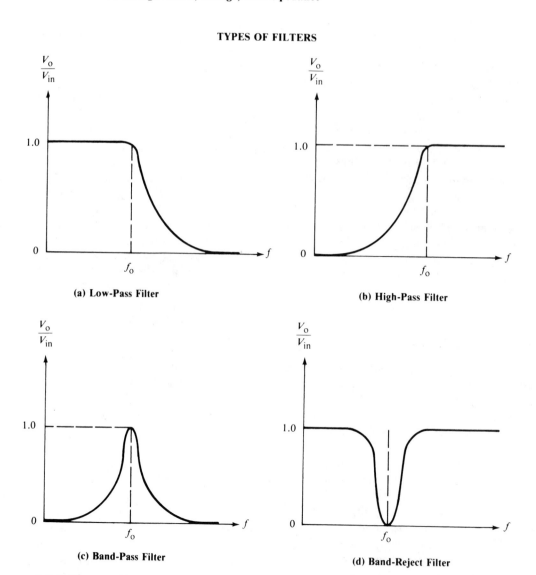

(a) Low-Pass Filter

(b) High-Pass Filter

(c) Band-Pass Filter

(d) Band-Reject Filter

FIGURE 19.1.   This section examines one of the major applications of ac circuits: filters.
Filter circuits are designed to pass sine waves in a certain range of frequencies while reject-
ing all others. They have wide-ranging application in audio, rf, and measurement systems.
The figure shows typical frequency response curves for the four major filter types.

filter that permits the amplitudes of high-
frequency tones to pass undiminished while attenuating the low-frequency tones. The base
control, on the other hand, is a low-pass filter
that passes tones of low frequencies while attenuating those of higher frequencies.

Radio and TV tuning knobs are adjustable
band-pass filters. Of all the rf signals received
by the antenna, the tuner will pass only one
while rejecting all others. Obviously, the
"band" of the filter must be narrow enough to
distinguish between two stations that are broadcasting at adjacent frequencies.

A common application of a band-reject

filter is to eliminate 60 Hz noise from audio systems. The ac line voltage operates at 60 Hz and comes right into an electronic system via the power supply. This 60 Hz signal may be picked up by electronic components of the circuit and superimposed on the signal voltage. If the signal frequencies are near 60 Hz, a 60 Hz bandreject filter may be included in the signal line to selectively eliminate this unwanted noise.

The design of filter circuits can be quite complicated, particularly in a specific application in which a particular frequency response is desired. The basic concepts of filter design, the basic characteristics of filters, and the terminology used to describe them are introduced here to serve as a foundation for more advanced study. The analysis begins with a discussion of the way in which filter behavior is described and the quantities used to measure it.

19.4 DESCRIBING FILTER CHARACTERISTICS

As indicated, the primary purpose of a filter circuit is to selectively pass sine waves with a certain range of frequencies. An ideal filter would have $V_o/V_{in} = 1$ for the frequencies that should pass and $V_o/V_{in} = 0$ for frequencies that should be rejected, but real filters are not so exact. Consequently, we need a way of expressing the relation V_o/V_{in} for sine waves of all possible frequencies. This information can be presented in two ways: by an equation or by a graph.

Voltage Transfer Functions

For simple filters, V_o/V_{in} for all frequencies can be expressed by an equation called the **voltage transfer function** of the circuit. The transfer function expresses how the filter will "transfer" an incoming sine wave with an amplitude V_{in} to its output, where its amplitude will be V_o. The transfer function, of course, includes the frequency of the sine wave because V_o/V_{in} depends on frequency.

The voltage divider equations given in Chapter 17 for the series RC circuit are examples of a voltage transfer function. If the applied voltage to a series RC circuit is called V_{in} and the capacitor voltage is taken as the output voltage V_o, then Equation 64 of Chapter 17 can be written as follows:

$$\frac{V_o}{V_{in}} = \frac{1}{\sqrt{1 + \omega^2 C^2 R^2}} \tag{1}$$

This expression is the voltage transfer function for a series RC circuit in which V_o is the capacitor voltage. Note that V_o/V_{in} depends on the frequency of the sine wave ω. If ω is large, V_o/V_{in} is small, approaching zero, and if ω is small, V_o/V_{in} approaches 1. Clearly, this behavior is a characteristic of a low-pass filter (Figure 19.2a).

The transfer function for the resistor voltage serving as V_o is given by Equation 61 of Chapter 17:

$$\frac{V_o}{V_{in}} = \frac{1}{\sqrt{1 + (1/\omega^2 C^2 R^2)}} \tag{2}$$

This expression has the opposite behavior. When ω is large, V_o/V_{in} approaches 1, and when ω is small, V_o/V_{in} is small, approaching zero. Clearly, this behavior is the characteristic of a high-pass filter (Figure 19.2b).

For relatively simple filters, therefore, we can calculate the voltage transfer function as we did for the series RC circuit. For more complex circuits, however, the equations are difficult to derive and are not a practical way of expressing the filter characteristics. In these cases, we use a graph.

Logarithmic Graphs

We have seen that the graph of a circuit's behavior as a function of frequency is termed its *frequency response*. For a filter, the frequency response graph is simply a graph of its voltage transfer function—that is, V_o/V_{in} as a function

VOLTAGE TRANSFER FUNCTION

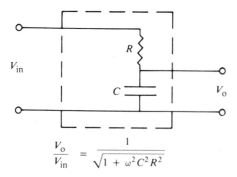

$$\frac{V_o}{V_{in}} = \frac{1}{\sqrt{1 + \omega^2 C^2 R^2}}$$

(a) Low-Pass Filter

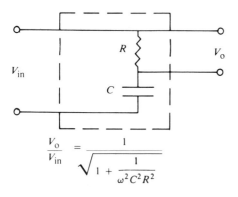

$$\frac{V_o}{V_{in}} = \frac{1}{\sqrt{1 + \dfrac{1}{\omega^2 C^2 R^2}}}$$

(b) High-Pass Filter

FIGURE 19.2. **The ac behavior of a filter is described by the ratio of the output voltage V_o to the input voltage V_{in} for all possible frequencies. For relatively simple circuits, this ratio can be found from an equation called the voltage transfer function. The figure gives the voltage transfer functions for a simple RC circuit used as a low-pass filter and as a high-pass filter.**

of frequency. Figure 19.3a shows a typical graph for Equation 1.

Note that the frequency axis in Figure 19.3 is logarithmic. The logarithmic scale permits the display of V_o/V_{in} over several decades of frequency. This wide range is necessary because frequency response data generally span many decades. Audio signals, for example, span the full range of frequencies distinguishable by the human ear: approximately 20 Hz to 20 kHz. Function generator outputs range from 0.1 Hz to 100 kHz or more.

The graphs of Figure 19.3 are termed **semilogarithmic** because only one axis is scaled logarithmically. If the vertical axis were also logarithmic, the graph would be called a **log–log graph**. In some cases log–log graphs are required to show more detailed behavior of V_o/V_{in}.

On a logarithmic scale, the unit separations are proportional to the logarithm of the quantity they represent. Below the frequency axis in Figure 19.3a is the logarithm of the frequency value. Note that:

$$\log 0.01 \text{ Hz} = \log 10^{-2} = -2$$
$$\log 0.1 \text{ Hz} = \log 10^{-1} = -1$$
$$\log 1 \text{ Hz} = \log 10^{0} = 0$$
$$\log 10 \text{ Hz} = \log 10^{1} = +1$$
$$\log 100 \text{ Hz} = \log 10^{2} = +2$$

Thus, a change of one major scale division on the logarithmic axis represents a frequency change of ten units.

Decibels

Because filters are designed to filter out sine waves of unwanted frequencies, a central question is, How well does a filter eliminate them? In mathematical terms, this question means, What is the value of V_o/V_{in}? Generally, we want this value to be extremely small—that is, 1/1000 or less! The quantity that is used to express this value is the **decibel** (db).

The decibel is used more generally to define the gain of a circuit. For an amplifier, the gain V_o/V_{in} is usually greater than 1. For a filter, however, the gain for unwanted frequencies is less than 1. Because the gain value can span several decades, it is defined in logarithmic

FREQUENCY RESPONSE GRAPHS

FIGURE 19.3. For more complex circuits, the frequency response is presented as a graph of V_o/V_{in} versus frequency, which is simply a graph of the voltage transfer function. Generally, the frequency axis is logarithmic in order to include a wide range of frequencies.

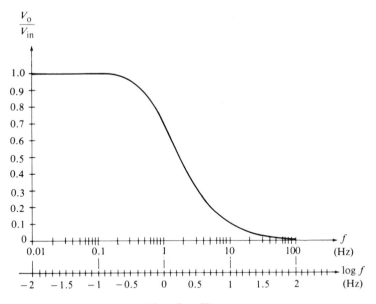

(a) Low-Pass Filter

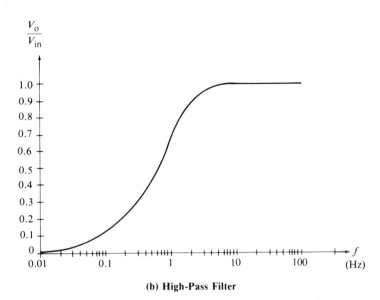

(b) High-Pass Filter

FILTER GAIN IN DECIBELS

$$\text{Gain (db)} = 20 \, \log_{10}\!\left(\frac{V_o}{V_{in}}\right)$$

$\dfrac{V_o}{V_{in}}$	Gain (db)
1.0	0
0.1	-20
0.01	-40
0.001	-60
0.0001	-80
0.00001	-100
—	—
—	—
—	—

FIGURE 19.4. The ability of a filter to reject sine waves of a given frequency is measured by the ratio V_o/V_{in}, which is generally quite small: 0.0001 or less. These small values of gain are expressed by the logarithmic quantity called the decibel (db). For each decrease of V_o/V_{in} by a factor of 10, the gain of the filter decreases by -20 db.

terms. Specifically, the gain in decibels is defined as:

$$\text{gain (db)} = 20 \, \log_{10}\!\left(\frac{V_o}{V_{in}}\right) \qquad (3)$$

When $V_o/V_{in} = 1$, the gain in decibels is:

$$\text{gain (db)} = 20 \, \log 1 = 0 \qquad (4)$$

When $V_o/V_{in} = 1/10$, it is:

$$\text{gain (db)} = 20 \, \log \frac{1}{10} = 20 \, \log(10^{-1})$$

$$= 20(-1) = -20 \qquad (5)$$

Figure 19.4 gives values of gain in decibels for several values of V_o/V_{in}. Note that each time the gain increases (or decreases) by a factor of 10, the gain in decibels increases (or de-creases) by 20. Thus, if a filter suppresses a given sine wave by 80 db, its amplitude has been reduced by four orders of magnitude—that is, $V_o/V_{in} = 10^{-4} = 0.0001$.

EXAMPLE 1: What is the change in decibels when a signal drops to half its initial power?

Solution: From the statement of the problem, we know that:

$$P_o = \frac{1}{2}P_{in}$$

or

$$\frac{P_o}{P_{in}} = \frac{1}{2}$$

Since power is proportional to voltage:

$$\frac{V_o^2}{V_{in}^2} = \frac{1}{2}$$

or

$$\frac{V_o}{V_{in}} = \frac{1}{\sqrt{2}} = 0.707$$

Substitution into Equation 3 gives:

$$\text{gain (db)} = 20 \, \log_{10}(0.707)$$

$$= 20(-0.151) = -3.0 \text{ db}$$

Thus, when the power of a signal drops by one-half, there is a decrease in gain of -3 db.

The vertical axis of a frequency response graph is often plotted in units of decibels rather than in values of V_o/V_{in}. Thus, the vertical axis is essentially logarithmic in values of V_o/V_{in}, but linear in values of decibels (Figure 19.5).

Note that the gain of the filter in Figure 19.5 is relatively flat up to 0.3 kHz, but de-creases, or rolls off, at higher frequencies. The

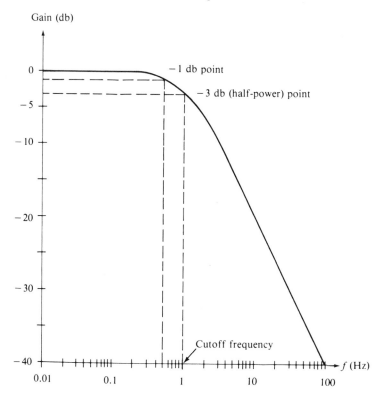

CUTOFF FREQUENCY

FIGURE 19.5. The vertical axis of a frequency response graph is often given in linear values of gain in decibels rather than in values of V_o/V_{in}. A critical frequency on such graphs is the cutoff frequency, for which the gain is -3 db ($V_o/V_{in} \approx 0.7$). This frequency is also termed the half-power point because $P_o/P_{in} \alpha (V_o/V_{in})^2 = 0.5$.

frequency at which the gain has decreased by -3 db is termed the **half-power point**. The half-power point is often used as an arbitrary measure of the **cutoff frequency** of the filter. That is, it is assumed that frequencies below the value at the half-power point pass essentially undiminished, though, in fact, at the half-power point $V_o \approx 0.7 V_{in}$. Remember that the -3 db criterion is arbitrary. We could be more demanding and use the -1 db point to determine the cutoff frequency.

EXAMPLE 2: What is the cutoff frequency in Figure 19.5 if the gain criterion is -1 db?

Solution: Using Equation 3, we find:

$$\text{gain (db)} = 20 \log\left(\frac{V_o}{V_{in}}\right) = 1 \text{ db}$$

$$\log\left(\frac{V_o}{V_{in}}\right) = -0.05$$

$$\frac{V_o}{V_{in}} = 0.89$$

That is, at the -1 db point, the output voltage V_o is about 90% of its original value V_{in}. Figure 19.5 shows that the corresponding frequency is **0.5 kHz**.

Bode Diagrams

One of the most important characteristics of a filter is how sharply it cuts off. In an ideal filter,

FIGURE 19.6.   Often the true frequency response of a filter is approximated by a Bode diagram,
which is simply an extrapolation
of the straight line segments that
intersect at the − 3 db frequency
f_o. The roll-off characteristic of the
filter, then, is the slope of the
curve, expressed as the change in
gain per decade of frequency (db/
decade) or per doubling of frequency (db/octave).

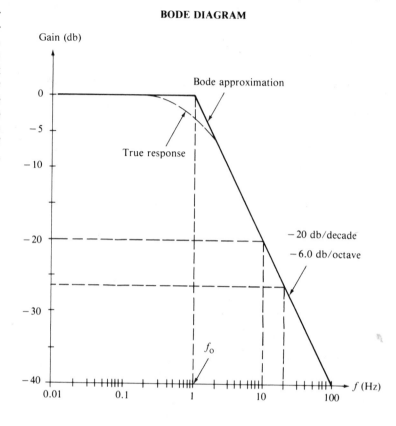

the cutoff would be vertical at the cutoff frequency f_o. For a low-pass filter, for example, all
sine waves with frequencies less than f_o would
have V_o/V_{in} = 1 and all sine waves with frequencies greater than f_o would have V_o/V_{in}
= 0.

   Real filters, of course, do not have a vertical cutoff, but are inclined, as shown in Figure
19.5. In general, the steeper the slope, the better the filter. The steepness of the slope is often
referred to as its roll-off characteristic.

   The roll-off characteristics of a filter are
most apparent in a Bode diagram of the frequency response. A **Bode diagram** is a simplification of the true frequency response, in which
the straight line segments are extrapolated to
their intersection, as shown in Figure 19.6.
Note that the intersection occurs at the fre-

quency f_o that corresponds to the − 3 db point.
This frequency is called the **corner frequency** f_o.

   The **roll-off characteristic** of the filter,
then, is simply the slope of the line for frequencies greater than f_o. This slope is generally expressed as the change in gain in decibels per
decade change of frequency, or **db/decade**.
Thus, in Figure 19.6, between 100 kHz and 10
kHz, the gain changes as follows:

$$\Delta \text{ gain} = -40 \text{ db} - (-20 \text{ db})$$
$$= -20 \text{ db}$$

Thus, the roll-off is:

   − 20 db/decade

   Another measure of the roll-off of the frequency response expresses the change in voltage

EXAMPLE 3

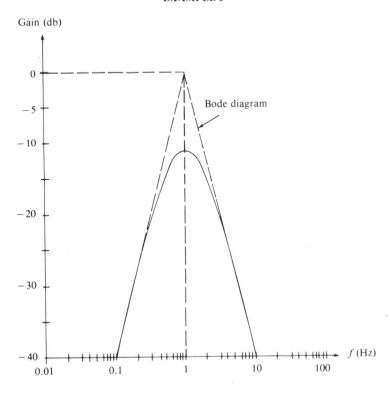

for each doubling in frequency, or **db/octave**. Thus, in Figure 19.6, between 20 kHz and 10 kHz, the gain is:

$$\Delta \text{ gain } = -26 \text{ db} - (-20 \text{ db})$$
$$= -6 \text{ db}$$

Thus, the roll-off is:

-6 db/octave

The Bode diagram for a filter is often used as a simple approximation of the true frequency response of a filter because it is easy to draw and generally contains the essential filter characteristics: the corner frequency and the roll-off.

EXAMPLE 3: What are the corner frequencies and the roll-off characteristics of the filter shown in Figure 19.7?

Solution: The filter circuit shown in Figure 19.7 is a band-pass filter. The Bode diagram is found by extending the slopes as shown by the dotted lines. The lower and upper cutoff frequencies are:

$f_1 = f_2 = $ **1.0 kHz**

The lower roll-off depends on the change in gain between 10 kHz and 1.0 kHz, which is:

$$\Delta \text{ gain } = -40 \text{ db } - (0 \text{ db})$$
$$= \mathbf{-40 \text{ db/decade}}$$

The upper roll-off depends on the change in gain between 1 kHz and 0.1 kHz, which is:

$$\Delta \text{ gain } = 0 - (-40 \text{ db})$$
$$= \mathbf{+40 \text{ db/decade}}$$

Now that we have ways of describing and measuring filter characteristics, the performance of various filter circuits can be analyzed. The analysis begins with the simple series RC filter and proceeds to relatively sophisticated active filters involving operational amplifiers.

19.5 SIMPLE RC FILTERS

The basic series RC filter is shown in Figure 19.8. This circuit can act as either a low-pass or a high-pass filter, depending on which voltage is selected as the output voltage.

Low-Pass Filter

For a low-pass filter, the output voltage is taken as the capacitor voltage. The filter's transfer function, then, is given by Equation 1:

$$\frac{V_o}{V_{in}} = \frac{1}{\sqrt{1 + \omega^2 C^2 R^2}}$$

To simplify the form of this expression, we will use the definition of f_o:

$$f_o = \frac{1}{2\pi CR} \qquad (6)$$

Recall that f_o is the value of f when:

$$\omega CR = 1 \qquad (7)$$

SIMPLE LOW-PASS FILTER

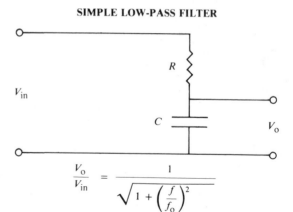

$$\frac{V_o}{V_{in}} = \frac{1}{\sqrt{1 + \left(\dfrac{f}{f_o}\right)^2}}$$

FIGURE 19.8. A simple low-pass filter is the capacitor voltage of a series RC circuit, as shown here. The voltage transfer function for the filter is more simply written in terms of the corner frequency f_o, which equals $1/(2\pi CR)$. Its roll-off characteristic is -20 db/decade (-6 db/octave).

Equation 7 allows us to rewrite Equation 1 for the **transfer function** of the simple low-pass filter as follows:

$$\frac{V_o}{V_{in}} = \frac{1}{\sqrt{1 + (f/f_o)^2}} \qquad (8)$$

When $f = f_o$, Equation 8 becomes:

$$\frac{V_o}{V_{in}} = \frac{1}{\sqrt{1 + 1}} = \frac{1}{\sqrt{2}} = 0.707$$

This value is also the value of V_o/V_{in} at the half-power point for which the gain is -3 db, as we saw in Example 1. Therefore, we conclude that the frequency f_o is the **corner frequency** for the simple low-pass filter.

EXAMPLE 4:   A simple low-pass RC filter has $C = 0.1 \ \mu\text{F}$ and $R = 1.6 \text{ k}\Omega$. What is the value of the corner frequency f_0 and the roll-off?

Solution: The corner frequency can be found from Equation 6:

$$f_0 = \frac{1}{2\pi CR} = \frac{1}{6.28 \times 0.1\ \mu F \times 1.6\ k\Omega}$$
$$= \textbf{1.0 kHz}$$

The roll-off can be found from the slope of the curve at frequencies that are *large* compared to f_0. Therefore, let us calculate the gain at 10 kHz and at 100 kHz from Equation 3 by substituting in the expression for V_0/V_{in} of Equation 1:

$$\text{gain (db)} = 20 \log\left(\frac{V_0}{V_{in}}\right)$$
$$= 20 \log\left[\frac{1}{\sqrt{1 + (f/f_0)^2}}\right] \quad (9)$$

At $f = 10$ kHz, Equation 9 gives:

$$\text{gain (db)} = 20 \log\left[\frac{1}{\sqrt{1 + (10\ kHz/1.0\ kHz)^2}}\right]$$
$$= 20 \log\left(\frac{1}{\sqrt{101}}\right) = -20\ db$$

At $f = 100$ kHz:

$$\text{gain (db)} = 20 \log\left[\frac{1}{\sqrt{1 + (100\ kHz/1.0\ kHz)^2}}\right]$$
$$= 20 \log\left(\frac{1}{\sqrt{10,001}}\right) = -40\ db$$

Thus, the roll-off is:

$$\text{roll-off} = -40\ db - (-20\ db)$$
$$= \textbf{-20 db/decade}$$

A similar calculation for a doubling of frequency — for example, from 50 kHz to 100 kHz — would give the roll-off as:

$$\text{roll-off} = \textbf{-6.0 db/octave}$$

Thus the **roll-off** of the simple low-pass RC filter is -20 db/decade or -6.0 db/octave.

The transfer function of Equation 1 for the values obtained in Example 4 is represented by the frequency response curve shown in Figure 19.3a. Note that the x axis is logarithmic in frequency, but the y axis is linear in gain.

The transfer function of Equation 9 can be graphed as a generalized Bode diagram that applies to any simple low-pass RC filter. This graph is shown in Figure 19.9. Note that the frequency axis is logarithmic for values of f/f_0 rather than frequency and that the gain axis is in decibels. Thus, Figure 19.9 is identical to Figure 19.3a if we give the corner frequency the value $f_0 = 1.0$ kHz and note that the gain axis is measured in decibels. Figure 19.9 can also be used to simplify the solution of any simple low-pass RC filter problem.

EXAMPLE 5: What is the gain in decibels of the filter in Example 4 at frequencies of 5 kHz and 50 kHz?

Solution: When $f = 5$ kHz, f/f_0 is:

$$\frac{f}{f_0} = \frac{5\ kHz}{1.0\ kHz} = 5$$

Figure 19.9 shows that the gain for $f/f_0 = 5$ is:

$$\text{gain} = -15\ db$$

When $f = 50$ kHz, f/f_0 is:

$$\frac{f}{f_0} = \frac{50\ kHz}{1.0\ kHz} = 50$$

FIGURE 19.9. The frequency response of any simple low-pass filter can be expressed by the generalized Bode diagram shown here. Note that the frequency axis is in units of f/f_o and that the gain axis is in decibels. This diagram can be used to solve any simple low-pass filter problem. See Example 5.

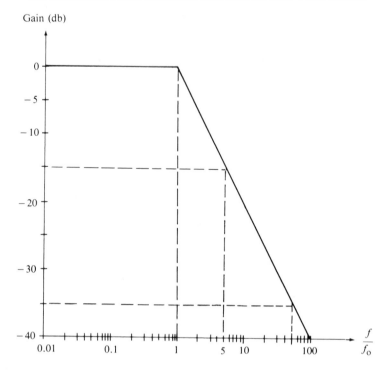

GENERALIZED BODE DIAGRAM FOR SIMPLE LOW-PASS FILTER

Figure 19.9 shows that the gain for $f/f_o = 50$ is:

$$\text{gain} = -35 \text{ db}$$

EXAMPLE 6: What values of R and C will yield a corner frequency of 20 kHz?

Solution: The corner frequency is given by Equation 6:

$$f_o = \frac{1}{2\pi RC} = 20 \text{ kHz}$$

We are free to choose a value of either R or C and then calculate the value of the other. Let us select:

$$C = 0.01 \ \mu\text{F}$$

Substitution then gives:

$$\frac{1}{2\pi R \times 0.01 \ \mu\text{F}} = 20 \text{ kHz}$$

$$R = \frac{1}{6.28 \times 0.01 \ \mu\text{F} \times 20 \text{ kHz}}$$
$$= 0.80 \text{ k}\Omega = \mathbf{800 \ \Omega}$$

One quantity that must be considered for the output voltage of a filter is the phase shift between the input and output voltages. In Problem 5 of Chapter 18, an expression was derived for the phase difference between the capacitor (output) voltage and the applied (input) voltage for a series RC circuit. It was given by:

$$\tan \phi_o = -\omega CR$$

Substituting in the expression for f_o, the

PHASE SHIFT FOR SIMPLE LOW-PASS FILTER

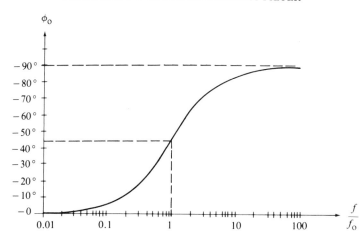

FIGURE 19.10. A generalized graph shows the phase shifts between V_o and V_{in} for a simple low-pass filter. Note that for $f \ll f_o$, ϕ_o approaches zero. When $f = f_o$, $\phi_o = -45°$, and for $f \gg f_o$, ϕ_o approaches $-90°$.

phase shift of a simple low-pass RC filter is found to be:

$$\tan \phi_o = -\frac{f}{f_o} \qquad (10)$$

The phase shift can be graphed as a generalized frequency response curve, as shown in Figure 19.10. Note that at the corner frequency, when $f = f_o$, the phase shift is $-45°$. At low frequencies, which pass undiminished, f is much smaller than f_o and the phase shift approaches zero. At high frequencies, which are strongly attenuated, f is much larger than f_o and the phase shift is $-90°$.

When using an RC filter in an actual circuit, the effects of components connected to the output must also be considered. Up to this point, the output has been assumed to be an open circuit. Therefore, V_o of the filter is essentially the filter's output open circuit voltage. V_o is also the actual output voltage if the circuit following the filter has a very high input resistance. This condition can be created by adding an op amp voltage follower, as shown in Figure 19.11. The voltage follower will have a very

LOW-PASS FILTER WITH VOLTAGE FOLLOWER

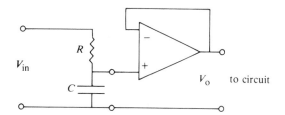

FIGURE 19.11. The simple low-pass filter has the frequency response characteristics described in the text provided that its output terminals look into a high impedance. Thus, it is necessary to insert a voltage follower between the filter and subsequent circuitry.

high input resistance, and its unity gain will not affect the magnitude of V_o.

If the circuit following the filter has a smaller resistance, shown as R_L in Figure 19.12a, then it must be included in the circuit analysis. The simplest way is to use Thevenin's theorem. Because the voltage across the resistor R_L is the same as the voltage across the capacitor C, we can reverse the positions of C and R_L, as shown in Figure 19.12b.

FIGURE 19.12. When a simple low-pass filter is connected to a load having a finite resistance R_L, loading effects on V_o must be determined. Using Thevenin's theorem as shown, we find that the net result is an increase in the corner frequency and a decrease in the output amplitudes by the voltage divider factor $R_L/(R + R_L)$.

LOADED LOW-PASS FILTER

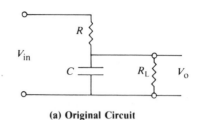

(a) Original Circuit

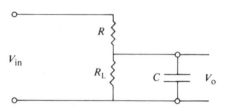

(b) Equivalent Circuit

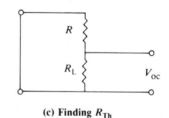

(c) Finding R_{Th}

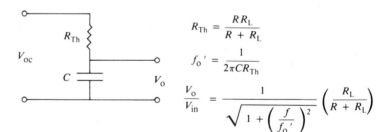

$$R_{Th} = \frac{RR_L}{R + R_L}$$

$$f_o' = \frac{1}{2\pi C R_{Th}}$$

$$\frac{V_o}{V_{in}} = \frac{1}{\sqrt{1 + \left(\frac{f}{f_o'}\right)^2}} \left(\frac{R_L}{R + R_L}\right)$$

(d) Thevenin Equivalent Circuit

The Thevenin resistance R_{Th} is found by short-circuiting the voltage source, as shown in Figure 19.12c. It becomes clear that R_{Th} is R_L in parallel with R:

$$R_{Th} = R \,||\, R_L = \frac{RR_L}{R + R_L} \qquad (11)$$

We can then find the open circuit voltage by neglecting the capacitor, as shown in Figure 19.12b. We find:

$$V_{oc} = \frac{R_L}{R + R_L} V_{in} \qquad (12)$$

Thus, the Thevenin equivalent network for the low-pass filter with a load is as shown in Figure 19.12d. This circuit is simply a low-pass filter in which R_{Th} is the series resistor and V_{oc} is the input voltage. Therefore, the filter's transfer function is given by Equation 1 with V_{oc} substituted for V_{in} and R_{Th} substituted for R:

$$\frac{V_o}{V_{oc}} = \frac{1}{\sqrt{1 + \omega^2 C^2 R_{Th}^2}}$$

or

$$\frac{V_o}{[R_L/(R + R_L)]V_{in}}$$

$$= \frac{1}{\sqrt{1 + \omega^2 C^2 [RR_L/(R + R_L)]^2}}$$

$$\frac{V_o}{V_{oc}} = \frac{R_L/(R + R_L)}{\sqrt{1 + \omega^2 C^2 [RR_L/(R + R_L)]^2}} \tag{13}$$

$$= \frac{R_L}{\sqrt{(R + R_L)^2 + \omega^2 C^2 R^2 R_L^2}} \tag{14}$$

Of more interest is the value of the **corner frequency** that is given by Equation 6:

$$f_o' = \frac{1}{2\pi CR_{Th}} = \frac{R + R_L}{2\pi CRR_L} \tag{15}$$

Hence, the Bode diagram is that of a simple low-pass filter with a corner frequency given by Equation 15. The effects of the load resistance become apparent when the equation for f_o' is rewritten as:

$$f_o' = \frac{1}{2\pi CR} \times \frac{R + R_L}{R_L} \tag{16}$$

When the load resistance R_L is large compared to R, the factor $(R + R_L)/R_L = 1$, and the circuit behaves as if R_L were an open circuit. As

R_L decreases toward R, the corner frequency increases by the factor:

$$\frac{R + R_L}{R_L} \tag{17}$$

For example, if $R_L = R$, then $(R + R_L)/R_L = 2$, and the corner frequency is given by:

$$f_o' = 2\left(\frac{1}{2\pi CR}\right)$$

which is twice the value without R_L.

Note also that the expression for the corner frequency can be substituted into Equation 14 for the **voltage transfer function** to yield:

$$\frac{V_o}{V_{in}} = \frac{1}{\sqrt{1 + (f/f_o)^2}} \times \frac{R_L}{R + R_L} \tag{18}$$

In the low-pass region, where $f \ll f_o$, the output voltage is:

$$V_o = V_{in}\left(\frac{R_L}{R + R_L}\right) \tag{19}$$

Hence, the amplitude of the output voltage is reduced by the voltage divider equation.

High-Pass Filter

For a high-pass filter, the resistor voltage is taken as V_o (Figure 19.13). The **transfer function** of the simple high-pass filter is:

$$\frac{V_o}{V_{in}} = \frac{1}{\sqrt{1 + (f_o/f)^2}} \tag{20}$$

where the expression for f_o has been substituted in.

SIMPLE HIGH-PASS FILTER

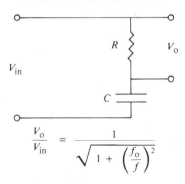

$$\frac{V_o}{V_{in}} = \frac{1}{\sqrt{1 + \left(\dfrac{f_o}{f}\right)^2}}$$

FIGURE 19.13. A simple high-pass filter is the resistor voltage of a series *RC* circuit, as shown. As for the low-pass filter, the voltage transfer function is more simply written in terms of f_o, the corner frequency. The roll-off is also 20 db/decade (3 db/octave). See Example 7.

As in the case of the low-pass filter, when $f = f_o$:

$$\frac{V_o}{V_{in}} = \frac{1}{\sqrt{1 + 1}} = 0.707 \tag{21}$$

which is the value at the half-power point. Therefore, f_o is also the **corner frequency** for the simple high-pass filter.

EXAMPLE 7: A simple high-pass filter has $C = 0.1\ \mu\text{F}$ and $R = 1.6\ \text{k}\Omega$. What is the value of the corner frequency f_o and the roll-off?

Solution: The corner frequency is found from Equation 6:

$$f_o = \frac{1}{2\pi CR} = \textbf{1.0 kHz}$$

which is the same in Example 4.

The roll-off can be found from the slope of the curve at frequencies that are *small*

compared to f_o. Therefore, we will calculate the gain at 0.1 kHz and at 0.01 kHz by substituting Equation 20 for V_o/V_{in} in Equation 3:

$$\text{gain (db)} = 20 \log \left[\frac{1}{\sqrt{1 + (f_o/f)^2}}\right] \tag{22}$$

At $f = 0.1$ kHz:

$$\text{gain (db)} = 20 \log \left[\frac{1}{\sqrt{1 + (1\ \text{kHz}/0.1\ \text{kHz})^2}}\right]$$

$$= 20 \log\left(\frac{1}{\sqrt{101}}\right) = -20\ \text{db}$$

At $f = 0.01$ kHz:

$$\text{gain (db)} = 20 \log \left[\frac{1}{\sqrt{1 + (1\ \text{kHz}/0.01\ \text{kHz})^2}}\right]$$

$$= 20 \log\left(\frac{1}{\sqrt{10.001}}\right) = -40\ \text{db}$$

Thus, the roll-off is:

$$\text{roll-off} = -20\ \text{db} - (-40\ \text{db})$$
$$= \textbf{20 db/decade}$$

A similar calculation for a doubled frequency gives:

$$\text{roll-off} = \textbf{6.0 db/octave}$$

Thus the **roll-off** of the simple high-pass *RC* filter is 20 db/decade or 6 db/octave.

The transfer function of Equation 22 is shown as a generalized Bode diagram in Figure 19.14. Compare this result with the Bode diagram for a low-pass filter, Figure 19.9.

The phase shift of the high-pass filter is the phase shift of the resistor voltage, which was found in Chapter 18, Problem 5 to be:

GENERALIZED BODE DIAGRAM FOR SIMPLE HIGH-PASS FILTER

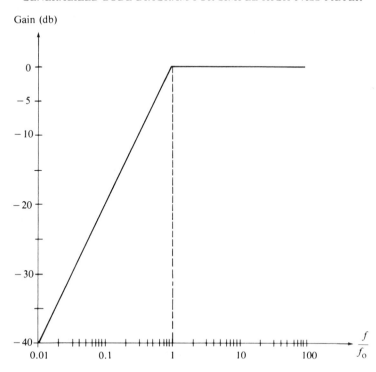

FIGURE 19.14. A generalized Bode diagram simplifies circuit calculations for the simple high-pass filter.

$$\tan \phi_o = \frac{1}{\omega CR} \tag{23}$$

Substituting in this value of f_o gives the **phase shift** of a simple high-pass RC filter:

$$\tan \phi_o = \frac{f_o}{f} \tag{24}$$

Figure 19.15 shows this phase shift as a frequency response graph. At the corner frequency, the phase shift is 45°; at low frequencies, which are strongly attenuated, the phase shift is 90°; and, at high frequencies, which pass undiminished, the phase shift is zero. Compare Figure 19.15 with the frequency response graph for a low-pass filter (Figure 19.10).

As in the case of the low-pass filter, the above results are valid only if the circuit following the filter has a high input resistance—for example, an op amp voltage follower. If the subsequent circuit has a smaller resistance, as shown in Figure 19.16a, then it must be included in the analysis. In this case, the equivalent resistance of the RC filter is $R \parallel R_L$, rather than simply R, as shown in Figure 19.16b:

$$R_{eq} = R \parallel R_L = \frac{RR_L}{R + R_L} \tag{25}$$

Thus, the voltage transfer function is:

$$\frac{V_o}{V_{in}} = \frac{1}{\sqrt{1 + 1/\omega^2 C^2 R_{eq}^2}} \tag{26}$$

$$= \frac{1}{\sqrt{1 + 1/\omega^2 C^2 [RR_L/(R + R_L)]^2}} \tag{27}$$

FIGURE 19.15. The generalized curve for the phase shift between V_o and V_{in} shows that for $f \ll f_o$, ϕ approaches 90°; when $f = f_o$, ϕ_o = 45°; and for $f \gg f_o$, ϕ approaches zero.

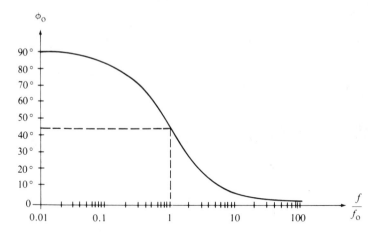

PHASE SHIFT FOR SIMPLE HIGH-PASS FILTER

The new **corner frequency** is:

$$f'_o = \frac{1}{2\pi CR_{eq}} = \frac{1}{2\pi CRR_L/(R + R_L)}$$

$$f'_o = \frac{1}{2\pi CR} \times \frac{R + R_L}{R_L} \qquad (28)$$

Note that this expression is identical to the corner frequency of the loaded low-pass filter. The additional factor $(R + R_L)/R_L$ serves to increase the corner frequency above that of the unloaded filter. The smaller the value of R_L, the greater the corner frequency. Substitution of Equation 28 into the **voltage transfer function** gives:

$$\frac{V_o}{V_{in}} = \frac{1}{\sqrt{1 + (f'_o/f)^2}} \qquad (29)$$

In the high-pass region where $f \gg f'_o$, Equation 29 reduces to:

$$V_o = V_{in} \qquad (30)$$

LOADED HIGH-PASS FILTER

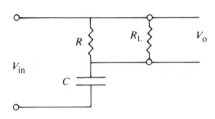

(a) Original Circuit

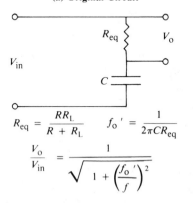

(b) Equivalent Circuit

FIGURE 19.16. Like the low-pass filter, if the high-pass filter has a load with a finite resistance R_L, then loading effects must be considered. Loading serves to increase the corner frequency, but leaves the amplitude of the input voltage for $f \gg f'_o$ unchanged.

which is identical to the voltage transfer function of the unloaded filter.

19.6 PASSIVE TWO-STAGE FILTERS

For many applications, a roll-off of 6 db/octave is not fast enough. For example, a signal voltage may occur at 100 Hz, and there may be a great deal of noise at 60 Hz. A high-pass filter with a corner frequency of 80 Hz or 90 Hz permits a significant amount of 60 Hz noise to get through. What is needed is a high-pass filter with a steeper roll-off to attenuate the 60 Hz noise more effectively.

The most straightforward method for increasing the roll-off is to connect two (or more) simple filters in series — that is, to filter the signal more than once. This method is not generally the most efficient, but it is examined here because the results provide a good introduction to more complex filter designs.

Two filters in series are shown schematically in Figure 19.17. Our task is to find the transfer function V_o / V_{in} for the net result of the composite filter. Suppose that the first circuit has a transfer function of:

$$\frac{V_{o_1}}{V_{in_1}} = H_1 \tag{31}$$

and the second has a transfer function of:

$$\frac{V_{o_2}}{V_{in_2}} = H_2 \tag{32}$$

where H_1 and H_2 represent the proper algebraic relationships involving ω, C, and R.

TWO-STAGE FILTER

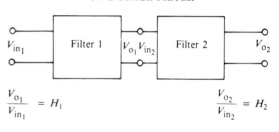

$$\frac{V_{o_1}}{V_{in_1}} = H_1 \qquad \frac{V_{o_2}}{V_{in_2}} = H_2$$

$$\frac{V_{o_2}}{V_{in_1}} = H_1 H_2$$

FIGURE 19.17. A roll-off greater than 6 db/octave can be achieved by adding more filter stages. The voltage transfer function, then, is simply the product of the voltage transfer functions for each stage. For this condition to be true, however, each stage must look into a high impedance, such as an op amp voltage follower.

In a two-stage filter, the output of the first filter serves as the input to the second filter. Therefore:

$$V_{o_1} = V_{in_2} \tag{33}$$

and Equation 32 becomes:

$$V_{o_2} = H_2 V_{o_1} \tag{34}$$

Substitution of Equation 31 for V_{o_1} gives:

$$V_{o_2} = H_1 H_2 V_{in_1} \tag{35}$$

or, simply:

$$V_o = H_1 H_2 V_{in} \tag{36}$$

This expression says that the transfer function for the composite filter is simply the product of the transfer functions H_1 and H_2 of the individual filters. Note that the product $H_1 H_2$ is equivalent to $H_2 H_1$. That is, it makes no difference in which order the circuits are connected because the output voltage is the same.

The connection of two filters must be done carefully in order that the two do not interact. Recall that the voltage transfer function assumes that the filter output looks into an open circuit. This arrangement can be achieved in practice in one of two ways. One is to separate the two filters by an op amp voltage follower. Because the follower has a very high input resistance, the voltage transfer function of the first filter will be accurately given by Equation 31. Also, the voltage follower will achieve the requirement of Equation 33 by faithfully transferring the output of the first filter to the input of the second.

This condition can also be approximated by selecting values of R and C for the second filter that yield an impedance that is large compared to that of the first filter. Then, the impedance seen by the first filter will be sufficiently large that any loading effects will be negligible, as the following examples demonstrate.

Low-Pass Filter

Two identical low-pass filters, separated by a voltage follower, are shown in Figure 19.18. The transfer function for each filter is given by Equation 8:

$$V_{o_{1,2}} = \frac{1}{\sqrt{1 + (f/f_o)^2}} \, V_{in_{1,2}}$$

where is is assumed that each filter has the same corner frequency $f_o = (1/2\pi RC)$. Therefore:

$$H_1 = H_2 = \frac{1}{\sqrt{1 + (f/f_o)^2}} \tag{37}$$

The product $H_1 H_2$ is:

$$H_1 H_2 = \frac{1}{1 + (f/f_o)^2} \tag{38}$$

so the **transfer function** for the two-stage low-pass filter is:

TWO-STAGE LOW-PASS FILTER

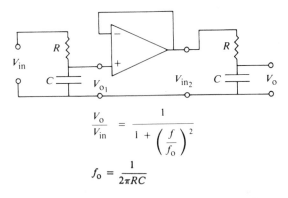

$$\frac{V_o}{V_{in}} = \frac{1}{1 + \left(\dfrac{f}{f_o}\right)^2}$$

$$f_o = \frac{1}{2\pi RC}$$

FIGURE 19.18. The design of a two-stage low-pass filter is shown. Note that the two filters are identical and have corner frequencies of $f_o = 1/(2\pi RC)$ and that they are separated by a high-impedance voltage follower.

$$V_o = \frac{1}{1 + (f/f_o)^2} \, V_{in} \tag{39}$$

The generalized Bode diagram for the two-stage low-pass filter is shown in Figure 19.19. It is similar to the Bode diagram for a single low-pass filter (Figure 19.9), but the roll-off in this case is 12 db/octave (40 db/decade).

Additional filter stages further increase the roll-off by 6 db/octave per stage. In other words, three stages will have a roll-off of 18 db/octave, four stages 24 db/octave, and so on.

The two-stage low-pass filter can also be achieved without an op amp, as the following example illustrates.

EXAMPLE 8: Without using voltage followers, design a two-stage low-pass filter with a corner frequency of 1.0 kHz.

BODE DIAGRAM FOR TWO-STAGE LOW-PASS FILTER

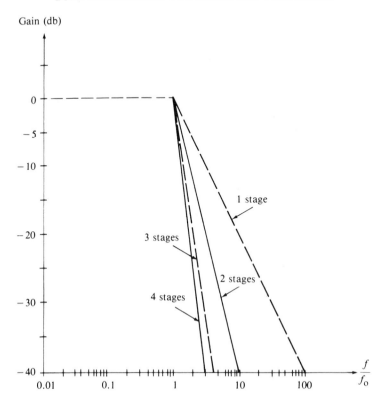

FIGURE 19.19. The generalized Bode diagram for the two-stage low-pass filter shows its roll-off to be – 12 db/octave, which is twice the value for a one-stage filter. Successive stages each increase the roll-off by – 6 db/octave per stage, as shown.

Solution: The corner frequency expression for the low-pass RC filter is given by Equation 6:

$$f_o = \frac{1}{2\pi R_1 C_1} = 1 \text{ kHz}$$

Let us select a relatively low value of resistance R for the first stage, for example:

$$R_1 = 1 \text{ k}\Omega$$

The necessary capacitance value can then be calculated from Equation 6:

$$\frac{1}{2\pi \times 1 \text{ k}\Omega \times C_1} = 1 \text{ kHz}$$

$$C_1 = \frac{1}{6.28 \times 1 \times 1} = 0.16 \ \mu\text{F}$$

For the second filter, we want the same corner frequency, $f_o = 1$ kHz, but a significantly larger impedance. Therefore, let us select:

$$R_2 = 100 \text{ k}\Omega$$

The corresponding capacitance value is:

EXAMPLE 8

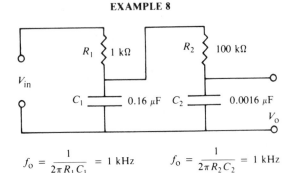

$$f_o = \frac{1}{2\pi R_1 C_1} = 1 \text{ kHz} \qquad f_o = \frac{1}{2\pi R_2 C_2} = 1 \text{ kHz}$$

FIGURE 19.20. A two-stage filter need not include an op amp voltage follower provided that the impedance of the second stage is large compared to that of the first stage.

$$\frac{1}{2\pi \times 100 \text{ k}\Omega \times C_2} = 1 \text{ kHz}$$

$$C_2 = \frac{1}{2\pi \times 100 \times 1} = 0.0016 \ \mu\text{F}$$

The complete circuit is shown in Figure 19.20.

$$H_1 H_2 = \frac{1}{1 + (f_o/f)^2} \tag{41}$$

and the **transfer function** for the two-stage high-pass filter is:

$$V_o = \frac{1}{1 + (f_o/f)^2} V_{in} \tag{42}$$

The generalized Bode diagrm for the two-stage high-pass filter is shown in Figure 19.22. Like the low-pass filter, this type has a roll-off of 12 db/octave (40 db/decade), and successive stages increase the roll-off by 6 db/octave per stage.

EXAMPLE 9: Without using a voltage follower, design a two-stage high-pass filter with a corner frequency of 1 kHz.

Solution: The solution to this problem is identical to that of Example 8, with a reversal of the roles of R and C for each stage. The circuit is shown in Figure 19.23.

High-Pass Filter

A two-stage high-pass filter is shown in Figure 19.21. The same method is used to find its transfer function as was used for the low-pass filter. The transfer function for each stage is given by Equation 20:

$$V_{o_{1,2}} = \frac{1}{\sqrt{1 + (f_o/f)^2}} V_{in_{1,2}}$$

where each stage has the same corner frequency $f_o = 1/2\pi RC$. Therefore:

$$H_1 = H_2 = \frac{1}{\sqrt{1 + (f_o/f)^2}} \tag{40}$$

Band-Pass Filter

A band-pass filter is simply a low-pass filter and a high-pass filter whose corner frequencies are selected to pass only a certain band of frequencies. Figure 19.24 shows a typical filter circuit with an op amp voltage follower between the two stages to eliminate loading. The transfer function of this circuit is the product of the transfer functions of the first-stage low-pass filter and the second-stage high-pass filter:

$$\frac{V_o}{V_{in}} = H_1 H_2$$

Substituting for H_1 and H_2 gives:

TWO-STAGE HIGH-PASS FILTER

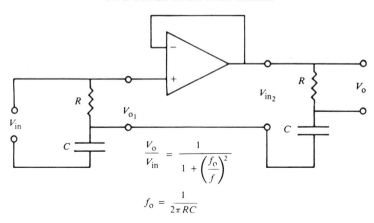

$$\frac{V_o}{V_{in}} = \frac{1}{1 + \left(\dfrac{f_o}{f}\right)^2}$$

$$f_o = \frac{1}{2\pi RC}$$

FIGURE 19.21. In a two-stage high-pass filter, as shown here, both filters have the same corner frequency and are separated by an op amp voltage follower.

BODE DIAGRAM FOR TWO-STAGE HIGH-PASS FILTER

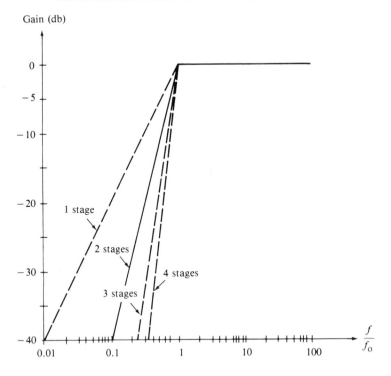

FIGURE 19.22. The generalized Bode diagram for the two-stage high-pass filter shows its roll-off to be 12 db/octave. Successive stages increase the roll-off by 6 db/octave per stage.

EXAMPLE 9

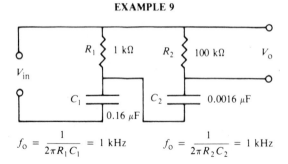

$$f_o = \frac{1}{2\pi R_1 C_1} = 1 \text{ kHz} \qquad f_o = \frac{1}{2\pi R_2 C_2} = 1 \text{ kHz}$$

FIGURE 19.23. The two-stage filter need not include an op amp voltage follower provided that the impedance of the second stage is large compared to that of the first stage.

$$\frac{V_o}{V_{in}} = \frac{1}{\sqrt{1 + (f/f_{LP})^2}} \times \frac{1}{\sqrt{1 + (f_{HP}/f)^2}} \quad (43)$$

where f_{LP} is the corner frequency of the low-pass filter and f_{HP} is the corner frequency of the high-pass filter. These corner frequencies are given by:

$$f_{LP} = \frac{1}{2\pi R_1 C_1}$$

$$f_{HP} = \frac{1}{2\pi R_2 C_2} \qquad (44)$$

The **width of the pass band** Δf_{PB} is the difference between the low and high corner frequencies:

$$\Delta f_{PB} = f_{LP} - f_{HP} \qquad (45)$$

The Bode diagram for a typical band-pass filter is shown in Figure 19.25. Note that the roll-offs of the leading and trailing edges are 6 db/octave. The roll-off can be increased, of

course, by adding more stages to each of the two filters.

EXAMPLE 10: Design a band-pass filter with the frequency response curve shown in Figure 19.25.

Solution: The basic circuit diagram for a band-pass filter is shown in Figure 19.24. Figure 19.25 gives the lower and upper corner frequencies as:

$$f_{LP} = 8.0 \text{ kHz}$$

$$f_{HP} = 5.0 \text{ kHz}$$

Using Equation 44, we can determine values for R_1, C_1, R_2, and C_2 as follows:

For $f_{LP} = 8 \text{ kHz} = \dfrac{1}{2\pi R_2 C_2}$

Let $R_2 = 10 \text{ k}\Omega$

Then $C_2 = \dfrac{1}{2\pi \times 10 \text{ k}\Omega \times 8 \text{ kHz}}$

$\qquad = 0.0020 \ \mu\text{F}$

For $f_{HP} = 0.5 \text{ kHz} = \dfrac{1}{2\pi R_1 C_1}$

Let $R_1 = 10 \text{ k}\Omega$

Then $C_1 = \dfrac{1}{2\pi \times 10 \text{ k}\Omega \times 5 \text{ kHz}}$

$\qquad = 0.0032 \ \mu\text{F}$

Band-Reject Filter

A band-reject filter is identical to a band-pass filter except that the corner frequency of the high-pass filter is *greater* than that of the low-pass filter. Therefore, the voltage transfer function is also given by Equation 43 and the corner frequencies by Equation 44.

The Bode diagram of a typical band-reject filter is shown in Figure 19.26. As in the case of

BAND-PASS FILTER

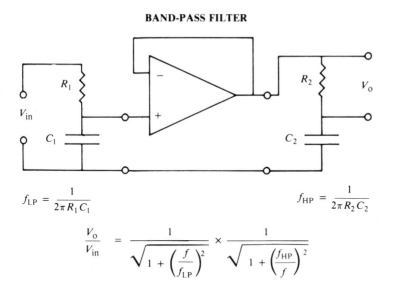

$$f_{LP} = \frac{1}{2\pi R_1 C_1}$$

$$f_{HP} = \frac{1}{2\pi R_2 C_2}$$

$$\frac{V_o}{V_{in}} = \frac{1}{\sqrt{1 + \left(\dfrac{f}{f_{LP}}\right)^2}} \times \frac{1}{\sqrt{1 + \left(\dfrac{f_{HP}}{f}\right)^2}}$$

FIGURE 19.24. A band-pass filter is simply a low-pass filter followed by a high-pass filter (or vice versa) in which the upper roll-off frequency is determined by the corner frequency of the low-pass filter f_{LP} and the lower roll-off frequency is determined by the corner frequency of the high-pass filter f_{HP}.

BODE DIAGRAM FOR BAND-PASS FILTER

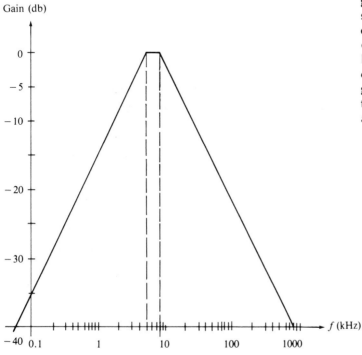

FIGURE 19.25. The Bode diagram for a typical band-pass filter shows the upper and lower roll-offs to be 6 db/octave, which is characteristic of the low- and high-pass filters that compose it. A circuit having the Bode diagram shown can be designed from the corner frequencies. See Example 10.

FIGURE 19.26. The circuit diagram and voltage transfer equations for a band-reject filter are identical to those for a band-pass filter except that the corner frequency of the high-pass filter is greater than that for the low-pass filter. The figure shows a typical Bode diagram. Note that roll-offs are 6 db/octave.

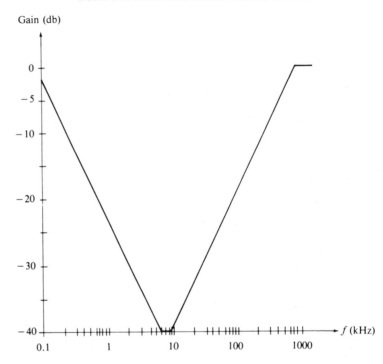

BODE DIAGRAM FOR BAND-REJECT FILTER

the band-pass filter, the roll-offs are 6 db/octave. Again, steeper roll-offs are obtainable with additional stages.

Q Factor

In many applications, we want to pass (or reject) only a single frequency or a very narrow range of frequencies. A common example is the tuning of a radio or TV for which an antenna picks up a wide range of radio frequencies and feeds them into the set's input circuit. The role of the tuning circuit is to permit only the frequency of a particular station to pass through to the amplifier. Because station frequencies are very close together, the pass band of the tuner must be very narrow to prevent stations at adjacent frequencies from getting through and being superimposed on the desired station.

A simple way of making a narrow band-pass filter is to make the corner frequencies of the high-pass and low-pass filters shown in Figure 19.25 identical, but increasing the roll-offs with additional stages. A narrow band-reject filter can be made in a similar way. A typical frequency response curve for a narrow band-pass filter with roll-offs of 24 db/octave is shown in Figure 19.27.

Narrow band filters are extremely important in ac and radio frequency circuits and, in general, the narrower they are, the better they function. Circuit designers often design very complex filters and go to great expense to make the roll-offs of the leading and trailing edges of the filter as steep and close together as possible.

The quality of a narrow band filter is measured, therefore, by the narrowness of its frequency response curve. This **quality factor** is given the symbol Q and is arbitrarily defined as

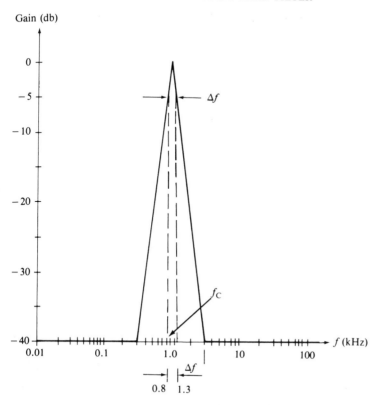

BODE DIAGRAM FOR NARROW BAND FILTER

FIGURE 19.27. Narrow band filters are extremely important circuits – for example, in the tuning circuits of radios and TVs. A measure of the narrowness of the frequency response is given by the Q factor, which is the center frequency f_C divided by the frequency width at the -3 db point, $\Delta f(-3\,db)$. See Example 11.

the central frequency of the filter f_C divided by the frequency width Δf at the -3 db point below maximum gain:

$$Q = \frac{f_C}{\Delta f(-3\,db)} \qquad (46)$$

The narrower the width Δf, the greater the value of Q and the narrower the response of the filter.

EXAMPLE 11: What is the Q of the filter whose frequency response curve is shown in Figure 19.27?

Solution: The center frequency of the filter is:

$$f_C = 1.0\ \text{kHz}$$

The frequency width at -3 db appears on the graph at:

$$\Delta f = 1.3\ \text{kHz} - 0.8\ \text{kHz} = 0.5\ \text{kHz}$$

Therefore:

$$Q = \frac{1.0\ \text{kHz}}{0.5\ \text{kHz}} = 2.0$$

Tuned Amplifier

In a narrow band filter, the corner frequencies of the low- and high-pass filter components are identical. According to the voltage transfer function for each filter, the gain at the corner

FIGURE 19.28. A narrow band filter can be turned into a tuned amplifier by adding gain to the voltage follower as shown. This gain will restore any signal losses due to the individual filters and increase the output for successive stages as well.

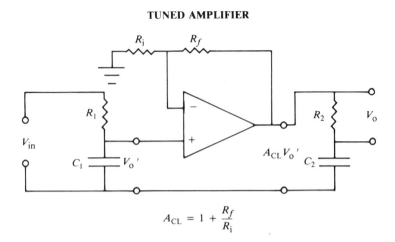

TUNED AMPLIFIER

$$A_{CL} = 1 + \frac{R_f}{R_i}$$

frequency is -3 db. Hence, there will be an attenuation of -6 db of the input signal voltage, even at the peak of the pass band. (Note that this feature is not shown in Figure 19.27.)

To restore the signal voltage to its incoming value or to increase it, an amplifier can be added to the circuit. A narrow band filter combined with an amplifier is called a **tuned amplifier**. The basic two-stage filter circuit shown in Figure 19.24 can be converted into a tuned amplifier by adding gain to the voltage follower that acts as a buffer between the two filters (Figure 19.28). That is, the voltage follower is converted to a noninverting amplifier whose closed loop gain A_{CL} is given by:

$$A_{CL} = 1 + \frac{R_f}{R_i} \qquad (47)$$

Voltage gains up to about 1000 are possible with a single stage of amplification, but additional gains can be achieved by adding more amplification stages.

19.7 OTHER FILTER CIRCUITS

The principles of specialized filter design can be quite complicated and they are beyond the

scope of this book. In most cases, designing a filter is unnecessary because we can buy a filter with almost any desired frequency characteristic. However, there are a number of basic filter configurations that appear frequently in ac circuits and generally form the basis of the more exotic filters, including active filters. These circuits are briefly introduced here so that they will be recognized if encountered.

Twin T Filter

The **twin T filter** circuit is shown in Figure 19.29. The reason for the name is obvious from the circuit diagram. The two resistors R and the capacitor $2C$ form one of the Ts, and the other two capacitors and the resistor $R/2$ form the second T. The importance of this circuit is that its output voltage V_o goes to zero at one frequency value. Hence, the twin T filter is a narrow band-reject, or notch, filter.

Bridge T Filter

The **bridge T filter** circuit is shown in Figure 19.30. This circuit is a T consisting of two equal capacitors C and a resistor R_1 that is "bridged" by a resistor R_2. This filter is also a notch filter,

TWIN T FILTER

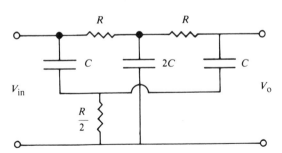

FIGURE 19.29. A common, somewhat complex, narrow band-reject filter is the twin T filter, shown. Its name comes from the "T" arrangement of resistors and capacitors. It has a very narrow notch.

BRIDGE T FILTER

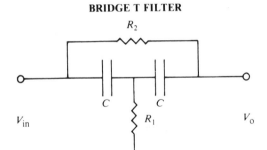

FIGURE 19.30. Another narrow band notch filter is the bridge T filter shown here. Its notch is not quite as narrow as that of the twin T, but it is much simpler to build and balance.

but its roll-off at the notch frequency is not as steep as that of a twin T.

Comparing the twin T to the bridge T shows some obvious trade-offs. The twin T is a much sharper filter, but it is much more complicated to build and balance than the bridge T. The selection of a filter for a given application involves balancing the desired value of Q against the cost of components.

19.8 ACTIVE FILTERS

Cascading simple filters is not a very practical method of achieving high values of roll-off or Q. Modern integrated circuits, particularly operational amplifiers, make it possible to achieve the same transfer functions with active filter designs that are much simpler to build and that require many fewer components.

An **active filter** is one that contains an external source of energy and thus can provide some voltage gain for the signal. The most common active filter designs utilize an operational amplifier whose external power supply provides the required energy to yield signal gain. The filter action is achieved by placing

various simple filter configurations in the feedback loop.

The full description of active filters is beyond the scope of this book. However, a general idea of what they look like and some simple tools to help in understanding their designs are given here.

Voltage Transfer Functions

The first point to make is that the voltage transfer function of a filter of a given order has the same general form, regardless of how it is achieved. For example, the transfer function of a passive second-order filter looks very much like the transfer function of an active second-order filter, even though their circuit diagrams are quite different, as illustrated in Figure 19.31.

Figure 19.31 lists the voltage transfer functions in s-notation for different two-stage passive filters and the more general form of the transfer function for a second-order filter. Note that the expressions are nearly the same except for the constant terms H and α (Greek letter alpha) in the general expressions.

The actual value of H is the gain of the

FIGURE 19.31. The voltage transfer function of a second-order filter is the same, regardless of how it is achieved. The table shows the voltage transfer functions for two-stage passive filters and the general form for any second-order filter. In the general form, H_o is the gain of the filter in the flat region and $\alpha = 1/Q$ is the inverse of the filter's Q value.

VOLTAGE TRANSFER FUNCTIONS

Type of Filter	Two-Stage Passive Filter	Generalized Second-Order Filter
Low-pass	$\dfrac{\omega_o^2}{s^2 + 2\omega_o s + \omega_o^2}$	$\dfrac{H_o \omega_o^2}{s^2 + \alpha \omega_o s + \omega_o^2}$
High-pass	$\dfrac{s^2}{s^2 + 2\omega_o s + \omega_o^2}$	$\dfrac{H_o s^2}{s^2 + \alpha \omega_o s + \omega_o^2}$
Band-pass	$\dfrac{\omega_o s}{s^2 + 2\omega_o s + \omega_o^2}$	$\dfrac{H_o \alpha \, \omega_o s}{s^2 + \alpha \omega_o s + \omega_o^2}$

filter in its flat region. Similarly, α is the inverse of the filter's Q value:

$$\alpha = \frac{1}{Q} \qquad (48)$$

The frequency ω_o is the corner frequency of the filter $\omega_o = 2\pi f_o$. Comparing the two-stage passive filter expressions to those of the general expression, we see that for the **low- and high-pass filters**:

$$H = 1 \qquad (49)$$
$$Q = \frac{1}{\alpha} = \frac{1}{2} \qquad (50)$$

and for the **band-pass filter**:

$$H = \frac{1}{2} \qquad (51)$$

$$Q = \frac{1}{\alpha} = \frac{1}{2} \qquad (52)$$

The particular values of H and α determine specific filter characteristics, but the general behavior is that of a second-order filter with a

roll-off of 12 db/octave provided that the placement of terms involving s and ω_o are those given in Figure 19.31. Similar expressions can be obtained for third-order filters with roll-offs of 18 db/octave, and so on.

With this general understanding, let us now examine the design of active second-order filters involving op amps. Just remember that the goal is to come up with circuit designs that have transfer functions comparable to those given in Figure 19.31.

General Operational Amplifier Filter Design

The general design of an active filter that includes an operational amplifier is shown in Figure 19.32. The five numbered boxes represent single circuit components: resistors, capacitors, and inductors. The arrows show the assumed direction of the currents. Note that the general structure is that of an inverting amplifier in which circuit components complete the negative feedback loop.

To analyze this circuit, an alternative form of Ohm's law is used:

ACTIVE FILTER

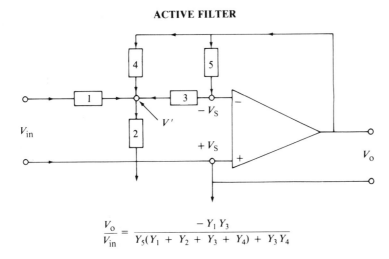

FIGURE 19.32. The most common way of making higher-order filters is to use an op amp as an active filter. The figure shows the general design of an active second-order filter in which boxes 1 through 5 represent single circuit elements: resistors, capacitors, and inductors. The voltage transfer function is expressed in terms of Y, the inverse of each component's impedance.

$$\frac{V_o}{V_{in}} = \frac{-Y_1 Y_3}{Y_5(Y_1 + Y_2 + Y_3 + Y_4) + Y_3 Y_4}$$

$$I = YV \tag{53}$$

$$\text{where} \quad Y = \frac{1}{R} \text{ for a resistor} \tag{54}$$

$$Y = sC \text{ for a capacitor} \tag{55}$$

$$Y = \frac{1}{sL} \text{ for an inductor} \tag{56}$$

The use of the inverted forms Y for impedance is for convenience only and does not affect the analysis or results. It merely simplifies the algebraic manipulations and the form of the result.

To begin the analysis, recall that one of the characteristics of an op amp is that the two input terminals, $+$ and $-$, are assumed to be at the same voltage. Therefore, since the $+$ input is at ground ($+V_S = 0$), we can also assume:

$$+V_S = 0 \tag{57}$$

$$-V_S = 0 \tag{58}$$

Applying Ohm's law as given by Equation 53, we can write the following equations for the currents through each of the five circuit elements:

$$I_1 = Y_1(V_{in} - V') \tag{59}$$

$$I_2 = Y_2 V' \tag{60}$$

$$I_3 = Y_3(-V') \tag{61}$$

$$I_4 = Y_4(V_o - V') \tag{62}$$

$$I_5 = Y_5 V_o \tag{63}$$

Another characteristic of the op amp is that it has a very high input impedance and, hence, is assumed to draw no input current. Thus, we must have:

$$I_5 = I_3 \tag{64}$$

By KCL, we must also have at the node:

$$I_4 + I_3 + I_1 = I_2 \tag{65}$$

Substituting in the values of the currents given by Equations 59 to 62 into Equation 64 gives an expression for the voltage transfer function of the circuit shown in Figure 19.32. This procedure yields:

$$\frac{V_o}{V_{in}} = \frac{-Y_1 Y_3}{Y_5(Y_1 + Y_2 + Y_3 + Y_4) + Y_3 Y_4} \tag{66}$$

This transfer function looks quite complicated, particularly because it involves the unknown quantities Y. However, it serves an extremely important role in converting the generalized circuit design of the active filter shown in Figure 19.31 into a real second-order filter.

The general strategy is to select values for the various Ys (and, hence, circuit components for Figure 19.32) that will yield the proper general form for the transfer function given in Figure 19.31. The following paragraphs illustrate the process.

Second-Order Low-Pass Filter

The general form for a second-order low-pass filter is given in Figure 19.31 as:

$$\frac{V_o}{V_{in}} = \frac{H_o \omega_o^2}{s^2 + \alpha \omega_o + \omega_o^2} \qquad (67)$$

Note that there is no s in the numerator. Therefore, we conclude that Y_1 and Y_3 in Equation 66 must be resistors:

$$Y_1 = \frac{1}{R_1} \qquad (68)$$

$$Y_3 = \frac{1}{R_3} \qquad (69)$$

In order to get s^2 in the denominator, the term Y_5 is surely a capacitor. (Note that the term $Y_3 Y_4$ cannot yield an s^2 because Y_3 has already been made a resistor.) Hence:

$$Y_5 = sC_5 \qquad (70)$$

If Y_5 is a capacitor, then Y_4 must be a resistor in order for us to have a non-s term in the denominator. Hence:

$$Y_4 = \frac{1}{R_4} \qquad (71)$$

Finally, the only other term, Y_2, must be a ca-

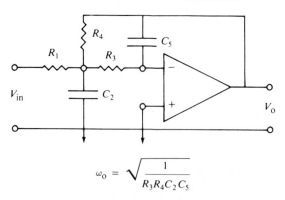

$$\omega_o = \sqrt{\frac{1}{R_3 R_4 C_2 C_5}}$$

FIGURE 19.33. Comparing the voltage transfer function for the active filter design shown in Figure 19.32 with that for a generalized second-order low-pass filter (Figure 19.31) yields the circuit design shown. Selection of values for the circuit components to achieve specific behavior is left for the specialist.

pacitor in order to produce the s^2 term in the denominator. Hence:

$$Y_2 = sC_2 \qquad (72)$$

Substitution of these terms into Equation 66 yields:

$$\frac{V_o}{V_{in}} = \frac{-(1/R_1 R_3 C_2 C_5)}{s^2 + \dfrac{s}{C_2}\left(\dfrac{1}{R_1} + \dfrac{1}{R_3} + \dfrac{1}{R_4}\right) + \dfrac{1}{R_3 R_4 C_2 C_5}} \qquad (73)$$

This equation will take the form of Equation 67 if the corner frequency ω_o is defined as:

$$\omega_o^2 = \frac{1}{R_3 R_4 C_2 C_5} \qquad (74)$$

The resulting circuit diagram for the filter is shown in Figure 19.33.

Selection of values for the five components to achieve a specific characteristic is a task for the design specialist and is beyond the scope of this book. The preceding discussion has given but a brief overview of the kind of analysis that is involved and has shown the general form for the circuit diagram.

Other Filter Designs

To complete this discussion, let us look at the generalized form for the other second-order filters: high-pass and band-pass. These filters are shown in Figure 19.34. The high-pass filter is comparable to the low-pass filter except that the resistor and capacitor are interchanged. The band-pass filter, however, has a distinct design.

Because the active band-pass filter design can also produce voltage gain, it is more commonly called a tuned amplifier. Of course, the addition of more amplifying stages to any of the designs will produce any required voltage gain.

OTHER ACTIVE FILTERS

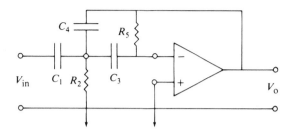

(a) High-Pass Filter

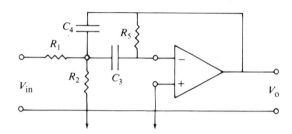

(b) Band-Pass Filter

FIGURE 19.34. The circuit design for an active, second-order high-pass filter is similar to that for a low-pass filter. The active, second-order band-pass filter is a distinctive design.

19.9 QUESTIONS AND PROBLEMS

1. What type of filter would you select for the following applications? Sketch the frequency response curve for each case.
   (a) To eliminate audio signals below 20 kHz in an ultrasound burglar alarm system.
   (b) To tune in a station broadcasting at 600 kHz on a radio.
   (c) To block noise in a European TV from the power line at 50 Hz.
   (d) To eliminate radio frequency noise above 100 kHz in an electronic measuring system.

2. Determine the gain in decibels of a circuit with the following ratios of input and output voltage:
   (a) $\dfrac{V_0}{V_{in}} = 0.67$

   (b) $\dfrac{V_0}{V_{in}} = 0.0038$

   (c) $\dfrac{V_0}{V_{in}} = 0.075$

   (d) $\dfrac{V_0}{V_{in}} = 0.11$

   (e) $\dfrac{V_0}{V_{in}} = 3.6$

3. Determine the gain in decibels and in values of V_0/V_{in} of the filter whose frequency response graph is shown in Figure 19.35 at the following frequencies: (a) 30 kHz, (b) 5.5 kHz, and (c) 200 kHz.

4. Find the frequencies at which the filter shown in Figure 19.36 has the following gains: (a) − 3 db, (b) − 15 db, and (c) − 30 db.

5. What is the cutoff frequency and the roll-off for the filters whose frequency response graphs are shown in Figures 19.35 and 19.36? Express the roll-off in both db/octave and db/decade.

6. Find the corner frequencies of low-pass filter

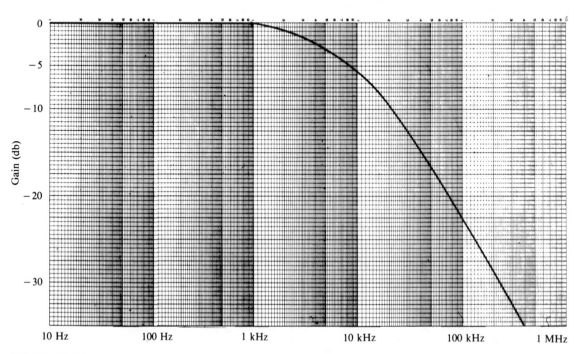

FIGURE 19.35.

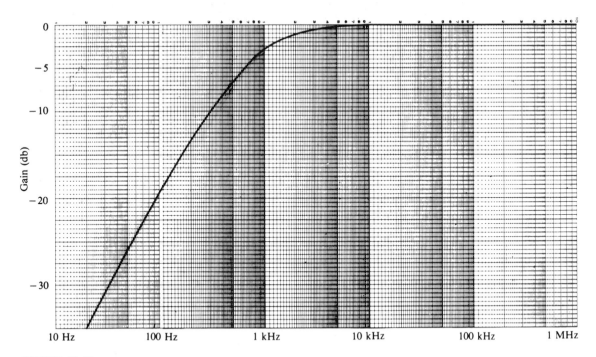

Gain (db)

10 Hz   100 Hz   1 kHz   10 kHz   100 kHz   1 MHz

FIGURE 19.36.

circuits with the following values of resistance and capacitance:

(a) $R = 1.5$ kΩ and $C = 0.01$ μF
(b) $R = 22$ kΩ and $C = 0.0015$ μF
(c) $R = 470$ kΩ and $C = 700$ pF

7. Design a low-pass filter with a corner frequency of 0.2 kHz and a roll-off of -20 db/decade. Give an expression for its voltage transfer function.

8. Design a high-pass filter with a corner frequency of 30 kHz and a roll-off of 6.0 db/octave. Give an expression for its voltage transfer characteristic.

9. Draw the frequency response curves for the circuits designed in Problems 7 and 8. Show the -3 db point and the Bode approximations.

10. Draw a graph of the output voltage phase angle versus frequency for the circuits designed in Problems 7 and 8.

11. For the filter circuits shown in Figure 19.37, determine the corner frquency when there is no load and when there is a load resistance of 4.7 kΩ.

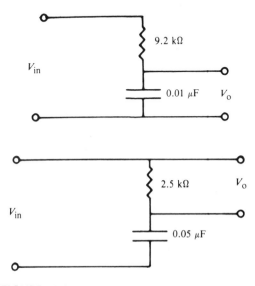

FIGURE 19.37.

12. What are the voltage transfer functions of the circuits shown in Figure 19.37 with and without the 4.7 kΩ load resistor?

13. Show how the circuits shown in Figure 19.37 can be redesigned to yield their unloaded corner frequency.

14. Design a low-pass filter with a corner frequency of 2.5 kHz and a roll-off of −12 db/octave without using a voltage follower. Give an expression for its voltage transfer function.

15. Design a high-pass filter with a corner fre-

quency of 15 kHz and a roll-off of 40 db/decade without using a voltage follower. Give an expression for its voltage transfer function.

16. Draw the Bode diagram for the circuits designed in Problems 14 and 15.

17. What are the Q factors for the narrow band filters whose frequency response graphs are given in Figure 19.38?

18. Design a tuned amplifier with a center frequency of 5 kHz, roll-off of 12 db/octave, and a gain of 100. Draw its Bode diagram.

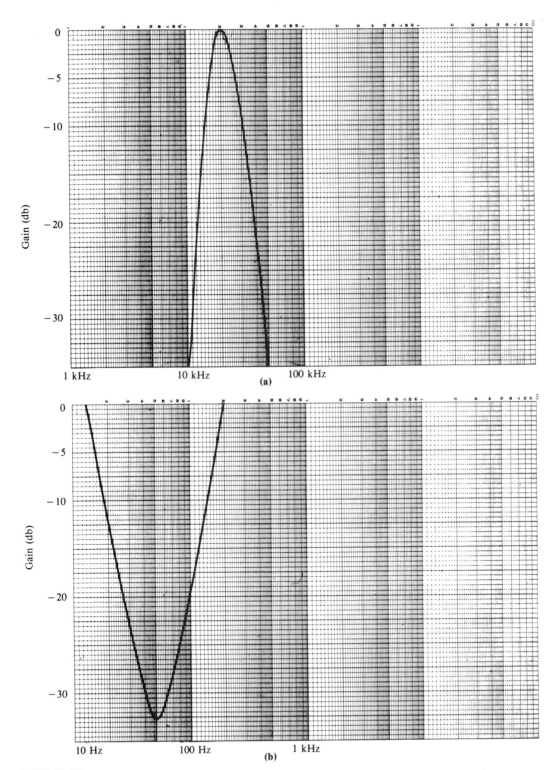

FIGURE 19.38.

INDEX